T0214186

Lecture Notes in Artificial Intelligence   11142

Subseries of Lecture Notes in Computer Science

LNAI Series Editors

Randy Goebel
  University of Alberta, Edmonton, Canada
Yuzuru Tanaka
  Hokkaido University, Sapporo, Japan
Wolfgang Wahlster
  DFKI and Saarland University, Saarbrücken, Germany

LNAI Founding Series Editor

Joerg Siekmann
  DFKI and Saarland University, Saarbrücken, Germany

More information about this series at http://www.springer.com/series/1244

Davide Ciucci · Gabriella Pasi
Barbara Vantaggi (Eds.)

Scalable Uncertainty Management

12th International Conference, SUM 2018
Milan, Italy, October 3–5, 2018
Proceedings

 Springer

Editors
Davide Ciucci ⓘ
University of Milano-Bicocca
Milan
Italy

Barbara Vantaggi ⓘ
Sapienza University of Rome
Rome
Italy

Gabriella Pasi ⓘ
University of Milano-Bicocca
Milan
Italy

ISSN 0302-9743 ISSN 1611-3349 (electronic)
Lecture Notes in Artificial Intelligence
ISBN 978-3-030-00460-6 ISBN 978-3-030-00461-3 (eBook)
https://doi.org/10.1007/978-3-030-00461-3

Library of Congress Control Number: 2018954666

LNCS Sublibrary: SL7 – Artificial Intelligence

© Springer Nature Switzerland AG 2018
This work is subject to copyright. All rights are reserved by the Publisher, whether the whole or part of the material is concerned, specifically the rights of translation, reprinting, reuse of illustrations, recitation, broadcasting, reproduction on microfilms or in any other physical way, and transmission or information storage and retrieval, electronic adaptation, computer software, or by similar or dissimilar methodology now known or hereafter developed.
The use of general descriptive names, registered names, trademarks, service marks, etc. in this publication does not imply, even in the absence of a specific statement, that such names are exempt from the relevant protective laws and regulations and therefore free for general use.
The publisher, the authors and the editors are safe to assume that the advice and information in this book are believed to be true and accurate at the date of publication. Neither the publisher nor the authors or the editors give a warranty, express or implied, with respect to the material contained herein or for any errors or omissions that may have been made. The publisher remains neutral with regard to jurisdictional claims in published maps and institutional affiliations.

This Springer imprint is published by the registered company Springer Nature Switzerland AG
The registered company address is: Gewerbestrasse 11, 6330 Cham, Switzerland

Preface

The series of annual conferences on Scalable Uncertainty Management (SUM) started in 2007, and it is dedicated to the management of large amounts of complex, uncertain, incomplete, or inconsistent information.

Managing uncertainty and inconsistency has been extensively explored in the field of artificial intelligence and statistics over a number of years. Recently, with the advent of massive amounts of data and knowledge from distributed, heterogeneous, and potentially conflicting sources, there has been an increasing interest in defining and applying formalisms capable of representing and managing uncertain and/or inconsistent data and knowledge. To meet the challenge of representing and manipulating large amounts of uncertain information, researchers draw from a wide range of different methodologies and uncertainty models.

While Bayesian methods remain the default choice in most disciplines, sometimes there is a need for more cautious and flexible approaches, and for more specific handling of incomplete or subjective information.

In fact, during the last fifty years, in areas like decision theory, artificial intelligence, or information processing, numerous approaches extending or orthogonal to the existing theory of probability and statistics have been successfully developed. These new approaches rely, for instance, on imprecise probabilities, fuzzy set theory, rough set theory, ordinal uncertainty representations, or even purely qualitative models.

The International Conference on Scalable Uncertainty Management (SUM) aims to provide a forum for researchers who are working on uncertainty management, in different communities and with different uncertainty models, to meet and exchange ideas. Previous SUM conferences have been held in Washington DC (2007), Naples (2008), Washington DC (2009), Toulouse (2010), Dayton (2011), Marburg (2012), Washington DC (2013), Oxford (2014), Québec City (2015), Nice (2016), and Granada (2017).

This volume contains contributions from the 12th SUM conference, which was held in Milan, Italy, during October 3–5, 2018. The conference attracted 37 submissions, of which 29 were accepted for publication (23 as regular and 6 as short papers) and presentation at the conference, based on peer reviews from three members of the Program Committee or by external reviewers.

In addition, the conference greatly benefited from the invited lectures by three world-leading researchers: Salem Benferhat, University of Artois; Georg Gottlob, University of Oxford; and Dominik Ślezak, University of Warsaw. To further embrace the aim of facilitating interdisciplinary collaboration and cross-fertilization of ideas, and building on the tradition of invited speakers at SUM, the conference featured eight tutorials, covering a broad set of topics related to uncertainty management. We thank Hassan Aït-Kaci, Didier Dubois, Salvatore Greco, Francesco Masulli, Agnès Rico, Grégory Smits, Fabio Stella, and Andrea Tettamanzi, for preparing and presenting these tutorials, two of which have a companion paper included in this volume.

We would like to thank all the authors, invited speakers, and tutorial speakers for their valuable contribution, and both the members of the Program Committee and the external reviewers for their detailed and critical assessment of the submissions. We are indebted to the Steering Committee for the suggestions and help given in setting up the conference and to Marco Viviani for his support in the organization. We are also very grateful to the University of Milano-Bicocca for hosting the conference, and to Springer for providing a grant of 500 euros for the best paper awards, and for the support of its staff in publishing this volume.

October 2018

Davide Ciucci
Gabriella Pasi
Barbara Vantaggi

Organization

Conference Chair

Davide Ciucci   University of Milano-Bicocca, Italy

Program Chairs

Gabriella Pasi   University of Milano-Bicocca, Italy
Barbara Vantaggi   Sapienza University of Rome, Italy

Steering Committee

Didier Dubois	IRIT-CNRS, France
Lluis Godo	IIIA-CSIC, Spain
Eyke Hüllermeier	Universität Paderborn, Germany
Anthony Hunter	University College London, UK
Henri Prade	IRIT-CNRS, France
Steven Schockaert	Cardiff University, UK
V. S. Subrahmanian	University of Maryland, USA

Web and Publicity Chair

Marco Viviani   University of Milano-Bicocca, Italy

Program Committee

Leila Amgoud	IRIT - CNRS, France
Christoph Beierle	University of Hagen, Germany
Salem Benferhat	University of Artois, France
Leopoldo Bertossi	Carleton University, Canada
Fernando Bobillo	University of Zaragoza, Spain
Gloria Bordogna	National Research Council (CNR), Italy
Federico Cerutti	Cardiff University, UK
Reynold Cheng	The University of Hong Kong, China
Olivier Colot	Lille 1 University, France
Fabio Cozman	University of São Paulo, Brazil
Alfredo Cuzzocrea	University of Trieste, Italy
Luis M. de Campos	University of Granada, Spain
Thierry Denoeux	University of Technology of Compiègne, France
Sébastien Destercke	CNRS, University of Technology of Compiègne, France
Didier Dubois	IRIT-CNRS, France

Zied Elouedi	Higher Institute of Management, University of Tunis, Tunisia
John Grant	Towson University, USA
Manuel Gómez-Olmedo	University of Granada, Spain
Anne Laurent	University of Montpellier 2, France
Sebastian Link	The University of Auckland, New Zealand
Thomas Lukasiewicz	University of Oxford, UK
Silviu Maniu	Paris-Sud University, France
Francesco Parisi	University of Calabria, Italy
Rafael Peñaloza	Free University of Bozen-Bolzano, Italy
Olivier Pivert	IRISA-ENSSAT, France
Henri Prade	IRIT - CNRS, France
Andrea Pugliese	University of Calabria, Italy
Steven Schockaert	Cardiff University, UK
Guillermo R. Simari	Universidad del Sur, Argentina
Umberto Straccia	National Research Council (CNR), Italy
Jef Wijsen	University of Mons, Belgium
Slawomir Zadrozny	Systems Research Institute, Polish Academy of Sciences, Poland

Additional Reviewers

Ma, Chenhao
Molinaro, Cristian
Parisi, Francesco
Pugliese, Andrea

Shan, Caihua
Smits, Grégory
Yan, Jing

Contents

Tutorials

A Crash Course on Generalized Possibilistic Logic 3
Didier Dubois and Henri Prade

Discrete Sugeno Integrals and Their Applications 18
Agnès Rico

Regular Papers

A Credal Extension of Independent Choice Logic 35
Alessandro Antonucci and Alessandro Facchini

Explaining the Most Probable Explanation 50
Raphaela Butz, Arjen Hommersom, and Marko van Eekelen

Fuzzification of Ordinal Classes. The Case of the HL7 Severity Grading 64
Federico Cabitza and Davide Ciucci

A Modular Inference System for Probabilistic Description Logics 78
*Giuseppe Cota, Fabrizio Riguzzi, Riccardo Zese, Elena Bellodi,
and Evelina Lamma*

Modeling the Dynamics of Multiple Disease Occurrence by Latent States . . . 93
*Marcos L. P. Bueno, Arjen Hommersom, Peter J. F. Lucas,
Mariana Lobo, and Pedro P. Rodrigues*

Integral Representations of a Coherent Upper Conditional Prevision
by the Symmetric Choquet Integral and the Asymmetric Choquet
Integral with Respect to Hausdorff Outer Measures 108
Serena Doria

Separable Qualitative Capacities 124
Didier Dubois, Francis Faux, Henri Prade, and Agnès Rico

An Approach Based on MCDA and Fuzzy Logic to Select Joint Actions 140
Abdelhak Imoussaten

Discovering Ordinal Attributes Through Gradual Patterns, Morphological
Filters and Rank Discrimination Measures 152
*Christophe Marsala, Anne Laurent, Marie-Jeanne Lesot, Maria Rifqi,
and Arnaud Castelltort*

Distribution-Aware Sampling of Answer Sets   164
  Matthias Nickles

Consequence-Based Axiom Pinpointing   181
  Ana Ozaki and Rafael Peñaloza

Probabilistic Semantics for Categorical Syllogisms of Figure II   196
  Niki Pfeifer and Giuseppe Sanfilippo

Measuring Disagreement Among Knowledge Bases   212
  Nico Potyka

On Enumerating Models for the Logic of Paradox Using Tableau   228
  Pilar Pozos-Parra, Laurent Perrussel, and Jean Marc Thévenin

On Instantiating Generalised Properties of Gradual
Argumentation Frameworks   243
  Antonio Rago, Pietro Baroni, and Francesca Toni

Lower and Upper Probability Bounds for Some Conjunctions
of Two Conditional Events   260
  Giuseppe Sanfilippo

Qualitative Probabilistic Relational Models   276
  Linda C. van der Gaag and Philippe Leray

Rule-Based Conditioning of Probabilistic Data   290
  *Maurice van Keulen, Benjamin L. Kaminski, Christoph Matheja,
  and Joost-Pieter Katoen*

Positional Scoring Rules with Uncertain Weights   306
  Paolo Viappiani

A New Measure of General Information on Pseudo Analysis   321
  Doretta Vivona and Maria Divari

A Formal Approach to Embedding First-Principles Planning
in BDI Agent Systems   333
  Mengwei Xu, Kim Bauters, Kevin McAreavey, and Weiru Liu

Short Papers

Imprecise Sampling Models for Modelling Unobserved Heterogeneity?
Basic Ideas of a Credal Likelihood Concept   351
  Thomas Augustin

Representation of Multiple Agent Preferences: A Short Survey   359
  Nahla Ben Amor, Didier Dubois, Henri Prade, and Syrine Saidi

Measuring and Computing Database Inconsistency via Repairs   368
  Leopoldo Bertossi

Scalable Bounding of Predictive Uncertainty in Regression Problems
with SLAC   373
  Arno Blaas, Adam D. Cobb, Jan-Peter Calliess, and Stephen J. Roberts

Predicting the Possibilistic Score of OWL Axioms Through Support
Vector Regression   380
  Dario Malchiodi, Célia da Costa Pereira, and Andrea G. B. Tettamanzi

Inferring Quantitative Preferences: Beyond Logical Deduction   387
  Maria Vanina Martinez, Lluis Godo, and Gerardo I. Simari

Handling Uncertainty in Relational Databases with Possibility Theory - A
Survey of Different Modelings   396
  Olivier Pivert and Henri Prade

An Argumentative Recommendation Approach Based
on Contextual Aspects   405
  Juan Carlos Lionel Teze, Lluis Godo, and Guillermo Ricardo Simari

Author Index   413

Tutorials

A Crash Course on Generalized Possibilistic Logic

Didier Dubois$^{(\boxtimes)}$ and Henri Prade

Institut de Recherche en Informatique de Toulouse (IRIT), CNRS & Université de Toulouse, 118 Route de Narbonne, 31062 Toulouse Cedex 9, France
{dubois,prade}@irit.fr

Abstract. This paper proposes a concise overview of the role of possibility theory in logical approaches to reasoning under uncertainty. It shows that three traditions of reasoning under or about uncertainty (set-functions, epistemic logic and three-valued logics) can be reconciled in the setting of possibility theory. We offer a brief presentation of basic possibilistic logic, and of its generalisation that comes close to a modal logic albeit with simpler more natural epistemic semantics. Past applications to various reasoning tasks are surveyed, and future lines of research are also outlined.

Keywords: Possibility theory · Possibilistic logic · Epistemic logic
Three-valued logic · Non-monotonic reasoning

1 Introduction

Uncertainty often pervades information and knowledge. For this reason, the handling of uncertainty in inference systems has been an issue for a long time in artificial intelligence (AI). As logical formalisms have dominated AI research for several decades, this problem was first tackled using modal logics or non-monotonic reasoning, or yet many-valued logics. Key-issues are reasoning under incomplete information (how to infer useful conclusions despite incompleteness?) and inconsistent information (how to protect useful conclusions from inconsistency?).

With the advent of Bayesian networks, Bayesian probabilities have become prominent at the forefront of AI methods, challenging the original supremacy of logical representation settings. However, Bayesian networks are in some sense miraculous, because, viewed as representing human information, they are neither incomplete nor inconsistent. The problem of dealing with incomplete probabilistic knowledge is then avoided, especially by Bayesian probability proponents, assuming that any state of knowledge can be accounted for by a unique probability distribution. However, due to the self-duality of probabilities, the Bayesian approach cannot distinguish between the lack of belief in a proposition and the belief in its negation, which is precisely what incomplete information is all about.

© Springer Nature Switzerland AG 2018
D. Ciucci et al. (Eds.): SUM 2018, LNAI 11142, pp. 3–17, 2018.
https://doi.org/10.1007/978-3-030-00461-3_1

The need for this distinction seems to have led to the emergence of ad hoc uncertainty calculi in early expert systems.

Besides, there has been a divorce between logic and probability in the early 20th century. Logic was then considered as a foundation for mathematics, while probability theory was found instrumental in representing statistical data. The school of subjective probability took it for granted that degrees of belief should be additive and self-dual, which led to paradoxes when trying to account for ignorance. Yet, logic and probability were originally motivated by the formalisation of some aspects of human reasoning. Boole, De Morgan, De Finetti used to attach probabilities to logical formulas and try to make deductions from them. But so doing, and in contrast with Bayesian networks, you run the risk of at worst assigning incompatible probabilities, or at best only deriving probability intervals rather than precise probabilities for conclusions. While this can be considered as a weakness of the logical approach, one may also claim that this state of facts is more faithful to human knowledge, which is often incomplete and sometimes inconsistent. There are now well-defined approaches to uncertainty that try to marry additivity and incompleteness [1]. These settings are characterized by the existence of a pair of dual measures for distinguishing between the support of a proposition and the lack of support of its negation, for instance Walley's imprecise probability theory, Shafer's evidence theory, and possibility theory, the latter first outlined by Shackle [2] in a very hostile scientific environment, later on taken up by Cohen [3] and Zadeh [4].

Possibility theory has a remarkable situation among the settings devoted to the representation of imprecise and uncertain information. First, it may be numerical or qualitative [5]. In the first case, possibility measures and the dual necessity measures can be regarded respectively as upper bounds and lower bounds of ill-known probabilities; they are also particular cases of Shafer plausibility and belief functions respectively. In fact, possibility measures and necessity measures constitute the simplest, non trivial, imprecise probability system. Second, when qualitative, possibility theory provides a natural approach to the grading of possibility and necessity modalities on finite ordinal scales. Especially, possibility theory has a logical counterpart, namely possibilistic logic [6], which remains close to classical logic. In this overview paper, we focus our attention on the unifying power of all-or-nothing possibility theory for epistemic reasoning. Then we briefly present possibilistic logic (PosLog), which handles degrees of certainty, and its recent extension called generalized possibilistic logic (GPL). Then we briefly review some application areas for PosLog and GPL.

2   Three Approaches to Handling Incomplete Knowledge

There seems to be three traditions for dealing with incomplete information and the ensuing representation of beliefs.

The Set-Function Tradition. Set-functions are used to assign degrees of beliefs to propositions represented by subsets of possible states of affairs, also called events: for instance, subjective probability adopts this point of view. More generally, if S represents a set of states of affairs and A is a subset thereof, the idea is to attach a degree of confidence $g(A)$ usually in the unit interval (but not always, for instance Spohn [7] uses integers). The main properties of the set-function g are monotonicity with inclusion ($g(B) \geq g(A)$ whenever $A \subseteq B$), and limit conditions $g(\emptyset) = 0, g(S) = 1$. Other examples of such confidence functions g are possibility and necessity measures, belief or plausibility functions, upper and lower probabilities [1].

The Multiple-Valued Logic Tradition. Very early in the 20th century, scholars in logic like Łukasiewicz realized that it is not always possible to declare that a given proposition is *true* or is *false*. Due to a lack of knowledge, only weaker qualifiers may be used, such as *possible, unknown,* or yet *indeterminate*. So the idea was to define the epistemic status of complex propositions from truth-tables extending the ones of classical logic, augmenting the set of truth-values with a new symbol lying between *true* and *false* and standing for *unknown* and the like, while sticking to the propositional syntax. This led to three-valued logics such as the ones of Łukasiewicz or Kleene. Of course, it was then natural to extend this view to more truth-values, even a continuum thereof. In the same spirit, Belnap [8] proposed to add a fourth truth-value standing for the idea of conflicting information about a proposition.

The Modal Logic Tradition. In this approach, the idea is to represent the belief modality at the syntactic level using a necessity symbol $\Box$ that prefixes propositions from a standard propositional language, and more generally, any formula, in fact. In this way, one can explicitly express that a proposition p is believed or known, writing $\Box p$, and we can make a clear syntactic distinction between the fact of not believing p (i.e., $\neg\Box p$) and believing its negation (i.e., $\Box\neg p$). This is the setting of epistemic or doxastic logics originally due to von Wright, Hintikka, and developed further by Fagin et al. [9]. These logics often adopt axioms (K, D) that ensure (i) the equivalence between knowing or believing a conjunction $\Box(p \land q)$ and believing each of the conjoints $\Box p$ land $\Box q$ (this is called the *adjunction* law), and (ii) that $\Box p$ is a stronger statement than $\Diamond p \equiv \neg\Box\neg p$ (expressing a lack of support for the negation of p). This approach makes a clear difference between belief and knowledge (viewed as true belief), and allows to construct very complex formulas involving nested modalities and the usual Boolean connectives.

These three traditions are supposed to address closely related issues: they try to propose tools to represent belief and uncertainty when information is lacking. But they look at odds with one another. The main claim of this paper is that possibility theory is instrumental to bridge these apparently unrelated approaches and to put them in a unified setting.

3 The Simplest Logic of Belief and Partial Ignorance

The simplest format to express beliefs is propositional logic (PL). Consider a propositional language $\mathcal{L} = \{p, q, r, \ldots\}$ built from atomic variables $\mathcal{A} =$

$\{a, b, c, \dots\}$ and usual connectives $\land, \lor, \neg, \dots$. A belief base is understood as a finite subset $\mathcal{K} \subset \mathcal{L}$ of propositions believed by an agent. In the usual semantics, a proposition p is evaluated on the set of interpretations of $\mathcal{L}$, say $S = \{s : \mathcal{A} \to \{0, 1\}\}$, assigning a truth-value to each variable. Let $[p]$ be the set of models of p, interpretations s for which p is true (which is denoted by $s \models p$), and $[\mathcal{K}]$ be the set of models of all $p \in \mathcal{K}$, i.e., $[\mathcal{K}] = \cap_{p \in \mathcal{K}}[p]$. Denote by $\mathcal{K} \vdash p$ the syntactic inference of p from $\mathcal{K}$, using inferences rules in PL and modus ponens, and by $\mathcal{K} \models p$ the semantic inference, defined by $[\mathcal{K}] \subset [p]$. It is well-known that $\mathcal{K} \models p \iff \mathcal{K} \vdash p$ (soundness and completeness).

3.1   Boolean Beliefs

The semantics in terms of interpretations is ontic in the sense that it tells whether propositions are true or false in each state of the world. However it does not match the idea that propositions in $\mathcal{K}$ represent beliefs. A belief should be evaluated with respect to an *epistemic state*, namely a *non-empty* set E of interpretations that are not ruled out by the agent. This is the simplest representation of incomplete information. Epistemic states are more or less informed: E is said to be more *specific* than E', if $E \subset E'$: E is more informed than E'. An epistemic state E is said to be fully informed if $E = \{s\}$ for some $s \in S$, and totally ignorant if $E = S$. In the tradition of epistemic logic, an agent is said to believe p if p is true in all interpretations the agent considers possible, i.e., in its epistemic state. This leads to an epistemic semantics for propositional logic, whereby propositions p are evaluated in epistemic states, namely:

$$E \models p \iff E \subseteq [p].$$

And we can define an epistemic semantic inference as $\mathcal{K} \models_e p \iff \forall E \neq \emptyset, E \subseteq [\mathcal{K}]$ implies $E \subseteq [p]$. However, there exists a least specific epistemic state for an agent whose belief base is $\mathcal{K}$ and this is $E = [\mathcal{K}]$. So the ontic and epistemic semantics lead to the same inference: $\mathcal{K} \models p \iff \mathcal{K} \models_e p$. But the epistemic semantics is potentially richer than the ontic one, as one may distinguish between the situation where $E \subseteq [\neg p]$ (the agent believes the negation of p) and the situation where $E \not\subseteq [p]$ (the agent does not believe p), which collapse if $E = \{s\}$. Clearly, the propositional language cannot grasp this distinction. To account for it at the syntactic level we must let the belief modality $\Box$ appear in the syntax, to mean that the agent believes p. In fact, the precise meaning of this modality can range from *the agent is informed that p* to *the agent knows that p*.

On this basis, we consider an epistemic language $\mathcal{L}_\Box$ built on top of $\mathcal{L}$:

- Atomic variables form the set $\mathcal{A}_\Box = \{\Box p : p \in \mathcal{L}\}$ (i.e., plain beliefs).
- $\mathcal{L}_\Box$ is the propositional language based on $\mathcal{A}_\Box$ with usual connectives $\land, \lor, \neg$.

In this language not only can we state that p is believed, but also that the agent ignores p, the latter being expressed by the formula $\neg \Box p \land \neg \Box \neg p$, or equivalently

$\Diamond p \land \Diamond \neg p$, where the formula $\Diamond p$ is short for $\neg \Box \neg p$. $\mathcal{L}_\Box$ can be viewed as the minimal language to express partial knowledge about the truth of propositions.

We define an epistemic semantics for the formulas ϕ, ψ in $\mathcal{L}_\Box$ by adapting the epistemic semantics of propositional logic with formulas in $\mathcal{L}$:

- $E \models \Box p$ if and only if $E \subseteq [p]$ (p is certainly true in the epistemic state E);
- $E \models \phi \land \psi$ if $E \models \phi$ and $E \models \psi$;
- $E \models \neg \phi$ if $E \models \phi$ is false.

A logic called MEL (Minimal Epistemic Logic) has been defined by Banerjee and Dubois [10] for reasoning about incomplete knowledge in $\mathcal{L}_\Box$. Axioms are as follows:

(PL) Axioms of propositional logic for $\mathcal{L}_\Box$-formulas
  (K) $\Box(p \to q) \to (\Box p \to \Box q)$
  (D) $\Box p \to \Diamond p$
(Nec) $\Box p$, for each $p \in \mathcal{L}$ that is a PL tautology, i.e. if $[p] = S$.

The inference rule is modus ponens. In other words, this is a two-tiered propositional logic with additional axioms, and we define the syntactic inference as $\mathcal{B} \vdash_{MEL} \phi$ if and only if $\mathcal{B} \cup \{K, D, Nec\} \vdash \phi$, where $\mathcal{B} \subset \mathcal{L}_\Box$ is a MEL base. This logic is sound and complete with respect to the epistemic semantics, defining as usual $\mathcal{B} \models_{MEL} \phi$ if and only if $E \models \psi, \forall \psi \in \mathcal{B}$ implies $E \models \phi$.

Note that the subjective language $\mathcal{L}_\Box$ is disjoint from the objective language $\mathcal{L}$. However, there is a clear embedding of propositional logic in MEL: Given a belief base $\mathcal{K}$ in PL, define the MEL base $\mathcal{K}_\Box = \{\Box p : p \in \mathcal{K}\}$. Then we have that $\mathcal{K}_\Box \vdash_{MEL} \Box q$ if and only if $\mathcal{K} \vdash q$.

3.2   Unifying Approaches to Reasoning with Incomplete Information

We are now in a position to reconcile the three traditions for reasoning with incomplete information.

Set Functions Underlying MEL. We have seen that at the semantic level incomplete information is represented by epistemic states $E \subseteq S$ representing mutually exclusive states of affairs, one of which is considered to be the real one by the agent with epistemic state E. Its characteristic function is a *possibility distribution* with values on $\{0, 1\}$ defined by $\pi(s) = \begin{cases} 1 \text{ if } s \in E; \\ 0 \text{ otherwise.} \end{cases}$

This representation leads us to build 2 set-functions (e.g., [5]):

- A *possibility function* $\Pi(A) = 1$ if and only if $E \cap A \neq \emptyset$ and 0 otherwise
- A *necessity function* $N(A) = 1$ if and only if $E \subseteq A$ and 0 otherwise.

Possibility and necessity functions are examples of confidence functions like probabilities, except that they are 2-valued. Their characteristic properties on finite sets are *maxitivity* for Π ($\Pi(A \cup B) = \max(\Pi(A), \Pi(B))$) and *minitivity* for N

$(N(A \cap B) = \max(N(A), N(B))$, which encodes the adjunction law), along with $N(S) = \Pi(S) = 1$, $N(\emptyset) = \Pi(\emptyset) = 0$.

It is clear that in a finite setting, an epistemic state is equivalent to a necessity function or to a possibility function, so that the semantics of MEL can be equivalently expressed in terms of these set functions. In particular, $\Box p$ means $N(p) = 1$ for some necessity function N, while $\Diamond p$ means $\Pi(p) = 1$ for some possibility function Π, and we have that $N(A) = 1 - \Pi(A^c)$, where A^c is the complement of A. Axiom Nec in MEL corresponds to $N(S) = 1$. Axiom D expresses the inequality $\Pi(A) \geq N(A)$. Axioms K and D ensure the minitivity axiom. So we can claim that MEL is the logic of all-or-nothing possibility theory.

MEL vs. Doxastic Logic. The close connection between MEL and modal epistemic logic S5 and doxastic logic KD45 is very clear, as MEL uses a fragment of their language. Moreover, all inferences that can be made in MEL can equally be made in KD45 as proved in [11]. However there are noticeable differences, reflecting the fact that MEL is in some sense a minimal uncertainty logic:

- MEL allows neither nested modalities nor plain (non-modal) propositional formulas.
- In MEL, formulas are evaluated on epistemic states while in KD45 formulas are evaluated on possible worlds (it makes sense to write $s \models \Box p$) via accessibility relations R on S (as $R(s) \subseteq [p]$).
- KD45 allows for very complex modal formulas that can be simplified using additional axioms (**4**: $\Box \phi \rightarrow \Box \Box \phi$; **5**: $\neg \Box \phi \rightarrow \Box \neg \Box \phi$), while MEL minimally augments the expressive power of PL. These additional introspection axioms **4** and **5** cannot be written in MEL.
- MEL has the deduction theorem because it is a two-tiered propositional logic. Modal logics often do not because axiom (Nec) in MEL is expressed by means of the necessitation rule $p \vdash \Box p$ (that cannot be written in MEL).

So MEL can be viewed as the poor man's epistemic or doxastic logic; we cannot see the difference here because axiom T: $\Box p \rightarrow p$ cannot be written in MEL (but we can include it by extending MEL to allow for objective formulas, see [11]).

While KD45 is supposed to model an agent reflecting on its own beliefs, MEL is rather dedicated to represent what an agent knows about the epistemic state of another agent. For instance $\Diamond p \wedge \Diamond \neg p$ should be read: the agent knows that the other agent ignores the truth value of p. Likewise, having $\Box p \vee \Box \neg p$ without having $\Box p$ nor $\Box \neg p$ means that the first agent knows that the other agent knows the truth value of p, while this first agent does not know what the second one believes. Note that this formula would be hard to interpret if it concerns one's own beliefs, as pointed out quite early in [12].

Regarding introspection, even if axioms **4** and **5** are absent from MEL, it is implicitly assumed in MEL that an agent can always say whether (s)he believes p, (s)he believes $\neg p$ or is ignorant about p (the formula $\Box p \vee \Box \neg p \vee (\Diamond p \wedge \Diamond \neg p)$ is a MEL tautology). In other words, an agent is supposed to know its own epistemic state (it can be described by a complete MEL base), which is how introspection is handled in MEL.

MEL vs. Three-Valued Logics. The 3-valued logic approach to reasoning with incomplete information can be captured in MEL [13]. For the sake of simplicity we restrict to Kleene logic, often used in incomplete databases, or logic programming. It has 3 truth values forming the set $\mathbb{V}_3 = \{0 < \frac{1}{2} < 1\}$, the same syntax as PL, with the same connectives semantically characterized as follows:

- *Negation*: $\mathbf{t}(\neg p) = 1 - \mathbf{t}(p)$;
- *Conjunction*: $\mathbf{t}(p \wedge q) = \min(\mathbf{t}(p), \mathbf{t}(q))$;
- *Disjunction*: $\mathbf{t}(p \vee q) = \max(\mathbf{t}(p), \mathbf{t}(q))$, by De Morgan laws;
- *Implication*: $\mathbf{t}(p \rightarrow_K q) = \max(1 - \mathbf{t}(p), \mathbf{t}(q))$ (using $p \rightarrow_K q \equiv \neg p \vee q$).

This logic has no tautologies, e.g., $\mathbf{t}(p \wedge \neg p) = \mathbf{t}(p \vee \neg p) = \frac{1}{2}$ when $\mathbf{t}(p) = \frac{1}{2}$.

In practice, the truth-value $\frac{1}{2}$ is often used to model the idea that the truth-value of an otherwise Boolean proposition is unknown. But *Unknown* is in opposition with *Known to be true* and *Known to be false*, not with *true* and *false*. The three-valued $\mathbf{t}(p)$ thus informs about the state of knowledge concerning the plain truth value $t(p) \in \{0,1\}$ of a Boolean proposition. In this view, the lack of tautologies in Kleene logic may look paradoxical since $p \wedge \neg p$ is known to be false regardless of our knowledge about p. The epistemic nature of Kleene truth-values led us to translate this logic into MEL [13]. Let $\mathcal{T}(\mathbf{t}(a) \in T)$ denote the translation into MEL of the atomic assertion $\mathbf{t}(a) \in T \subseteq \mathbb{V}_3$. We define, in agreement with the intended meaning of Kleene truth-values:

$$\mathcal{T}(\mathbf{t}(a) = 1) = \Box a \text{ (certainty of } a\text{)}; \qquad \mathcal{T}(\mathbf{t}(a) = 0) = \Box \neg a = \mathcal{T}(\mathbf{t}(\neg a) = 1);$$
$$\mathcal{T}(\mathbf{t}(a) = \tfrac{1}{2}) = \Diamond a \wedge \Diamond \neg a \text{ (ignorance)};$$
$$\mathcal{T}(\mathbf{t}(a) \geq \tfrac{1}{2}) = \Diamond a \text{ (possibility of } a\text{)}; \qquad \mathcal{T}(\mathbf{t}(a) \leq \tfrac{1}{2}) = \Diamond \neg a.$$

The translation of composite formulas in Kleene logic can be done recursively. In particular, for conjunction and disjunction we just have $\mathcal{T}(\mathbf{t}(a \vee b) = 1) = \Box a \vee \Box b$ and $\mathcal{T}(\mathbf{t}(a \wedge b) = 1) = \Box a \wedge \Box b$.

Asserting a formula p in Kleene logic comes down to the statement $\mathbf{t}(p) = 1$. A knowledge base $\mathcal{K}$ in Kleene logic can be put in conjunctive normal form (without simplifying the terms $a \vee \neg a$). The translation $\mathcal{T}(\mathbf{t}(\mathcal{K}) = 1)$ into MEL consists in the same conjunction of clauses with modality $\Box$ in front of literals. As a consequence, the translation process reaches a small fragment of the MEL language, namely $\mathcal{L}_\Box^K := \Box a | \Box \neg a | \phi \vee \psi | \phi \wedge \psi$ where $\Box$ only appears in front of literals and no negation is allowed to prefix the $\Box$ symbol. This result clarifies the limited expressive power of Kleene logic.

- Knowledge can be expressed about atomic variables only. In particular, truth-functionality is a trivial matter as it just says that $a \vee b$ means $\Box a \vee \Box b$. We can never express $\Box(a \vee b)$ using Kleene logic.
- The lack of tautologies in Kleene logic is explained, as $\Box a \vee \Box \neg a$ (translation of $\mathbf{t}(a \vee \neg a) = 1$) is not a tautology! More generally, replacing literals ℓ by $\Box \ell$ in a tautology of propositional logic does not yield a tautology in MEL.
- In $\mathcal{L}_\Box^K$, only partial models can be expressed by conjunctions of plain beliefs, that is "rectangular" subsets of S of the form $[\wedge_{a \in A^+} a \wedge \wedge_{a \in A^-} \neg a]$, where A^+ and A^- are disjoint subsets of atomic formulas.

These results extend to many three-valued logics where the third truth-value refers to ignorance (like Łukasiewicz 3-valued logic). They can be translated into MEL and these translations are theorem-preserving; see [13].

4   Gradual Beliefs in Possibility Theory

In this section, we consider beliefs can be a matter of degree. We first show that in order to satisfy the adjunction law, we must rely on the graded version of possibility theory. Then we describe possibilistic logic, that augments propositional logic with certainty weights attached to formulas. We generalize this approach by extending the MEL framework with certainty and possibility weights. We then survey the applications of these formalisms.

4.1   Degrees of Accepted Belief

In order to relate accepted beliefs to graded beliefs, one should extract accepted beliefs from a confidence function g. A natural way of proceeding is to define a belief as a proposition in which an agent has enough confidence. So we should be in a position to define a positive belief threshold β such that A is a belief if and only if $g(A) \geq \beta > 0$. However the adjunction law (accepted beliefs are closed under conjunction) leads to enforce the following property

Accepted belief postulate: If $g(A) \geq \beta$ and $g(B) \geq \beta$ then $g(A \cap B) \geq \beta$.

This requirement is very strong as (along with the monotonicity of g) it enforces the equality $g(A \cap B) = \min(g(A), g(B))$, that is g is a graded necessity measure, still denoted by N. Letting $\iota : S \to [0,1]$ be the function defined by $\iota(s) = N(S \setminus \{s\})$ (the degree of belief that the actual state of affairs is not s), it is clear that

$$N(A) = \min_{s \notin A} \iota(s).$$

The value $1 - \iota(s)$ can be interpreted as the degree of plausibility $\pi(s)$ of state s, where π is the membership function of a fuzzy epistemic state $\tilde{E}$ that represents a possibility distribution [5]. The set function $\Pi : 2^S \to [0,1]$ such that

$$\Pi(A) = 1 - N(A^c) = \max_{s \in A} \pi(s)$$

represents the degree of plausibility of A, measuring to what extent A is not totally ruled out by the agent. This setting is the one of possibility theory that seems to be the only one that accounts for the notion of accepted belief.

Possibility theory was proposed by L.A. Zadeh in the late 1970's for representing uncertain pieces of information expressed by fuzzy linguistic statements [4], and later developed in an artificial intelligence perspective [5]. Formally speaking, the proposal is quite similar to the one made some thirty years before by the economist Shackle [2], who had considered a non probabilistic view of uncertainty based on the idea of degree of potential surprise, which can be modelled as $N(A^c)$. Namely the more you believe in A^c, the more surprising you find the occurrence of A. The degree $N(A)$ was explicitly used by Cohen [3] under the name "Baconian probability" capturing the idea of provability. Especially if you can prove A with some confidence, you cannot at the same time prove its negation, which makes condition $\min(N(A), N(A^c)) = 0$ natural. So, condition $N(A) > 0$ expresses that A is *prima facie* an accepted belief, its absolute value expressing the strength of acceptance. Such Baconian probabilities, viewed as shades of certainty, are claimed to be more natural than probabilities in legal matters. Deciding whether someone is guilty cannot be based on statistics, nor on subjective probabilities: you must prove guilt using convincing arguments. About a decade later, Spohn [7] introduced ordinal conditional functions, now called ranking functions, as a basis for a dynamic theory of epistemic states. Ranking functions κ are a variant of potential surprize, taking values on the non-negative integers, that is $g_\kappa(A) = 2^{-\kappa(A^c)}$ is a degree of necessity. The theory of ranking functions and possibility theory can be developed in parallel, even if they were independently devised [14].

4.2   Graded Possibilistic Logic

Possibilistic logic, PosLog for short, has been developed for about thirty years [6,15]. Basic possibilistic logic has been first introduced in AI as a tool for the logical handling of uncertainty in a qualitative way. A basic possibilistic logic formula is a pair (p, α) made of a classical logic formula p associated with a certainty level $\alpha \in (0, 1]$, viewed as a lower bound of a *necessity measure*, i.e., (p, α) is understood as $N(p) \geq \alpha$. A Poslog base is a conjunction of weighted formulas in PosLog. Formulas of the form $(p, 0)$ do not contain any information and are not part of the language. The interval $[0, 1]$ can be replaced by any linearly ordered scale.

The axioms of PosLog [6] are those of propositional logic where each axiom schema is now supposed to hold with maximal certainty, i.e., is assigned level 1. PosLog has two inference rules:

- if $\beta \leq \alpha$ then $(p, \alpha) \vdash (p, \beta)$ (certainty weakening);
- $(\neg p \lor q, \alpha), (p, \alpha) \vdash (q, \alpha), \forall \alpha \in (0, 1]$ (modus ponens).

We may equivalently use certainty weakening with the PosLog counterpart of the resolution rule: $(\neg p \lor q, \alpha), (p \lor r, \alpha) \vdash (q \lor r, \alpha), \forall \alpha \in (0, 1]$.

Using certainty weakening, the following inference rule is valid:

$$(\neg p \lor q, \alpha), (p \lor r, \beta) \vdash (q \lor r, \min(\alpha, \beta)) \text{ (weakest link resolution)}.$$

The idea that in a reasoning chain, the certainty level of the conclusion is the smallest of the certainty levels of the formulas involved in the premises is at the basis of the syntactic approach proposed by Rescher [16] for plausible reasoning, and would date back to Theophrastus, an Aristotle's follower.

An interesting feature of possibilistic logic is its ability to deal with inconsistency. Indeed a PosLog base Γ has an inconsistency level $incl(\Gamma)$ defined as the least degree such that the set of formulas with a strictly greater weight is consistent. When $incl(\Gamma) > 0$, the propositional part of Γ is inconsistent but the consequences of the (consistent) maximal subset of formulas above the inconsistency level are said to be *non-trivial*.

A possibilistic logic base is semantically equivalent to a possibility distribution that restricts the set of interpretations (w.r.t. the propositional language) that are more or less compatible with the base. Instead of an ordinary subset of models as in classical logic, we have a fuzzy set of models, since the violation of a formula by an interpretation becomes a matter of degree. A PosLog formula (p, α) encodes the statement $N(p) \geq \alpha$. Its semantics is given by the possibility distribution $\pi_{(p,\alpha)}$ defined by:

$$\pi_{(p,\alpha)}(s) = 1 \text{ if } s \models p \text{ and } \pi_{(p,\alpha)}(s) = 1 - \alpha \text{ if } s \models \neg p$$

The underlying idea is that any model of p should be fully possible, and that a counter-model of p is all the less possible as p is more certain, i.e., as α is higher. It can be easily checked that $\pi_{(p,\alpha)}$ is the least informative (i.e., maximizing possibility degrees) possibility distribution whose associated necessity measure N satisfies $N(p) \geq N_{(p,\alpha)}(p) = \alpha$. We write $\pi \models (p, \alpha)$ instead of $N(p) \geq \alpha$ to denote the satisfaction by epistemic models.

A PosLog knowledge base $\Gamma = \{(p_i, \alpha_i), i = 1, \ldots, m\}$, corresponding to the conjunction of PosLog formulas (p_i, α_i), is semantically associated with the possibility distribution: $\pi_\Gamma(s) = \min_{i=1}^{m} \pi_{(p_i,\alpha_i)}(s)$. Thus, π_Γ is the least informative possibility distribution that is a model of each weighted formula (p_i, α_i) in Γ. Due to the min-decomposability of necessity measures, $N(\bigwedge_i p_i) \geq \alpha \Leftrightarrow \forall i, N(p_i) \geq \alpha$, and then any possibilistic propositional base can be put in clausal form. Possibilistic logic with the inference rules recalled above is sound and complete with respect to this semantics.

4.3 Generalized Possibilistic Logic

In basic possibilistic logic, only conjunctions of possibilistic logic formulas are allowed. But since (p, α) is semantically interpreted as $N(p) \geq \alpha$, a possibilistic formula can be manipulated as a propositional formula that is true (if $N(p) \geq \alpha$) or false (if $N(p) < \alpha$). Then possibilistic formulas can be combined with all propositional connectives, including disjunction and negation. This is *generalized possibilistic logic* (GPL) [17]. GPL is a two-tiered propositional logic, in which propositional formulas are encapsulated by weighted modal operators interpreted in terms of uncertainty measures from possibility theory. Let $\Lambda_k = \{0, \frac{1}{k}, \frac{2}{k}, \ldots, 1\}$ with $k \in \mathbb{N} \setminus \{0\}$ be a finite set of certainty degrees, and let $\Lambda_k^+ = \Lambda_k \setminus \{0\}$. The language of GPL, $\mathcal{L}_{\mathbf{N}}^k$, with $k + 1$ certainty levels is built on top of the propositional language $\mathcal{L}$ as follows:

– If $p \in \mathcal{L}$ and $\alpha \in \Lambda_k^+$, then $\mathbf{N}_\alpha(p) \in \mathcal{L}_{\mathbf{N}}^k$.
– If $\varphi \in \mathcal{L}_{\mathbf{N}}^k$ and $\psi \in \mathcal{L}_{\mathbf{N}}^k$, then $\neg\varphi$ and $\varphi \wedge \psi$ are also in $\mathcal{L}_{\mathbf{N}}^k$.

Here we use the notation $\mathbf{N}_\alpha(p)$, instead of (p, α), emphasizing the closeness with modal logic. So, an agent asserting $\mathbf{N}_\alpha(p)$ has an epistemic state π such that $N(p) \geq \alpha > 0$. Hence $\neg\mathbf{N}_\alpha(p)$ stands for $N(p) < \alpha$, which, given the finiteness of the set of considered certainty degrees, means $N(p) \leq \alpha - \frac{1}{k}$ and thus $\Pi(\neg p) \geq 1 - \alpha + \frac{1}{k}$. Let $\nu(\alpha) = 1 - \alpha + \frac{1}{k}$. Then, $\nu(\alpha) \in \Lambda_k^+$ iff $\alpha \in \Lambda_k^+$, and $\nu(\nu(\alpha)) = \alpha, \forall \alpha \in \Lambda_k^+$. Thus, we can write $\mathbf{\Pi}_\alpha(p) \equiv \neg\mathbf{N}_{\nu(\alpha)}(\neg p)$. So, in GPL, like in MEL (retrieved by identifying $\square p$ with $\mathbf{N}_1(p)$ and $\lozenge p$ with $\mathbf{\Pi}_1(p)$) one can distinguish between the absence of sufficient certainty that p is true $(\neg\mathbf{N}_\alpha(p))$ and the stronger statement that p is somewhat certainly false $(\mathbf{N}_\alpha(\neg p))$.

The semantics of GPL is as in Poslog defined in terms of normalized possibility distributions over propositional interpretations, where possibility degrees are limited to Λ_k. A model of a GPL formula $\mathbf{N}_\alpha(p)$ is any Λ_k-valued possibility distribution π such that $N(p) \geq \alpha$, where N is the necessity measure induced by π, and then the standard definition for $\pi \models \varphi_1 \wedge \varphi_2$ and $\pi \models \neg\varphi$. As usual, π is called a model of a set of GPL formulas Γ, written $\pi \models \Gamma$, if π is a model of each formula in Γ. We write $\Gamma \models \phi$, for Γ a set of GPL formulas and ϕ a GPL formula, iff every model of Γ is also a model of ϕ. Note that a formula in GPL will not always have one least specific possibility distribution that satisfies it. For instance, the set of possibility distributions satisfying the disjunction '$\mathbf{N}_\alpha(p) \vee \mathbf{N}_\alpha(q)$' no longer has a unique least informative model as it is the case for conjunctions in PosLog. In fact, there are two of them: $\pi_{(p,\alpha)}$ and $\pi_{(q,\alpha)}$. The soundness and completeness of the following axiomatization of GPL holds with respect to the above semantics [17]:

(PL) Axioms of propositional logic for $\mathcal{L}_N^k$-formulas
(K) $\mathbf{N}_\alpha(p \to q) \to (\mathbf{N}_\alpha(p) \to \mathbf{N}_\alpha(q))$
(N) $\mathbf{N}_1(\top)$
(D) $\mathbf{N}_\alpha(p) \to \mathbf{\Pi}_1(p)$
(W) $\mathbf{N}_{\alpha_1}(p) \to \mathbf{N}_{\alpha_2}(p)$, if $\alpha_1 \geq \alpha_2$

with modus ponens as the only inference rule.

Note in particular that when α is fixed we get a fragment of the modal logic KD. See [11] for a survey of previous studies on the links between modal logics and possibility theory. When $k = 1$, GPL with a value scale Λ_1 coincides with MEL. Figure 1 recaps the links between propositional logic, its extensions PosLog and MEL, and GPL that generalizes both. Note that in MEL, we have $\mathbf{\Pi}_1(p) \equiv \neg\mathbf{N}_1(\neg p)$, whereas in GPL we only have $\mathbf{\Pi}_1(p) \equiv \neg\mathbf{N}_{\frac{1}{k}}(\neg p)$ if $k > 1$.

GPL and MEL are suitable for reasoning about the revealed beliefs of an external agent (and not introspection). It captures the idea that while a consistent epistemic state of an agent about the world is represented by a normalized possibility distribution over possible worlds, the meta-epistemic state of another agent about the former's epistemic state is a family of possibility distributions. Standard inference in GPL is co-NP complete, but other kinds of inference (e.g., using specificity criterion) can be more complex; see [17] for a detailed study.

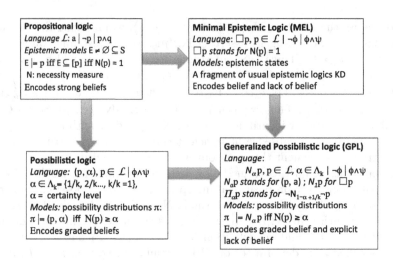

Fig. 1. Comparison of belief logics in possibility theory

5 Applications to Reasoning and Decision

PosLog and GPL proved to be useful for modeling some forms of commonsense reasoning [15].

Reasoning with Exceptions. A rule having exceptions "if p then q, generally", denoted by $p \rightsquigarrow q$, is understood formally in possibility theory as the constraint $\Pi(p \wedge q) > \Pi(p \wedge \neg q)$ on a possibility distribution π describing the normal course of things. Any finite set $\Delta = \{p_i \rightsquigarrow q_i, i = 1, \cdots, n\}$ of conditional statements is represented by a set constraints of the form $\Pi(p_i \wedge q_i) > \Pi(p_i \wedge \neg q_i)$. The family of possibility distributions on S compatible with these constraints, if not empty, possesses a maximal element according to the principle of minimal specificity, e.g., [18]. This principle assigns to each interpretation s the highest possibility level (yielding a well-ordered partition of S) without violating the constraints. This defines a unique complete plausibility preorder on S. Let $E_1, \ldots, E_k$ be the obtained partition. A possibility distribution π_Δ can be defined on S by $\pi_\Delta(s) = \frac{k+1-i}{k}$ if $s \in E_i, i = 1, \ldots, k$. Note that this numerical scale is arbitrary. Namely the range of π is used as an ordinal scale.

Each default $p_i \rightsquigarrow q_i \in \Delta$ can be turned into a possibilistic clause of the form $(\neg p_i \vee q_i, N(\neg p_i \vee q_i))$, where N is computed from π_Δ induced by the set of constraints corresponding to the conditional knowledge base. We thus obtain a possibilistic logic base Γ_Δ encoding the generic knowledge embedded in Δ. As shown in [19], non trivial inference of q from $\Gamma_\Delta \cup \{p\}$ using possibilistic logic turns out to be equivalent to inferring $p \rightsquigarrow q$ from Δ using rational closure in the sense of Lehmann and Magidor [20]. As for preferential inference [21], it can be captured in GPL since its semantics in terms of possibilistic logic reads: Δ implies $p \rightsquigarrow q$ if and only if $\forall \pi$, for which $\Pi(p_i \wedge q_i) > \Pi(p_i \wedge \neg q_i), \forall i = 1, \ldots, k$, we have that $\Pi(p \wedge q) > \Pi(p \wedge \neg q)$. It just needs a proper encoding of such statements in GPL. In [17], we encode this constraint as $\bigvee_{i=1}^{k} \mathbf{N}_{\frac{i}{k}}(\neg p \vee q) \wedge \neg \mathbf{N}_{\frac{i}{k}}(\neg p \vee \neg q)$.

Prioritized Constraints. Basic possibilistic logic can also be used for representing preferences. Then, each logic formula (p, α) represents a goal p to be reached with some priority level α (rather than a statement that is more or less believed) [22]. Beyond PosLog, interpretations (corresponding to the different alternatives) can be compared in terms of vectors acknowledging the satisfaction or the violation of the formulas associated with the different goals, using suitable order relations. Thus, partial orderings of interpretations can be obtained [23].

Non-monotonic Logic Programming. Another remarkable application of GPL is its capability to encode answer set programs, adopting a three-valued scale $\Lambda_2 = \{0, 1/2, 1\}$. In this case, we can discriminate between propositions we are fully certain of and propositions we consider only more plausible than not. This is sufficient to encode non-monotonic ASP rules (with negation as failure) within GPL and lay bare their epistemic semantics. For instance, an ASP rule of the form $a \leftarrow b \wedge \mathrm{not}\, c$, where "not" denotes negation as failure, is encoded as $\mathbf{N}_1(b) \wedge \boldsymbol{\Pi}_1(\neg c) \rightarrow \mathbf{N}_1(a)$ in GPL. See [17] for theoretical results, and [24] for the GPL encoding of Pearce equilibrium logic [25].

Reasoning About Ignorance. *All that is known is p* means that p is known but no q such that $q \models p$ is known. This can be expressed in MEL by the formula $OKp \equiv \Box p \wedge \bigwedge_{w \models p} \Diamond p_w$ where p_w is a propositional formula whose only ontic model is w [10]. It is clear that $E \models \bigwedge_{w \models p} \Diamond p_w$ if and only if $[p] \subseteq E$, so that the only epistemic model of OKp is $[p]$. In terms of set-functions it corresponds to the guaranteed possibility $\Delta(A) = \min_{w \in A} \pi(A)$ [5]. The expression of OKp can be related to the Moebius transform of a belief function in the sense of Shafer [10]. In GPL, one can similarly define a formula whose only epistemic model is a possibility distribution π [17].

6 Perspectives Beyond GPL

As per the above discussions, GPL is a rather versatile tool for knowledge representation and reasoning that has simpler and more natural semantics than epistemic logic, and can handle degrees of belief. On this basis, several lines of research can be envisaged, some of which were already investigated in the recent past

- *Comparative certainty logic.* Rather than using weighted formulas, one might wish to represent at a syntactic level statements of the form "*p is more certain than q*". This kind of statements can be to some extent captured by GPL. The idea is to define a partial certainty ordering on a propositional belief base, and to induce a partial order on the propositional language that completes it, via suitable inference (see [17,26] for examples of such logics). This topic is related to Lewis logics of comparative possibility [27].
- *Symbolic possibilistic logic.* Instead of using weights from a totally ordered scale, one may use pairs (p, x) where x is a symbolic entity that stands for an unknown weight. Then we can model the situation where only a partial

ordering between ill-known weights is known. See [28] for this approach. It differs from comparative certainty logic [26], because the latter does not rely on the principle of minimal specificity at work in possibilistic logic.

- *The logic of capacities.* The connection between epistemic logic and possibility theory can be extended to general capacities (monotonic set-functions). The corresponding modal logics are then non-normal, only classical, and the semantics in terms of capacities is close to neighborhood semantics. This enlarged logical framework seems to capture some inconsistency-tolerant logics (like Belnap logic [8]). See for instance [29].
- *Multisource logic.* Not only can we attach certainty weights to propositional formulas, but we can attach to them sources (agents) that supply this information. So we can extend possibilistic formulas with labels describing groups of agents, and extend the corresponding inference machinery [30], so as to compute for each proposition which agents believe it and to what extent.
- *Multiagent epistemic reasoning.* Finally, it would be of interest to apply GPL to problems where an agent's decisions are based on what agents mutually know about each other's knowledge. It would mean extending GPL to nested modalities labelled by agents, and compare to the state of the art in multiagent modal logics [9]. A question to address is whether accessibility relations of usual epistemic logics are needed or not to solve such problems. This is a topic of interest, not explored yet.

References

1. Dubois, D., Prade, H.: Formal representations of uncertainty. In: Bouyssou, D.E. (ed.) Decision-Making - Concepts and Methods, pp. 85–156. Wiley, Hoboken (2009). Chapter 3
2. Shackle, G.: Expectation in Economics. Cambridge University Press, Cambridge (1949)
3. Cohen, L.: The Probable and the Provable. Clarendon Press, Oxford (1977)
4. Zadeh, L.: Fuzzy sets as a basis for a theory of possibility. Fuzzy Sets Syst. **1**, 3–28 (1978)
5. Dubois, D., Prade, H.: Possibility theory: qualitative and quantitative aspects. In: Gabbay, D., Smets, P. (eds.) Handbook of Defeasible Reasoning and Uncertainty Management Systems, vol. 1, pp. 169–226. Kluwer Academic, Dordrecht (1998)
6. Dubois, D., Lang, J., Prade, H.: Possibilistic logic. In: Gabbay, C.D., Hogger, J., Robinson, D.N. (eds.) Handbook of Logic in Artificial Intelligence and Logic Programming, vol. 3, pp. 439–513. Oxford University Press, Oxford (1994)
7. Spohn, W.: The Laws of Belief. Oxford University Press, Oxford (2012)
8. Belnap, N.D.: A useful four-valued logic. In: Epstein, G. (ed.) Modern Uses of Multiple-Valued Logic, pp. 8–37. Reidel, Dordrecht (1977)
9. Fagin, R., Halpern, J.Y., Moses, Y., Vardi, M.: Reasoning About Knowledge. MIT Press, Cambridge (1995)
10. Banerjee, M., Dubois, D.: A simple logic for reasoning about incomplete knowledge. Int. J. Approx. Reason. **55**(2), 639–653 (2014)
11. Banerjee, M., Dubois, D., Godo, L., Prade, H.: On the relation between possibilistic logic and modal logics of belief and knowledge. J. Appl. Non Class. Log. **27**(3–4), 206–224 (2017)

12. Halpern, J.Y., Moses, Y.: Toward a theory of knowledge and ignorance: preliminary report. In: Apt, K. (ed.) Logics and Models of Concurrent Systems, vol. 13, pp. 459–476. Springer, Heidelberg (1985). https://doi.org/10.1007/978-3-642-82453-1_16

13. Ciucci, D., Dubois, D.: A modal theorem-preserving translation of a class of three-valued logics of incomplete information. J. Appl. Non Class. Log. **23**, 321–352 (2013)

14. Dubois, D., Prade, H.: Qualitative and semi-quantitative modeling of uncertain knowledge - a discussion. In: Beierle, C., Brewka, G., Thimm, M. (eds.) Computational Models of Rationality, Essays Dedicated to Gabriele Kern-Isberner on the Occasion of her 60th Birthday, pp. 280–296. College Publications (2016)

15. Dubois, D., Prade, H.: Possibilistic logic - an overview. In: Siekmann, J.H. (ed.) Computational Logic. Handbook of the History of Logic, vol. 9, pp. 283–342. Elsevier, Amsterdam (2014)

16. Rescher, N.: Plausible Reasoning. Van Gorcum, Amsterdam (1976)

17. Dubois, D., Prade, H., Schockaert, S.: Generalized possibilistic logic: foundations and applications to qualitative reasoning about uncertainty. Artif. Intell. **252**, 139–174 (2017)

18. Benferhat, S., Dubois, D., Prade, H.: Possibilistic and standard probabilistic semantics of conditional knowledge bases. J. Log. Comput. **9**(6), 873–895 (1999)

19. Benferhat, S., Dubois, D., Prade, H.: Practical handling of exception-tainted rules and independence information in possibilistic logic. Appl. Intell. **9**(2), 101–127 (1998)

20. Lehmann, D., Magidor, M.: What does a conditional knowledge base entail? Artif. Intell. **55**(1), 1–60 (1992)

21. Kraus, S., Lehmann, D., Magidor, M.: Nonmonotonic reasoning, preferential models and cumulative logics. Artif. Intell. **44**(1–2), 167–207 (1990)

22. Lang, J.: Possibilistic logic as a logical framework for min-max discrete optimisation problems and prioritized constraints. In: Jorrand, P., Kelemen, J. (eds.) FAIR 1991. LNCS, vol. 535, pp. 112–126. Springer, Heidelberg (1991). https://doi.org/10.1007/3-540-54507-7_10

23. Ben Amor, N., Dubois, D., Gouider, H., Prade, H.: Preference modeling with possibilistic networks and symbolic weights: a theoretical study. In: Proceedings of ECAI 2016, pp. 1203–1211 (2016)

24. Dubois, D., Prade, H., Schockaert, S.: Stable models in generalized possibilistic logic. In: Proceedings of KR 2012, pp. 519–529 (2012)

25. Pearce, D.: Equilibrium logic. Ann. Math. Artif. Intell. **47**, 3–41 (2006)

26. Touazi, F., Cayrol, C., Dubois, D.: Possibilistic reasoning with partially ordered beliefs. J. Appl. Log. **13**(4), 770–798 (2015)

27. Lewis, D.: Counterfactuals. Blackwell Publishers, Worcester (1986)

28. Cayrol, C., Dubois, D., Touazi, F.: Symbolic possibilistic logic: completeness and inference methods. J. Log. Comput. **28**(1), 219–244 (2018)

29. Ciucci, D., Dubois, D.: A two-tiered propositional framework for handling multi-source inconsistent information. In: Antonucci, A., Cholvy, L., Papini, O. (eds.) ECSQARU 2017. LNCS (LNAI), vol. 10369, pp. 398–408. Springer, Cham (2017). https://doi.org/10.1007/978-3-319-61581-3_36

30. Belhadi, A., Dubois, D., Khellaf-Haned, F., Prade, H.: Multiple agent possibilistic logic. J. Appl. Non Class. Log. **23**(4), 299–320 (2013)

Discrete Sugeno Integrals and Their Applications

Agnès Rico[(⊠)]

ERIC, Université Claude Bernard Lyon 1, 43 bld du 11 novembre,
69100 Villeurbanne, France
agnes.rico@univ-lyon1.fr

Abstract. This paper is an overview of the discrete Sugeno integrals and their applications when the evaluation scale is a totally ordered set. The various expressions of the Sugeno integrals are presented. Some major characterisation results are recalled: results based on characteristic properties and act-based axiomatisation. We discuss the properties of a preference relation modelling by a Sugeno integral. We also present its power expression to represent a dataset and its interpretation with a set of if-then rules.

Keywords: Sugeno integrals · Qualitative data representation
If-then rules · Decision under uncertainty
Multi-criteria decision making

1 Introduction

In the setting of multi-criteria decision making or in decision under uncertainty the use of numerical data is not always natural. For instance, by lack of time or lack of precision there is uncertainty pervading the data collection. An illustration is that we cannot ask too many questions to the users in order to obtain information and even if the users answer on a real interval scale, making an average with these data is a questionable process. So it is natural to search for qualitative aggregation operators, i.e., operators involving only order.

The problems concerning multi-criteria decision making or the decision under uncertainty can be modeled in a similar way, even if there is at least one difference: The number of criteria present in multi criteria decision making is finite but the number of states of nature in decision under uncertainty is infinite even if they play the same role. In this paper, we consider a finite number of states of nature.

- Decision making under uncertainty is modeled by a 4-tuple $(N, X, A, \succeq)$ where
  - $N = \{1, \cdots, n\}$ is the set of possible states of nature;
  - X is the set of possible consequences of acts;
  - A is the sets of acts, i.e., the functions $x : N \to X$;

© Springer Nature Switzerland AG 2018
D. Ciucci et al. (Eds.): SUM 2018, LNAI 11142, pp. 18–32, 2018.
https://doi.org/10.1007/978-3-030-00461-3_2

- $\succeq$ is a preference relation on A. We assume that $\succeq$ is a complete preorder. The 4-tuple is equipped with:
- a utility function that rates the consequences: $u : X \to U$ where U is a totally ordered set;
- A monotonic set function, named capacity, $\mu : 2^N \to V$ where V is a totally ordered scale, represents the knowledge of the decision maker concerning the states of nature.

U and V are supposed to be included in the same totally ordered set L. Hence it is possible to compare $\mu(S)$ and $u(x)$ for each state of nature S and each act x. In order to simplify the notations we assume that U and V are the same set L.

- A multi-criteria decision-making problem is a 3-tuple $(C, A, \succeq)$ where
  - $C = \{1, \cdots, n\}$ is the set of criteria;
  - A is the set of objects or alternatives which are evaluating according to the criterion;
  - $\succeq$ is the relation of preference on A.

Each criterion i is equipped with an evaluation scale X_i, so an act is represented by an element of $\Pi_{i=1}^n X_i$. We assume the existence of n utility functions $u_i : X_i \to L$ where L is totally ordered set. In order to simplify the notations, we assume that an alternative is represented by a n-vector in L^n. The capacity $\mu : 2^C \to L$ represents the importance of the subsets of criteria.

Sugeno integrals introduced in [15] is an ordinal counterpart of the classical Lebesgue integral. They are commonly used as qualitative aggregation operators [14] or as qualitative preference functional in multi criteria decision making or decision under uncertainty. They calculate a global evaluation of alternatives or objects according to some local evaluations. For many years, many authors have studied its properties [2]. A major result is that the Sugeno integral is the only solution for aggregating scores only using min and max [23]. For instance, in the framework of multi-criteria decision making the properties can be read in [22] and in [18] in the context of decision under uncertainty.

This paper is an overview of the discrete Sugeno integrals and their applications when the evaluation scale is a totally ordered set. We start with the various expressions of the Sugeno integrals. All of them are based on a capacity. The capacities are used in uncertainty modeling [12,13], multiple criteria aggregation [9,17] or in game theory [11]. They qualitatively represent the importance of sets of criteria or the likelihood of sets of possible states of nature. The complexity of the Sugeno integrals is not good because of this capacity; the number of coefficients present in the Sugeno integral expression grows exponentially with the number of criteria involved in the model. A main difficulty is to identify these coefficients. In order to decrease this number we consider the Moebius transform of the capacity which contains only the minimal information to construct the capacity. Section 3 deals with particular Sugeno integrals before presenting different characterisations of the Sugeno integrals in Sect. 4. There are two families of characterisation results, those based on characteristic properties and those based on acts axiomatisation. Section 5 is devoted to the representation of the qualitative data with Sugeno integrals. We present how to elicit Sugeno integrals compatible with a dataset and the link with version space learning. An elicited capacity contains properties concerning the subsets of criteria involved. This part also presents the representation of Sugeno integrals in terms of if-then rules involving thresholds [8,10,19]. Then we briefly present a logical representation for the Sugeno integrals in the setting of possibilistic logic. The conclusion is a quick overview of some extensions of Sugeno integrals in order to increase its power of expression.

2   The Discrete Sugeno Integral

We use the notations of multi criteria decision making where objects are evaluated according to some criteria. $C = \{1, \cdots, n\}$ is the set of criteria. 2^C is the power set, the evaluation scale L is a totally ordered set with a top denoted 1 and a bottom denoted 0. We assume that L is equipped with an involutive negation denoted $1-$. An object is represented by a vector $x = (x_1, \ldots, x_n) \in L^n$ where x_i is the evaluation of x according to the criterion i. $\bar{\alpha}$ will denote the constant vector equal to α everywhere. We also need the following notation: If $x, y \in L^n$ are objects and $A \subseteq C$ is a subset of criteria, xAy is the object or alternative equal to x on A and y on $\bar{A}$ where $\bar{A}$ is the complement of A.

The respective weights of the subsets of criteria are represented by a capacity or fuzzy measure.

Definition 1. *A capacity (or fuzzy measure) is a set function $\mu : 2^C \to L$ such that $\mu(\emptyset) = 0$, $\mu(C) = 1$ and $A \subseteq B$ implies $\mu(A) \leq \mu(B)$.*

Two capacities play a particular role: the necessity measure N which are capacities such that $N(A \cap B) = \min(N(A), N(B))$; and the possibility measures Π which are capacities such that $\Pi(A \cup B) = \max(\Pi(A), \Pi(B))$. $\Pi(\{i\})$ will be denoted π_i. We have $\Pi(A) = \max_{i \in A} \pi_i$ forall $A \subseteq C$. In multi-criteria decision making π_i is the importance of the criterion i. In case of decision under uncertainty π_i represents the plausibility of the state i.

The conjugate capacity of a capacity μ is defined by: $\mu^c(A) = 1 - \mu(\bar{A})$, where $\bar{A}$ is the complement of A. Note that $\Pi^c = N$.

The Sugeno integral was first defined in [15,16].

Definition 2. *The Sugeno integral of $x \in L^n$ with respect to a capacity μ is $S_\mu(x) = \max_{\alpha \in L} \min(\alpha, \mu(x \geq \alpha))$, where $\mu(x \geq \alpha) = \mu(\{i \in C | x_i \geq \alpha\})$.*

The Sugeno integral can be expressed in various ways [15,22–24] useful to study its properties:

- $S_\mu(x) = \max_{A \subseteq C} \min(\mu(A), \min_{i \in A} x_i)$,
- $S_\mu(x) = \min_{A \subseteq C} \max(\mu(\bar{A}), \max_{i \in A} x_i)$

- If we consider x and $(\cdot)$ the permutation on $C = \{1, \cdots, n\}$ such that $x_{(1)} \leq \cdots \leq x_{(n)}$ then $S_\mu(x) = \max_{i=1}^n \min(x_{(i)}, \mu(A_{(i)}))$ where $A_{(i)} = \{(i), \cdots, (n)\}$.
- Given a capacity μ, for all $\alpha > 0, \alpha \in L$, let $\mu_\alpha : 2^C \to \{0, 1\}$ be the Boolean capacity defined by $\mu_\alpha(A) = \begin{cases} 1 & \text{if } \mu(A) \geq \alpha \\ 0 & \text{otherwise.} \end{cases}$ μ_α is named a α-cut of μ. The capacity μ can be reconstructed from the μ_α's as follows: $\mu(A) = \max_{\alpha > 0} \min(\alpha, \mu_\alpha(A))$ and $S_\mu(u) = \max_{\alpha > 0} \min(\alpha, S_{\mu_\alpha}(u))$.

The Sugeno integral can be interpreted as a weight median:

$$S_\mu(x) = median[x_1, \cdots, x_n, \mu(A_{(2)}), \cdots, \mu(A_{(n)})].$$

Using this property, let us present an example in order to illustrate that the capacity models the attitude of the decision maker.

Example 1. Let us consider $\alpha, \beta \in L$, $\beta < \alpha$ and the object $\bar{\alpha} A \bar{\beta}$. We have $S_\mu(\bar{\alpha} A \bar{\beta}) = median(\alpha, \beta, \mu(A))$.

- If $\mu(A)$ is small enough, i.e., $\mu(A) \leq \beta < \alpha$ then $S_\mu(\bar{\alpha} A \bar{\beta}) = \beta$. In this case the decision maker does not have much confidence and the bad outcome occurs.
- If $\mu(A)$ is great enough, i.e., $\mu(A) \geq \alpha > \beta$ then $S_\mu(\bar{\alpha} A \bar{\beta}) = \alpha$. The decision maker does have confidence and the good outcome occurs.
- If $\mu(A)$ is mild, then $S_\mu(\bar{\alpha} A \bar{\beta}) = \mu(A)$.

The Sugeno integral definition uses a capacity which implies to know 2^n terms. We can reduce this complexity defining the Sugeno integrals with the qualitative Moebius transform of the capacity. The Moebius transform $\mu_\#$ of a capacity μ is the set function defined as follows:

$$\mu_\#(T) = \begin{cases} \mu(T) & \text{if } \mu(T) > \max_{x \in T} \mu(T \backslash \{x\}) \\ 0 & \text{otherwise.} \end{cases}$$

The Moebius transform contains the minimal information to construct the capacity μ: $\mu(A) = \max_{T \subseteq A} \mu_\#(T)$. The sets T satisfying $\mu_\#(T) > 0$ are named the focal sets of μ. The set of focal sets of μ will be denoted by $\mathcal{F}(\mu)$. The focal sets play a central role in the applications for instance in the if-then rules (see Sect. 5.2). Note that the Moebius transform of a possibility measure is its possibility distribution: $\Pi_\#(A) = \pi(s)$ if $A = \{s\}$ and 0 otherwise.

The focal sets are sufficient to calculate the Sugeno integral:

$$S_\mu(x) = \max_{A \subseteq C} \min(\mu_\#(A), \min_{i \in A} x_i) \tag{1}$$

The focal sets of the conjugate μ^c of μ are also sufficient to calculate the Sugeno integral [10]:

$$S_\mu(x) = \min_{T \in \mathcal{F}(\mu^c)} \max(1 - \mu_\#^c(T), \max_{i \in T} x_i). \tag{2}$$

3 Sugeno Integrals with Respect to Possibility Measure or Necessity Measure

This section is devoted to the link between the Sugeno integrals and the Sugeno integrals with respect to a possibility measure or to a necessity measure.

There are two elementary aggregation schemes on a finite totally ordered chain:

- The pessimistic one $\min_{i=1}^{n} x_i$ which is very demanding since in order to obtain a good evaluation an object needs to satisfy all the criteria.
- The optimistic one, $\max_{i=1}^{n} x_i$ which is very loose since one fulfilled criterion is enough to obtain a good evaluation.

These two aggregation schemes can be slightly generalised by means of importance levels or priorities $\pi_i \in L$, on the criteria i, thus yielding weighted minimum and maximum [1]:

$$MIN_\pi(x) = \min_{i=1}^{n} \max(1-\pi_i, x_i) = S_N(x); \quad MAX_\pi(x) = \max_{i=1}^{n} \min(\pi_i, x_i) = S_\Pi(x).$$

In these aggregation operations, a fully important criterion can alone bring the whole global score to 0 ($MIN_\pi(x)$) or to 1 ($MAX_\pi(x)$), according to the chosen attitude (resp. pessimistic or optimistic). If $\pi = 1$ everywhere then the Sugeno integral with respect to N is the minimum and the Sugeno integral with respect to Π is the maximum.

Example 2. Let $L = [0,1]$ be the evaluation scale, $\{1,2,3\}$ be the set of criteria and π be the possibility distribution. We consider two alternatives x and y and we calculate MIN_π and MAX_π:

	1 2 3	MIN_π	MAX_π
π	1 1 0.6		
x	1 1 0	0.4	1
y	1 0 0	0	1

We can see that x violates the least important criterion 3 only, and it makes the global evaluation positive according to the pessimistic criterion, while y is ruled out because it violates one essential criterion. But the optimistic evaluation puts the two decisions on a par because they both satisfy one important criterion.

Any qualitative capacity is a lower possibility measure [36]: $\mu(A) = \min_{\Pi \in C_q(\mu)} \Pi(A)$ where $C_q(\mu) = \{\pi : \Pi(A) \geq \mu(A) \forall A \subseteq C\}$ is the possibilistic core of the capacity μ. A Sugeno integral is both a lower possibility and an upper necessity integral [37]:

Theorem 1. $S_\mu(x) = \min_{\pi \in C_q(\mu)} S_\Pi(x)$ *and* $S_\mu(x) = \max_{\pi \in C_q(\mu^c)} S_N(x)$.

Note that if a capacity is defined by very few possibility distributions it may simplify the calculation of Sugeno integrals.

4   Different Characterisations of the Sugeno Integrals

In decision under uncertainty and in multi criteria decision making, an interesting question is Can I use Sugeno integrals to model a given aggegation function or a given preference relation? This section presents characterisation results in order to be able to answer this question. In these results comonotone vectors play a major role, let us recall their definition. Two objects or alternatives $x, y \in L^n$ are said to be comonotone if for every $i, j \in \{1, \cdots, n\}$, if $x_i < x_j$ then $y_i \leq y_j$.

4.1   Characteristic properties

This section presents properties that allow a functional to be a Sugeno integral.

Theorem 2. *Let $I : L^C \to L$. There is a capacity μ such that $I(x) = S_\mu(x)$ for every $x \in L^C$ if and only if the following properties are satisfied*

1. *Comonotonic maxitivity: $I(\max(x, y)) = \max(I(x), I(y))$, for any comonotone $x, y \in L^C$.*
2. *Weak minitivity: $I(\min(\alpha, x)) = \min(\alpha, I(x))$, for every $\alpha \in L$ and $x \in L^C$ (min-homogeneity).*
3. *$I(\mathbf{1}_C) = 1$.*

Most older formulations of this theorem [29,31] add an assumption of increasing monotonicity of the functional I (if $x \geq y$ then $I(x) \geq I(y)$) to the three conditions (1–3). This condition turned out to be reduntant. The conditions (1–3) are necessary and sufficient [30]. They can be equivalently replaced by conditions (1'–3') below.

1'. Comonotone minitivity: $I(\min(x, y)) = \min(I(x), I(y))$, for any comonotone $x, y \in L^C$.
2'. Weak maxitivity $I(\max(\beta, x)) = \max(\beta, I(x))$, for every $\beta \in L$ and $x \in L^C$ (max-homogeneity).
3'. $I(\mathbf{0}_C) = 0$.

The existence of these two equivalent characterisations is due to the possibility of writing Sugeno integral in conjunctive and disjunctive forms equivalently The three conditions (1–3) presented above can be replaced by:

1". I is an increasing function,
2". I is idempotent $S_\mu(\bar{\alpha}) = \alpha$,
3". I is non compensatory $S_\mu(\bar{\alpha}A\bar{0}), S_\mu(\bar{1}A\bar{\alpha}) \in \{S_\mu(\bar{1}A\bar{0}), \alpha\}$.

4.2   Act-Based Axiomatisation of Sugeno Integral

Now we consider a preference relation $\succeq$ on a set of acts. This section deals with the axioms that allow a preference relation to be representable by a Sugeno integral. The axioms proposed in [12] are as follows:

A1. *Totality:* $\succeq$ is a non-trivial total preorder, i.e., it is transitive and complete, and $x \succ y$ for some acts.

WP3. *Weak compatibility with constant acts:* $\forall A \subseteq C, \forall \alpha, \beta \in L, \forall x, \alpha \geq \beta$ implies $\bar{\alpha} A x \succeq \bar{\beta} A x$.

RCD. *Restricted conjunctive dominance:* For any acts x, y and any constant act $\bar{\alpha}$, $\bar{\alpha} \succ y$ and $x \succ y$ imply $\bar{\alpha} \wedge x \succ y$.

Upper-bounding the expectation of an act x better than another act y by a constant value that is better than the utility of act y still yields an act better than y.

RDD. *Restricted max-dominance:* For any acts x, y and any constant act $\bar{\alpha}$, $y \succ \bar{\alpha}$ and $y \succ x$ imply $y \succ \bar{\alpha} \vee x$.

If an act y is preferred to an act x and also to the constant act $\bar{\alpha}$ if the bad consequences of x are upgraded to α, y is still preferred to x.

Note that axioms $A1$ and $WP3$ entail Pareto-dominance: if $x \geq y$ then $x \succeq y$ (see Lemma 4 in [12]). Moreover, **RCD** and **RDD** make sense for one-shot decisions, i.e., without repetition, making the compensation of bad results by good ones impossible.

We recall here the main result about this axiomatization for decision under uncertainty.

Theorem 3 ([12]). *The following propositions are equivalent:*

- $(L^C, \succeq)$ *satisfies A1, plus WP3, RCD, RDD.*
- *there exists a finite chain L of preference levels, an L-valued monotonic set-function μ, and an L-valued utility function u on X, such that $x \succeq y$ if and only if $S_\mu(u(x)) \geq S_\mu(u(y))$.*

In the above theorem RCD and RDD can be replaced by a non-compensation axiom:

$$\textbf{Axiom NC:} \begin{cases} \bar{1}A\bar{\beta} \sim \bar{\beta} \;\text{ or }\; \bar{1}A\bar{\beta} \sim \bar{1}A\bar{0} \;\textbf{(DNC)} \\ \text{and} \\ \bar{\alpha}A\bar{0} \sim \bar{\alpha} \;\text{ or }\; \bar{\alpha}A\bar{0} \sim \bar{0} \;\textbf{(CNC)} \end{cases}$$

For a binary act, this axiom reflects the fact that $S_\mu(\bar{\alpha}A\bar{\beta})$ equals either α, β, or the likelihood of A, $\mu(A)$, or the likelihood of $\overline{A}$, $\mu(\overline{A})$.

Axiom **NC** also reflects the fact that Sugeno integral is decomposable in terms of medians [33].

- If the **RCD** axiom is replaced with the **CD** axiom in Theorem 3: For any acts x, y, z, $z \succ y$ and $x \succ y$ imply $z \wedge x \succ y$, then the capacity is a possibility measure.
- If the **RDD** axiom is replaced with the **DD** axiom in Theorem 3: For any acts x, y, z, $y \succ z$ and $y \succ x$ imply $y \succ z \vee x$, then the capacity is a necessity measure.

There exists another way to characterise the Sugeno integral with respect to a possibility measure. This method is based on the pessimistic and optimistic substitute profiles [34].

Definition 3. *Let π be a possibility distribution and σ be a permutation on the criteria such that: $\pi_{\sigma(1)} \leq \cdots \leq \pi_{\sigma(n)}$. A pessimistic profile $x^{\sigma,-}$ and an optimistic profile $x^{\sigma,+}$ are defined: $\forall i = 1, \ldots, n$,*

$$x_i^{\sigma,-} = \min_{j:\pi_j \geq \pi_{\sigma(i)}} x_j = \overset{i}{\underset{j=1}{\min}} x_{\sigma(j)}; \quad x_i^{\sigma,+} = \max_{j:\pi_j \geq \pi_{\sigma(i)}} x_j = \overset{i}{\underset{j=1}{\max}} x_{\sigma(j)}. \quad (3)$$

Clearly, $x^{\sigma,-}$ and $x^{\sigma,+}$ are respectively obtained by taking the worst and the best evaluation at each level of π. Observe that only the ordering of elements i induced by π on $\mathcal{C}$ is useful in the definition of the pessimistic and optimistic profiles associated with x. Hence we obtain the following characterisation [34]:

Theorem 4. *The following propositions are equivalent:*

- $\succeq$ *satisfies A1, plus WP3, RCD, RDD and the axiome $\Pi 4$: There exists a permutation σ on $\mathcal{C}$ such that for all $A \subseteq C, \bar{1}A\bar{0} \sim (\bar{1}A\bar{0})^{\sigma,+}$.*
- *there exists a possibility measure Π, such that $x \succeq y$ if and only if $S_\Pi(x) \geq S_\Pi(y)$.*

We obtain a similar theorem for the necessity measure replacing the optimistic profiles by the pessimistic profile,i.e., replacing the axiom $\Pi 4$ by the axiom: There exists a permutation σ on $\mathcal{C}$ such that for all $A \subseteq C, \bar{1}A\bar{0} \sim (\bar{1}A\bar{0})^{\sigma,-}$.

Another act-based characterisation uses the link between the Sugeno integral model and the non compensatory model. In a non compensatory model: For each $x \in X$ we have a subset $A_i^x \subseteq X_i$ containing the levels on attribute i that are satisfactory for x. In order to be at least as good as x an alternative must have evaluations that are satisfactory for x on a subset of attributes interpreted as sufficiently important to warrant preference on x.

When the preference relation $\succeq$ is a weak order that has a numerical representation hence the Sugeno integral and the non compensatory model are equivalent (see [9] fore more details). This is a useful link to prove the following characterisation [9,19]:

Theorem 5. *A binary relation $\succeq$ on X has a representation in the discrete Sugeno integral model if and only if it is a weak order satisfying*

- *the order-denseness condition: there is a countable set $Y \subseteq X$ that is dense in X for $\succeq$.*
- *strongly 2-graded axiom: for all criterion i for all $x, y, z, w \in X$ and all*
$$a_i \in X_i, \begin{cases} x \succeq z \\ y \succeq w \\ z \geq w \end{cases} \Rightarrow \begin{cases} a_i\{i\}x \succeq z \\ or \\ x_i\{i\}y \succeq w \end{cases}.$$

For each criterion i, the strongly 2-graded axiom can be interpreted as follows: If y_i is better than x_i w.r.t $w \in X$ there does not exist an element in X_i that could be worse than x_i.

The Sugeno integral fails to satisfy strict monotonicity. We can find x and y such that $x > y$ and $S_\mu(x) = S_\mu(y)$. More precisely the Sugeno integral satisfies strict monotonicity if and only if there exists a unique important criteria according to which the objects are compared. This lack of discrimination is due to the drowning effect. If we consider two objects with identical good evaluations (resp. bad), the maximum (the minimum) give the same global evaluation. In this case this effect can be fixed using the leximin and the leximax. The leximin procedure (resp. the leximax procedure) consists in ordering both vectors in increasing (resp. decreasing) order and then lexicographically comparing them.

When looking at the Sugeno integral expression, we can refine in different ways: using the leximin, the leximax. We can also refine the capacity considering big-stepped mass functions (see [32] for details of the procedure).

5   Representation of Qualitative Data

5.1   Elicitation of the Sugeno Integral

We consider a dataset and we want to elicit Sugeno integrals to represent it [6, 7]. A piece of data (x, α) is a vector $x \in L^n$ and $\alpha \in L$ is the global evaluation of x. (x, α) is compatible with a Sugeno integral S_μ if and only if $S_\mu(x) = \alpha$. A dataset $(x^k, \alpha_k)_k$ is said to be compatible if there exists at least one family of Sugeno integrals compatible with all the data. An interesting result is: a dataset is compatible if each pair of data (x^i, α_i) and (x^j, α_j) are compatible. Hence if there exists a family of Sugeno integrals compatible with the dataset, this family is obtained calculating the intersection of families obtained for each piece of data. If the dataset is not compatible we look for a partition where each set is a compatible dataset. For instance, the partition is obtained using simulated annealing algorithm. This procedure is based on the theoretical results presented in [27]. The major one is the set of the capacities compatible with a piece of data (x, α):

Proposition 1. $\{\mu | S_\mu(x) = \alpha\} = \{\mu | \check{\mu}_{x,\alpha} \le \mu \le \hat{\mu}_{x,\alpha}\}$ *where* $\check{\mu}_{x,\alpha}$ *and* $\hat{\mu}_{x,\alpha}$ *are capacities defined by*

$$\check{\mu}_{x,\alpha}(A) = \begin{cases} \alpha \text{ if } \{i | x_i \ge \alpha\} \subseteq A \\ 0 \text{ otherwise} \end{cases} \quad and \quad \hat{\mu}_{x,\alpha}(A) = \begin{cases} \alpha \text{ if } A \subseteq \{i | x_i > \alpha\} \\ 1 \text{ otherwise.} \end{cases}$$

$\check{\mu}_{x,\alpha}$ is a necessity measure with the focal sets A_α and C. $\hat{\mu}_{x,\alpha}(A)$ is a possibility measure with respect to the possibility distribution $\hat{\pi}_{x,\alpha}(i) = 1$ if $x_i \le \alpha$ and α otherwise.

The set of compatible capacities with the dataset $(x^k, \alpha_k)_k$ is $\{\mu | \max_k \check{\mu}_{x^k,\alpha_k} \le \mu \le \min_k \hat{\mu}_{x^k,\alpha_k}\}$ where $\check{\mu}_{x^k,\alpha_k}$ and $\hat{\mu}_{x^k,\alpha_k}$ are the capacities associated to (x^k, α_k). This set of solutions can be empty so it is worth noticing that in order to compare $\max_k \check{\mu}_{x^k,\alpha_k}$ and $\min_k \hat{\mu}_{x^k,\alpha_k}$ it is not necessary to calculate their values and to compare them on each subset of criteria. It is proved in [27] that the set of compatible capacities is not empty if and only if for all $\alpha_k < \alpha_l$ we have $\{i \in C | x_i^l \ge \alpha_l\} \not\subseteq \{i \in C | x_i^k > \alpha_k\}$.

Example 3. Let us present the illustration proposed in [6]: the evaluation of the subjective mental workload of stewards, stewardesses or cabin chiefs in a NASA-TLX setting.

The data used were collected during a series of five rotations of planes pertaining to a big European airline company. Overall, the data set contains 840 pairs (x, α) where x is a 6-component vector corresponding to the evaluation of the six NASA-TLX criteria and α is a global corresponding evaluation of their mental workload. All the values are in $[0, 1]$. In practice, the participants were invited to estimate the criteria and their global mental workload by drawing a cross on a bounded segment. A simulated annealing algorithm gave the following results. From the 840 data, 811 satisfy the representability condition (the global evaluation is between the minimum and the maximum of the locals evaluations). The 29 outliers were not further taken in consideration. We obtained a partition with 30 subsets. The larger subset contains 168 pieces of data, 9 subsets have more than 30 elements and 6 subsets have less than 2 elements.

Note that the way we calculate a family of Sugeno integrals when there is a new piece of data taken into account is similar to the evolution of the version space learning [6]. More precisely the elicitation of the Sugeno integral can be encoded as a bipolar variant of the problem of version space learning:

- $\mathcal{X}$ is the n-tuple $x = (x_1, \cdots, x_n)$
- $\mathcal{S}_\mu$: x associated to a global evaluation α is a positive and negative example with a level α.
- The hypotheses space $\mathcal{H}$ is the space of Sugeno integrals defined on 2^C.

A piece of data (x, α) is viewed as two triples $(x, 0, \alpha)$ and $(x, 1, \alpha)$. x is both guaranteed to be possible and is not impossible at level α. α indicates how far the piece of data is from an extreme case. Piece of data $(x, 1)$ is totally possible and non guaranteed to be impossible at level α; piece of data $(x, 0)$ is impossible and non guaranteed to be possible. The bounds of the Sugeno integrals compatible with a dataset can be interpreted as the guaranteed possibility and the non impossibility that the new piece of data is near the reference piece of data for the compatible set considered.

5.2   If-then Rules Representation

This section deals with the equivalence between the Sugeno integral and a set of if-then rules.

The inequality $S_\mu(x) \geq \gamma$ is equivalent to $\exists T \in \mathcal{F}(\mu)$ such that $\forall i \in T, x_i \geq \gamma$; hence each focal T of μ with level $\mu_\#(T)$ corresponds to the selection rule:

$$R_T^s: \text{If } x_i \geq \mu_\#(T) \text{ for all } i \in T \text{ then } S_\mu(x) \geq \mu_\#(T).$$

Symmetrically [10], the inequality $S_\mu(x) \leq \alpha$ is equivalent to $\exists F \in \mathcal{F}(\mu^c)$ with $\mu^c(F) \geq 1 - \alpha$ s.t. $\forall x_i \in F \ x_i \leq \alpha$; hence each focal set of the conjugate μ^c corresponds to the elimination rule:

R_F^e: If $x_i \leq 1 - \mu_\#^c(F)$ for all $i \in F$ then $S_\mu(x) \leq 1 - \mu_\#^c(F)$.

More precisely, [19], the Sugeno integral is equivalent to a set of single-threshold rules. This set of rules corresponds to the focal sets of the capacity.

Example 4. This result was applied with algal and pollution [10]. Samples were collected on retention basins in the Eastern suburbs of Lyon before groundwater seepage. The waters contain many pollutants (like heavy metals, pesticides, hydrocarbons, PCB, ...) and our aim is to assess their impact on the water ecosystem health.

First, algal growth (C) was measured as a global indicator of algal health with standardized biossay (NF EN ISO 8692), then bioassays more specific of different metabolic pathways were carried out: chlorophyll fluorescence (F) as phosynthesis indicator and two enzymatic activities, Alkaline phosphatase Activity (APA) and Esterase Activity (EA) as nutrients metabolism indicators. Assays were performed after 24 hours exposure to samples collected during 7 different rainfall events and for different periods of the year for dry weather. With all of these local evaluations the expert can give an evaluation concerning the impact of the pollutants. After the elicitation of the Sugeno integral corresponding to the expert, the rules obtained include pieces of knowledge familiar to experts in the application area.

The representation with if-then rules permits to understand that the Sugeno integral can be encoded by means of a possibilistic logic base with positive clauses $\{[\wedge_{j \in T} x_j, \mu_\#(T)], T \text{ focal set}\}$ (see [8] for more details). Note that the Sugeno integral can also be written in a N-base using the transformation between a Δ-base and a N base. In such representation for each criterion i the predicates are $p_i(\gamma)$ with the convention $p_i(\gamma)$ is true if $x_i \geq \gamma$. We consider a simple first order language where the above predicates are the atoms, the connective are negation, conjunction and disjunction. Let us present an example.

Example 5. We consider three criteria $C = \{1, 2, 3\}$ and the evaluation scale $L = \{1, \lambda_2, \lambda_1, 0\}$ with $1 > \lambda_2 > \lambda_1 > 0$. We consider a Sugeno integral with respect to μ defined by $\mu(\{1\}) = \mu(\{2\}) = \lambda_1$, $\mu(\{3\}) = \lambda_2$, $\mu(\{1,2\}) = \lambda_1$, $\mu(\{2,3\}) = \lambda_2$, $\mu(\{1,3\}) = \mu(\{1,2,3\}) = 1$. The qualitative Moebius transform is: $\mu_\#(\{1\}) = \mu_\#(\{2\}) = \lambda_1$, $\mu_\#(\{3\}) = \lambda_2$ and $\mu_\#(\{1,3\}) = 1$.

It corresponds to the following Δ-possibilistic logic base:

$$B_\uparrow^\Delta(\mu) = \{[p_1(1) \wedge p_3(1), 1], [p_3(\lambda_2), \lambda_2], [p_1(\lambda_1), \lambda_1], [p_2(\lambda_1), \lambda_1]\}.$$

Hence the Sugeno integral can be interpreted as follows:

- An object that satisfies properties 1 and 3 to degree 1 is fully satisfactory.
- An object that satisfies property 3 to at least level λ_2 is satisfactory at least to degree λ_2.
- An object that satisfies any of the properties 1, 2 to at least level λ_1 is satisfactory at least to degree λ_1.

6   Conclusion

This survey presents the Sugeno integral as a useful tool for decision under uncertainty and multicriteria decision making when the evaluation scale is a finite totally ordered set. In order to increase the expressive power of Sugeno integrals we can consider various types of scales (negative, lattices) or we can consider the Sugeno integral as a particular case of more general aggregation functions. We expect that their study entails a better knowledge about it.

Let us conclude this survey with an overview of some extensions.

The Sugeno integrals can be considered as particular cases of families of aggregation functions. Noticing that the Sugeno integrals are idempotent lattice polynomial functions [22], a natural extension is to consider that the evaluation scale is a bounded lattice [2,33]. We can also consider aggregation functions satisfying a part of the properties characterizing the Sugeno integrals. For instance, we can cite the Level-dependant Sugeno integrals [4] and the Universal integrals [5]. Another extension is to consider a bipolar scale when decisions are evaluated by sets of positive and negative arguments [38]. As mentioned above a currently used hypothesis is to assume that there is a common evaluation scale for all criteria. In order to remove this one we can use a Sugeno utility function which is a combination of a Sugeno integral and a unary order preserving map on each criterion: $S_{\mu,\varphi}(x) = \max_{I \in \mathcal{F}(\mu)} \min(\mu(I), \min_{i \in I} \varphi_i(x_i))$ for all $x \in X = \Pi_{i=1} X_i$ where μ is a capacity, X_i is the evaluation scale on the criterion i with a top 1_{X_i} and a bottom 0_{X_i}, $\varphi = (\varphi_1, \cdots, \varphi_n)$ and for all $i \in C$, $\varphi_i : X_i \to L$ is order preserving with $\varphi_i(0_{X_i}) = 0$ and $\varphi_i(1_{X_i}) = 1$. The conjunction or the disjunction of Sugeno utility functions modelises the ordinal functions representable with a set of multi-threshold rules [26,35].

Another extension is to change the action of the weights of criteria on the local evaluations before aggregating them. The Sugeno integrals correspond to the saturation levels scheme, i.e., the values of the scale are reduced all the more as criteria are less important. There are two others possible actions:

- Softening thresholds: π_i makes local evaluation all the less demanding as the criterion is less important.
- Drastic thresholds: if x_i is higher than the threshold it is considered fully satisfied but it is all the more severely decreased if the criterion is important.

The extension of these schemes defines the other integrals named qualitative integrals [37].

The last extension we mention here is the desintegrals [39]. They are used when the evaluation scale for each criterion is decreasing, i.e., 0 is better score than 1 but the scale of the global evaluation is increasing. There are two methods: Reversing the local evaluation and then aggregate; aggregating the local score and then reverse the global evaluation.

Acknowledgements. Dear Didier thank you very much for your comments.

References

1. Dubois, D., Prade, H.: Weighted minimum and maximum operations. Inf. Sci. **39**, 205–210 (1986)
2. Couceiro, M., Marichal, J.-L.: Associative polynomial functions over bounded distributive lattices. Order **28**, 1–8 (2011)
3. Halas, R., Mesiar, R., Pocs, J., Torra, V.: A note on some algebraic properties of discrete Sugeno integrals. Fuzzy Sets Syst. (2018)
4. Meziar, R., Mesiarova-Zemankova, A., Ahmad, K.: Level-dependent Sugeno integral. IEEE Trans. Fuzzy Syst. **17**, 167–172 (2009)
5. Klement, E.P., Mesiar, R., Spizzichino, F., Stupnanova, A.: Universal integrals based on copulas. Fuzzy Optim. Decis. Mak. **13**, 273–286 (2014)
6. Prade, H., Rico, A., Serrurier, M.: Elicitation of sugeno integrals: a version space learning perspective. In: Rauch, J., Raś, Z.W., Berka, P., Elomaa, T. (eds.) ISMIS 2009. LNCS (LNAI), vol. 5722, pp. 392–401. Springer, Heidelberg (2009). https://doi.org/10.1007/978-3-642-04125-9_42
7. Prade, H., Rico, A., Serrurier, M., Raufaste, E.: Elicitating sugeno integrals: methodology and a case study. In: Sossai, C., Chemello, G. (eds.) ECSQARU 2009. LNCS (LNAI), vol. 5590, pp. 712–723. Springer, Heidelberg (2009). https://doi.org/10.1007/978-3-642-02906-6_61
8. Dubois, D., Prade, H., Rico, A.: The logical encoding of Sugeno integrals. Fuzzy Sets Syst. **241**, 61–75 (2014)
9. Bouyssou, D., Marchant, T., Pirlot, M.: A conjoint measurement approach to the discrete Sugeno integral. In: Brams, S.J. (ed.) The Mathematics of Preference, Choice and Order: Essays in Honor of Peter C. Studies in Choice and Welfare, pp. 85–109. Springer, Heidelberg (2009). https://doi.org/10.1007/978-3-540-79128-7_6
10. Dubois, D., Durrieu, C., Prade, H., Rico, A., Ferro, Y.: Extracting decision rules from qualitative data using sugeno integral: a case-study. In: Destercke, S., Denoeux, T. (eds.) ECSQARU 2015. LNCS (LNAI), vol. 9161, pp. 14–24. Springer, Cham (2015). https://doi.org/10.1007/978-3-319-20807-7_2
11. Schmeidler, D.: Cores of exact games. J. Math. Anal. Appl. **40**(1), 214–225 (1972)
12. Dubois, D., Prade, H., Sabbadin, R.: Qualitative decision theory with Sugeno integrals. In: Grabisch, M. (ed.) Fuzzy Measures and Integrals Theory and Applications, pp. 314–322. Physica Verlag, Heidelberg (2000)
13. Dubois, D., Prade, H., Sabbadin, R.: Decision-theoretic foundations of qualitative possibility theory. Eur. J. Oper. Res. **128**, 459–478 (2001)
14. Grabisch, M., Labreuche, C.: A decade of application of the Choquet and Sugeno integrals in multi-criteria decision aid. Ann. Oper. Res. **175**, 247–286 (2010)
15. Sugeno, M.: Theory of fuzzy integrals and its applications. Ph.D. thesis, Tokyo Institute of Technology (1974)
16. Sugeno, M.: Fuzzy measures and fuzzy integrals: a survey. In: Gupta, M.M., et al. (eds.) Fuzzy Automata and Decision Processes, pp. 89–102. North-Holland, Amsterdam (1977)
17. Grabisch, M.: The application of fuzzy integrals in multicriteria decision making. Europ. J. Operat. Res. **89**(3), 445–456 (1996)
18. Dubois, D., Marichal, J.-L., Prade, H., Roubens, M., Sabbadin, R.: The use of the discrete Sugeno integral in decision making: a survey. Int. J. Uncertain. Fuzziness Knowl.-Based Syst. **9**, 539–561 (2001)

19. Greco, S., Matarazzo, B., Slowinski, R.: Axiomatic characterization of a general utility function and its particular cases in terms of conjoint measurement and rough-set decision rules. Europ. J. Oper. Res. **158**, 271–292 (2004)
20. Rico, A., Grabisch, M., Labreuche, C., Chateauneuf, A.: Preference modeling on totally ordered sets by the Sugeno integral. Discret. Appl. Math. **147**, 113–124 (2005)
21. Murofushi, T., Sugeno, M.: Fuzzy measures and fuzzy integrals. In: Grabisch, M., Murofushi, T., Sugeno, M. (eds.) Fuzzy Measures and Integrals - Theory and Applications, pp. 3–41. Physica Verlag (2000)
22. Marichal, J.-L.: An axiomatic approach of the discrete Sugeno integral as a tool to aggregate interacting criteria in a qualitative framework. IEEE Trans. Fuzzy Syst. **9**(1), 164–172 (2001)
23. Marichal, J.-L.: On Sugeno integral as an aggregation function. Fuzzy Sets Syst. **114**, 347–365 (2000)
24. Dubois, D., Prade, H.: Qualitative possibility functions and integrals. In: Pap, E. (ed.) Handbook of Measure Theory, vol. 2, pp. 1469–1521. Elsevier (2002)
25. Dubois, D., Prade, H., Rico, A., Teuheux, B.: Generalized qualitative Sugeno integrals. Inf. Sci. 1–17 (2017)
26. Brabant, Q., Couceiro, M., Dubois, D., Prade, H., Rico, A.: Extracting decision rules from qualitative data via Sugeno utility functional. In: Proceeding of IPMU (2018)
27. Rico, A., Grabisch, M., Labreuche, C., Chateauneuf, A.: Preference modelling on totally ordered sets by the Sugeno integral. Discret. Appl. Math. **147**, 113–124 (2005)
28. Murofushi, T., Sugeno, M.: Null sets with respect to fuzzy measures. In: Proceedings of the 3rd International Fuzzy System Association World Congress (ISFA 1989), pp. 172–175 (1989)
29. Ralescu, D., Sugeno, M.: Fuzzy integral representation. Fuzzy Sets Syst. **84**, 127–133 (1996)
30. Rico, A.: Modélisation des Préférences pour l'Aide à la Décision par l'Intégrale de Sugeno, Ph. D. thesis, Université Paris I Sorbonne (2002)
31. de Campos, L.M., Bolaños, M.J.: Characterization and comparison of Sugeno and Choquet integrals. Fuzzy Sets Syst. **52**, 61–67 (1992)
32. Dubois, D., Fargier, H.: Making discrete Sugeno integrals more discriminant. Int. J. Approx. Reason. **50**, 880–898 (2009)
33. Marichal, J.L.: Weighted lattice polynomials. Discret. Math. **309**(4), 814–820 (2009)
34. Dubois, D., Rico, A.: Axiomatisation of Discrete Fuzzy Integrals with Respect to Possibility and Necessity Measures. In: Torra, V., Narukawa, Y., Navarro-Arribas, G., Yañez, C. (eds.) MDAI 2016. LNCS (LNAI), vol. 9880, pp. 94–106, 147, 195–197. Springer, Cham (2016). https://doi.org/10.1007/978-3-319-45656-0_8
35. Couceiro, M., Dubois, D., Prade, H., Rico, A.: Enhancing the expressive power of sugeno integrals for qualitative data analysis. In: Kacprzyk, J., Szmidt, E., Zadrożny, S., Atanassov, K.T., Krawczak, M. (eds.) IWIFSGN/EUSFLAT -2017. AISC, vol. 641, pp. 534–547. Springer, Cham (2018). https://doi.org/10.1007/978-3-319-66830-7_48
36. Dubois, D., Prade, H., Rico, A.: Qualitative capacities as imprecise possibilities. In: van der Gaag, L.C. (ed.) ECSQARU 2013. LNCS (LNAI), vol. 7958, pp. 169–180. Springer, Heidelberg (2013). https://doi.org/10.1007/978-3-642-39091-3_15

37. Dubois, D., Prade, H., Rico, A.: Residuated variants of Sugeno integrals: Towards new weighting schemes for qualitative aggregation methods. Inf. Sci. **329**, 765–781 (2016)
38. Greco, S., Rindone, F.: Bipolar fuzzy integrals. Fuzzy Sets Syst. **220**, 21–33 (2013)
39. Dubois, D., Prade, H., Rico, A.: Qualitative integrals and desintegrals: how to handle positive and negative scales in evaluation. In: Proceedings of IPMU, pp. 347–355 (2012)

Regular Papers

A Credal Extension of Independent Choice Logic

Alessandro Antonucci$^{(\boxtimes)}$ and Alessandro Facchini

Istituto Dalle Molle di Studi Sull'Intelligenza Artificiale (IDSIA),
Lugano, Switzerland
{alessandro,alessandro.facchini}@idsia.ch

Abstract. We propose an extension of Poole's *independent choice logic* based on a relaxation of the underlying independence assumptions. A *credal* semantics involving multiple joint probability mass functions over the possible worlds is adopted. This represents a conservative approach to probabilistic logic programming achieved by considering *all* the mass functions consistent with the probabilistic facts. This allows to model tasks for which independence among some probabilistic choices cannot be assumed, and a specific dependence model cannot be assessed. Preliminary tests on an object ranking application show that, despite the loose underlying assumptions, informative inferences can be extracted.

Keywords: Probabilistic logic programming · Imprecise probabilities
PSAT · Independence

1 Introduction

Probabilistic logic programming (PLP) is an emerging area in AI research aiming to develop reasoning tools for relational domains under uncertainty [9]. PLP can be either intended as a combination of logic programs with probabilistic statements, or a part of the wider current interest in probabilistic programming [4] specialised to highly structured probability spaces. After some early proposals such as Probabilistic Horn Abduction [26] and Independent Choice Logic [27], Distribution Semantics [28] and Probabilistic Datalog [14], PLP is currently subject of intense research as well as the theoretical basis for many real-world applications [8].

Most of these theories make assumptions such as the program acyclicity and the mutual independence of the probabilistic facts. This leads to the specification of a single probability mass function over the possible words (*least models*) associated to the logic program. In more recent times, several approaches are trying to go beyond some of these assumptions, but without giving away independence. In some of these cases [7,22], this leads to the adoption of a *credal* semantics, i.e., a PLP does not define a single mass function but a set of joint mass functions over the least models. In this paper we propose a different path consisting

© Springer Nature Switzerland AG 2018
D. Ciucci et al. (Eds.): SUM 2018, LNAI 11142, pp. 35–49, 2018.
https://doi.org/10.1007/978-3-030-00461-3_3

in a generalisation of Poole's independent choice logic which keeps the acyclicity condition on programs, but relaxes the independence assumptions. Such an extension can be intended as a credal, conservative, semantics considering *all* the joint mass functions over the least models being also consistent with the probabilistic facts. To introduce our motivations for such proposal, consider the following example.

Example 1. Two urns contain red, green and blue balls. Let $\mu_1 = [0.60; 0.30; 0.10]$ and $\mu_2 = [0.20; 0.35; 0.45]$ denote the normalised proportions of the colours in the urns.[1] Assuming that a ball is randomly drawn from each urn, the joint probability over the nine possible outcomes is the left matrix in Table 1. The matrix on the right corresponds to a different situation in which a ball is randomly drawn from the first urn, while the second ball is required to have a colour different from that of the first. To achieve that, an unbiased coin is flipped to decide which one of the two permitted colours should be picked. Both joint mass functions are consistent with the above marginal probabilities (i.e., the sums of the rows/columns are the values in μ_1 and μ_2).

Table 1. Two joint mass functions over two ternary variables sharing the same marginal probabilities. Dark grey cells corresponds to a query of interest (first ball non-red, second ball non-green).

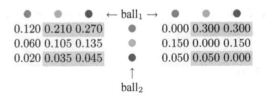

This trivial example makes clear that marginal probabilities are in general not sufficient to uniquely determine a joint mass function unless additional assumptions, such as independence or an explicit dependence, are made. If no further assumptions can be made, the most conservative approach corresponds to consider *all* the joint mass functions consistent with the constraints and computing inferences with respect to the multiple probabilistic specification. This is clarified by the following example.

Example 2. With the same setup as in Example 1, we want to compute the probability of having the first ball not red and the second not green. Assuming independence, this query has probability $(1 - \mu_1(r))(1 - \mu_2(g)) = 0.56$. The same result can be obtained by summing the (consistent) dark grey values in the matrix on the left of Table 1. Under the coin-based dependence relation, the result becomes

---

[1] We use semicolons to separate the elements of an array and commas to separate the two bounds of an interval.

instead 0.65 *(sum of the dark grey values in the matrix on the right). With no dependence/independence assumptions, the probability of the query can be only said to belong to the interval* $[0.5, 0.7]$. *These values are the maximum and the minimum obtained from a linear programming task involving the nine joint probabilities as optimisation variables and the consistency with the marginals (together with non-negativity and normalization) as linear constraints.*

The feasible region of the linear programming task in the above example is a convex set of (joint) probability mass functions, i.e., a *credal set* [3,20]. Inferences based on credal sets are conservatively intended as the computation of the lower/upper bounds of the query with the mass function varying in the credal set. The interval estimate is also called *imprecise probability*. When coping with relational domains, adding independence statements leading to a unique, "precise", specification might be not always a tenable assumption, and the imprecise-probabilistic technique we present in this paper can be regarded as the most conservative approach to the modeling of a condition of *ignorance* about the dependencies among the atoms in a program.

After reviewing some necessary background concepts in Sect. 2, our proposal is first introduced by an example in Sect. 3 and then formalised as a theory in Sect. 4. An inference algorithm is derived in Sect. 5, while the results of some preliminary experiments are reported in Sect. 6. Conclusions and outlooks are in Sect. 7.

2   Background

We review here some background information about logic programming and its probabilistic extension.

Logic Programming. A *term* t is defined as being either a constant or a variable. An *atom* $r(t_1, \ldots, t_k)$ is thence obtained by applying a relational symbol r to a sequence of terms $t_1, \ldots, t_k$. We identify Boolean propositional variables with 0-ary predicates. An atom a or its negation $\neg a$ is called a *literal*. A *clause* has form

$$a_0 \leftarrow a_1, \ldots, a_m, \neg a_{m+1}, \ldots, \neg a_n, \tag{1}$$

where a_i are atoms for each $i = 0, \ldots, n$, a_0 is called the *head* of the clause, while the other atoms are the *body*. *Facts* are clauses with empty body. If a clause is not a fact is called a *rule*. A *ground* atom (literal, clause) is an atom (literal, clause) that does not contain any variable. The *grounding* of a clause is a clause obtained by uniformly replacing constants for the variables in the considered clause. A *logic program* $\mathbf{P}$ is a finite set of clauses. The *Herbrand base* of a program is the set of all ground instances of atoms in the program. A *query* Q for a program $\mathbf{P}$ is a set of ground literals whose positive part belong to the Herbrand base of $\mathbf{P}$. A logic program is *acyclic* if there is an assignment of a positive integer to each element of the Herbrand base such that for every grounding of a rule the number assigned to the head is greater then each number assigned to an element of the body. An *interpretation* ι of a program $\mathbf{P}$ is a function assigning a truth value to each member of the Herbrand base for $\mathbf{P}$. An interpretation ι is said to be a stable model for $\mathbf{P}$ if for every ground atom a, a is true in ι if and only if a is a fact in $\mathbf{P}$ or it is the head of the grounding of a rule in $\mathbf{P}$ whose base is true in ι. A negation $\neg a$ is true in ι if and only if the atom a is not true in ι.

The following classical result [2] is used in the rest of the paper.

Proposition 1. *An acyclic logic program has a unique stable model.*

Probabilistic Logic Programming. There are different ways of representing probabilistic information with logic programming. A possible approach is to annotate or extend clauses with probabilities (see, e.g., [21,24]). Another, nowadays very popular, approach consists in adding to a logic program *independent* probabilistic alternatives representing mutually independent random events with a finite number of different outcomes, such as tossing a coin or rolling a die. Within this latter stream, various languages have been proposed such as PRISM [28], ICL [27], pD [15], LPAD [30] or ProbLog [10]. Despite the differences in their syntactical presentation, under Sato's distribution semantics all these formalisms are in general comparable in their expressive power [29]. In this work we focus on Poole's ICL. The choice of this formalism, rather than another among the languages listed above, is due to the fact that its explicit set-theoretical formulation of probabilistic choices will simplify the presentation of our approach. ICL syntax and semantics are introduced here below.

Independent Choice Logic (ICL). An ICL theory is a triple $\langle \mathbf{P}, \mathbf{C}, \mu \rangle$ such that:

- $\mathbf{P}$ is an acyclic logic program.
- $\mathbf{C}$, called the *choice space*, is a family of non-empty sets of ground atoms. The elements of $\mathbf{C}$ are called *alternatives*, and the elements of each alternative *atomic choices*. Atomic choices from the same or different alternatives cannot unify with each other, nor with the head of any clause in $\mathbf{P}$.
- μ specifies a probability mass function over each alternative, i.e., $\mu : \bigcup_{C \in \mathbf{C}} C \to [0, 1]$ with $\sum_{a \in C} \mu(a) = 1$ for each $C \in \mathbf{C}$.

The following example demonstrates the ICL syntax.

Example 3. Consider the setup of Example 1 under the assumption of independence between the colours of the two balls (Table 1 left). Let a_i^j denote the Boolean variable true if and only if the colour j has been drawn from the i-th urn. The example corresponds to an ICL theory over the ground atoms $\{a_i^j\}_{i=1,2}^{j=r,g,b}$. The logical program is empty, i.e., $\mathbf{P} = \emptyset$, as the logical constraints over the atoms can be directly embedded in the choice space $\mathbf{C}$. Such a space contains two alternatives, i.e., $\mathbf{C} := \{C_1, C_2\}$, with $C_i = \{a_i^r, a_i^g, a_i^b\}$ for each $i = 1, 2$. Finally, the probabilities are directly obtained from the marginals, i.e., $\mu(C_i) = \mu_i$ for each $i = 1, 2$.

ICL semantics is defined in terms of *possible words*. A *total choice* $\mathfrak{c}$ for the choice space $\mathbf{C}$ is a choice function selecting exactly one atomic choice from each alternative in $\mathbf{C}$, and we associate to $\mathfrak{c}$ a possible world $w_{\mathfrak{c}}$. For each total choice $\mathfrak{c}$ the program $\mathbf{P} \cup \{a \leftarrow \mid a \in \text{image}(\mathfrak{c})\}$ is acyclic, and therefore by Proposition 1 it has a unique stable model $\iota_{\mathfrak{c}}$. We thence say that an atom a is true at $w_{\mathfrak{c}}$, written $w_{\mathfrak{c}} \models a$, if a is true in $\iota_{\mathfrak{c}}$, and states $w_{\mathfrak{c}} \models \neg a$ if $w_{\mathfrak{c}} \not\models a$. Given a query Q, we write $w \models Q$ whenever $w \models q$, for any member of $q \in Q$.

As alternatives are assumed to be independent, a ICL theory defines a unique mass function μ' over the collection Ω of all the possible worlds as follows:

$$\mu'(w_{\mathfrak{c}}) := \prod_{a \in \text{image}(\mathfrak{c})} \mu(a) \cdot \prod_{a \notin \text{image}(\mathfrak{c})} (1 - \mu(a)). \tag{2}$$

Example 4. The ICL theory in Example 3 has nine possible worlds corresponding to the elements of $C_1 \times C_2$. For each w, the associated probability assigned by μ as in Eq. (2) reproduces the value in Example 1. The query in Example 2 is $Q := \{\neg a_1^r, \neg a_2^g\}$ and it is true only in four possible worlds. The corresponding success probability is $\mu'(Q) = \sum_{w \models Q} \mu'(w) = 0.210 + 0.270 + 0.035 + 0.045$ as in Example 2.

3   A Motivating Example

Consider a possibility space determined by the fact that, on a working day, a person called Andrea is using her car or not, she is working late or not, and that in Milan, the city where she is living, it is raining or not. We denote by r the atomic fact that it rains in Milan, by c the fact that she uses the car, and by w the fact that she is working late. Hence, the elements of the possibility space we consider correspond to the eight possible worlds associated with r, c, w in Table 2, where $\mathbf{t}$ denotes true and $\mathbf{f}$ false.

Let us compute the probability that Andrea will hang out with friends under the following assumptions.

- The probability of rain in Milan is 0.1, that for Andrea using her car 0.5 and that for her working late 0.2.
- If Andrea is with her car or it is raining, she will visit her parents.
- If she is neither visiting her parents nor working late, then she will be hanging out with friends.

Let p denote the fact that Andrea visits her parents, and h the fact associated to her hanging out with friends. In a PLP dialect such as ICL, if we introduce symbols nr, nc and nw to denote the complementary atomic facts associated respectively to r, c and w, the example corresponds to a theory with:

- $\mathbf{C} = \{C_1, C_2, C_3\}$, with alternatives $C_1 = \{r, nr\}$, $C_2 = \{c, nc\}$, $C_3 = \{w, nw\}$;
- $\mu(r) = 0.1$, $\mu(nr) = 0.9$, $\mu(c) = 0.5$, $\mu(nc) = 0.5$, $\mu(w) = 0.2$, $\mu(nw) = 0.8$;
- and $\mathbf{P} = \{p \leftarrow c, p \leftarrow r, h \leftarrow \neg p, nw\}$.

Table 2. A possibility space as a set of possible worlds, and nine mass functions over it.

	ω_1	ω_2	ω_3	ω_4	ω_5	ω_6	ω_7	ω_8
r (Rain in Milan)	t	t	t	t	f	f	f	f
c (Andrea using her car)	t	t	f	f	t	t	f	f
w (Andrea working late)	t	f	t	f	t	f	t	f
$\mu'(\omega_i)$	0.01	0.04	0.01	0.04	0.09	0.36	0.09	0.36
$\mu^{(1)}(\omega_i)$	0.1	0.0	0.0	0.0	0.0	0.4	0.1	0.4
$\mu^{(2)}(\omega_i)$	0.1	0.0	0.0	0.0	0.1	0.3	0.0	0.5
$\mu^{(3)}(\omega_i)$	0.0	0.1	0.0	0.0	0.0	0.4	0.2	0.3
$\mu^{(4)}(\omega_i)$	0.0	0.1	0.0	0.0	0.2	0.2	0.0	0.5
$\mu^{(5)}(\omega_i)$	0.0	0.0	0.1	0.0	0.1	0.4	0.0	0.4
$\mu^{(6)}(\omega_i)$	0.0	0.0	0.1	0.0	0.0	0.5	0.1	0.3
$\mu^{(7)}(\omega_i)$	0.0	0.0	0.0	0.1	0.2	0.3	0.0	0.4
$\mu^{(8)}(\omega_i)$	0.0	0.0	0.0	0.1	0.0	0.5	0.2	0.2

Hence, for instance, $\omega_i \models c$ if and only if $\omega_i \not\models nc$ if and only if $\omega_i \not\models \neg c$, with $i \in \{1, \ldots, 8\}$. It also holds that $\omega_i \models p$, for $i \in \{1, \ldots, 6\}$ and $\omega_i \models \neg p$, for $i = 7, 8$. Similarly $\omega_8 \models h$ and $\omega_i \models \neg h$ for $i \neq 8$. Since alternatives are assumed to be independent, a unique probability mass function μ' (see Table 2) is defined over $\Omega := \{\omega_1, \ldots, \omega_8\}$ and, for instance, the success probability of the query h is $\mu'(\omega_8) := 0.9 \cdot 0.5 \cdot 0.8 = 0.36$. Yet, it is easy to identify realistic situations in which such independence assumptions are violated. E.g., Andrea using her car might be not independent of raining in Milan. While the modelling of a deterministic dependence (e.g., $c \leftarrow r$) or a probabilistic influence (e.g., $\mu(c \mid r) = 0.95$ and $\mu(c \mid nr) = 0.45$) can be described in standard PLPs, a condition of complete ignorance about the relations between two or more atomic facts requires a generalisation of the semantics. For instance, if we do not make any assumption about the (in)dependence relations among the possible choices corresponding to the three variables in our example, the whole set of mass functions consistent with the marginals is the convex hull of the eight mass functions $\{\mu^{(i)}\}_{i=1}^8$ in Table 2 and the probability of the query h can only be said to belong to the range $[0.2, 0.5]$. This interval shrinks to $[0.32, 0.40]$ if no assumptions can be made about the independence between choices on elements of C_1 and choices on elements of C_2, but both are assumed to be independent with respect to choices on C_3.

In the next section we formalise these ideas in a general framework, called *credal choice logic* (CCL), which provides an extension of Poole's independent choice logic in which the independence condition is relaxed.

4   Credal Choice Logic

Syntax. From a syntactical point of view, the idea pursued in this paper is to consider elements of the choice space not as independent alternatives but as *independent families*, each including possibly correlated alternatives. This is formally stated in the following definition.

Definition 1. *A CCL theory $\mathcal{T}$ is a triple $\langle \mathbf{P}, \mathcal{C}, \mu \rangle$ where:*

- **P** *is an acyclic logic program.*
- $\mathcal{C} = \{\mathbf{C}_1, \ldots, \mathbf{C}_k\}$ *is a family of choice spaces (i.e., a set of sets of non-empty sets of ground atoms). Alternatives and atomic choices are intended as in ICL. In particular we assume that:*
  1. *for each choice space $\mathbf{C} \in \mathcal{C}$, no atomic choice in $\bigcup \mathbf{C}$ should unify with the head of any clause in **P**;*
  2. *for each pair $\mathbf{C}, \mathbf{C}' \in \mathcal{C}$ with $\mathbf{C} \neq \mathbf{C}'$, the sets of atomic choices $\bigcup \mathbf{C}$ and $\bigcup \mathbf{C}'$ are disjoint.*
- μ *specifies a probability mass function over each alternative of each choice space, i.e., $\mu : \bigcup \bigcup \mathcal{C} \to [0,1]$ with $\sum_{a \in C} \mu(a) = 1$, for every $C \in \mathbf{C}$, and every $\mathbf{C} \in \mathcal{C}$.*

Contrary to ICL, in a CCL theory the alternatives in a given choice space are not assumed to be independent. Consequently there could be a choice space $\mathbf{C}$ with alternatives $C_1, C_2 \in \mathbf{C}$ such that $C_1 \neq C_2$ but $C_1 \cap C_1 \neq \emptyset$. On the other hand, any ICL theory can be formalised as a CCL theory $\langle \mathbf{P}, \mathcal{C}, \mu \rangle$ whose choice spaces $\mathbf{C} \in \mathcal{C}$ are singletons.

Example 5. Dropping the independence assumption about the relation between *Andrea going by car and raining in Milan in the example in the previous section transforms the corresponding ICL theory in a CCL theory $\mathcal{T}_{\text{friends}} := \langle \mathbf{P}, \{\mathbf{C}_1, \mathbf{C}_2\}, \mu \rangle$, where $\mathbf{C}_1 = \{C_1, C_2\}$, $\mathbf{C}_2 = \{C_3\}$, while **P** and μ are defined as in the original example. The fact fact that alternatives C_1 and C_2 belong to the same choice space means that the choice among elements of C_1 (namely between c and nc) is not assumed to be independent from the choice among elements of C_2 (namely between r and nr). On the other hand the fact that the alternative C_3 does not belong to the same choice space as C_1 and C_2 means that the choice between w and nw is independent of the two previous choices.*

Semantics. As in ICL, the CCL semantics is defined in terms of possible words. A total choice for a family $\mathcal{C}$ of choice spaces is a choice function $\mathfrak{c}_i$ on $\bigcup \mathcal{C}$ such that, for C and C' belonging to the same choice space $\mathbf{C}$, the following coherence condition is satisfied:

$$\mathfrak{c}_i(C) = \mathfrak{c}_i(C'), \tag{3}$$

whenever $\mathfrak{c}_i(C) \in C \cap C'$. Hence a total choice selects coherently (in the sense of Eq. (3)) exactly one atomic choice from each alternative of every choice space of $\mathcal{C}$.

Now, consider the program $\mathbf{P}(\mathfrak{c}) := \mathbf{P} \cup \{a \leftarrow \mid \varphi \in \mathsf{image}(\mathfrak{c})\}$. From the disjointness conditions, $\mathbf{P}(\mathfrak{c})$ is necessarily acyclic and therefore by Proposition 1 it has a unique stable model $I_{(\mathfrak{c})}$. As for ICL, for each total choice $\mathfrak{c}$ we define a corresponding possible world $\omega_{\mathfrak{c}}$ and the associated notion of being true in it.

In the case of ICL, the probability for a possible world is given as in Eq. (2) by the product of the probabilities of the atomic choices true in it and one minus the probabilities of the atomic choices false in it. The distribution semantics we defined for CCL acts similarly but on sets of probabilities.

To illustrate the idea, notice that a total choice $\mathfrak{c}$ restricted to a choice space $\mathbf{C}_i$ determines a partial selection $\mathfrak{c}_i$. Such partial selection can be identified with all total choices extending it, and thus with the collection $E_{\mathfrak{c}_i}$ of possible world in which all elements in the image of $\mathfrak{c}_i$ are true. Such collection represents the possible worlds an agent cannot tell apart if she is given only the partial information provided by $\mathfrak{c}_i$. Let Ω_i be the set of all collections associated with $\mathbf{C}_i$. Assume μ_i is a probability mass function over Ω_i, and define $\mu_i(a) := \sum_{\mathfrak{c}_i : a \in \mathsf{image}(\mathfrak{c}_i)} \mu_i(E_{\mathfrak{c}_i})$, for $a \in \bigcup \mathbf{C}_i$. We say that μ_i *agrees* with μ on $\mathbf{C}_i$ if $\mu_i(a) = \mu(a)$, for every $a \in \bigcup \mathbf{C}_i$. Hence, call $\mathcal{M}_i$ the set of all probability distributions over Ω_i agreeing with μ on $\mathbf{C}_i$.

Each possible world $\omega_{\mathfrak{c}} \in \Omega$ can be identified with the unique sequence $(E_{\mathfrak{c}_i} : i \in \{1, \ldots, k\}) \in \times_{i \in \{1, \ldots, k\}} \Omega_i$ such that $\{\omega_{\mathfrak{c}}\} = \bigcap_{i \in \{1, \ldots, k\}} E_{\mathfrak{c}_i}$. For $\mu_i \in \mathcal{M}_i$ with $i \in \{1, \ldots, k\}$, let $\mu' := \prod_{i \in \{1, \ldots, k\}} \mu_i$ be the mass function on Ω defined by:

$$\mu'(\omega_{\mathfrak{c}}) = \prod_{i \in \{1, \ldots, k\}} \mu_i(E_{\mathfrak{c}_i}). \tag{4}$$

We can therefore associate with a CCL theory $\mathcal{T}$ a joint credal set $\mathcal{M}_{\mathcal{T}}$ obtained as the following set of factorising mass functions:

$$\left\{ \prod_{i_{i \in \{1, \ldots, k\}}} \mu_i \mid \mu_i \in \mathcal{M}_i, i \in \{1, \ldots, k\} \right\}. \tag{5}$$

It is immediate to verify the following result.

Proposition 2. *Let $\mathcal{T}$ be a CCL theory. $\mathcal{M}_{\mathcal{T}}$ is the largest closed convex set of probability mass functions μ' on Ω that agree with μ, that is $\mu'(a) = \mu(a)$ for every $a \in \bigcup \bigcup \mathcal{C}$.*

In the imprecise-probability jargon, the condition in Eq. 5 is called *strong independence* between the variables associated to the different choice spaces [3]. Accordingly, we call the credal set $\mathcal{M}_{\mathcal{T}}$ *strong extension* of the marginals specified by μ. In particular, we have that for $\mu_i \in \mathcal{M}_i$ there is $\mu' \in \mathcal{M}_{\mathcal{T}}$ such that:

$$\mu_i(E_{\mathfrak{c}_i}) = \sum_{\omega \in E_{\mathfrak{c}_i}} \mu'(\omega) = \mu'(\mathsf{image}(\mathfrak{c}_i)). \tag{6}$$

Given a CCL theory $\mathcal{T}$, a question is therefore whether or not a probability mass function over the possible worlds consistent with the marginals exists. By

Eq. (5) it is enough to answer such question when $\mathcal{C}$ contains a single choice space. But this can be shown by induction on the number of alternatives. Hence, the following holds.

Proposition 3. *The credal set* $\mathcal{M}_T$ *of a CCL theory* T *is non-empty.*

Example 6. Consider the CCL theory $T_{\text{friends}} := \langle \mathbf{P}, \{\mathbf{C}_1, \mathbf{C}_2\}, \mu \rangle$ *in Example 5. There are four partial choices* c_1^j *with respect to* $\mathbf{C}_1$ *and thus four collections* $E_{c_1^j}$ *of possible worlds and, similarly, two partial choices* c_2^j *with respect to* $\mathbf{C}_2$ *and two collections* $E_{c_2^j}$ *of possible worlds (see Table 3). We have that* $\mathcal{M}_1$ *is the collection of all probability mass functions* $\mu_1 : \{E_{c_1^1}, \dots, E_{c_1^4}\} \to [0,1]$ *such that* $\mu(r) = \mu_1(E_{c_1^1}) + \mu_1(E_{c_1^2})$, $\mu(nr) = \mu_1(E_{c_1^3}) + \mu_1(E_{c_1^4})$, $\mu(c) = \mu_1(E_{c_1^1}) + \mu_1(E_{c_1^3})$ *and* $\mu(nc) = \mu_1(E_{c_1^2}) + \mu_1(E_{c_1^4})$. *On the other hand,* $\mathcal{M}_2$ *is given by taking* $\mu_2 : \{E_{c_2^1}, E_{c_2^2}\} \to [0,1]$ *defined simply as* $\mu_2(E_{c_2^1}) = \mu(w)$ *and* $\mu_2(E_{c_2^2}) = \mu(nw)$. *Hence* $\mathcal{M}_{T_{\text{friends}}} = \{\mu' = \mu_1 \cdot \mu_2 \mid \mu_1 \in \mathcal{M}_1, \mu_2 \in \mathcal{M}_2\}$, *where* $\mathcal{M}_1$ *is the convex hull of* $\mu_1^{(1)} = [0.0; 0.1; 0.4; 0.5]$ *and* $\mu_1^{(2)} = [0.1; 0.0; 0.5; 0.4]$, *while* $\mathcal{M}_2$ *has a single element* $\mu_2 = [0.2; 0.8]$.

Table 3. Classes (of possible worlds) for $\mathcal{M}_1$ and $\mathcal{M}_2$.

(i,j)	$(1,1)$	$(1,2)$	$(1,3)$	$(1,4)$	$(2,1)$	$(2,2)$
$E_{c_i^j}$	$\{\omega_1, \omega_2\}$	$\{\omega_3, \omega_4\}$	$\{\omega_5, \omega_6\}$	$\{\omega_7, \omega_8\}$	$\{\omega_1, \omega_3, \omega_5, \omega_7\}$	$\{\omega_2, \omega_4, \omega_6, \omega_8\}$
image(c_i^j)	$\{r, c\}$	$\{r, nc\}$	$\{nr, c\}$	$\{nr, nc\}$	$\{w\}$	$\{nw\}$

5   Inference

In the previous section, by relaxing the independence assumptions, we introduced CCL as a generalisation of ICL. Here we discuss how to compute inferences in CCL.

Given a CCL theory T, the CCL semantics leads to the specification of a credal set $\mathcal{M}_T$ over Ω called the *strong extension* of T. Given a *query* Q (see definition in Sect. 2) for the program $\mathbf{P}$ in T, inference is intended as the computation of the lower $\underline{\mu}(Q)$ and upper $\overline{\mu}(Q)$ bounds of the success probability of the query with respect to the strong extension, i.e.,

$$\underline{\mu}(Q) := \min_{\mu(\Omega) \in \mathcal{M}_T} \sum_{\omega \models Q} \mu(\omega), \qquad (7)$$

and similarly, with the maximum replacing the minimum for the upper success probability $\overline{\mu}(Q)$. As $\mathcal{M}_T$ is non-empty (Proposition 3), Eq. (7) always returns proper inferences. Moreover, being induced by a finite number of linear constraints over the joint probabilities (non-negativity, normalisation, and consistency with the marginal probabilistic facts), the convex set $\mathcal{M}_T$ has only a finite number of extreme points and Eq. (7) can be regarded as a linear programming task, whose solution corresponds to an extreme point of $\mathcal{M}_T$. E.g., the interval $[0.2, 0.5]$ obtained in Sect. 3 for the query h under a complete relaxation of the independence assumptions can be regarded as a CCL inference over a strong extension corresponding to the convex hull of the mass functions $\{\mu^{(i)}\}_{i=1}^8$ in Table 2. Such a brute-force approach cannot be used in general as the number of extreme points might be exponentially large with respect to the input size. In the rest of this section we show that even if, as expected, inference in CCL is NP-hard, these inferences can be reduced to a classical task for which dedicated solver and algorithms have been developed.

To do that, let us first denote as CCL_α the decision task associated to an inference, i.e., given a $\alpha \in [0, 1]$, deciding whether or not $\alpha \in [\underline{\mu}(Q), \overline{\mu}(Q)]$. In case the theory contains a single choice space, CCL_α can be reduced to *probabilistic satisfiability* (PSAT). PSAT is a generalisation of the classical *satisfiability* (SAT) decision problem, in which each element of a finite set of Boolean formulas is paired with a probabilistic assessment [25]. The task is to decide whether or not the assessments are consistent, i.e., to determine whether or not a joint mass function over the variable assigning the given probability to each formula exists. A PSAT instance is thus a set $\{P(\varphi_i) = \alpha_i \mid i < n\}$ where each φ_i is a Boolean formula and α_i is the associated probability. As the problem generalises SAT for propositional calculus it is NP-hard. The NP-completeness of the task has been proved in [17] and, unlike SAT, PSAT remains NP-complete even if each clause does not contains more than two literals.

We first describe how PSAT can be used to solve CCL_α. Let $\langle \mathbf{P}, \mathcal{C} = \{\mathbf{C}\}, \mu \rangle$ be a CCL theory with a single choice space. First, we convert $\mathbf{P}$ to an equivalent Boolean formula $\varphi_{\mathbf{P}}$ (see, e.g., [19] for some conversions), and encode the choice space $\mathbf{C}$ as a conjunction of XOR:

$$\varphi_{\mathbf{C}} := \bigwedge_{C \in \mathbf{C}} \bigoplus_{a \in C} a. \tag{8}$$

The class of models of $\varphi_{\mathbf{C}} \wedge \varphi_{\mathbf{P}}$ corresponds to the stable models of the original CCL theory. The associated PSAT instance is therefore the set $\Sigma_\alpha := \{P(\varphi_{\mathbf{C}} \wedge \varphi_{\mathbf{P}}) = 1\} \cup \{P(a) = \mu(a) \mid a \in \bigcup \mathbf{C}\} \cup \{P(\bigwedge Q) = \alpha\}$. Clearly $\alpha \in [\underline{\mu}(Q), \overline{\mu}(Q)]$ if and only if Σ_α is satisfiable. With a small abuse of notation, we denote as $\mathrm{SAT}(\mathrm{CCL}_\alpha)$ a Boolean variable true if and only if Σ_α is satisfiable.

The computation of $[\underline{\mu}(Q), \overline{\mu}(Q)]$ can be therefore achieved by iterating the above reduction for different values of α according to a bracketing scheme that identifies both the bounds (Fig. 1). The procedure is basically a bisection method that recognises UNSAT values of α as outer approximations (red points in Fig. 1) and SAT values as inner approximations (blue points in Fig. 1). To solve PSAT instances, solvers such of those developed by [6,12] can be used. To achieve a precision ϵ in the estimates, the number of calls of these solvers is $O(\log \epsilon^{-1})$.

A key point for the bracketing algorithm is the computation of an inner (i.e., SAT) value $\mu_I(Q)$ (e.g., black point in Fig. 1). We achieve this by translating it into the inference task of computing in ProbLog the marginal probability

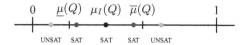

Fig. 1. Inner and outer approximations of CCL inferences. (Color figure online)

of the query given some specific evidence. More precisely, we first translate $\mathcal{T}$ into ProbLog by distinguishing different occurrences of the same atomic choices and adding, for each duplicated atomic choice, a clause of the form $f \leftarrow a, a'$, where f is some fixed new symbol. Reading f as denoting the false, the newly introduced clauses mimics the coherence conditions on total choices defined on CCL theories. Then, we modify the probabilistic assignment μ to each atomic choice a in such a way that the marginal probability of a given the evidence $f = \mathbf{f}$ (i.e. f is false) coincide with $\mu(a)$. Hence, the value $\mu_I(Q)$ corresponds to the ProbLog computation of the marginal probability for the query Q given the evidence $f = \mathbf{f}$. Such task can be reduced to weighted model counting [11].

The above procedure should be adopted when coping with CCL theories having a single choice space as in the case of object ranking (see Sect. 6). For the general case, we suggest the following (outer) approximation scheme:

$$\underline{\mu}(Q) \geq \sum_{\omega_c \models Q} \mu(\omega_c) = \sum_{\omega_c \models Q} \prod_{i=1}^{k} \mu(\mathrm{image}(c_i)), \tag{9}$$

where the inequality follows from Eq. (7) by simply swapping the minimum and the sum, while the equality with the term on the right-hand side follows from the factorisation in Eq. (4) and then by applying Eq. (6).

Example 7. Consider the CCL theory $\mathcal{T}_{\mathsf{friends}}$ from Example 5. We calculate the lower success probability of query h. As the theory contains multiple choice spaces, we should use Eq. (9). However notice that $\omega_i \models h$ exactly when $i = 8$ (see Sect. 3) and therefore:

$$\underline{\mu}(\{h\}) = \underline{\mu}(\omega_8) = \underline{\mu}_1(\{nr, nc\}) \cdot \underline{\mu}_2(\{nw\}),$$

which gives the numerical value 0.32, as expected from Sect. 3. Analogously we get 0.40 for the corresponding upper success probability.

6 Empirical Analysis

In this section we report the results of a very first empirical analysis of the approach to CCL inference described in the previous section. Unlike most of the statistical models based on credal sets [3], CCL has no parameters directly affecting the *imprecision* level of the inferences (i.e., the difference between the upper and the lower probability of a query). A first important question is therefore whether or not, our relaxation of the standard independence assumptions in PLP, allows to obtain non-vacuous inferences from a query. Another point is whether or not available PSAT algorithms are able to solve instances induced by CCL theories.[2]

To achieve that we consider *object ranking*, that is the task of deriving a complete ranking over a set of n objects from a data set $\mathcal{D}$ of complete rankings (if features are also considered, the term preference/label learning is used instead) [16]. We assume rankings not directly available: only the n^2 *marginal* counts reporting how many times a certain object gets a certain rank are available. Note that, in general, complete information about the rankings cannot be recovered from these counts (Fig. 2).

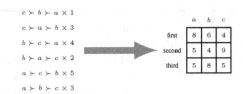

Fig. 2. From complete rankings to marginal counts.

Consider for instance a horse race. Call r_j the unary predicate denoting the property "ending the race in j-th position", and h_i the constant associated to the i-th horse with $i, j = 1, \ldots, n$. The ground atom $r_j(h_i)$ means "horse i ended the race in j-th position". Under our assumptions, we can learn from the data the (marginal) probabilities $\alpha_{i,j} := \mu(r_j(h_i))$. Object ranking can be based on a joint mass function over these n^2 ground toms. This function should reproduce the marginal probabilities being also consistent with the obvious logical constraints ("one and only one horse ends the race in j-th position", and "one and only one position is reached by the i-th horse at the end of the race"). These constraints are encoded in the following formula:

$$\varphi_{\mathsf{ranking}} := \bigwedge_{j=1}^{n} \bigoplus_{i=1}^{n} r_j(h_i) \wedge \bigwedge_{i=1}^{n} \bigoplus_{j=1}^{n} r_j(h_i), \tag{10}$$

where $\oplus$ is the exclusive disjunction (XOR). Algebraically, the task corresponds therefore to the specification of a joint mass function over n^2 Boolean variables, i.e., 2^{n^2} probabilities. Only $n!$ probabilities are non-zero as soon as we impose the constraints in Eq. (10), while the n^2 marginal probability induce an equal number of (linear) constraints. Thus, we might have non-unique specifications even with four objects only.

Consistently with what above, although complete rankings over n objects are available for training, we only learn the n^2 marginal probabilities. This is done by

---

[2] Here we use the solver [12], freely available at http://psat.sourceforge.net.

smoothing the frequencies with a Laplace prior of equivalent size two. The corresponding CCL theory is defined as the program-free CCL theory $\mathcal{T}_{\text{ranking}} :=$ $\langle \emptyset, \mathcal{C}, \mu \rangle$ with a single choice space, i.e., $\mathcal{C} = \{\{C_1, \ldots, C_n\} \cup \{C'_1, \ldots, C'_n\}\}$ with $C_\ell = \{r_1(h_\ell), \ldots, r_n(h_\ell)\}$ and $C'_\ell = \{r_\ell(h_1), \ldots, r_\ell(h_n)\}$, and by stating $\mu(r_j(h_i)) := \alpha_{i,j}$ for each $i, j = 1, \ldots, n$. This eventually induces PSAT instances with the same probabilistic facts and logical constraints obtained by a CNF (*conjunctive normal form*) conversion of Eq. (10).[3]

As a first benchmark, we use four classical UCI datasets (Vehicle, Stock, Glasses, and Bodyfat) with $n = 4, 5, 6, 7$. We evaluate pairwise preferences between pairs of objects. We write $i' \succ i''$ to denote the fact that object i' has a higher ranking than object i''. To do that, we extend the theory $\mathcal{T}_{\text{ranking}}$ by adding to the (initially empty) program $\mathbf{P}$, the following clauses:

$$\left\{ q \leftarrow r_{j'}(h_{i'}), r_{j''}(h_{i''}) \mid \begin{smallmatrix} j',j''=1,\ldots,n \\ s.t. \quad j'>j'' \end{smallmatrix} \right\}, \tag{11}$$

where q is a new symbol. Hence, the query corresponding to the property $i' \succ i''$ will be given by $Q_{i' \succ i''} := \{q\}$. For this query we compute both $\underline{\mu}(Q_{i' \succ i''})$ and $\overline{\mu}(Q_{i' \succ i''})$ and, on the basis of these values, we decide whether $i' \succ i''$ or $i'' \succ i'$. Note that when coping with imprecise, interval-valued, inferences, a condition of indecision between the two options can be also observed.

This is the case if $[\underline{\mu}(Q_{i' \succ i''}), \overline{\mu}(Q_{i' \succ i''})]$ overlaps the decision threshold .5, otherwise a clear preference is returned. This decision (denoted as CCL) is compared against the one based on the original data with the complete rankings regarded here as a ground truth and the one based on the marginal probabilities treated as independent (denoted as ICL). The ICL accuracy on the queries is evaluated separately on the pairs on which CCL is determinate and the ones on which CCL is indeterminate (i.e., indecision is returned). These results are in Fig. 3. The separation between these two accuracies is clear: CCL becomes undecided on the tasks on which a less conservative approach would be less accurate, thus providing a more robust approach to the inferences.

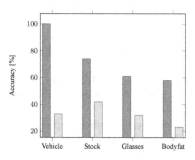

Fig. 3. ICL accuracy for preference queries on instances in which CCL is determinate (blue) and indeterminate (red). (Color figure online)

---

[3] A XOR rewrites as a disjunction together with negations of pairwise conjunctions.

7   Conclusions

In the last years, several works in the probabilistic logic tradition have proposed formalisms to explicitly deal with independence, see e.g. [1,5,18,23]. Our approach differs in the sense that, rather than attempting to combine probabilistic logics and probabilistic networks, is closer to logic programming. Indeed, we have introduced CCL, a conservative extension of Poole's ICL in which both independent and non independent choices can be modelled. In the proposed setting, a theory specifies the (credal) set of all probability mass functions over least models compatible with the marginals on atomic choices. In guise of example, we applied CCL to object ranking and have shown how to infer the lower and upper probabilities of a query. In future work, on top of presenting more complex applications and deeper experiments of the proposed formalism, we plan to compare our work with related approaches such as for instance the one discussed by [13] in the context of probabilistic databases.

References

1. Andersen, K.A., Hooker, J.N.: Bayesian logic. Decis. Support Syst. **11**(2), 191–210 (1994)
2. Apt, K.R., Bezem, M.: Acyclic programs. New Gener. Comput. **9**(3), 335–363 (1991)
3. Augustin, T., Coolen, F., de Cooman, G., Troffaes, M.: Introduction to Imprecise Probabilities. Wiley, Hoboken (2014)
4. Carpenter, B., et al.: Stan: a probabilistic programming language. J. Stat. Softw. **20**, 1–37 (2016)
5. Cozman, F., de Campos, C., da Rocha, J.C.: Probabilistic logic with independence. Int. J. Approx. Reason. **49**(1), 3–17 (2008)
6. Cozman, F., di Ianni, L.: Probabilistic satisfiability and coherence checking through integer programming. Int. J. Approx. Reason. **58**, 57–70 (2015)
7. Cozman, F., Mauá, D.: On the semantics and complexity of probabilistic logic programs. J. Artif. Intell. Res. **60**, 221–262 (2017)
8. De Raedt, L.: Applications of probabilistic logic programming. In: International Conference on Inductive Logic Programming (2015)
9. De Raedt, L., Kimmig, A.: Probabilistic (logic) programming concepts. Mach. Learn. **100**(1), 5–47 (2015)
10. De Raedt, L., Kimmig, A., Toivonen, H.: ProbLog: a probabilistic Prolog and its application in link discovery. In: International Joint Conference on Artificial Intelligence, pp. 2462–2467 (2007)
11. Fierens, D., et al.: Inference and learning in probabilistic logic programs using weighted Boolean formulas. Theory Pract. Log. Program. **15**(3), 358–401 (2015)
12. Finger, M., De Bona, G.: Probabilistic satisfiability: logic-based algorithms and phase transition. In: International Joint Conference on Artificial Intelligence, pp. 528–533 (2011)
13. Flesca, S., Furfaro, F., Parisi, F.: Consistency checking and querying in probabilistic databases under integrity constraints. J. Comput. Syst. Sci. **80**(7), 1448–1489 (2014)

14. Fuhr, N.: Probabilistic datalog: a logic for powerful retrieval methods. In: International ACM SIGIR Conference on Research and Development in Information Retrieval, pp. 282–290. ACM (1995)
15. Fuhr, N.: Probabilistic datalog: implementing logical information retrieval for advanced applications. J. Assoc. Inf. Sci. Technol. **51**(2), 95–110 (2000)
16. Fürnkranz, J., Hüllermeier, E.: Preference learning: an introduction. In: Fürnkranz, J., Hüllermeier, E. (eds.) Preference Learning, pp. 1–17. Springer, Heidelberg (2010). https://doi.org/10.1007/978-3-642-14125-6_1
17. Georgakopoulos, G., Kavvadias, D., Papadimitriou, C.: Probabilistic satisfiability. J. Complex. **4**(1), 1–11 (1988)
18. Haenni, R., Romeijn, J.W., Wheeler, G., Williamson, J.: Probabilistic Logics and Probabilistic Networks, vol. 350. Springer, Dordrecht (2010). https://doi.org/10.1007/978-94-007-0008-6
19. Janhunen, T.: Representing normal programs with clauses. In: European Conference on Artificial Intelligence, pp. 358–362. IOS Press (2004)
20. Levi, I.: The Enterprise of Knowledge: An Essay on Knowledge, Credal Probability, and Chance. MIT press, Cambridge (1983)
21. Lukasiewicz, T.: Probabilistic logic programming. In: European Conference on Artificial Intelligence, pp. 388–392 (1998)
22. Lukasiewicz, T.: Probabilistic description logic programs. Int. J. Approx. Reason. **45**(2), 288–307 (2007)
23. Michels, S., Hommersom, A., Lucas, P.J., Velikova, M.: A new probabilistic constraint logic programming language based on a generalised distribution semantics. Artif. Intell. **228**, 1–44 (2015)
24. Ng, R., Subrahmanian, V.S.: Probabilistic logic programming. Inf. Comput. **101**(2), 150–201 (1992)
25. Nilsson, N.J.: Probabilistic logic. Artif. Intell. **28**(1), 71–87 (1986)
26. Poole, D.: Probabilistic Horn abduction and Bayesian networks. Artif. Intell. **64**(1), 81–129 (1993)
27. Poole, D.: The independent choice logic for modelling multiple agents under uncertainty. Artif. Intell. **94**(1), 7–56 (1997)
28. Sato, T.: A statistical learning method for logic programs with distribution semantics. In: International Conference on Logic Programming, pp. 715–729 (1995)
29. Vennekens, J., Verbaeten, S.: Logic programs with annotated disjunctions. Technical report CW 368, K.U.Leuven (2003)
30. Vennekens, J., Verbaeten, S., Bruynooghe, M.: Logic programs with annotated disjunctions. In: Demoen, B., Lifschitz, V. (eds.) ICLP 2004. LNCS, vol. 3132, pp. 431–445. Springer, Heidelberg (2004). https://doi.org/10.1007/978-3-540-27775-0_30

Explaining the Most Probable Explanation

Raphaela Butz[1,2(✉)], Arjen Hommersom[2,3], and Marko van Eekelen[2,3]

[1] Institute for Computer Science, TH Köln, Cologne, Germany
raphaela.butz@th-koeln.de
[2] Department of Computer Science, Open University of the Netherlands, Heerlen,
The Netherlands
[3] ICIS, Radboud University, Nijmegen, The Netherlands

Abstract. The use of Bayesian networks has been shown to be powerful
for supporting decision making, for example in a medical context. A particularly useful inference task is the most probable explanation (MPE),
which provides the most likely assignment to all the random variables
that is consistent with the given evidence. A downside of this MPE solution is that it is static and not very informative for (medical) domain
experts. In our research to overcome this problem, we were inspired by
recent research results on augmenting Bayesian networks with argumentation theory. We use arguments to generate explanations of the MPE
solution in natural language to make it more understandable for the
domain expert. Moreover, the approach allows decision makers to further explore explanations of different scenarios providing more insight
why certain alternative explanations are considered less probable than
the MPE solution.

Keywords: Bayesian networks · Most probable explanation
Argumentation theory

1 Introduction

With the increasing availability of data, machine learning approaches are more
and more used to analyse these datasets and to build models, such as Bayesian
networks (BNs) [9]. While these BNs provide a graph-structure of the direct
dependences between random variables, they are in practice hard to interpret
for domein experts. For example, two random variables which are uncondition-
ally independent may become dependent when observing a third variable (a
process which is called *explaining away*). Another example is that an arc in a
BN does not necessarily imply a causal dependence. These types of properties
of a BN makes the representation and reasoning with BN sometimes counter-
intuitive and the interpretation of the results difficult in practice [11]. Another
issue with using statistical models learned from (observational) data in general
is that it is hard to combine the knowledge from these networks with other types

© Springer Nature Switzerland AG 2018
D. Ciucci et al. (Eds.): SUM 2018, LNAI 11142, pp. 50–63, 2018.
https://doi.org/10.1007/978-3-030-00461-3_4

of knowledge, for example from intervention studies or other types of domain knowledge. While some work has been done for statistical matching of different BN (see e.g. [2]), these methods are quite limited compared to the heterogeneous types of knowledge that are available in practice.

Argumentation-based inference provides an attractive alternative for building models and developing decision support systems, as they model human-like reasoning in a natural manner (see e.g. [4,5]). A second advantage compared to statistical models is that logic is *modular*, which allows the user to (more) easily incorporate knowledge to an existing knowledge base. For example, the Semantic Web has shown that knowledge bases can become extremely large, and sensible reasoning is still possible [3]. Nevertheless, first-order logic lacks a sense of 'strength', that it, the representation does not provide means to decide which inference is more probable than others [16]. Moreover, it is unclear how the knowledge derived from data, represented in a BN, could be exploited in a logic-based formalism.

Recently, researchers have considered using argumentation for explaining Bayesian network inference, e.g. [11,12,15,16]. In brief, this line of research aims at explaining a posterior probability of a random variable by the most likely chain of inference from evidence. This is attractive, because of the nature of argumentation, the resulting reasoning is easier to understand than Bayesian network inference. However, logical structure can still easily be misinterpreted by domain experts. For this reason, in the remainder of this paper, we will focus on generating a semi-structured natural language from a Bayesian network knowledge base through an argumentation formalism. In particular, we aim to generate explanations about the dependences between variables.

This paper addresses the problem of explaining scenarios which are consistent with the evidence. Therefore we start with the particular inference problem in Bayesian networks to find the most likely value combination for all variables to get the highest probability given the evidence. This instantiation of a Bayesian network is called most probable explanation (MPE). Most other approaches to explain a BN are focusing on one or several variables of interest, hence the domain expert has to choose variables he want to look at. However, we believe that it is worthwhile to look at the complete scenario to get an overview before deciding which variables should be focused on. Consider for example Fig. 1, which we will use as a running example in this paper. It involves a Bayesian network for the diagnosis of diabetes type 2. In this BN the variable diabetes indicates whether a person has this disease and the variable *quality_of_beta_cells* represents how well the insulin production is working. Furthermore, *food_intake* indicates if someone has recently eaten and finally, the variable *glycaemia* shows which blood sugar level a person has. In this BN, it was computed that the MPE of having *glycaemia(hyper)* is: *diabetes(false)*, *food_intake(true)*, *quality_of_beta_cells(high)*. In the following, we address the question how the method proposed in this paper can help to better explain this conclusion to a domain expert.

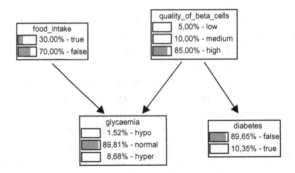

Fig. 1. Example Bayesian network for the diagnosis of diabetes

We will start with the scenario given by the MPE solution and start to deploy more scenarios in an interactive manner. In each scenario the domain expert can argue with the MPE. For example, one could disagree with the explanation of the MPE to structure a new explanation. Such an interactive approach can provide valuable feedback to the domain expert. To accomplish this all, we propose three principles for generating explanations of MPE. The first principle is build a logical knowledge base which maps the dependences represented in the BN. The second principle is to incorporate strength of the dependences in the Bayesian network. The third principle is to use an interactive representation of the knowledge to the domain expert for exploration, to give more insight into the knowledge base and ultimately into the BN.

This paper is structured as follows. In Sect. 2 the necessary preliminaries with respect to BNs, MPE inference, and natural language generation (NLG) are presented. In Sect. 3, we will discuss how the knowledge base is constructed and how the method can be used to explore the BN. In Sect. 4 it is shown, how the MPE is explained through natural language. In Sect. 5, we will discuss related work. Finally, in Sect. 6, we conclude and discuss future research.

2 Preliminaries

In this section, we provide the necessary background for Bayesian networks and natural language generation that will be used in subsequent sections.

2.1 Bayesian Networks

BNs are a type of probabilistic graphical models, which represent the conditional independence structure of a probabilistic model in the form of a parameterized directed acyclic graph [6]. The nodes represent random variables and the directed edges model the absence of probabilistic independences. The joint probability distribution of the graph is defined by the conditional probability of every node given their parents:

$$P(Y) = \prod_{i=0}^{n} P(Y_i \mid \pi(Y_i)) \tag{1}$$

where $Y = \{Y_0, Y_1, \ldots, Y_n\}$, and $\pi(Y_i)$ represent the parents of Y_i in the graph. In the following, we will assume that each random variable is discrete.

The independences between random variables given evidence can be read from the graph structure of the BNs, by means of a d-separation criterion. Consider for example the structure $Y_1 \rightarrow Y_2 \rightarrow Y_3$. Without evidence, Y_1 and Y_3 are dependent; however, if there is evidence on Y_2, then this 'blocks' the information between Y_1 and Y_3. Generally, two sets of nodes X and Y are d-separated by a set of nodes Z $(X \perp Y \mid Z)$ if every path from X to Y is blocked given Z [6].

2.2 Most Probable Explanation

A typical task for a BN is to compute a MAP (maximum a-posteriori), for example to decide which is the most likely diagnosis given the available evidence: e.g., diabetes is a-priori unlikely, but may become almost certain after observing insulin autoantibodies in the blood.

The most probable explanation (MPE) is a special case of MAP. The MPE computes the most probable instantiation of all nodes which are not observed, instead of choosing one or a set of variables of interest as in MAP. The most probable assignment for all non-evidence nodes $Y_1 = y_1, \ldots, Y_m = y_m$ is computed by maximizing $P(Y_1 = y_1, \ldots, Y_m = y_m \mid E)$, whereas E is the evidence, $Y_1, \ldots, Y_m$ are the non-evidence nodes and $y_1, \ldots, y_m$ their values [6].

2.3 Natural Language Generation

Natural language generation (NLG) is the deliberate construction of natural language to meet defined communication goals [10]. These goals define the purpose of the communication process. Most NLG Systems using Reiters [10] work as base. Reiter is dividing the NLG process from a non-linguistic input to a automatically produced textual output into three steps: *document planning* (also *content planning*), *microplanning* (also *sentence planning*) and *surface realisation*. In the document planning step the information which should be communicated is determined. In addition it is decided how the information is structured to make coherent text. The document plan is then refined in the second step, which is microplanning. In microplanning the document plan is converted into lexical content through the individual structuring of each sentence to give the text a smoother flow. Therefore conjunctions, pronominalisation, and discourse markers can be used. In the last step (realisation), all specifications are converted into actual text and the correctness of the language is verified. These steps can be achieved by using simple boilerplates, syntactic processing through linguistic formalisms, and theories up to the usage of AI to get a more flexible natural language feel.

3 Building the Knowledge Base

In this paper, we present a framework consisting of different steps. An overview of these steps in given in Fig. 2. Building the knowledge base is the first major step of this approach, which extracts the knowledge from a Bayesian network into a probability tree. The second major step is the representation of the knowledge base in natural language, described in Sect. 4. Note that this framework is quite modular, so this step could be replaced with another representation, such as formal logic. It can be extended with further modules, by extending for example the microplanning step as well. The same terms hold for the building of the knowledge base. For example, in Sect. 3.2 we present a pruning strategy for the argumentative trees that we will use; however, this could be omitted if the BN is relatively simple.

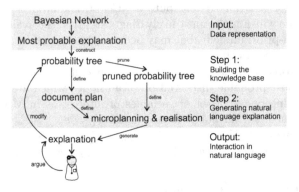

Fig. 2. Overview of the methodology

For the construction of the knowledge base, we assume that two key properties are essential for a good understanding of MPE solution in a Bayesian network:

Following a Chain of Deduction. A chain of deduction, meaning that conclusions are drawn from immediate previous conclusions, is easy to follow and to argue with. For example, if someone disagrees with one reasoning step, the chain generated by this reasoning step onward can be replaced for different steps of reasoning. Arguing this chain therefore directly allows for an interactive approach. With the construction of chains the knowledge base can be built. To achieve this, the chain starts with the evidence and further knowledge is added by iteratively drawing conclusions from the evidence and previous conclusions.

Reasoning with the Strongest Dependence. To build a logical knowledge base we assume that stronger probabilistic dependences will explain the MPE better, as each of the reasoning is more plausible. Therefore the concluding process should always follow the most likely, hence strongest, next dependence.

3.1   Argumentative Probability Tree

Probability trees are regularly used to represent probability distributions, for example in semantics for the language *CP-logic* [13]. However, the construction and use of it in our argumentative probability tree is different and described further below.

Fig. 3. Argumentative probability tree, a: With the MPE, b: MPE with one alternative, c: MPE and two alternative branches

To construct the argumentative probability tree, we start with the evidence, since this knowledge is certain. In the running example *glycaemia(hyper)* was observed, so this variable is used as root. Then the most likely dependence to another variable is searched, which is in our example *food_intake(true)* with the probability of 0.9194, hence *food_intake(true)* is attached to *glycaemia(hyper)* as child. Given *glycaemia(hyper)* and *food_intake(true)* the strongest dependence is searched again. This continues until no dependent variables are left. Then one chain of deduction and branch of the tree is finished as shown in Fig. 3a. The joint probability $P(Y_1 = y_1, \ldots, Y_n = y_n \mid e)$ of the variable-value pairs given the evidence will be used as the strength for a chain. In the example, the strength of the first chain is 0.55. Note that this reasoning strategy generates the MPE solution, which is {*diabetes(false), food_intake(true), quality_of_beta_cells(high)*}. This result matches the intuitive notion that the strongest dependences lead to the most likely solution. Therefore, every other chain with the same variable-value combination will not be a better explanation, and therefore will have a lower joint probability. In a normal probability tree, there are chains which lead to the same MPE solution, by exchanging the order of random variables. In the argumentative probability tree this cannot occur, because the order of variable-value pairs is important.

The argumentative probability tree can be extended by disagreeing with the MPE branch. This will lead to a new chain of deduction as a new branch in the probability tree. For example, if the domain expert is disagreeing with *diabetes(false)*, because she thinks that *diabetes* is *true*, another chain is attached to the tree. To attach this chain, *diabetes(false)* is replaced with *diabetes(true)* which only has the probability of 0.2694 given *glycaemia(hyper)* and *food_intake-(true)*. Given *glycaemia(hyper)*, *food_intake(true)* and *diabetes(true)*, *quality_of_ beta_cells* is then replaced with *low*, because this

---

Algorithm 1. Dynamic construction of the argumentative probability tree

---

```
 1  buildBranch(BN, parent, tree);
 2  for node in BN do
 3      if node is not in tree.parents then
 4      |   continue;
 5      end
 6      for value in node do
 7          if node[value] ≥ child[value] then
 8          |   child = node;
 9          end
10      end
11  end
12  tree.add(parent,child);
13  tree = tree.buildBranch(BN, child, tree);
14  return tree;

15  buildMPE(BN, e);
16  tree = tree.addRoot(e);
17  tree = buildBranch(BN, e, tree);

18  argueMPE(BN, argue, tree);
19  tree = buildBranch(BN, argue, tree);
```

---

value is with probability 0.5589 more likely than *quality_of_beta_cells(high)*, which is down to 0.225. The extended tree is shown in Fig. 3b.

As another example, the domain expert could also disagree with *food_intake-(true)*, because she knows that the patient has not eaten, and furthermore argues that *quality_of_beta_cells(low)* is a better explanation for *glycaemia(hyper)*. Then *quality_of_beta_cells(low)* is attached to the root with the probability of 0.1578. Given *glycaemia(hyper)* and *quality_of_beta_cells(low)* the final dependent variable is *diabetes* with the most probable value *true*, having a probability of 0.90. Because the evidence in *quality_of_beta_cells* is now fixed to a value, *food_intake* becomes independent. The result is shown in 3c. The general dynamic process of the construction of the tree is shown in Algorithm 1.

This chains or branches in the argumentative probability tree can also be presented as a set of rules. An argument A derived from the tree has a structure as follows:

$$A : P(child_1, \ldots, child_n \mid e) =$$
$$child_1 \leftarrow e : P(child_1|e) \tag{2}$$
$$child_n \leftarrow e \wedge child_1 \wedge \ldots \wedge child_{n-1} : P(child_n|e, child_1, \ldots, child_{n-1})$$

where P is the probability distribution, e is the observed evidence, and $child_1 \ldots child_n$ are all children of the evidence in one branch of the tree.

For example, the A_{MPE} from our running example is constructed by means of deductions obtained from the rules in the following equation:

$A_{\text{MPE}} : 0.55 =$

$food\_intake(true) \leftarrow glycaemia(hyper) : 0.9194,$

$diabetes(false) \leftarrow food\_intake(true) \wedge glycaemia(hyper) : 0.7306,$ (3)

$quality\_of\_beta\_cells(high) \leftarrow$

$\quad food\_intake(true) \wedge glycaemia(hyper) \wedge diabetes(false) : 0.827$

This rule-based representation shows, that every node in the argumentative probability tree is explained by its parents, because every branch is a chain of deduction; hence every node in the chain is explained by its former nodes. As one can imagine this representation is getting more complex with every child, which is added. Therefore we aim for a natural language approach to explain the knowledge. Nevertheless this rule-based representation can be used to link the knowledge base to other logical knowledge bases, such as ontologies to enrich the information.

3.2   Pruned Argumentative Probability Tree

In this paper, we aim to explain a whole scenario in the form of the MPE solution or alternatives of that. However, explaining every parent for every node can result in long and complex explanations, especially in larger BNs. Therefore pruning the tree can be a solution, such that information which is not important for the scenario does not have to be explained. In Sect. 3.1, d-separated nodes are already omitted. In this section we will filter for parent nodes, which have a significant impact on its child nodes. Only such parents will be used in the explanation.

To calculate the threshold of significance we use the likelihood ratio, which has been used before in related literature [11,12]. Typically, binary-valued variables are used. Here we generalize this to multinomial distributions. Suppose that we are interested in the effect of a set of nodes $\mathbf{x}$ on $Y = y_i$. We define the strength as the likelihood ratio, i.e.

$$s(y_i, \mathbf{x}) = \frac{P(y_i \mid \mathbf{x})}{\sum_{j \neq i} P(y_j \mid \mathbf{x})}$$

One reasonable threshold th is $th = 1$, such that all node pairs with a strength $th \leq s(y_i, \mathbf{x})$ are not explained, because there it is unlikely that y_i follows from $\mathbf{x}$. For example, the strength of $food\_intake(true)$ given $glycaemia(hyper)$ is 3.81, whereas $quality\_of\_beta\_cells(high)$ given $glycaemia(hyper)$ has the strength 0.72, therefore $glycaemia(hyper)$ is not considered in the explanation of $quality\_of\_beta\_cells(high)$.

A related method to calculate the threshold of significance was recently introduced by Kyrimi et al. [7]. The main difference is that Kyrimi et al. aims at finding the most significant evidence for a variable of interest. However, in our case, evidence is given, and the most related variables are selected, which can be seen as the reverse problem. However, we also borrow the idea by Kyrimis [7] to perform a conflict analysis. This analysis aims to examine if every premise in a rule has the same direction in creating the overall change in a variable of interest, i.e., a positive or negative influence to the conclusion. We will explain a node with significant parent nodes which are contributing to the observed value of the node. For example, in the diabetes example $glycaemia(hyper)$ and $food\_intake(true)$ cause the same direction of change when they are observed on all other variables, for example $quality\_of\_beta\_cells(high)$ is decreasing. A different direction of change is caused by the observation $diabetes(false)$, because the probability of $quality\_of\_beta\_cells(high)$ is now increased. Since the overall change of $quality\_of\_beta\_cells$ is $high$, it is only explained by $diabetes(false)$ as well. The pruned argument structure for the MPE A_{PrMPE} is therefore the following:

$$
\begin{aligned}
A_{\mathrm{PrMPE}} : 0.55 = \\
food\_intake(true) \leftarrow glycaemia(hyper) : 0.9194, \\
diabetes(false) \leftarrow food\_intake(true) : 0.7306, \\
quality\_of\_beta\_cells(high) \leftarrow diabetes(false) : 0.827
\end{aligned}
\tag{4}
$$

4   Generating Natural Language Explanations

The communication goals, which we want to meet with our NLG, are related to the principles we proposed in Sect. 1. The first goal is to explain the dependences in a BN and the second, that the strength of these dependences should be verbalized. The third principle is only indirectly achieved with NLG, because the interactive arguing process does not change the NLG. To achieve the steps defined by Reiter [10] we are first defining a document plan, then microplanning and realisation are merged to one step by using a formal grammar.

4.1   Document Plan

The communication goals are derived from the main principles, used in the construction of the argumentative probability tree. Therefore it is straightforward to derive a document plan from the tree as well. Hence the resulting document plan shown in Fig. 4, looks like a generalized argumentative probability tree. The important nodes are emphasised in the document plan. Important nodes are nodes which are the first child of the evidence or where alternative branches are introduced. The first child in the MPE argumentation is important, because it is always the best explanation for the evidence. At the beginning the domain expert only has the explanation for the scenario described in the MPE. If the domain expert starts to argue with the explanation, MPE alternatives will be constructed. The start of an alternative argumentation should be represented in the natural language as well. The same can be said if the argument is not starting with the most likely child. Every one of this occurrences should be phrased differently in natural language to emphasise changes in the deduction of a chain.

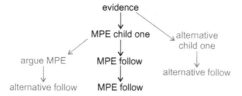

Fig. 4. The document plan derived from the argumentative probability tree

4.2   Microplanning and Realisation

To get the impression of natural language, boilerplates are constructed for the microplanning and the realisation of the text. That is, these two steps are merged since static boilerplates have no need for a further grammar definition. The boilerplates are described in Backus-Naur form (BNF) shown below. The nodes of the argumentative probability tree are displayed as constants in the BNF and have to be replaced in a defined phrasing for every specific node-value pair. These constants are displayed between #. This grammar can be used to describe different scenarios, but only if the domain expert is arguing with the MPE, only then new branches are attached in the argumentative probability tree and in the document plan, as well. This changes are then considered in the grammar, through the document plan.

⟨*explanation*⟩	::=	The observed evidence #e# ⟨*MPE child(1)*⟩
	|	#e# ⟨*Alternative*⟩
⟨*MPE child(1)*⟩	::=	is best explained by #child(1)# ⟨*MPE follow*⟩
	|	is best explained by #child(1)# ⟨*Argue MPE*⟩
	|	is best explained by #child(1)#
⟨*MPE follow*⟩	::=	⟨*Sentences MPE*⟩ ⟨*MPE follow*⟩
	|	⟨*Sentences MPE*⟩ ⟨*Argue MPE*⟩ | ⟨*Sentences MPE*⟩
⟨*Alternative*⟩	::=	could ⟨*probability*⟩ indicate #child(1)# ⟨*Alternative follow*⟩
	|	could ⟨*probability*⟩ indicate #child(1)#
⟨*Argue MPE*⟩	::=	Given #parents# #child(n)#, is alternatively ⟨*probability*⟩ ⟨*Alternative follow*⟩
	|	Given #parents#, #child(n)# is alternatively ⟨*probability*⟩
⟨*Alternative follow*⟩	::=	⟨*Sentences Alternative*⟩ ⟨*Alternative follow*⟩
	|	⟨*Sentences Alternative*⟩
⟨*Sentences MPE*⟩	::=	Given #parents#, #child(n)# is ⟨*probability*⟩
	|	With the description of the previous sentence #child(n)# will most probable be ⟨*probability*⟩
	|	With the description of the previous sentence #child(n)# is ⟨*probability*⟩
	|	With the prior explanation #child(n)# is ⟨*probability*⟩
	|	In the pre-established combination #child(n)# is ⟨*probability*⟩
⟨*Sentences Alternative*⟩	::=	In that situation #child(n)# is ⟨*probability*⟩
	|	With the description of the previous sentence #child(n)# is ⟨*probability*⟩
⟨*probability*⟩	::=	⟨*#s(0.0-0.2)#*⟩ | ⟨*#s(0.21-0.3)#*⟩ | ⟨*#s(0.31-0.4)#*⟩ | ⟨*#s(0.41-0.5)#*⟩ | ⟨*#s(0.51-0.6)#*⟩
	|	⟨*#s(0.61-0.7)#*⟩ | ⟨*#s(0.71-0.8)#*⟩ | ⟨*#s(0.81-0.9)#*⟩ | ⟨*#s(0.91-1)#*⟩

⟨*#s(0.0-0.2)#*⟩	::=	highly unlikely;	⟨*#s(0.21-0.3)#*⟩	::=	very unlikely
⟨*#s(0.31-0.4)#*⟩	::=	unlikely;	⟨*#s(0.41-0.5)#*⟩	::=	not plausible
⟨*#s(0.0-0.2)#*⟩	::=	highly unlikely;	⟨*#s(0.21-0.3)#*⟩	::=	very unlikely
⟨*#s(0.31-0.4)#*⟩	::=	unlikely;	⟨*#s(0.41-0.5)#*⟩	::=	not plausible
⟨*#s(0.51-0.6)#*⟩	::=	plausible;	⟨*#s(0.61-0.7)#*⟩	::=	possible
⟨*#s(0.71-0.8)#*⟩	::=	likely;	⟨*#s(0.81-0.9)#*⟩	::=	very likely
⟨*#s(0.91-1)#*⟩	::=	highly likely			

According to this BNF, the MPE from the diabetes example is explained as followed: "The observed evidence hyperglycaemia is best explained by food intake. Given hyperglycaemia and food intake, not having diabetes is likely. Given not having diabetes is likely, high quality of beta cells is very likely". If the domain expert argues that diabetes could be true the explanation changes to: "The observed hyperglycaemia is best explained by food intake. Given hyperglycaemia and food intake, diabetes is alternatively very unlikely. In that situation low quality of beta cells is plausible." If the domain expert argues that hyperglycaemia can be explained with a low quality of beta cells the explanation is: "Hyperglycaemia could highly unlikely indicate low quality of beta cells. With very unlikely low quality of beta cell, diabetes is highly likely".

4.3   Extended Example

In this section we will present a small but realistic example on a breast cancer dataset. The dataset contains nodes about x-ray properties and is intended for diagnostic purposes. The focus in this paper concentrates on explaining an MPE, therefore we will not discuss the medical background or how the BN was created. For further information to the data set, see [1]. However, it is notable that this network contains 16 nodes, which means that MPEs are quite hard to comprehend.

To compute an exemplary MPE, we choose the scenario that no imaging has been done yet, and select a number of observations that can be done in clinical practice, in particular, we take the observations that *Age(35–49)*, *SkinRetract(Yes)*, *NippleDischarge(No)* and *Location(UpInQuad)*. Given this evidence, the MPE solution that we find is: *AD(No)*, *BC(No)*, *BreastDensity(medium)*, *FibrTissue-Dev(Yes)*, *LymphNodes(no)*, *Margin(Ill-defined)*, *Mass(No)*, *MC(No)*, *Metastasis-(no)*, *Shape(Other)*, *Size(1 cm)*, *Spiculation(Yes)*. The explanation with our method is: *"The observed evidence age 35–49 and no nipple discharge is best explained by no architectural distortions. Given age 35–49, no nipple discharge and no architectural distortions, no metastasis is very likely. With the description of the previous sentence no lymph nodes will most probable be highly likely. Since no lymph nodes is highly likely and age 35–49 and no nipple discharge is given, no microcalcifications is very likely. Given age 35–49, no nipple discharge, no architectural distortions, no metastasis, no lymph nodes and no microcalcifications, breast cancer is very unlikely. With the description of the previous sentence size smaller than 1 cm will most probable be highly likely. In the pre-established combination other shape is highly likely. With the prior explanation no mass will most probable be highly likely. Given skin retract and UpInQuad location, fibrous tissue development is likely. Since fibrous tissue development is likely and skin retract and UpInQuad location is given, ill-defined margin is likely. With the description of the previous sentence spiculation is highly likely. Given no architectural distortions, no nipple discharge, size smaller than 1 cm, other shape and medium breast density is plausible".*

We believe that this is a reasonable explanation that could be understandable for physician, although clearly not trivial. The explanation shows, which other nodes are important for the explanation of a node, in this scenario. More specific one can read that the observed nodes *Age(35–49)* and *NippleDischarge(No)* are best explained by *AD(No)*, whereas the observed nodes *SkinRetract(Yes)* and *Location(UpInQuad)* is explained by *FibrTissueDev(Yes)*. On can derive, that the observed nodes are not highly dependent. The explanation shows that there are two different sub-chains of deduction in this scenario. Hence the first part showing the first sub-chain of deduction and the second part showing the smaller sub-chain *SkinRetract(Yes)*, *Location(UpInQuad)*, *FibrTissueDev(Yes)*, *Margin(Ill-defined)* and *Spiculation (Yes)*. Nevertheless the last explained node *BreastDensity(medium)* is explained with a subset of nodes from the first sub-chain. This means that breast density can be explained by parts of the first chain but does not belong in it. Being the last node also means, that the node breast density is not as important for the explanation of this scenario under the given evidence.

5   Related Work

In 1988–1999 research has been done in explaining BNs, however this research was manly focusing to explain BN models in general, how they where generated and how inferences could be made. This research has been summarized in Lacave et al. [8]. More recent research, including this paper, is focusing on the dynamic behavior of BNs, explaining posteriors and certain variables of interest or in this case, the MPE solution.

Several approaches for combining formal argumentation with Bayesian networks have been proposed in literature [11,12,15,16]. Their methods are based on building inference rules from variable-value pairs. The approach of Vreeswijk et al. [15] uses a multi-agent system to decide if a inference rule is supporting a logical argument. Williams et al. [16] uses argumentation theory to decide which arguments are justified for a particular patient in order to explain predictions of the Bayesian network. In Timmer et al. [11] the approach from Williams et al. is refined and in [12] a so called support graph is introduced. This support graph reduces the amount of rules extracted from the BN by only considering variables that are not conditionally independent, given the variable of interest, i.e., the variables which are in the Markov blanket of the variable of interest. Furthermore Timmer et al. are not only showing one argument to explain a variable of interest, but showing arguments derived from different non-blocking paths in the network (so-called support chains). Therefore the user can decide which argument is best for explaining the variable of interest. In this paper, we take a different perspective: instead of explaining a particular marginal probability, we aim to use argumentation for exploring scenarios which are consistent with the evidence.

There are also other approaches for explaining a variable of interest given evidence in Bayesian networks. Yap et al. [17] introduced a method to explain the variable of interest by capturing how variable interactions in a BN lead to inferences, independently of the evidence, just using variables needed to predict the behavior of the variable of interest. Vlek et al. [14] are providing a text form report for different scenarios, consistent with the evidence, regarding a case in legal evidence. The report estimates the probability of each chosen scenario being likely, to present a global perspective on the case. In Kyrimi et al. [7], variables of interest are not explained by all variables, but only from variables having an significant impact on it. To achieve this Kyrimi et al. are computing the impact of the evidence, and all variables in the Markov blanket of the variable of interest. The motivation of Vlek et al. [14] is similar to ours, because they also explain a scenario. However, Vlek et al. are focusing on one variable of interest and are searching for the most probable values for the other variables. Our approach however is focusing on a scenario given the evidence, without highlighting certain variables, to get an overview over the scenario. This is also the difference to Kyrimi et al. [7]: even though we make use of their methods to prune the explanation, we prune the explanation of a variable in only one chain of deduction.

6   Conclusion

The main contribution of this paper is a new method to explain BNs based on scenarios starting with the MPE. From the MPE further scenarios can be explained to explore the BN. To this end a knowledge base representing the dependences and the strength of dependences is drawn from the Bayesian network with a argumentative probability tree. With this tree several desirable properties are achieved. In particular, the tree structures the order of dependences to get a chain of reasoning to build the knowledge base. Through the strength attached to every dependence, the tree provides the information how likely each dependence is as well. With the mapping of the argumentative probability tree to natural language, domain experts can argue with the explanation and explore the BN to get more insights.

In future work, we are going to evaluate this framework by comparing it to other frameworks in different contexts. In addition, we would further like to exploit the modularity of this framework. For example, in follow-up work we would like to study if arguments can semantically be enriched through linking them to ontologies. Subsequently, arguments could be linked to further data sources to present even more specific information, relevant for the domain expert. Furthermore we will attempt to put state-of-the-art NLG research to use, to get a more flexible natural language generation, where for example node-value pairs can be given a more natural appearance in an explanation.

References

1. Amirkhani, H., Rahmati, M., Lucas, P.J., Hommersom, A.: Exploiting experts knowledge for structure learning of Bayesian networks. IEEE Trans. Pattern Anal. Mach. Intell. **39**(11), 2154–2170 (2017)
2. Endres, E., Augustin, T.: Statistical matching of discrete data by Bayesian networks. In: Conference on Probabilistic Graphical Models, pp. 159–170 (2016)
3. Fensel, D., et al.: Towards LarKC: a platform for web-scale reasoning. In: 2008 IEEE International Conference on Semantic Computing, pp. 524–529. IEEE (2008)
4. Fox, J., Glasspool, D., Grecu, D., Modgil, S., South, M., Patkar, V.: Argumentation-based inference and decision making-a medical perspective. IEEE Intell. Syst. **22**(6) (2007)
5. Hommersom, A., Lucas, P.J.F. (eds.): Foundations of Biomedical Knowledge Representation: Methods and Applications. LNCS (LNAI), vol. 9521. Springer, Cham (2015). https://doi.org/10.1007/978-3-319-28007-3
6. Korb, K.B., Nicholson, A.E.: Bayesian Artificial Intelligence, 2nd edn. CRC Press Inc., Boca Raton (2010)
7. Kyrimi, E., Marsh, W.: A progressive explanation of inference in 'hybrid' Bayesian networks for supporting clinical decision making. In: Proceedings of the Eighth International Conference on Probabilistic Graphical Models, pp. 275–286 (2016)
8. Lacave, C., Díez, F.J.: A review of explanation methods for Bayesian networks. Knowl. Eng. Rev. **17**(2), 107–127 (2002)
9. Pearl, J.: Probabilistic Reasoning in Intelligent Systems: Networks of Plausible Inference. Morgan Kaufmann, Burlington (1988)
10. Reiter, E., Dale, R.: Building Natural Language Generation Systems. Studies in Natural Language Processing. Cambridge University Press, New York (2000)
11. Timmer, S., Meyer, J., Prakken, H., Renooij, S., Verheij, B.: Inference and attack in Bayesian networks. In: Hindriks, K., De Weerdt, M., Van Riemsdijk, B., Warnier, M. (eds.) Proceedings of the 25th Benelux Conference on Artificial Intelligence (BNAIC 2013), pp. 199–206. Delft University Press (2013)
12. Timmer, S.T., Meyer, J.J.C., Prakken, H., Renooij, S., Verheij, B.: A two-phase method for extracting explanatory arguments from Bayesian networks. Int. J. Approx. Reason. **80**, 475–494 (2017)
13. Vennekens, J., Denecker, M., Bruynooghe, M.: CP-logic: a language of causal probabilistic events and its relation to logic programming. Theory Pract. Logic Program. **9**(3), 245–308 (2009)
14. Vlek, C.S., Prakken, H., Renooij, S., Verheij, B.: A method for explaining Bayesian networks for legal evidence with scenarios. Artif. Intell. Law **24**(3), 285–324 (2016)
15. Vreeswijk, G.A.W.: Argumentation in Bayesian belief networks. In: Rahwan, I., Moraïtis, P., Reed, C. (eds.) ArgMAS 2004. LNCS (LNAI), vol. 3366, pp. 111–129. Springer, Heidelberg (2005). https://doi.org/10.1007/978-3-540-32261-0_8
16. Williams, M., Williamson, J.: Combining argumentation and Bayesian nets for breast cancer prognosis. J. Logic Lang. Inf. **15**(1), 155–178 (2006)
17. Yap, G.E., Tan, A.H., Pang, H.H.: Explaining inferences in Bayesian networks. Appl. Intell. **29**(3), 263–278 (2008)

Fuzzification of Ordinal Classes. The Case of the HL7 Severity Grading

Federico Cabitza[1,2] and Davide Ciucci[1(✉)]

[1] Università degli Studi di Milano-Bicocca, Milan, Italy
{cabitza,ciucci}@disco.unimib.it
[2] IRCCS Istituto Ortopedico Galeazzi, Milan, Italy

Abstract. Despite the vagueness and uncertainty that is intrinsic in any medical act, interpretation and decision (including acts of data reporting and representation of relevant medical conditions), still little research has focused on how to explicitly take this uncertainty into account. In this paper, we focus on a general and wide-spread HL7 terminology, which is grounded on a traditional and well-established convention, to represent severity of health conditions (e.g., pain, visible signs), ranging from *absent* to *very severe* (as a matter of fact, different versions of this standard present minor differences, like 'minor' instead of 'mild', or 'fatal' inst ead of 'very severe'). Our aim is to provide a fuzzy version of this terminology. To this aim, we conducted a questionnaire-based qualitative research study involving a relatively large sample of clinicians to represent numerically the five different labels of the standard terminology: absent, mild, moderate, severe and very severe. Using the collected values we then present and discuss three different possible representations that address the vagueness of medical interpretation by taking into account the perceptions of domain experts. In perspective, our hope is to use the resulting fuzzifications to improve machine learning approaches to medicine.

Keywords: Fuzzy sets · Interval-valued fuzzy sets · Linguistic labels
Ground truth

1 Introduction

Medical data are intrinsically uncertain and during the years several attempts to deal with the vagueness of medical terms have been made. We recall the use of fuzzy sets to represent medical terms [13,20] or to model patient records [24], the fuzzy version of the Arden markup language [21] and several fuzzy ontology applications to medicine [10,17].

In this paper, we are interested in a very common terminology to represent severity of health conditions and symptoms in medical documents, which has been recently adopted also by the HL7 Fast Healthcare Interoperability

© Springer Nature Switzerland AG 2018
D. Ciucci et al. (Eds.): SUM 2018, LNAI 11142, pp. 64–77, 2018.
https://doi.org/10.1007/978-3-030-00461-3_5

Resources (FHIR) framework, i.e., the most widely adopted standards framework for the representation of health data on the Internet and in digital health applications[1].

According to this terminology doctors can report health conditions in terms of five ordinal categories, namely: *absent, mild, moderate, severe* and *very severe*. Although these categories are used extensively and on a daily basis by most medical doctors around the world in most forms, charts and reports (even paper-based ones), their meaning has never been established univocally and, more importantly from the computational point of view, quantitatively [19]. As a matter of fact, no standardizing body nor single doctor can establish what, say, *moderate* really means in objective terms [1], nor determine that the transition from a *mild* condition to a *moderate* one is like passing from a *moderate* one to a *severe* condition: a standard terminology to describe severity is just a set of available values, in which only a total order relation is defined.

Ordinal scales are very common in medicine [11,15] and on their basis doctors can understand each other and make critical decisions despite their seeming arbitrariness and loosely defined semantics; ordinal values like those mentioned above are also extensively used to annotate medical records, and to some extent report a written interpretation of other medical data, like laboratory results and medical images. For this reason severity labels are increasingly used in *ground truthing*, that is the preparation of training and test data sets for the definition and evaluation of predictive models. Of course these terms are subject to personal views, contextual situations, interpretation of evidence, etc.: in a word, they are intrinsically *fuzzy*.

The scope of the present work is thus to represent severity categories as fuzzy sets or their extensions. We will consider these categories as linguistic labels and assign them different types of fuzzy sets with domain on numerical scales according to a human-centered study. In so doing, we can get a representative, yet approximate, model to map ordinal categories to numerical values (on a scale [0–100] or [0–25], where the lower bound represents absence of perceivable signs of the condition of medical interest and the upper bound its strongest expression), and *vice versa*.

The data set we used to define this mapping is a collection of intervals for each category/label, provided by domain experts (i.e., medical doctors) by means of an ad-hoc Web-based questionnaire, administered during an on line survey. We present and discuss several ways to aggregate these values in order to obtain some kind of *fuzzification* of the severity conditions. This approach is different from existing approaches to fuzzify ordinal scales such as [18,22], where the fuzzification process is done automatically by assigning a fuzzy number to each label and then applied to a case study. Here, our aim is to fuzzify the ordinal scale starting from the collected data.

There is a twofold interest in this fuzzification: representing the intrinsic uncertainty of the severity description and then, the possibility to apply fuzzy techniques. Just to make two examples, the obtained fuzzy sets could be inserted

---

[1] http://hl7.org/fhir/ValueSet/condition-severity.

in the fuzzy version of the Arden markup language [21], which provides a way to use fuzzy sets as a data type; or they could represented in fuzzy OWL [3] in order to be used in ontology based medical applications.

The rest of the paper is organized as follows: Sect. 2 presents the methods used to collect the perceptions from a sample of domain experts (doctors); Sect. 3 presents the fuzzification methods. Finally, Sect. 4 discusses the findings of this case study and the generalizability of the methods reported in the previous sections to other application domains.

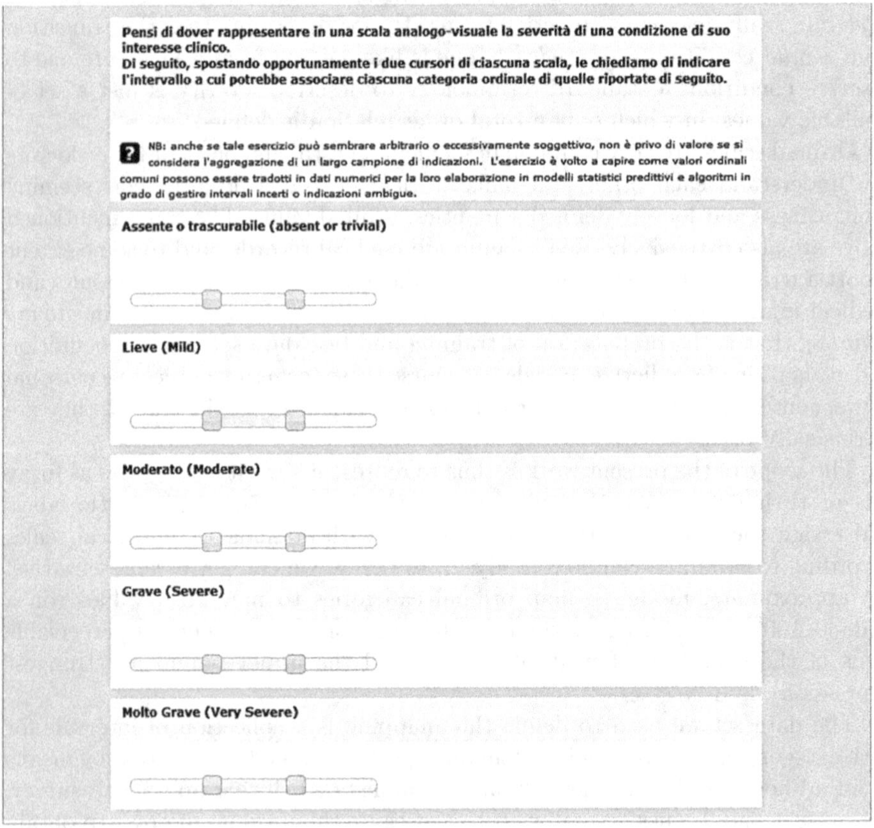

Fig. 1. The first page of the on-line questionnaire that we administered to the sample of clinicians to collect their perception on severity categories (original text in Italian).

2   The Data Set

To collect data on the subjective perception of the quantitative meaning of the categories (each denoted by a specific label) of the severity HL7 ordinal scale, we designed a closed-ended two-page questionnaire to be administered on line in a Computer-Assisted-Web-self-Interview (CAWI) configuration. The first page of this questionnaire (depicted in Fig. 1) asked the respondents to express each level of severity of the original 5-item HL7 scale (i.e., absent, mild, moderate, severe, and very severe) into a Visual Analogue Scale (VAS). A VAS is a measurement instrument that has been devised and introduced in health care to try to measure characteristics that appear or are easily perceived as continuous but that cannot be directly measured easily, like pain, and by which to overcome the intrinsically discrete nature of ordinal categorizations [8].

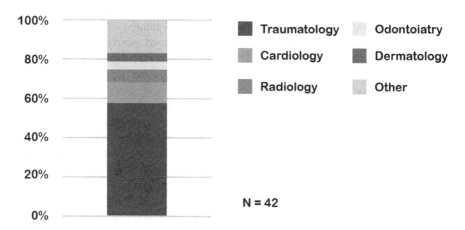

Fig. 2. Stacked bar chart representing the composition of the sample of respondents involved in this study. The majority of the sample were trauma and orthopedic surgeons, the rest of the sample is relatively varied, as also shown by the 'other' category, which is the second one for numerosity and encompasses (among the others) two neurologists, one endocrinologist and one rheumatologist. This suggests that, despite the relatively small sample, this is sufficiently heterogeneous not to consider the responses limited to a specific medical discipline.

To this aim, we associated each item with a 2-cursor range slider control. By moving each of the two independent cursors the respondents could thus create an *inner interval*, comprised *within* the two cursors, encompassing all those numerical values that they felt could represent the ordinal category properly. The interface was designed so that initially the slider control could be seen in the central position depicted in Fig. 1 and respondents would want to move the cursors to set the new intervals and, in doing so, "see" the overlap that they deem useful to report between the categories. This overlap was neither promoted nor prevented, as the cursors could be moved freely along each range slider with the only constraint that the 'lower' extreme cursor could never be moved to the right of the 'higher' extreme cursor, and vice versa. Moreover, the respondents could get only an approximate idea of the numerical values that were associated with the position of the cursors (and in fact this association was not mentioned in the task description, reported at the top of Fig. 1, but only in the help section), since the range was intended to be on a strict analogue scale, with no explicit nor numerical anchor. That notwithstanding, VASs are common representational tools most potential respondents were very familiar with for its wide adoption in clinical practice, as said above, and this suggests that respondents performed the task effortlessly. The second page of the questionnaire was intended to collect a few data on the respondent's professional profile (which was intended to be anonymous), namely her medical specialty.

At the end of November 2017, we invited 97 clinicians by email to fill in the two-page questionnaire. Most respondents worked as clinicians and surgeons at the Orthopedic Institute Galeazzi (IOG), which is one of the largest teaching hospitals in Italy specialized in the study and treatment of musculoskeletal disorders; at IOG almost 5,000 surgeries are performed yearly, mostly arthroplasty (hip and knee prosthetic surgery) and spine-related procedures. After two weeks since this first invitation we sent a gentle reminder and one week later we definitely closed the survey. Response rate was moderately high, especially in light of the very busy daily schedule of the involved prospective respondents, the anonymity of the survey and the lack of incentives: indeed slightly less than half of the potential respondents accepted the invitation and filled in the online questionnaire: thus we collected 42 questionnaires by as many respondents (see Fig. 2). When we analyzed the responses, some questionnaires were found filled in with seemingly random data and were discarded: then the final dataset contained 298 datapoints, corresponding to 149 intervals (see Fig. 3) by 32 different respondents. Moreover, the questionnaires completed in each and every item were 27.

Some respondents contacted us after doing the CAWI to warn us that they had found it difficult to move the cursors of the range slider controls on mobile and multi-touch devices like smart phones and tablets. Although we did not collect information on the device used during the CAWI, we can consider that several people could have tried to fill in questionnaire from their smart phones: this could account for some of the "dirtiness" we detected in the original data set (like improbable interval extremes and empty cells). In any case, to our knowledge no study has so far involved more than thirty domain experts to have them represent the quantitative "meaning" (onto a numerical 0–100 range) for the ordinal categories they use in their reports and records on a daily basis.

Thus, the original data set contained the lower and upper extremes of the five ordinal categories expressing increasing levels of severity for all of the survey respondents, that is a 32×10 matrix of data points on the severity dimension, ranging from 0 to 100. From this data set of coordinates of interval extremes we computed a new one, which included all of the data points between the extremes (with the granularity of the hundredth). This was the data set that we used for the fuzzification, which required both the binning (to reduce the effects of minor observation errors) and the weighting of each coordinate on the basis of the interval width.

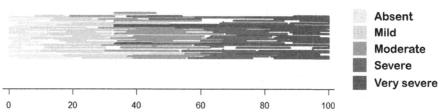

Fig. 3. Diagram showing the data set at a glance. Different questionnaires are represented along the vertical dimension; intervals related to different severity categories are represented in different hues along the horizontal 0–100 continuum.

Table 1. An extract of the dataset, for each severity level the min and max values are shown.

Absent or trivial	Mild	Moderate	Severe	Very severe
3–20	23–40	39–55	56–76	83–100
0–18	18–36	37–58	61–81	82–100
2–15	17–37	39–61	63–83	84–97
23–58	44–78	55–93	60–91	71–97
0–9	10–30	30–53	54–77	78–100
7–7	30–30	56–56	67–67	95–95

An extract of this dataset is reported in Table 1. Both from Fig. 2 and Table 1, it can be seen that in the majority of cases, each level is represented as an interval, not just a coordinate point, and these intervals can overlap. Also, significant differences can exist between different doctors.

3 Alternative Fuzzification Methods

Starting from the collected data, we are now going to define three different fuzzifications of the five severity conditions.

3.1 Using Fuzzy Sets

Let us consider the linguistic variable [23] *Severity condition* with values $V = \{$ *absent, mild, moderate, severe, very severe*$\}$. In this section, we give two different semantics to each term in V by means of a fuzzy set in the universe $U = [0, 100]$ or $U = [0, 25]$. In the following sub-section, the semantic will be given by an interval-fuzzy set.

Using the Extremes. As a first solution, only the extremes of each interval assigned by a doctor to each term is used. As an example, let us consider the extract given in Table 1 and specifically the value of *absent* in the first row. We are going to use the two values 3 and 20 to define a fuzzy set for *absent*.

For each term v in V and each value x in the range $[0-100]$ we simply count how many times x appears in the column corresponding to v as one of the extremes of the interval. In this way, only one fuzzy set is needed for each term $v \in V$ (in Sect. 3.2, we will discuss the separate use of the two extremes). The resulting fuzzy sets are shown in Fig. 4.

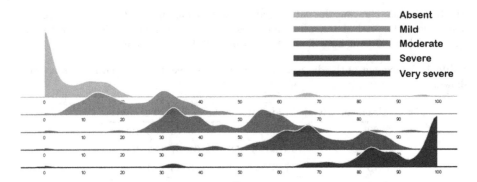

Fig. 4. Joy plot depicting the fuzzification before normalization to 1 obtained by assigning one fuzzy set to each severity category. (A joy plot is a recent diagram that arranges together a vertical succession of area charts, whose lines are sufficiently close together so that their peaks overlap the shapes behind and give a sort of 3D effect. The name, oddly enough, derives to the iconic cover of an album by Joy Division from 1979 [16].)

Let us notice that these are fuzzy sets but not fuzzy numbers [9] contrary to what is often used to represent linguistic terms. Indeed, the fuzzy sets in Fig. 4 have more than one peak. Formally, there does not exist a value x such that in the interval $(-\infty, x)$ the fuzzy set is ascending and descending in the interval (x, ∞). However, as argued in [12], it should not be mandatory to use convex and normalized fuzzy sets if non-convex and non-normalized ones are a more natural and better choice.

The advantages of this representation is to somehow take into account the notion of interval. Especially, in the "middle" cases of Mild, Moderate and Severe, it can be easily seen that we have two peaks, which can be considered as a fuzzy number each, representing the extreme of the interval.

Using the Whole Interval. By using only the extreme values, the information hidden in the interval is lost. Thus, here for each interval we use all the values it contains. For example, let us consider again the first interval in Table 1: $[3, 20]$. In order to obtain a fuzzy set as semantics for *Absent*, all the values $\{3, 4, 5 \ldots, 20\}$ are taken into account.

Data points have been splitted in 25 bins by diving each coordinate in the 0–100 range by 4 and taking the integer part of the quotient. This was done to account for the fact that respondents could not choose the exact coordinate but just move a cursor that was 1/14 wide with respect to the range lenght; thus, considering the middle point of the cursor as an approximate point the respondents might have referred to drop the cursor and select the coordinate of choice, we decided to divide the 0–100 range in 25 bins. Considering the full granularity of the original data (in hundredths) would result in a too noisy representation.

Moreover, we also weighted the interval data by dividing each data point associated by a respondent with a category by the total number of data points in the same interval, that is its width. In so doing, each data point has been weighted by the intended accuracy of the respondent. That is, if a doctor for a given term selected an interval $[x, y]$, we weighted it by a factor $\frac{1}{(y-x+1)}$. For instance, a data point associated by a respondent to the 'mild' category within a 2-bin wide interval got a double weight in comparison to the same data point chosen by another respondent who created a 4-bin wide interval for the same category, as we consider the former indication as at a higher *resolution* (so to speak).

Example 1. Let us consider once more the first row in Table 1. After data binning, the new intervals look likes as in Table 2. Then, to build the fuzzy set corresponding to *Absent*, the values in the range [0–5] with weight 1/6 are considered.

Table 2. Dataset transformation from the interval [0–100] to the interval [0–25]

	Absent or trivial	Mild	Moderate	Severe	Very severe
Original [0–100]	3–20	23–40	39–55	56–76	83–100
Bin [0–25]	0–5	5–10	9–13	14–19	20–25

The two above intervention of feature engineering and the normalisation to 1 resulted in a smoothing of the original distributions, into the ones depicted in Fig. 5.

In this case the resulting fuzzy sets are more regular and most of them can be considered as a fuzzy number. Due to this regularity, we could think to approximate these sets with standard shaped fuzzy sets, such as triangular or Gaussian, in order to ease representation and computation. Of course, we lose, at least from a visual standpoint, the information of the extremes of the interval.

3.2   Using Interval Valued Fuzzy Sets

Another possibility to use all the available information contained in the interval is to directly use the interval in the definition of the fuzzy set. Thus, in this section we propose to use an interval–valued fuzzy set [4,14].

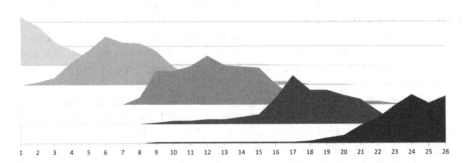

Fig. 5. Joyplot obtained using the whole interval, after data binning and weighting.

Definition 1. *An* interval–valued fuzzy set *on a universe U is a mapping I : $U \mapsto L([0,1])$ where $L([0,1])$ is a partially order set defined as $\{(x,y) : (x,y) \in [0,1]^2 \quad and \quad x \leq y\}$.*

From the available data, we can define an interval valued fuzzy set similarly as it was done in Subsect. 3.1 but considering that the extremes represent two fuzzy sets instead of only one. Thus, given a point u in the universe $U = [0,100]$, and a term $t \in V$, $I_t(u)$ is the interval $[x,y]$ with x the number of times (normalized to 1) u has been selected as lower bound for l and y the number of times (normalized to 1) u has been selected as upper bound for l. So, for instance, $I_{Mild}(30) = [0.2, 0.8]$ represents the fact that 30 has been selected as a lower extreme for *Mild* in proportion of 0.2 of all the values selected for the *Mild* lower extreme and in a proportion 0.8 as an upper extreme.

4   Discussion and Conclusion

This paper addresses the fuzzication of a common terminology, which is also adopted by the HL7 framework in the digital health domain, to characterise health conditions, the appearance of medical signs and other expressions of medical relevance, as these are perceived by either the medical doctors or the patients themselves (e.g., in the so called Patient Reported Outcome measures [2]).

This allows to consider a single medical condition as something that can be characterized with multiple categories instead of only one, that is to avoid the limit of having just one category (e.g., mild) considered as true (1) while all of the other labels are false (0) and to adopt a more fuzzy representation, where multiple categories can have a truth value between 0 and 1. This can be motivated for a number of reasons: first, medical conditions are intrinsically ambiguous. Fuzzifying their computer-readable representation can help raise awareness of the variability of severity perceptions by different health practitioners, even of the same condition, to be expressed symbolically. Health practitioners are

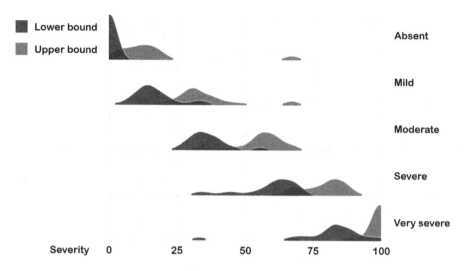

Fig. 6. Joy plot showing the responses collected, grouped by severity level (horizontal lines) and kind of extreme of the intervals proposed by the respondents: blue for the lower extreme and red for the higher extreme. As expected variability is lower for the whole range extremes, the one associated with the lower extreme of the absent category and the higher extreme of the very severe category, although both ceiling and flooring effects were not severe. (Color figure online)

increasingly called to produce health data of high quality (both in terms of reliability and validity) to represent medical events and conditions faithfully on electronic medical records. While we agree that doctors should continue to use textual categories (including codes and numbers) they feel more confident with (like the HL7 ones), we aim to raise awareness on the intrinsic ambiguity of these labels, and on the need to be wary of any clear-cut mapping between each label and a set of numerical values: converting ordinal categories like the severity ones into numbers, in order to treat them as scalar values (e.g., to calculate the average pain, or the average severity of a condition, within a sample of patients, also for inference statistical aims) is a serious methodological mistake. In this paper we also showed how different clinicians (that is different domain experts) intend the same textual labels differently, when invited to reflect on their meaning in numerical terms. This means that also using labels as univocal tokens in advanced statistical techniques, like the ones employed in machine/statistical learning and in the definition of predictive models, can be harmful. The same patient could be associated with a 'mild' label by a doctor, and a 'severe' label by another doctor, and this even if either doctors intend to characterize the very same condition, which can be represented by the same numerical value on a 0–100 continuum. This observation regards the phenomenon of inter-rater reliability that, although widely known in the medical ambit [6], is still little known and considered in most of the fields of applied computer science [5].

For this reason our approach is grounded on an empirical and human-centered approach, that is on the subjective perceptions of domain experts for whom the ordinal categories to be fuzzified are meaningful according to the context, right in virtue of their descriptive power and despite their ambiguity. Our study confirms this ambiguity and proposes a way to leverage it, instead of "sweeping it under the carpet" or underrating its potential to undermine the real-world performance of predictive models. To this regard, we are aware that the data collection phase is critical in getting reliable data to fuzzify. As said in Sect. 2, our on-line questionnaire allowed the respondents to set a different interval on a continuous range for each ordinal category of the severity HL7 terminology; moreover all of the ranges were displayed in a single page, as depicted in Fig. 1. This allowed (and perhaps fostered) overlaps to occur not only between the representations of the same category across different respondents (as expected), but also between representations of different categories (usually consecutive categories) by the *same* respondent, as this happened in the 30% of cases. A different approach would have requested respondents to set the five different intervals without the possibility to see inter-category overlap at a glance, that is each interval in a distinct page of the questionnaire. However, this would not have prevented overlaps, nor made them less frequent, as also the opposite effect could be induced, in case respondents had intended the representational task as requiring to avoid or at least minimize interval overlaps.

In an alternative design, the questionnaire could have requested respondents to set only 4 points within a single severity range, so as to identify 5 intervals univocally and without inner overlaps[2].

These different design setups would have yielded different data sets and hence different outputs of the fuzzification methods described in this paper, but would likely have little impact on these latter ones. The adopted setup was aimed at keeping the questionnaire short and simple and at not inducing the surreptitious idea that each respondent should define clear-cut boundaries between severity categories on a continuum: this was left possible, by aligning the cursors with each other vertically, but it was not imposed by design.

Finally, it is noteworthy to say the fuzzification methods proposed and discussed in this paper have been applied to the traditional 5-item severity terminology only as a proof of the concept: we chose this terminology because it is common to many health conditions, used in most medical specialties, and it has also been recently adopted by the HL7 standard developing organization and hence it is nowadays widespread in most digital health applications. However, these fuzzification methods can be applied to *any* ordinal terminology, and not only to those specific of the medical domain.

---

[2] This latter design would have allowed to compute the inter-rater agreement for each severity coordinate, as each observer would have associated each coordinate to one and only one severity category. In our setup this is not feasible unless spurious labels for category overlaps are introduced, e.g. mild-and-moderate, besides the five regular ones, making the reliability assessment more laborious.

Indeed, in this paper we have provided elements to consider fuzzification as a convenient way to convert single ordinal labels into numbers, which are the representation of choice of many predictive models[3], and vice versa, as a way to transform single numerical coordinates on a range representing the variability of a dimension of interest (like severity) into corresponding fuzzy values, which are associated to one or more ordinal classes. These coordinates could be the output of regressive models, which de-fuzzification could transform into sets (usually pairs) of labels, each associated with a fuzzy truth value. In so doing, the fuzzy representation could convey a better idea of the intrinsic uncertainty of the prediction itself, and make this latter one more meaningful in the eyes of the practitioners involved (although also less clear-cut and more ambiguous, paradoxically).

The advantage of this approach lies in the fact just mentioned above: the mapping is made on the basis of the perceptions of a heterogeneous sample of domain experts, in our case, clinicians. If perceptions are collected from the experts who annotated a ground truth data set, this mapping could optimally represent the implicit meaning that group of people, as a collective, attach to the annotation labels, and hence to the classes the machine learning have to work with. Even if the perceptions are not collected from the same group of people involved in the observations and the annotations, the opportune selection of the sample (e.g.,through stratified random sampling) could guarantee a certain degree of representativeness and bring forth reasonable and meaningful mappings.

Future work should be aimed at testing the above conjecture that fuzzification could yield to either more accurate predictions than in multi-class classification, or more meaningful ones for the domain experts involved or both.

Acknowledgements. The authors would like to thank Giorgio Pilotti and Gabriele Caldara, two Master students of the Master Degree in Data Science, who have conceived and realised a preliminary version of the charts depicted in Figs. 5 and 6, respectively. Figure 3 was developed after an intuition of Pietro de Simoni, another student from the same master degree course. The authors are also grateful to Prof. Giuseppe Banfi for advocating the survey at IOG and to all of the anonymous clinicians who spontaneously participated in the research by playing the game of reporting severity categories on a traditional VAS.

References

1. Atkinson, T.M., et al.: What do 'none', 'mild', 'moderate', 'severe', and 'very severe' mean to patients with cancer? Content validity of PRO-CTCAE response scales. J. Pain Symptom Manag. **55**(3), e3–e6 (2018)
2. Black, N.: Patient reported outcome measures could help transform healthcare. BMJ: Brit. Med. J. **346**, f167 (2013)

---

[3] Indeed, since many machine learning algorithms cannot operate on label data directly, usual feature engineering tasks consist in the transformation of ordinal (categorical) data into numbers (i.e., Integer Encoding), or in applying specific encoding schema to create dummy variables or binary features for each value of a specific nominal categorical attribute (One-Hot Encoding), like severity [7].

3. Bobillo, F., Straccia, U.: Fuzzy ontology representation using OWL 2. Int. J. Approx. Reason. **52**(7), 1073–1094 (2011)

4. Bustince, H., Montero, J., Pagola, M., Barrenechea, E., Gomez, D.: A survey of interval valued fuzzy sets, Chap. 22, pp. 489–515. Wiley-Blackwell (2008)

5. Cabitza, F., Ciucci, D., Rasoini, R.: A giant with feet of clay: on the validity of the data that feed machine learning in medicine. In: Cabitza, F., Batini, C., Magni, M. (eds.) Organizing for the Digital World. LNISO, vol. 28, pp. 121–136. Springer, Cham (2019). https://doi.org/10.1007/978-3-319-90503-7_10

6. Cabitza, F., Locoro, A., Laderighi, C., Rasoini, R., Compagnone, D., Berjano, P.: The elephant in the record: on the multiplicity of data recording work. Health Inform. J. (2018, to be published)

7. Coates, A., Ng, A.Y.: The importance of encoding versus training with sparse coding and vector quantization. In: Proceedings of the 28th International Conference on Machine Learning, ICML 2011, pp. 921–928 (2011)

8. Crichton, N.: Visual analogue scale (VAS). J. Clin. Nurs. **10**(5), 706–6 (2001)

9. Dijkman, J., van Haeringen, H., de Lange, S.: Fuzzy numbers. J. Math. Anal. Appl. **92**(2), 301–341 (1983)

10. El-Sappagh, S., Elmogy, M.: A fuzzy ontology modeling for case base knowledge in diabetes mellitus domain. Eng. Sci. Technol. Int. J. **20**(3), 1025–1040 (2017)

11. Forrest, M., Andersen, B.: Ordinal scale and statistics in medical research. Br. Med. J. (Clin. Res. Ed.) **292**(6519), 537–538 (1986)

12. Garibaldi, J.M., John, R.I.: Choosing membership functions of linguistic terms. In: The 12th IEEE International Conference on Fuzzy Systems, FUZZ-IEEE 2003, St. Louis, Missouri, USA, 25–28 May 2003, pp. 578–583. IEEE (2003)

13. Godo, L., de Mántaras, R.L., Sierra, C., Verdaguer, A.: Milord the architecture and the management of linguistically expressed uncertainty. Int. J. Intell. Syst. **4**, 471–501 (1989)

14. Guijun, W., Xiaoping, L.: The applications of interval-valued fuzzy numbers and interval-distribution numbers. Fuzzy Sets Syst. **98**(3), 331–335 (1998)

15. Jakobsson, U.: Statistical presentation and analysis of ordinal data in nursing research. Scand. J. Caring Sci. **18**(4), 437–440 (2004)

16. Kosara, R.: Joy plots, May 2017. https://eagereyes.org/blog/2017/joy-plots, http://archive.is/Ui0NN. Accessed 9 May 2018

17. Lee, C.S., Wang, M.H., Hsu, C.Y., Chen, Z.W.: Type-2 fuzzy set and fuzzy ontology for diet application. In: Sadeghian, A., Mendel, J., Tahayori, H. (eds.) Advances in Type-2 Fuzzy Sets and Systems. STUDFUZZ, vol. 301, pp. 237–256. Springer, New York (2013). https://doi.org/10.1007/978-1-4614-6666-6_15

18. Li, Q.: A novel Likert scale based on fuzzy sets theory. Expert Syst. Appl. **40**(5), 1609–1618 (2013)

19. Salomon, J.A.: Reconsidering the use of rankings in the valuation of health states: a model for estimating cardinal values from ordinal data. Popul. Health Metr. **1**(1), 12 (2003)

20. Sanchez, E.: Medical applications with fuzzy sets. In: Jones, A., Kaufmann, A., Zimmermann, H.J. (eds.) Fuzzy Sets Theory and Applications. ASIC, vol. 177, pp. 331–347. Springer, Dordrecht (1986). https://doi.org/10.1007/978-94-009-4682-8_16

21. Vetterlein, T., Mandl, H., Adlassnig, K.P.: Fuzzy Arden syntax: a fuzzy programming language for medicine. Artif. Intell. Med. **49**(1), 1–10 (2010)

22. Vonglao, P.: Application of fuzzy logic to improve the Likert scale to measure latent variables. Kasetsart J. Soc. Sci. **38**(3), 337–344 (2017)
23. Zadeh, L.: The concept of a linguistic variable and its application to approximate reasoning I. Inf. Sci. **8**(3), 199–249 (1975)
24. Żywica, P.: Modelling medical uncertainties with use of fuzzy sets and their extensions. In: Medina, J., Ojeda-Aciego, M., Verdegay, J.L., Perfilieva, I., Bouchon-Meunier, B., Yager, R.R. (eds.) IPMU 2018. CCIS, vol. 855, pp. 369–380. Springer, Cham (2018). https://doi.org/10.1007/978-3-319-91479-4_31

A Modular Inference System for Probabilistic Description Logics

Giuseppe Cota[1](✉)🆔, Fabrizio Riguzzi[2]🆔, Riccardo Zese[1]🆔, Elena Bellodi[2]🆔, and Evelina Lamma[1]🆔

[1] Dipartimento di Ingegneria, University of Ferrara,
Via Saragat 1, 44122 Ferrara, Italy
{giuseppe.cota,fabrizio.riguzzi,riccardo.zese,elena.bellodi,
evelina.lamma}@unife.it
[2] Dipartimento di Matematica e Informatica, University of Ferrara,
Via Saragat 1, 44122 Ferrara, Italy

Abstract. While many systems exist for reasoning with Description Logics knowledge bases, very few of them are able to cope with uncertainty. BUNDLE is a reasoning system, exploiting an underlying non-probabilistic reasoner (Pellet), able to perform inference w.r.t. Probabilistic Description Logics. In this paper, we report on a new *modular* version of BUNDLE that can use other OWL (non-probabilistic) reasoners and various approaches to perform probabilistic inference. BUNDLE can now be used as a standalone desktop application or as a library in OWL API-based applications that need to reason over Probabilistic Description Logics. Due to the introduced modularity, BUNDLE performance now strongly depends on the method and OWL reasoner chosen to obtain the set of justifications. We provide an evaluation on several datasets as the inference settings vary.

Keywords: Probabilistic Description Logic · Semantic Web
Reasoner · OWL Library

1 Introduction

The aim of the Semantic Web is to make information available in a form that is understandable and automatically manageable by machines. In order to realize this vision, the W3C has supported the development of a family of knowledge representation formalisms of increasing complexity for defining ontologies, called OWL (Web Ontology Language), that are based on Description Logics (DLs). Many inference systems, generally called reasoners, have been proposed to reason upon these ontologies, such as Pellet [23], Hermit [22] and Fact++ [24].

Nonetheless, modeling real-world domains requires dealing with information that is incomplete or that comes from sources with different trust levels. This motivates the need for the uncertainty management in the Semantic Web, and many proposals have appeared for combining probability theory with OWL languages, or with the underlying DLs [4,8,12,14,15]. Among them, in [18,26] we

© Springer Nature Switzerland AG 2018
D. Ciucci et al. (Eds.): SUM 2018, LNAI 11142, pp. 78–92, 2018.
https://doi.org/10.1007/978-3-030-00461-3_6

introduced the DISPONTE semantics for probabilistic DLs. DISPONTE borrows the distribution semantics [20] from Probabilistic Logic Programming, that has emerged as one of the most effective approaches for representing probabilistic information in Logic Programming languages. Examples of probabilistic reasoners that perform inference under DISPONTE are BUNDLE [18,19,26], TRILL and TRILL^P [26,27]. The first one is implemented in Java, whereas the other two are written in Prolog to exploit Prolog's backtracking facilities during the search of all the possible justifications.

In order to perform probabilistic inference over DISPONTE knowledge bases (KBs), it is necessary to find the covering set of justifications and this is accomplished by a non probabilistic reasoner. The first version of BUNDLE was able to execute this search by exploiting only the Pellet reasoner [23].

In this paper, we propose a new version of BUNDLE which is modular and allows one to use different OWL reasoners and different approaches for justification finding. In particular, it embeds Pellet, Hermit, Fact++ and JFact as OWL reasoners, and three justification generators, namely GlassBox (only for Pellet), BlackBox and OWL Explanation. The introduced modularity has two main advantages with respect to BUNDLE's previous version. First, it allows one to "plug-in" a new OWL API-based reasoner in a very simple manner. Second, the framework can be easily extended by including new concrete implementations of algorithms for justification finding.

In this modular version, BUNDLE performance will strongly depend on the sub-system employed to build the set of justifications for a given query. To evaluate it we ran several experiments on different real-world and synthetic datasets.

The paper is organized as follows: Sect. 2 briefly introduces DLs, while Sect. 3 illustrates the justification finding problem. Sections 4 and 5 present DISPONTE and the theoretical aspects of inference in DISPONTE KBs respectively. The description of BUNDLE is provided in Sect. 6. Finally, Sect. 7 shows the experimental evaluation and Sect. 8 concludes the paper.

2 Description Logics

An ontology describes the concepts of the domain of interest and their relations with a formalism that allows information to be processable by machines. The *Web Ontology Language* (OWL) is a family of knowledge representation languages for authoring ontologies or knowledge bases. OWL 2 [25] is the last version of this language and since 2012 it became a W3C recommendation.

Descriptions Logics (DLs) provide a logical formalism for knowledge representation. They are useful in all the domains where it is necessary to represent information and to perform inference on it, such as software engineering, medical diagnosis, digital libraries, databases and Web-based informative systems. They possess nice computational properties such as decidability and (for some DLs) low complexity [1].

There are many different DL languages that differ in the constructs that are allowed for defining concepts (sets of individuals of the domain) and roles (sets of pairs of individuals). The $\mathcal{SROIQ}(\mathbf{D})$ DL is one of the most common fragments; it was introduced by Horrocks et al. in [7] and it is of particular importance because it is semantically equivalent to OWL 2.

Let us consider a set of *atomic concepts* $\mathbf{C}$, a set of *atomic roles* $\mathbf{R}$ and a set of individuals $\mathbf{I}$. A *role* could be an atomic role $R \in \mathbf{R}$, the inverse R^- of an atomic role $R \in \mathbf{R}$ or a complex role $R \circ S$. We use $\mathbf{R}^-$ to denote the set of all inverses of roles in $\mathbf{R}$. Each $A \in \mathbf{A}$, $\bot$ and $\top$ are concepts and if $a \in \mathbf{I}$, then $\{a\}$ is a concept called *nominal*. If C, C_1 and C_2 are concepts and $R \in \mathbf{R} \cup \mathbf{R}^-$, then $(C_1 \sqcap C_2)$, $(C_1 \sqcup C_2)$ and $\neg C$ are concepts, as well as $\exists R.C$, $\forall R.C$, $\geq nR.C$ and $\leq nR.C$ for an integer $n \geq 0$.

A *knowledge base* (KB) $\mathcal{K} = (\mathcal{T}, \mathcal{R}, \mathcal{A})$ consists of a TBox $\mathcal{T}$, an RBox $\mathcal{R}$ and an ABox $\mathcal{A}$. An RBox $\mathcal{R}$ is a finite set of *transitivity axioms* $Trans(R)$, *role inclusion axioms* $R \sqsubseteq S$ and *role chain axioms* $R \circ P \sqsubseteq S$, where $R, P, S \in \mathbf{R} \cup \mathbf{R}^-$. A *TBox* $\mathcal{T}$ is a finite set of *concept inclusion axioms* $C \sqsubseteq D$, where C and D are concepts. An *ABox* $\mathcal{A}$ is a finite set of *concept membership axioms* $a : C$ and *role membership axioms* $(a, b) : R$, where C is a concept, $R \in \mathbf{R}$ and $a, b \in \mathbf{I}$.

A KB is usually assigned a semantics using interpretations of the form $\mathcal{I} = (\Delta^{\mathcal{I}}, \cdot^{\mathcal{I}})$, where $\Delta^{\mathcal{I}}$ is a non-empty *domain* and $\cdot^{\mathcal{I}}$ is the *interpretation function* that assigns an element in $\Delta^{\mathcal{I}}$ to each individual a, a subset of $\Delta^{\mathcal{I}}$ to each concept C and a subset of $\Delta^{\mathcal{I}} \times \Delta^{\mathcal{I}}$ to each role R. The mapping $\cdot^{\mathcal{I}}$ is extended to complex concepts as follows (where $R^{\mathcal{I}}(x, C) = \{y | \langle x, y \rangle \in R^{\mathcal{I}}, y \in C^{\mathcal{I}}\}$ and $\#X$ denotes the cardinality of the set X):

$$\top^{\mathcal{I}} = \Delta^{\mathcal{I}} \qquad\qquad \bot^{\mathcal{I}} = \emptyset$$
$$\{a\}^{\mathcal{I}} = \{a^{\mathcal{I}}\} \qquad\qquad (\neg C)^{\mathcal{I}} = \Delta^{\mathcal{I}} \setminus C^{\mathcal{I}}$$
$$(C_1 \sqcup C_2)^{\mathcal{I}} = C_1^{\mathcal{I}} \cup C_2^{\mathcal{I}} \qquad (C_1 \sqcap C_2)^{\mathcal{I}} = C_1^{\mathcal{I}} \cap C_2^{\mathcal{I}}$$
$$(\exists R.C)^{\mathcal{I}} = \{x \in \Delta^{\mathcal{I}} | R^{\mathcal{I}}(x) \cap C^{\mathcal{I}} \neq \emptyset\} \qquad (\forall R.C)^{\mathcal{I}} = \{x \in \Delta^{\mathcal{I}} | R^{\mathcal{I}}(x) \subseteq C^{\mathcal{I}}\}$$
$$(\geq nR.C)^{\mathcal{I}} = \{x \in \Delta^{\mathcal{I}} | \#R^{\mathcal{I}}(x, C) \geq n\} \qquad (\leq nR.C)^{\mathcal{I}} = \{x \in \Delta^{\mathcal{I}} | \#R^{\mathcal{I}}(x, C) \leq n\}$$
$$(R^-)^{\mathcal{I}} = \{(y, x) | (x, y) \in R^{\mathcal{I}}\} \qquad (R_1 \circ \ldots \circ R_n)^{\mathcal{I}} = R_1^{\mathcal{I}} \circ \ldots \circ R_n^{\mathcal{I}}$$

$\mathcal{SROIQ}(\mathbf{D})$ also permits the definition of datatype roles, which connect an individual to an element of a datatype such as integers, floats, etc.

A query Q over a KB $\mathcal{K}$ is usually an axiom for which we want to test the entailment from the KB, written as $\mathcal{K} \models Q$.

Example 1. Consider the following KB "Crime and Punishment"

Nihilist $\sqsubseteq$ GreatMan	$\exists$killed.$\top \sqsubseteq$ Nihilist
(raskolnikov, alyona) : killed	(raskolnikov, lizaveta) : killed

This KB states that if you killed someone then you are a nihilist and whoever is a nihilist is a "great man" (TBox). It also states that Raskolnikov killed Alyona and Lizaveta (ABox). The KB entails the query $Q = $ raskolnikov : GreatMan (but are we sure about that?).

3   Justification Finding Problem

Here we discuss the problem of finding the covering set of justifications for a given query. This non-standard reasoning service is also known as *axiom pinpointing* [21] and it is useful for tracing derivations and debugging ontologies. This problem has been investigated by various authors [2,6,9,21]. A justification corresponds to an *explanation* for a query Q. An explanation is a subset of logical axioms $\mathcal{E}$ of a KB $\mathcal{K}$ such that $\mathcal{E} \models Q$, whereas a justification is an explanation such that it is minimal w.r.t. set inclusion. Formally, we say that an explanation $\mathcal{J} \subseteq \mathcal{K}$ is a justification if for all $\mathcal{J}' \subset \mathcal{J}$, $\mathcal{J}' \not\models Q$, i.e. $\mathcal{J}'$ is not an explanation for Q. The problem of enumerating all justifications that entail a given query is called axiom pinpointing or justification finding. *The set of all the justifications for the query Q is the covering set of justifications for Q.* Given a KB $\mathcal{K}$, the covering set of justifications for Q is denoted by ALL-JUST$(Q, \mathcal{K})$.

Below, we provide the formal definitions of justification finding problem.

Definition 1 (Justification finding problem).
Input: *A knowledge base $\mathcal{K}$, and an axiom Q such that $\mathcal{K} \models Q$.*
Output: *The set* ALL-JUST$(Q, \mathcal{K})$ *of all the justifications for Q in $\mathcal{K}$.*

There are two categories of algorithms for finding a single justification: glass-box algorithms [9] and black-box algorithms. The former category is reasoner-dependent, i.e. a glass-box algorithm implementation depends on a specific reasoner, whereas a black-box algorithm is reasoner-independent, i.e. it can be used with any reasoner. In both cases, we still need a reasoner to obtain a justification.

It is possible to incrementally compute all justifications for an entailment by using Reiter's Hitting Set Tree (HST) algorithm [17]. This algorithm repeatedly calls a glass-box or a black-box algorithm which builds a new justification. To avoid the extraction of already found justifications, at each iteration the extraction process is performed on a KB from which some axioms are removed by taking into account the previously found justifications. For instance, given a KB $\mathcal{K}$ and a query Q, if the justification $\mathcal{J} = \{E_1, E_2, E_3\}$ was found, where E_is are axioms, to avoid the generation of the same justification, the HST algorithm tries to find a new justification on $\mathcal{K}' = \mathcal{K} \setminus E_1$. If no new justification is found the HST algorithm backtracks and tries to find another justification by removing other axioms from $\mathcal{J}$, one at a time.

4   Probabilistic Description Logics

DISPONTE [18,26] applies the distribution semantics [20] to Probabilistic Description Logic KBs.

In DISPONTE, a *probabilistic knowledge base $\mathcal{K}$* is a set of certain axioms or probabilistic axioms. Certain *axioms* take the form of regular DL axioms. *Probabilistic axioms* take the form

$$p :: E$$

where $p \in [0, 1]$ and E is a DL axiom. $p :: E$ means that we have degree of belief p in axiom E.

DISPONTE associates independent Boolean random variables to the DL axioms. The set of axioms that have the random variable assigned to 1 constitutes a *world*. The probability of a world w is computed by multiplying the probability p_i for each probabilistic axiom E_i included in the world by the probability $1 - p_i$ for each probabilistic axiom E_i not included in the world.

Below, we provide some formal definitions for DISPONTE.

Definition 2 (Atomic choice). *An atomic choice is a couple (E_i, k) where E_i is the ith probabilistic axiom and $k \in \{0, 1\}$. The variable k indicates whether E_i is chosen to be included in a world $(k = 1)$ or not $(k = 0)$.*

Definition 3 (Composite choice). *A composite choice κ is a consistent set of atomic choices, i.e., $(E_i, k) \in \kappa$, $(E_i, m) \in \kappa$ implies $k = m$ (only one decision is taken for each axiom).*

The probability of composite choice κ is

$$P(\kappa) = \prod_{(E_i,1) \in \kappa} p_i \prod_{(E_i,0) \in \kappa} (1 - p_i)$$

where p_i is the probability associated with axiom E_i, because the random variables associated with axioms are independent.

Definition 4 (Selection). *A selection σ is a total composite choice, i.e., it contains an atomic choice (E_i, k) for every probabilistic axiom of the theory. A selection σ identifies a theory w_σ called a world: $w_\sigma = \mathcal{C} \cup \{E_i | (E_i, 1) \in \sigma\}$, where $\mathcal{C}$ is the set of certain axioms.*

$P(w_\sigma)$ is a probability distribution over worlds. Let us indicate with $\mathcal{W}$ the set of all worlds. The probability of Q is [18]:

$$P(Q) = \sum_{w \in \mathcal{W}: w \models Q} P(w)$$

i.e. the probability of the query is the sum of the probabilities of the worlds in which the query is true.

Example 2. Let us consider the knowledge base and the query $Q = $ raskolnikov : GreatMan of Example 1 where some of the axioms are probabilistic:

$E_1 = 0.2 ::$ Nihilist $\sqsubseteq$ GreatMan $C_1 = \exists$killed.$\top \sqsubseteq$ Nihilist

$E_2 = 0.6 ::$ (raskolnikov, alyona) : killed $E_3 = 0.7 ::$ (raskolnikov, lizaveta) : killed

Whoever is a nihilist is a "great man" with probability 0.2 (E_1) and Raskolnikov killed Alyona and Lizaveta with probability 0.6 and 0.7 respectively (E_2 and E_3). Moreover there is a certain axiom (C_1). The KB has eight worlds and Q is true in three of them, corresponding to the selections:

$$\{\{(E_1, 1), (E_2, 1), (E_3, 1)\}, \ \{(E_1, 1), (E_2, 1), (E_3, 0)\}, \ \{(E_1, 1), (E_2, 0), (E_3, 1)\}\}$$

The probability is $P(Q) = 0.2 \cdot 0.6 \cdot 0.7 + 0.2 \cdot 0.6 \cdot (1 - 0.7) + 0.2 \cdot (1 - 0.6) \cdot 0.7 = 0.176$.

5 Inference in Probabilistic Description Logics

It is often infeasible to find all the worlds where the query is true. To reduce reasoning time, inference algorithms find, instead, explanations for the query and then compute the probability of the query from them. Below we provide the definitions of DISPONTE explanations and justifications, which are tightly intertwined with the previous definitions of explanation and justification for the non-probabilistic case.

Definition 5 (DISPONTE Explanation). *A composite choice ϕ identifies a set of worlds $\omega_\phi = \{w_\sigma | \sigma \in \mathcal{S}, \sigma \supseteq \phi\}$, where $\mathcal{S}$ is the set of all selections. We say that ϕ is an explanation for Q if Q is entailed by every world of ω_ϕ.*

Definition 6 (DISPONTE Justification). *We say that an explanation γ is a justification if, for all $\gamma' \subset \gamma$, γ' is not an explanation for Q.*

A set of explanations $\boldsymbol{\Phi}$ is *covering* Q if every world $w_\sigma \in \mathcal{W}$ in which Q is entailed is such that $w_\sigma \in \bigcup_{\phi \in \boldsymbol{\Phi}} \omega_\phi$. In other words a covering set $\boldsymbol{\Phi}$ identifies all the worlds in which Q succeeds.

Two composite choices κ_1 and κ_2 are *incompatible* if their union is inconsistent. For example, $\kappa_1 = \{(E_i, 1)\}$ and $\kappa_2 = \{(E_i, 0)\}$ are incompatible. A set K of composite choices is *pairwise incompatible* if for all $\kappa_1 \in K$, $\kappa_2 \in K$, $\kappa_1 \neq \kappa_2$ implies that κ_1 and κ_2 are incompatible. The *probability of a pairwise incompatible set of composite choices* K is $P(K) = \sum_{\kappa \in K} P(\kappa)$.

Given a query Q and a covering set of pairwise incompatible explanations $\boldsymbol{\Phi}$, the probability of Q is [18]:

$$P(Q) = \sum_{w_\sigma \in \omega_\Phi} P(w_\sigma) = P(\omega_\Phi) = P(\boldsymbol{\Phi}) = \sum_{\phi \in \boldsymbol{\Phi}} P(\phi) \tag{1}$$

where ω_Φ is the set of worlds identified by the set of explanations $\boldsymbol{\Phi}$.

Example 3. Consider the KB and the query $Q = $ raskolnikov : GreatMan of Example 2. We have the following covering set of pairwise incompatible explanations: $\boldsymbol{\Phi} = \{\{(E_1, 1), (E_2, 1)\}, \{(E_1, 1), (E_2, 0), (E_3, 1)\}\}$. The probability of the query is $P(Q) = 0.2 \cdot 0.6 + 0.2 \cdot 0.4 \cdot 0.7 = 0.176$.

Unfortunately, in general, explanations (and hence justifications) are not pairwise incompatible. The problem of calculating the probability of a query is therefore reduced to that of finding a covering set of justifications and then transforming it into a covering set of pairwise incompatible explanations.

We can think of using justification finding algorithms for non-probabilistic DLs to find the covering set of non-probabilistic justifications, then consider only the probabilistic axioms and transform the covering set of DISPONTE justifications into a pairwise incompatible covering set of explanations from which it is easy to compute the probability.

Example 4. Consider the KB and the query Q = raskolnikov : GreatMan of Example 2. If we use justification finding algorithms by ignoring the probabilistic annotations, we find the following non-probabilistic justifications: $\mathcal{J} = \{\{E_1, C_1, E_2\}, \{E_1, C_1, E_3\}\}$. Then we can translate them into DISPONTE justifications: $\boldsymbol{\Gamma} = \{\{(E_1, 1), (E_2, 1)\}, \{(E_1, 1), (E_3, 1)\}\}$. Note that $\boldsymbol{\Gamma}$ is not pairwise incompatible, therefore we cannot directly use Eq. (1). The solution to this problem will be shown in the following section.

6 BUNDLE

The reasoner BUNDLE [18,19] computes the probability of a query w.r.t. DISPONTE KBs by first computing all the justifications for the query, then converting them into a pairwise incompatible covering set of explanations by building a Binary Decision Diagram (BDD). Finally, it computes the probability by traversing the BDD. A BDD for a function of Boolean variables is a rooted graph that has one level for each Boolean variable. A node n has two children corresponding respectively to the 1 value and the 0 value of the variable associated with the level of n. When drawing BDDs, the 0-branch is distinguished from the 1-branch by drawing it with a dashed line. The leaves store either 0 or 1.

Given the set $\boldsymbol{\Phi}$ of all DISPONTE explanations for a query Q, we can define the Disjunctive Normal Form Boolean formula $f_{\boldsymbol{\Phi}}$ representing the disjunction of all explanations as $f_{\boldsymbol{\Phi}}(\mathbf{X}) = \bigvee_{\phi \in \boldsymbol{\Phi}} \bigwedge_{(E_i, 1)} X_i \bigwedge_{(E_i, 0)} \overline{X_i}$. The variables $\mathbf{X} = \{X_i \mid p_i :: E_i \in \mathcal{K}\}$ are independent Boolean random variables with $P(X_i = 1) = p_i$ and the probability that $f_{\boldsymbol{\Phi}}(\mathbf{X})$ takes value 1 gives the probability of Q.

BDDs perform a Shannon's expansion of the Boolean function $f_{\boldsymbol{\Phi}}$ that makes the disjuncts, and hence the associated explanations, mutually exclusive, i.e. pairwise incompatible.

Given the BDD, we can use function PROBABILITY described in [10] to compute the probability. This dynamic programming algorithm traverses the diagram from the leaves to the root and computes the probability of a formula encoded as a BDD.

Example 5 (Example 2 cont.). Let us consider the KB and the query of Example 2. If we associate random variables X_1 with axiom E_1, X_2 with E_2 and X_3 with E_3, the BDD representing the set of explanations is shown in Fig. 1. By applying function PROBABILITY [10] to this BDD we get

$$\text{PROBABILITY}(n_3) = 0.7 \cdot 1 + 0.3 \cdot 0 = 0.7$$
$$\text{PROBABILITY}(n_2) = 0.6 \cdot 1 + 0.4 \cdot 0.7 = 0.88$$
$$\text{PROBABILITY}(n_1) = 0.2 \cdot 0.88 + 0.8 \cdot 0 = 0.176$$

and therefore $P(Q) = \text{PROBABILITY}(n_1) = 0.176$, which corresponds to the probability given by DISPONTE.

BUNDLE uses implementations of the HST algorithm to incrementally obtain all the justifications. However, *the first version was able to use only a glass-box approach which was dependent on the Pellet reasoner* [23].

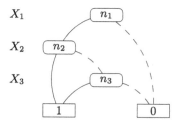

Fig. 1. BDD representing the set of explanations for the query of Example 2.

In the following, we illustrate the modifications introduced in the new modular version of BUNDLE.

Figure 2 shows the new architecture of BUNDLE. The main novelties are the adoption of the OWL Explanation library[1] [6] and of the BlackBox approach offered by OWL API. Thanks to them, BUNDLE is now reasoner-independent and it can exploit different OWL reasoners.

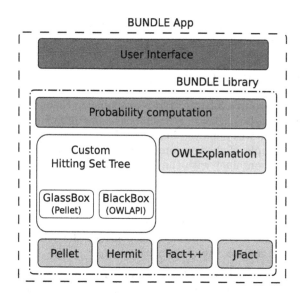

Fig. 2. Software architecture of BUNDLE.

Modularity is therefore realized in two directions: (1) support of different OWL reasoners: Pellet 2.5.0, Hermit 1.3.8.413 [22], Fact++ 1.6.5 [24], and JFact 4.0.4[2]; (2) three different strategies for finding a justification, which are:

---

[1] https://github.com/matthewhorridge/owlexplanation.
[2] http://jfact.sourceforge.net/.

GlassBox. A glass-box approach which depends on Pellet. It is a modified version of the `GlassBoxExplanation` class contained in the Pellet Explanation library.

BlackBox. A black-box approach offered by the OWL API[3] [5]. The OWL API is a Java API for the creation and manipulation of OWL 2 ontologies.

OWL Explanation. A library that is part of the OWL Explanation Workbench [6]. The latter also contains a Protégé plugin, underpinned by the library, that allows Protégé users to find justifications for entailments in their OWL 2 ontologies.

All reasoners can be paired with the BlackBox and OWL Explanation methods, while only Pellet can exploit the GlassBox method.

To find all justifications of a given query with the GlassBox and BlackBox approaches an implementation of the HST algorithm is used, which is a modified version of the `HSTExplanationGenerator` class of the OWL API. We modified this class in order to support annotated axioms (DISPONTE axioms are OWL axioms annotated with a probability). OWL Explanation, instead, already contains an HST implementation and a black-box approach that supports annotated axioms.

BUNDLE can be used as standalone desktop application or, in this new version, as a library.

6.1   Using BUNDLE as Application

BUNDLE is an open-source software and is available on Bitbucket, together with its manual, at https://bitbucket.org/machinelearningunife/bundle.

A BUNDLE image was deployed in Docker Hub. Users can start using BUNDLE with just a couple of docker commands. All they have to do is pull the image and start the container with the commands:

```
sudo docker pull giuseta/bundle:3.0.0
sudo docker run -it giuseta/bundle:3.0.0 bash
```

A bash shell of the container then starts and users can use BUNDLE by running the command `bundle`. For instance, if we consider the KB and the query of Example 2, the user can ask the query with:

```
bundle -instance http://www.semanticweb.org/
    crime_and_punishment#raskolnikov,http://www.
    semanticweb.org/crime_and_punishment#GreatMan file:
    examples/crime_and_punishment.owl
```

6.2   Using BUNDLE as Library

BUNDLE can also be used as a library. The library is set up as a Maven application and published on Maven Central[4].

---

[3] http://owlcs.github.io/owlapi/.

[4] With groupId `it.unife.endif.ml`, artifactId `bundle` and version 3.0.0.

Once the developer has added BUNDLE dependency in the project's POM file, the probability of the query can be obtained in just few lines:

```
1  Bundle reasoner = new Bundle();
2  reasoner.setRootOntology(rootOntology);
3  reasoner.setReasonerFactory(new JFactFactory());
4  reasoner.init();
5  QueryResult result = reasoner.computeQuery(query);
```

where `rootOntology` and `query` are objects of the classes `OWLOntology` and `OWLAxiom` of the OWL API library respectively.

Line 3 shows that the developer can inject the preferred OWL API-based reasoner and perform probabilistic inference without modifying BUNDLE.

7   Experiments

We performed three different tests to compare the possible configurations of BUNDLE, which depend on the reasoner and the justification search strategy chosen, for a total of 9 combinations. In the first test we compared all configurations on four different datasets, in order to highlight which combination reasoner/strategy show the best behavior in terms of inference time. In the second one, we considered KBs of increasing size in terms of the number of probabilistic axioms. Finally, in the third test, we asked queries on a synthetic dataset of increasing size. The last two experiments were targeted to investigate the scalability of the different configurations. All tests were performed on the HPC System Marconi[5] equipped with Intel Xeon E5-2697 v4 (Broadwell) @ 2.30 GHz, using 8 cores for each test.

Test 1. The first test considers 4 real world KBs of various complexity as in [27]: (1) **BRCA** [11], which models the risk factors of breast cancer; (2) an extract of **DBPedia**[6] [13], containing structured information from Wikipedia, usually those contained in the information box on the righthand side of pages; (3) **Biopax level 3**[7] [3], which models metabolic pathways; (4) **Vicodi**[8] [16], which contains information on European history and models historical events and important personalities.

We used a version of the DBPedia and Biopax KBs without the ABox and a version of BRCA and Vicodi with an ABox containing 1 individual and 19 individuals respectively. For each KB we added a probability annotation to each axiom. The probability values were randomly assigned. We randomly created 50 subclass-of queries for all the KBs and 50 instance-of queries for BRCA and Vicodi, following the concepts hierarchy of the KBs, ensuring each query had at least one explanation.

---

[5] http://www.hpc.cineca.it/hardware/marconi.
[6] http://dbpedia.org/.
[7] http://www.biopax.org/.
[8] http://www.vicodi.org/.

Table 1 shows the average time in seconds to answer queries with different BUNDLE configurations. Bold values highlight the fastest configuration for each KB. With the exception of DBPedia, the best results are obtained by Pellet with the GlassBox approach, corresponding to the configuration of the previous non-modular version of BUNDLE. However, the use of OWL Explanation library with Pellet shows competitive results. For BioPax and Vicodi KBs, the BlackBox approach with Fact++ wasn't able to return a result (cells with "crash").

Table 1. Average time (in seconds) for probabilistic inference with all possible configurations of BUNDLE over different datasets (Test 1). For BioPax and Vicodi KBs Fact++/BlackBox wasn't able to return a result due to an internal error.

Reasoner	Method	Subclass-of queries				Instance-of queries	
		BioPax	DBPedia	BRCA	Vicodi	BRCA	Vicodi
Pellet	GlassBox	**0.501**	0.416	**0.85**	**0.393**	**1.654**	**0.42**
Pellet	BlackBox	1.779	0.484	1.488	0.667	5.671	0.804
Pellet	OWLExp	0.768	0.937	1.051	0.772	2.564	0.687
Hermit	BlackBox	4.281	2.192	7.68	1.968	29.944	2.416
Hermit	OWLExp	2.304	2.216	3.373	1.739	10.645	2.17
Fact++	BlackBox	crash	**0.254**	0.586	crash	3.368	crash
Fact++	OWLExp	1.568	1.077	0.934	0.667	2.532	1.183
JFact	BlackBox	1.757	0.501	1.974	0.726	7.273	0.812
JFact	OWLExp	1.072	1.248	2.036	0.869	3.47	1.291

Test 2. The second test was performed following the approach presented in [11] on the BRCA KB ($\mathcal{ALCHF}(D)$, 490 axioms). To test BUNDLE, we randomly generated and added an increasing number of subclass-of probabilistic axioms. The number of these axioms was varied from 9 to 16, and, for each number, 100 different consistent ontologies were created. Although the number of additional axioms, they may cause an exponential increase of the inference complexity (please see [11] for a detailed explanation).

Finally, an individual was added to every KB, randomly assigned to each simple class that appeared in the probabilistic axioms, and a random probability was attached to it. Complex classes contained in the conditional constraints were split into their components, e.g., the complex class *PostmenopausalWoman-TakingTestosterone* was divided into *PostmenopausalWoman* and *WomanTakingTestosterone*. Finally, we ran 100 probabilistic queries of the form $a : C$ where a is the added individual and C is a class randomly selected among those that represent women under increased and lifetime risk such as *WomanUnderLifetimeBRCRisk* and *WomanUnderStronglyIncreasedBRCRisk*, which are at the top of the concept hierarchy.

Table 2 shows the execution time averaged over the 100 queries as a function of the number of probabilistic axioms. For each size, bold values indicate the best configuration. The BlackBox approaches are much slower on average than the others. The best performance is shown by Pellet/GlassBox until size 12, and by Pellet/OWLExp, Fact++/OWLExp and JFact/OWLExp from size 13.

Table 2. Average execution time (ms) for probabilistic inference with different configurations of BUNDLE on versions of the BRCA KB of increasing size (Test 2).

Reasoner	Method	9	10	11	12	13	14	15	16
Pellet	GlassBox	**1.360**	**1.076**	16.149	**16.448**	7.157	14.895	9.884	7.889
Pellet	BlackBox	19.747	16.406	51.258	42.258	42.309	65.690	63.269	55.006
Pellet	OWLExp	3.520	3.271	18.554	17.951	7.333	**8.848**	8.811	7.350
Hermit	BlackBox	31.871	26.373	72.473	62.064	66.664	77.378	79.785	57.745
Hermit	OWLExp	6.518	6.380	24.245	23.666	15.211	23.697	18.462	15.694
Fact++	BlackBox	3.718	2.846	20.880	18.879	8.479	19.221	15.837	10.411
Fact++	OWLExp	1.829	1.618	14.254	16.871	5.776	13.384	**8.224**	**6.640**
JFact	BlackBox	5.570	4.483	25.897	22.272	15.319	28.686	24.082	16.120
JFact	OWLExp	1.748	1.509	**13.366**	16.823	**2.267**	13.747	8.591	7.294

Test 3. In the third test we artificially created a set of KBs of increasing size of the following form:

$$(E_{1,i}) \; 0.6 :: B_{i-1} \sqsubseteq P_i \sqcap Q_i \quad (E_{2,i}) \; 0.6 :: P_i \sqsubseteq B_i \quad (E_{3,i}) \; 0.6 :: Q_i \sqsubseteq B_i$$

where $n \geq 1$ and $1 \leq i \leq n$. The query $Q = B_0 \sqsubseteq B_n$ has 2^n explanations, even if the KB has a size that is linear in n.

We increased n from 2 to 10 in steps of 2 and we collected the running time, averaged over 50 executions. Table 3 shows, for each n, the average time in seconds that the systems took for computing the probability of the query Q (in bold the best time for each size). We set a timeout of 10 min for each query, so the cells with "–" indicate that the timeout occurred. This experiment confirms what already suggested by *Test 2*, i.e. the best results in terms of scalability are provided by the OWL Explanation method paired with any reasoner except Hermit. Thanks to this library the new version of BUNDLE is able to beat the first version (corresponding to Pellet/GlassBox), by reaching a larger dataset size.

Table 3. Average time (in seconds) for probabilistic inference with different configurations of BUNDLE on a synthetic dataset (Test 3). "–" means that the execution timed out (600 s).

Reasoner	Method	2	4	6	8	10
Pellet	GlassBox	0.404	**0.673**	**3.651**	–	–
Pellet	BlackBox	0.558	1.217	4.868	456.71	–
Pellet	OWLExp	0.972	1.957	4.45	**13.459**	**52.084**
Hermit	BlackBox	2.800	13.965	117.886	–	–
Hermit	OWLExp	2.307	8.507	37.902	185.158	–
Fact++	BlackBox	**0.248**	1.026	5.96	487.091	–
Fact++	OWLExp	0.815	1.708	4.282	15.331	76.313
JFact	BlackBox	0.405	1.178	4.895	497.745	–
JFact	OWLExp	0.946	1.878	4.258	17.547	78.831

8   Conclusions

In this paper, we presented a modular version of BUNDLE, a system for reasoning on Probabilistic Description Logics KBs that follow DISPONTE. Modularity allows one to pair 4 different OWL reasoners with 3 different approaches to find query justifications. In addition, BUNDLE can now be used both as a standalone application and as a library. We provided a comparison between the various configurations reasoner/approach over different datasets, showing that Pellet paired with GlassBox or any reasoner (except Hermit) paired with the OWLExplanation library achieve the best results in terms of inference time on a probabilistic ontology. In the future, we plan to study the effects of glass-box or grey-box methods for collecting explanations.

Acknowledgement. This work was supported by the "GNCS-INdAM".

References

1. Baader, F., Horrocks, I., Sattler, U.: Description logics, chap. 3, pp. 135–179. Elsevier, Amsterdam (2008)
2. Baader, F., Peñaloza, R., Suntisrivaraporn, B.: Pinpointing in the description logic $\mathcal{EL}^+$. In: Hertzberg, J., Beetz, M., Englert, R. (eds.) KI 2007. LNCS (LNAI), vol. 4667, pp. 52–67. Springer, Heidelberg (2007). https://doi.org/10.1007/978-3-540-74565-5_7
3. Demir, E., Cary, M.P., Paley, S., Fukuda, K., Lemer, C., Vastrik, I., Wu, G., D'Eustachio, P., Schaefer, C., Luciano, J.: The BioPax community standard for pathway data sharing. Nat. Biotechnol. **28**(9), 935–942 (2010)
4. Ding, Z., Peng, Y.: A probabilistic extension to ontology language OWL. In: 37th Hawaii International Conference on System Sciences (HICSS-37 2004), CD-ROM/Abstracts Proceedings, 5–8 January 2004, Big Island, HI, USA. IEEE Computer Society (2004)

5. Horridge, M., Bechhofer, S.: The OWL API: a Java API for OWL ontologies. Semant. Web **2**(1), 11–21 (2011)
6. Horridge, M., Parsia, B., Sattler, U.: The OWL explanation workbench: a toolkit for working with justifications for entailments in OWL ontologies (2009)
7. Horrocks, I., Kutz, O., Sattler, U.: The even more irresistible $\mathcal{SROIQ}$. In: Proceedings of the Tenth International Conference on Principles of Knowledge Representation and Reasoning, vol. 6, pp. 57–67. AAAI Press (2006). http://dl.acm.org/citation.cfm?id=3029947.3029959
8. Jaeger, M.: Probabilistic reasoning in terminological logics. In: Doyle, J., Sandewall, E., Torasso, P. (eds.) 4th International Conference on Principles of Knowledge Representation and Reasoning, pp. 305–316. Morgan Kaufmann (1994)
9. Kalyanpur, A.: Debugging and repair of OWL ontologies. Ph.D. thesis, The Graduate School of the University of Maryland (2006)
10. Kimmig, A., Demoen, B., De Raedt, L., Costa, V.S., Rocha, R.: On the implementation of the probabilistic logic programming language ProbLog. Theory Pract. Log. Prog. **11**(2–3), 235–262 (2011)
11. Klinov, P., Parsia, B.: Optimization and evaluation of reasoning in probabilistic description logic: towards a systematic approach. In: Sheth, A., Staab, S., Dean, M., Paolucci, M., Maynard, D., Finin, T., Thirunarayan, K. (eds.) ISWC 2008. LNCS, vol. 5318, pp. 213–228. Springer, Heidelberg (2008). https://doi.org/10.1007/978-3-540-88564-1_14
12. Koller, D., Levy, A.Y., Pfeffer, A.: P-CLASSIC: a tractable probabilistic description logic. In: Kuipers, B., Webber, B.L. (eds.) Fourteenth National Conference on Artificial Intelligence and Ninth Innovative Applications of Artificial Intelligence Conference, AAAI 1997, 27–31 July 1997, Providence, Rhode Island, pp. 390–397. AAAI Press/The MIT Press (1997)
13. Lehmann, J., Isele, R., Jakob, M., Jentzsch, A., Kontokostas, D., Mendes, P.N., Hellmann, S., Morsey, M., van Kleef, P., Auer, S., Bizer, C.: DBpedia - a large-scale, multilingual knowledge base extracted from Wikipedia. Semant. Web **6**(2), 167–195 (2015)
14. Lukasiewicz, T.: Expressive probabilistic description logics. Artif. Intell. **172**(6–7), 852–883 (2008)
15. Lutz, C., Schröder, L.: Probabilistic description logics for subjective uncertainty. In: Lin, F., Sattler, U., Truszczynski, M. (eds.) 12th International Conference on Principles of Knowledge Representation and Reasoning (KR 2010), pp. 393–403. AAAI Press, Menlo Park (2010)
16. Nagypál, G., Deswarte, R., Oosthoek, J.: Applying the semantic web: the VICODI experience in creating visual contextualization for history. Lit. Linguist. Comput. **20**(3), 327–349 (2005)
17. Reiter, R.: A theory of diagnosis from first principles. Artif. Intell. **32**(1), 57–95 (1987)
18. Riguzzi, F., Bellodi, E., Lamma, E., Zese, R.: Probabilistic description logics under the distribution semantics. Semant. Web **6**(5), 447–501 (2015). https://doi.org/10.3233/SW-140154
19. Riguzzi, F., Bellodi, E., Lamma, E., Zese, R.: Reasoning with probabilistic ontologies. In: Yang, Q., Wooldridge, M. (eds.) 24th International Joint Conference on Artificial Intelligence (IJCAI 2015), pp. 4310–4316. AAAI Press, Palo Alto (2015)
20. Sato, T.: A statistical learning method for logic programs with distribution semantics. In: Sterling, L. (ed.) ICLP 1995, pp. 715–729. MIT Press (1995)

21. Schlobach, S., Cornet, R.: Non-standard reasoning services for the debugging of description logic terminologies. In: Gottlob, G., Walsh, T. (eds.) Proceedings of the Eighteenth International Joint Conference on Artificial Intelligence, IJCAI 2003, Acapulco, Mexico, 9–15 August 2003, pp. 355–362. Morgan Kaufmann Publishers Inc., San Francisco (2003)

22. Shearer, R., Motik, B., Horrocks, I.: HermiT: a highly-efficient OWL reasoner. In: OWL: Experiences and Direction, vol. 432, p. 91 (2008)

23. Sirin, E., Parsia, B., Cuenca-Grau, B., Kalyanpur, A., Katz, Y.: Pellet: a practical OWL-DL reasoner. J. Web Semant. **5**(2), 51–53 (2007)

24. Tsarkov, D., Horrocks, I.: FaCT++ description logic reasoner: system description. In: Furbach, U., Shankar, N. (eds.) IJCAR 2006. LNCS (LNAI), vol. 4130, pp. 292–297. Springer, Heidelberg (2006). https://doi.org/10.1007/11814771_26

25. W3C: OWL 2 web ontology language, December 2012. http://www.w3.org/TR/2012/REC-owl2-overview-20121211/

26. Zese, R.: Probabilistic semantic web: reasoning and learning, studies on the semantic web, vol. 28. IOS Press, Amsterdam (2017). https://doi.org/10.3233/978-1-61499-734-4-i, http://ebooks.iospress.nl/volume/probabilistic-semantic-web-reasoning-and-learning

27. Zese, R., Bellodi, E., Riguzzi, F., Cota, G., Lamma, E.: Tableau reasoning for description logics and its extension to probabilities. Ann. Math. Artif. Intell. **82**(1–3), 101–130 (2018). https://doi.org/10.1007/s10472-016-9529-3

Modeling the Dynamics of Multiple Disease Occurrence by Latent States

Marcos L. P. Bueno[1,2,5(✉)], Arjen Hommersom[1,3], Peter J. F. Lucas[1,4], Mariana Lobo[2], and Pedro P. Rodrigues[2]

[1] Institute for Computing and Information Sciences, Radboud University Nijmegen, Nijmegen, The Netherlands
{mbueno,arjenh,peterl}@cs.ru.nl

[2] CINTESIS - Centre for Health Technology and Services Research, Porto, Portugal
{marianalobo,pprodrigues}@med.up.pt

[3] Faculty of Management, Science and Technology, Open University, Heerlen, The Netherlands

[4] Leiden Institute of Advanced Computer Science, Leiden University, Leiden, The Netherlands

[5] Department of Computer Science, Federal University of Uberlândia, Uberlândia, Brazil

Abstract. The current availability of large volumes of health care data makes it a promising data source to new views on disease interaction. Most of the times, patients have multiple diseases instead of a single one (also known as multimorbidity), but the small size of most clinical research data makes it hard to impossible to investigate this issue. In this paper, we propose a latent-based approach to expand patient evolution in temporal electronic health records, which can be uninformative due to its very general events. We introduce the notion of *clusters of hidden states* allowing for an expanded understanding of the multiple dynamics that underlie events in such data. Clusters are defined as part of hidden Markov models learned from such data, where the number of hidden states is not known beforehand. We evaluate the proposed approach based on a large dataset from Dutch practices of patients that had events on comorbidities related to atherosclerosis. The discovered clusters are further correlated to medical-oriented outcomes in order to show the usefulness of the proposed method.

Keywords: Machine learning · Unsupervised learning
Hidden Markov model · Clustering · Electronic health records
Multimorbidity

1 Introduction

With the availability of large volumes of health care data, additional data sources become available for understanding disease interaction, in particular in multimorbidity research, i.e. when multiple diseases occur in people [2,16,19]. Influenced by factors such as the population aging, multimorbidity is the rule, not the

© Springer Nature Switzerland AG 2018
D. Ciucci et al. (Eds.): SUM 2018, LNAI 11142, pp. 93–107, 2018.
https://doi.org/10.1007/978-3-030-00461-3_7

exception in patients. The small size of most clinical research data makes it hard to impossible to investigate this issue, thus leaving room for alternative data sources that could support, e.g., the discovery of previously unnoticed disease interactions. Recently, machine learning techniques applied to large electronic health records (in the order of billion data points) have been able to provide accurate predictions [18], which shows that it is possible to take advantage of such datasets, despite their great differences compared to traditional clinical data.

In spite of its volume-related advantages, health care data are noisy, incomplete, and mostly not completely research-tailored, making analysis hard. For example, a patient visit to their general practitioner is often represented as a single event containing a main diagnosis, which can be very generic telling no more than which chronic or non-chronic disease was involved. Hence, more specific data such as symptoms and signs are often not available in patient data. As a consequence, a deeper understanding of patient situation beyond such episodic information is very challenging to be obtained, which could potentially help understand how the involved diseases interact. Uncertainty also plays a central role because future events are typically not completely determined by the current patient status. Much research has been dedicated to the analysis of health care data, but it tends to focus on managerial aspects such as patient flow, hospital resources, etc. [3,12] more often than on understanding diseases dynamics [7,14].

In this paper, we hypothesize that latent information next to the diagnostic data can increase our understanding of disease dynamics. By using as a basis hidden Markov models (HMMs) [17], multiple latent states can be associated to a given diagnostic event (where an event could be a visit due to, e.g., type 2 diabetes mellitus or a myocardial infarction). Based on this, we introduce the notion of *clusters of hidden states*, where a cluster contains all the states that produce the same observation (i.e. the same event). Although apparently simplistic, states within a cluster can have quite different dynamics in terms of transitioning patterns (i.e. how a state can be reached by/left from). By looking at these transition patterns, we will be able to give multiple meanings to each event, which sheds light on the influence of such event in the whole care process, as well as on the comorbidity interaction. Besides the structural differences of states within a cluster, we show that these states are associated in different ways to medical outcomes. The identification of latent information has been shown valuable for gaining a better understanding of health care data [6,7], although we pursue a different angle on what to cluster than previous research.

The contributions of this paper are as follows. We first define the notion of clusters of states from the perspective of electronic health records. This is followed by the identification of general transition patterns that might emerge in clusters of hidden states. We then introduce a case study based on data collected from Dutch practices amounting to 32,227 patients that had visits related to atherosclerosis. Atherosclerosis is a medical condition that can be seen as an umbrella to many other diseases, thus it is suitable for illustrating clusters and the role of their states in real-world data. Once an HMM is learned from the atherosclerosis data, we provide application-oriented interpretation to the clusters of states by looking at a medical outcome (the number of total diseases that were registered in patients) correlated to states of clusters.

2 Multimorbidity Event Data

2.1 Representation

In the considered electronic health records (EHRs), patient visits to their general practitioner are recorded such that each patient visit is assigned a diagnosis code. This diagnosis code can be related, e.g., to a chronic condition (e.g. due to diabetes mellitus) or not (e.g. a fracture). The time interval between two any visits is arbitrary. Next to the diagnosis data, additional data might be available, such as medication prescription and lab exams.

In general, let us consider that there are n possible diagnoses, each one represented by a random variable X_i taking values on $\{0, 1\}$. The full set of diagnosis variables is denoted by $\mathbf{X} = \{X_1, \ldots, X_n\}$. As a patient visit in the considered EHRs will be often assigned a single diagnosis code (sometimes called the main diagnosis), each visit will be represented by an instantiation of $\mathbf{X}$, such that $X_i = 1$ and $X_1 = \cdots = X_{i-i} = X_{i+1} = \ldots = X_n = 0$, where X_i corresponds to the main diagnosis. An alternative representation would be using a single variable taking values on a domain with n values, which could be seen as the state space of a Markov chain. However, we prefer using individual diagnosis variables because it is more general and flexible enough for easily allowing one to add more information into events. This could be the case, for example, when a secondary diagnosis is also available, or when one wants to lump together multiple visits into a single event (e.g. based on a pre-determined time interval or number of visits, if such knowledge is available).

2.2 Modeling

Multimorbidity event data tends to be fine grained, in the sense that each event will likely reflect information limited to the current patient visit. This differs, e.g., from longitudinal clinical trials [21], which are often characterized by repeated measurements of symptoms and signs associated to one or more conditions. As a consequence, data from such clinical trials normally allows for a more complete assessment of patient evolution, as opposed to multimorbidity event data. This suggests that unmeasured patient information could be searched for in such data in the form of latent information, such that when combined with observable data could provide a richer characterization of patients.

Hidden Markov models are often used to capture the interaction between observable and latent variables in a sequential process. In the multimorbidity context, the diagnosis variables $\mathbf{X}$ would be the observable variables, and we assume that there is a latent variable S for representing the hidden or latent states, with domain $\mathrm{val}(S) = \{s_1, \ldots, s_K\}$. The hidden states would, thus, compensate for the mentioned difficulties present in temporal EHRs. In order to fully specify an HMM over a time horizon $\{0, \ldots, T\}$, we specify a factorization of its joint distribution:

$$P(\mathbf{S}^{(0:T)}, \mathbf{X}^{(0:T)}) = P(S^{(0)}, \mathbf{X}^{(0)}) \prod_{t=0}^{T-1} P(S^{(t+1)} \mid S^{(t)}) P(\mathbf{X}^{(t+1)} \mid S^{(t+1)}) \quad (1)$$

By assuming that the probabilistic interaction between the observables is mediated by the states, we factorize the observables (also known as the emission distribution) as:

$$P(\mathbf{X}^{(t)} \mid S^{(t)}) = \prod_{i=1}^{n} P(X_i^{(t)} \mid S^{(t)}) \quad (2)$$

where $t = 0, \ldots, T$.

3   Identifying Transition Patterns

3.1   Clusters of States

The events constructed from multimorbidity data imply that in order to fully comply with the data concerning n diagnoses, the hidden states should be constrained to emit one out of n different observations at each moment. In spite of this apparent simplicity, the underlying process being modeled could still be quite complex (e.g. by having multiple stages at different moments). In order to properly capture such distribution, more states could be needed, which can lead to the situation where multiple states are associated to the same diagnosis (e.g. if one decides to model more states than observable variables). From these considerations, we define a *cluster of states* as a set of states that have the same emission distribution.

3.2   Transition Patterns

Modeling state transitions in a probabilistic way, e.g. as in Markov chains, implies that a state can often be reached by different ways and can lead to different future states. When clusters of states are considered, such dynamics are further enriched, because such past-present-future transitioning can occur by multiple ways. For example, consider two states s_i and s_j belonging to a cluster C, as shown in Fig. 1. This suggests that s_i will likely be reached earlier for the first time than s_j, and it also suggests that both states can lead to quite different incoming and outgoing states. Of course, such multiple *roles* of a given diagnosis (represented by the cluster C) stem from the complexity of the underlying process, where the a given diagnosis could be associated to different medical outcomes when one looks at the whole care process. For example, the cluster states could be associated to different levels of severity or worsening of patient health that could happen at different moments.

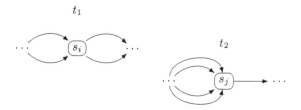

Fig. 1. Cluster of states $C = \{s_i, s_j\}$, where s_i can be reached from two states and transition to two states, while s_j can be reached from four states and can transition to a single state.

In order to better understand the roles of states in clusters, we discuss transition patterns that might arise. This characterization involves states and transitions from and to them, and is provided at a high level, because it is intuitively unfeasible to anticipate all the possible ways by which the states of clusters can interact without having an actual model at hand.

Internal Patterns. A state is associated to an *internal transition pattern* if most of the probability mass of its incoming and outgoing probabilities associates to states from the same cluster. The most trivial internal pattern occurs when a state has a loop probability close to 1, which we call a *recurrent pattern*. A more formal description is that a state s has a recurrent pattern if s has a transition probability $P(s \rightarrow s) \geq \alpha$, where α will typically be close to 1.

A more complex internal pattern would occur when there is a cycle involving two or more states from the same cluster. In this case, at any moment it is very likely that the system (e.g. a patient) is switching between the same diagnosis represented by different states. We call such patterns *internal feedback patterns*.

External Patterns. By taking a closer look at the internal feedback patterns, one would probably conclude that it does not seem to make sense to have different states within a cluster if they switch only among themselves. It would probably suffice to have a single state with a recurrent pattern instead. By opposition, we consider *external feedback patterns*, in which there are states from multiple clusters.

In the context of multimorbidity, these patterns would mean that transitions could involve different diagnoses, as opposed to internal patterns. Hence, if a cluster is involved in both an internal and an external pattern, then the same diagnosis could lead to different multimorbidity future events. In other words, the same diagnosis could play distinct roles.

Example 1. Suppose two clusters of states $C1 = \{s_1, s_2\}$ and $C2 = \{s_3, s_4, s_5\}$, where $C1$ and $C2$ are associated to two different diagnosis codes, as shown in Fig. 2. It holds that state s_1 is involved in a recurrent pattern due to its high self-transition probability (for $\alpha = 0.95$). States s_4 and s_5 are involved in an internal feedback pattern, while states s_2 and s_3 are involved in an external feedback pattern.

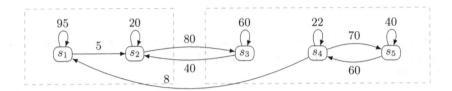

Fig. 2. An example with two clusters of states $C1$ (left) and $C2$ (right) for depicting transition patterns.

4   Case Study

In order to illustrate the value of the proposed methods, we consider the Primary Care Database from the NIVEL institute (Netherlands Institute for Health Services Research), a Dutch institute that maintains routinely electronic health records from health care providers to monitor health in Dutch patients [1]. In the NIVEL data, patient visits are assigned an ICPC code (International Classification of Primary Care) indicating a diagnosis for the visit. Each patient visit is assigned an ICPC code (International Classification of Primary Care), which indicates the diagnosis for the visit.

4.1   Variables and Observations

Atherosclerosis is a cardiovascular condition that has complex associations to a number of other conditions. Although in the literature atherosclerosis has been known to be associated to chronic diseases like diabetes [8], there is still active research on its implications and associations [13, 15, 20]. In our data preprocessing steps, we first selected ICPC codes related to atherosclerosis, then groups of codes that refer to a given medical symptom or condition were built based on medical experts. As a result, each group of codes gave rise to a variable (i.e. an observable), as shown in Table 1. The variables constructed based on Table 1 can be seen as comorbidities that might occur in patients with atherosclerosis.

In order to construct the event data from the raw NIVEL data, we first ordered the raw data in ascending dates. Then, whenever a patient visit having as diagnosis one of the ICPC codes from Table 1 was found, a new observation was created, where the variable associated to the ICPC code would have a 1 and the remaining variables would have zeros. The visits that were not associated to any of such ICPC codes were ignored.

Table 1. ICPC codes related to atherosclerosis, and their mapping into variables of the model.

ICPC code, description	Variable (model)
K02.00, Pressure/tightness of heart	*Angina*
K74.00, Angina pectoris	
K74.02, Stable angina pectoris	
K76.01, Coronary sclerosis	
K75.00, Acute myocardial infarction	*Myocardial infarction*
K76.02, Previous myocardial infarction (> 4 weeks earlier)	
K89.00, Transient cerebral ischemia/TIA	*Cerebrovascular accident*
K90.00, Cerebrovascular accident	
K90.03, Cerebral infarct	
K92.01, Intermittent claudication	*Claudication*
K99.01, Aortic aneurysm	*Aortic aneurysm*
K91.00, Atherosclerosis	*Atherosclerosis*

4.2   Sample

We considered a sample of 32,227 patients that had visits between 1st of January, 2003 and 31st of December, 2011. To be included, a patient must have had at least one visit related to one of the diagnoses listed in Sect. 4.1. The data construction procedure previously discussed resulted in a dataset with 216,580 observations, where the average number of observations per patient is 6.7 (SD = 10.9), 11,932 patients have only one observation, whereas 20,295 have two or more.

4.3   Number of Hidden States

In order to select an appropriate number of states when learning HMMs, the Akaike Information Criterion (AIC) shown in Eq. 3 was minimized. Models are evaluated by increasing number of states until the addition of states does not improve the score substantially, which is an indication of model overfitting.

$$\text{AIC}(M) = 2\log p - 2\log(\hat{L}) \tag{3}$$

where M is a candidate model, p is the number of parameters of M, and $\hat{L}$ is the log-likelihood of M based on maximum likelihood estimates of the parameters.

During the learning of HMMs, the expectation-maximization algorithm is used, which is quite sensitive to its initial parameters, especially with larger number of states. In order to reduce such effect, the best initial model was selected out of 30 candidates randomly generated.

4.4   Clinical Interpretation of Clusters

If clusters of states are identified in the learned model, one would expect that states within a cluster are indeed necessary, i.e. they should not be replaced by a single state, at the cost of, e.g., worsening model fit. States can be distinguished based on the transition patterns in which they are involved, which provides a dynamics-based description of their differences. Moreover, states of a cluster can be distinguished at a medical level by looking at associations with other data available in patient data. In this case study, we consider as a medical outcome the total number of distinct diagnoses that were registered for each patient (which might include other events than those listed in Sect. 4.1), which provides an approximation to the number of diseases that have occurred in the patient. Such result can indicate medical significance to the cluster states.

Let us consider a state j and a patient i. We first compute the chances that this patient is in such state at some instant t based on the full observations of the patient:

$$\gamma_i^t(j) = P\left(S^{(t)} = j \mid \mathbf{X}^{(0:T_i)}\right) \tag{4}$$

where T_i refers to the last observation of patient i. When the patient has more than one observation, this will result in a sequence of probabilities for a state j. As we will associate the states to the total number of diseases, the average of such probabilities is taken:

$$\overline{\gamma}_i(j) = \frac{1}{T_i + 1} \sum_{t=0}^{T_i} P\left(S^{(t)} = j \mid \mathbf{X}^{(0:T_i)}\right) \tag{5}$$

Once the quantities in Eq. 5 are computed, they are grouped based on the total number of diseases. Then, the average of such quantities is taken per group. Grouping is used to adjust for the near-zero probabilities that might occur in $\overline{\gamma}_i(j)$. As a result, pairs with number of diseases and group averages are obtained, which we use for computing associations (e.g. the Pearson correlation coefficient).

5   Experimental Results

5.1   Model Dimension

Figure 3 shows the model selection scores, which served as a basis for selecting an HMM with 9 states as the suitable model. All the states of the model were associated to fully deterministic emission distributions, such that only one diagnosis variable had a probability equal to 1 in each state, while the other variables had probabilities equal to zero. This means that the property that models should produce events with only one active variable (representing the main diagnosis) was met.

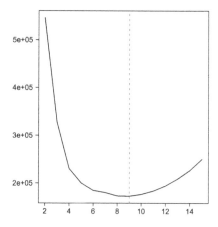

Fig. 3. Model selection scores. X axis: number of hidden states, Y axis: AIC score. The vertical line indicates the number of states where the AIC was minimal.

5.2   Clusters

Figure 4 shows the learned HMM, where each state is named according to the observable that is active (i.e. the observable that has probability equal to 1). Figure 4 shows that three non-unitary clusters were obtained, suggesting that visits associated to angina, myocardial infarction and cerebrovascular accident were suitably represented by 2 states each. Intuitively, it is relevant to model a visit to, e.g., angina by means of 2 different states, hence such a could lead to two different patient courses. As expected, determining which of the two states a visit is associated to depends, e.g., on what is known so far about the patient in terms of past visits.

5.3   Transition Patterns

Figure 4 shows the state transitions of the learned HMM. For each cluster, there is clearly a state that will very likely take a self-transition, which are CVA6, Angina7 and MI3. Thus, such states produce internal patterns as recurrent transition patterns. The HMM shows external patterns as well. In particular, angina seems to be a central event in this model: when moving from the two other clusters, it is more likely that this transition will reach angina (i.e. Angina5). Once in angina, a transition to the other clusters is also possible, with probability larger than 0.05. Hence, such external patterns can be thought of as feedback transition patterns.

5.4   Clinical Interpretation of Clusters

The average probabilities defined in Eq. 5 are summarized in Fig. 5. The histograms suggest that within each cluster there are states that are substantially

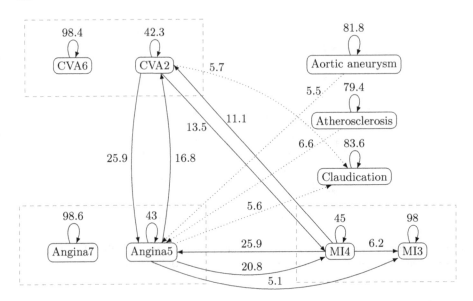

Fig. 4. Clusters of hidden states. Arcs denote state transitions, with labels indicating probability (in %). For the sake of visualization, only transitions with probability greater than or equal to 5% are shown and some transitions were shown dashed.

more prevalent than others, and such separation is more or less uniform depending on the cluster. In general, the recurrent states were usually more likely than their counterparts, which suggests that patients likely had several visits due to the same diagnosis before a diagnosis associated to a different comorbidity was registered.

Figure 6 shows the total number of diseases in patients against the group probabilities. Visual inspection shows that up to 50 diagnoses the trend is substantially more stable than that of all the groups; indeed, around 97% of the patients had at most 50 diagnoses, thus we focus on such groups for obtaining a better understanding of the general trend.

We first note that whenever a patient has a given event (e.g. a CVA), then it holds that the patient will be in one of the states of the corresponding cluster (e.g. either in CVA 2 or CVA 6 states). Figure 6 suggests that, in general, the states of clusters are correlated to the number of diseases in different ways. For the CVA case, patients with only a few diseases are more likely in state CVA 6 (internal patterns) rather than CVA 2 (external patterns). However, as the number of diseases increases, the chances to be in CVA 6 decreases while the chances to be in CVA 2 increases, although such trends occur at different paces. Analogously, for an MI event, it is likely the patient will be in state MI 3 (internal patterns) if the case involves only a few diseases, but a probability decrease is expected for cases involving more diseases. On the other hand, not much can be said about MI 4, as the correlation is very low. Intuitively, one would indeed

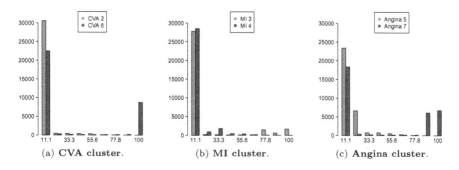

(a) **CVA cluster.**   (b) **MI cluster.**   (c) **Angina cluster.**

Fig. 5. Histograms of average probabilities of states (in %). X axis: average probability of state j in patient i (i.e. $\overline{\gamma}_i(j)$), Y axis: number of patients.

expect that with more diseases patients will transit more between the clusters, which partially explains the trends of the CVA and MI clusters.

As opposed to the previous clusters, Fig. 6 suggests that the Angina cluster has a less straightforward dynamics. In this cluster, both of its states become more prevalent as the number of diseases increases (up to 50), which might suggest the increasing importance of angina that might work as a proxy for the comorbidities considered in this paper, as well as for other chronic and non-chronic events not explicitly considered.

5.5 Are the Clusters Needed? A Comparison to Markov Chains

The need for the clusters learned in the HMM can be assessed by comparing the model fit of the HMM with that of a Markov chain (MC). The state space of such MC is $\mathbf{X}$, i.e., the six comorbidities listed in Sect. 4.1, hence learning this MC amounts to estimating the initial and transition probabilities involving the variables in $\mathbf{X}$. This comparison can illustrate whether the multiple states associated to a given comorbidity (in this paper, the multiple states of CVA, MI and Angina) are indeed necessary for delivering a better model.

Table 2 shows the AIC scores computed for the 9-state HMM and for the MC, which indicates a superior model fit for the HMM. Besides such advantage, with the MC it is no longer possible to identify that the occurrence of a certain event such as angina, can be correlated to different patient characteristics (we used in this paper the total amount of diseases, but other medical outcomes could be devised as well).

Table 2. AIC scores of the HMM and the Markov chain learned from the multimorbidity data. The smaller the AIC, the better the model fit is.

Model	State clusters	AIC
9-state HMM	3 clusters	172,942.8
Markov chain	No clusters	185,013.5

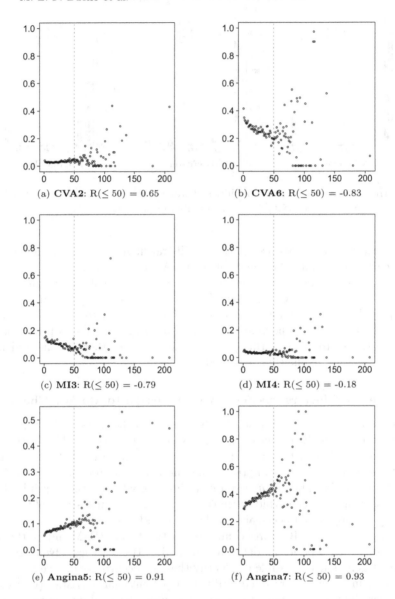

Fig. 6. Association of cluster states to clinical outcome. X axis: number of distinct diagnoses, Y axis: group probability. The vertical line is drawn at $X = 50$. R indicates the Pearson coefficient, calculated considering only the groups with at most 50 diagnoses (which amounts to 97% of all the patients).

6   Related Work

The notion of clustering states in hidden Markov models has not been investigated so far to the best of our knowledge. A related approach is clustering applied to timed automata [5,22], where state sequences are clustered based on their distance by means of hierarchical clustering methods. Based on Bayesian HMMs that use topic modeling, clustering of patient journeys has been proposed [6], which uses the full set of events associated to unstable angina. By opposition, in our case the clusters are determined based on the states, which shifts the focus towards the dynamics that involve states within clusters. Despite their differences, our methods and those from the literature share the goal of moving towards explainable artificial intelligence [4,11], as we aimed not only to obtain a model with suitable fit, but also to understand more about the patient situation by looking at the structure of the HMM. An example in our case is the deterministic emissions, which can facilitate interpreting models like HMMs to a great extent, at the same time obeying constraints of the multimorbidity problem.

In the context of electronic health records of multimorbidity, a cohort of the NIVEL data used in this paper had been used for learning graphical models based on Bayesian networks, in static [9] and temporal [10] contexts. In those cases, however, the goal was to model differences in practices, hospitals, or regions, without taking into account latent variables.

7   Conclusions

In this paper we proposed a modeling methodology for representing multimorbidity data collected from Dutch practices as single-event observations. Due to the fine-grained nature of such data, we used latent variable-based models, namely HMMs, for extracting additional information that are not directly measured. By using such models, we showed that clusters of states could be discovered, which are states associated to the same observation (or diagnosis, in the case of multimorbidity data), which can, however, be reached from and lead to different transitions. Based on this, we defined the notion of transition patterns.

For the experiments, we considered data concerning variables that are associated to atherosclerosis. The learned model had 9 states, in which clusters involving angina, myocardial infarction and cerebrovascular accident were identified. This suggests that these diagnoses are too complex to be managed by a single latent state, hence a model with better fit is obtained when such diagnoses are allowed to be each represented by multiple states (or roles), as we did with the used HMMs.

Suggestions for future work include investigating the effect of modeling medication and lab exams, which are available to some patients in the NIVEL data. These could be added into the model as inputs (i.e. covariates), for example, which would allow to capture switching regimes for the transitions. Regarding the methodological aspect, it might be worth to extend the learning algorithm to impose deterministic model emissions. Although we were able to obtain a model that satisfied this problem requirement, it might not be always the case that unconstrained learning would be suitable, as there might not be enough data for learning enough states. As a consequence, non-deterministic emissions could be obtained, which although could lead to better model fit than their deterministic counterpart, would violate the problem property that a single event should be produced at each moment.

Acknowledgments. This work has been partially funded by NWO (Netherlands Organisation for Scientific Research), project Careful (62001863), by FAPEMIG, and by NORTE 2020 (project NanoSTIMA). Project "NORTE-01-0145-FEDER-000016" (NanoSTIMA) is financed by the North Portugal Regional Operational Programme (NORTE 2020), under the PORTUGAL 2020 Partnership Agreement, and through the European Regional Development Fund (ERDF).

References

1. NIVEL Primary Care Database. https://www.nivel.nl/en/dossier/nivel-primary-care-database. Accessed 30 Apr 2018
2. Barnett, K., Mercer, S., Norbury, M., Watt, G., Wyke, S., Guthrie, B.: Epidemiology of multimorbidity and implications for health care, research, and medical education: a cross-sectional study. Lancet **380**, 37–43 (2012). https://doi.org/10.1016/S0140-6736(12)60240-2
3. Côté, M.J., Stein, W.E.: A stochastic model for a visit to the doctors office. Math. Comput. Model. **45**(3), 309–323 (2007). https://doi.org/10.1016/j.mcm.2006.03.022
4. Gunning, D.: Explainable Artificial Intelligence (XAI) (2016). http://www.darpa.mil/program/explainable-artificial-intelligence
5. Hammerschmidt, C.A., Verwer, S., Lin, Q., State, R.: Interpreting finite automata for sequential data. In: Interpretable Machine Learning for Complex Systems: NIPS 2016 Workshop Proceedings (2016)
6. Huang, Z., Dong, W., Wang, F., Duan, H.: Medical inpatient journey modeling and clustering: a Bayesian hidden Markov model based approach. In: AMIA Annual Symposium Proceedings, pp. 649–658 (2015)
7. Huang, Z., Dong, W., Bath, P., Ji, L., Duan, H.: On mining latent treatment patterns from electronic medical records. Data Min. Knowl. Discov. **29**(4), 914–949 (2015). https://doi.org/10.1007/s10618-014-0381-y
8. Hyvärinen, M., et al.: The impact of diabetes on coronary heart disease differs from that on ischaemic stroke with regard to the gender. Cardiovasc. Diabetol. **8**(1), 17 (2009). https://doi.org/10.1186/1475-2840-8-17
9. Lappenschaar, M., Hommersom, A., Lucas, P.J., Lagro, J., Visscher, S.: Multilevel Bayesian networks for the analysis of hierarchical health care data. Artif. Intell. Med. **57**(3), 171–183 (2013). https://doi.org/10.1016/j.artmed.2012.12.007
10. Lappenschaar, M., et al.: Multilevel temporal Bayesian networks can model longitudinal change in multimorbidity. J. Clin. Epidemiol. **66**(12), 1405–1416 (2013). https://doi.org/10.1016/j.jclinepi.2013.06.018
11. Liu, S., Wang, X., Liu, M., Zhu, J.: Towards better analysis of machine learning models: a visual analytics perspective. Vis. Inform. **1**(1), 48–56 (2017). https://doi.org/10.1016/j.visinf.2017.01.006
12. Marshall, A., Vasilakis, C., El-Darzi, E.: Length of stay-based patient flow models: recent developments and future directions. Health Care Manag. Sci. **8**(3), 213–220 (2005). https://doi.org/10.1007/s10729-005-2012-z

13. Mytton, O.T., et al.: Association between intake of less-healthy foods defined by the United Kingdom's nutrient profile model and cardiovascular disease: a population-based cohort study. PLOS Med. **15**(1), 1–17 (2018). https://doi.org/ 10.1371/journal.pmed.1002484

14. Najjar, A., Reinharz, D., Girouard, C., Gagn, C.: A two-step approach for mining patient treatment pathways in administrative healthcare databases. Artif. Intell. Med. **87**, 34–48 (2018). https://doi.org/10.1016/j.artmed.2018.03.004

15. O'Donovan, G., Lee, I., Hamer, M., Stamatakis, E.: Association of "weekend warrior" and other leisure time physical activity patterns with risks for all-cause, cardiovascular disease, and cancer mortality. JAMA Internal Med. **177**(3), 335–342 (2017). https://doi.org/10.1001/jamainternmed.2016.8014

16. Prados-Torres, A., et al.: Multimorbidity patterns in primary care: interactions among chronic diseases using factor analysis. PLOS ONE **7**(2), 1–12 (2012). https://doi.org/10.1371/journal.pone.0032190

17. Rabiner, L.R.: A tutorial on hidden Markov models and selected applications in speech recognition. Proc. IEEE **72**, 257–286 (1989)

18. Rajkomar, A., et al.: Scalable and accurate deep learning with electronic health records. NPJ Digit. Med. **1**(1), 18 (2018). https://doi.org/10.1038/s41746-018-0029-1

19. Sinnige, J., Korevaar, J.C., Westert, G.P., Spreeuwenberg, P., Schellevis, F.G., Braspenning, J.C.: Multimorbidity patterns in a primary care population aged 55 years and over. Family Pract. **32**(5), 505–513 (2015)

20. Stamatakis, E., de Rezende, L.F.M., Rey-López, J.P.: Sedentary behaviour and cardiovascular disease. In: Leitzmann, M., Jochem, C., Schmid, D. (eds.) Sedentary Behaviour Epidemiology, pp. 215–243. Springer, Cham (2018). https://doi.org/10. 1007/978-3-319-61552-3_9

21. Wijkstra, J., et al.: Treatment of unipolar psychotic depression: a randomized, double-blind study comparing imipramine, venlafaxine, and venlafaxine plus quetiapine. A. Psych. Scandinavica **121**(3), 190–200 (2009)

22. Zhang, Y., Lin, Q., Wang, J., Verwer, S.: Car-following behavior model learning using timed automata. IFAC-PapersOnLine **50**(1), 2353–2358 (2017). 20th IFAC World Congress. https://doi.org/10.1016/j.ifacol.2017.08.423

Integral Representations of a Coherent Upper Conditional Prevision by the Symmetric Choquet Integral and the Asymmetric Choquet Integral with Respect to Hausdorff Outer Measures

Serena Doria[✉] [ID]

Department of Engineering and Geology,
University G.d'Annunzio, Chieti-Pescara, Italy
s.doria@dst.unich.it

Abstract. Complex decisions in human decision-making may arise when the Emotional Intelligence and Rational Reasoning produce different preference ordering between alternatives. From a mathematical point of view, complex decisions can be defined as decisions where a preference ordering between random variables cannot be represented by a linear functional. The Asymmetric and the Symmetric Choquet integrals with respect to non additive-measures have been defined as aggregation operators of data sets and as a tool to assess an ordering between random variables. They could be considered to represent preference orderings of the conscious and unconscious mind when a human being make decision. Sufficient conditions are given such that the two integral representations of a coherent upper conditional prevision by the Asymmetric Choquet integral and the Symmetric Choquet integral with respect to Hausdorff outer measures coincide and linearity holds.

Keywords: Coherent upper conditional prevision
Asymmetric Choquet integral · Symmetric Choquet integral
Hausdorff outer measures

1 Introduction

In the last years non-additive measures have been introduced with the aim to represent partial knowledge with more flexible models which are closer to human and artificial thought. The concept of Emotional Intelligence and its role in decision-making abilities has been investigated [2].

Emotional Intelligence is defined as the capability of individuals to recognize their own emotions and those of others, discern between different feelings and label them appropriately, use emotional information to guide thinking and behavior, and manage emotions to adapt to environments or achieve goals.

© Springer Nature Switzerland AG 2018
D. Ciucci et al. (Eds.): SUM 2018, LNAI 11142, pp. 108–123, 2018.
https://doi.org/10.1007/978-3-030-00461-3_8

Emotions, defined as instinctive or intuitive feelings distinguished from reasoning or knowledge, are the way to reach the unconscious, the part of the mind which is inaccessible to the conscious mind but which affects behavior and emotions; they function the same way as the unconscious and are the means to decode it.

In [18] one of the most comprehensive descriptions of the structure and functioning of the unconscious has been produced with the purpose to account for the non-logical aspects of human thought, as they can be detected in serious mental disorders. The author drew a distinction between the logical conscious thought, structured on the categories of time and space, which he defined *asymmetrical thought*, and the unconscious thought, which he defined *symmetrical thought*.

Both types of thoughts combine in the different experiences of human thinking thus yielding to a bi-logic asset. Emotions do not involve thinking. Thinking is a process and develops over time, it is something that happens while feeling is experienced as something that does not happen but simply is. The experience of feeling the emotions thus cannot be immediately reduced to thought and thinking. Feeling and thinking represent a constitutive conflict of human beings since they are of bi-logical nature. Such constitutive conflict can become pathological if the contrast or conflict between these ways of being symmetric and asymmetric, that is conscious or unconscious, becomes too strong. The conscious and unconscious are two different modes of being, asymmetric and in becoming the first, symmetric and static the second. The fundamental principles of the unconscious are the *generalization principle* and the *symmetric principle*. The generalization principle claims that the unconscious treats any single thing as it were a member of a class which contains other members. In turn, this class is considered as a subclass of another more general class, and so on. According to the symmetric principle, in the realm of the unconscious, every relation is symmetric (as in the mathematical sense of this term).

Complex decisions in human decision-making may arise when the Emotional Intelligence and Rational Reasoning produce different preference ordering between alternatives.

From a mathematical point of view, complex decisions can be defined as decisions where a preference ordering $\succ$ between random variables cannot be represented by a linear functional, that is there exist no linear functional Γ such that

$$X \succ Y \Leftrightarrow \Gamma(X) > \Gamma(Y) \text{ and } X \approx Y \Leftrightarrow \Gamma(X) = \Gamma(Y)$$

Example 1. *Let Ω be a non empty set and let $\boldsymbol{B} = \{B_1, B_2\}$ be a partition of Ω. Let consider the class $K = \{X_1, X_2, X_3\}$ of bounded random variables defined on Ω by*

random variables	B_1	B_2
X_1	0.3	0.3
X_2	0.7	0.0
X_3	0.0	0.7

The preference ordering $X_1 \succ X_2$ and $X_2 \approx X_3$ cannot be represented by the linear functional $\Gamma(X) = \sum_{i=1}^{2} X(B_i)\mu(B_i)$ since there exists no probability measure μ such that the following system has solution:

$$\begin{cases} X_1 \succ X_2 \\ X_2 \approx X_3 \end{cases} \Leftrightarrow \begin{cases} 0.3\mu(B_1) + 0.3\mu(B_2) > 0.7\mu(B_1) + 0.0\mu(B_2) \\ 0.7\mu(B_1) + 0.0\mu(B_2) = 0.0\mu(B_1) + 0.7\mu(B)_2. \end{cases}$$

The given ordering is represented by a coherent lower conditional prevision defined as the Choquet integral with respect to a coherent lower conditional probability $\underline{\mu}$ such that $\underline{\mu}(B_1) = \underline{\mu}(B_2) = 0$. We can observe that if only the random variables X_1 and X_2 are considered then the preference ordering $X_1 \succ X_2$ can be represented by a linear functional.

The previous example put in evidence the necessity to introduce non-linear functionals to represent preference orderings, to investigate equivalent random variables and to manage null-events.

A normalized monotone set function μ, also called non-additive measure or fuzzy measure, represents a functional Γ defined on a class L of random variables if $\Gamma(X) = \int X d\mu$.

The Asymmetric and the Symmetric Choquet integrals ([3,6]) with respect to non additive-measures have been defined as aggregation operators of data sets or as a tool to assess an ordering between random variables. According to the theory developed in [18] they could be considered as two different tools to represent preference orderings of the conscious and unconscious mind when a human being make decision.

Applications in decision making of the Symmetric and Asymmetric Choquet integrals on finite domains have been investigated in [15].

An interesting problem is to determine if fuzzy integrals always represent non-linear operators and when different fuzzy integrals coincide ([19]).

Coherent upper and lower previsions are examples of non-linear functionals.

In recent papers ([7,8]) a model of coherent upper conditional prevision and probability ([25,26]), based on Hausdorff outer measures is introduced.

Let (Ω, d) be a metric space where Ω is a set with positive and finite Hausdorff outer measure in its Hausdorff dimension. Let **B** be a partition of Ω. Let $B \in$ **B** be a set with positive and finite Hausdorff outer measure in its Hausdorff dimension s. The s-dimensional Hausdorff outer measure h^s is the Hausdorff measure *associated* to the coherent conditional prevision $P(X|B)$. The model permits to manage conditioning events with zero probability.

The necessity to propose a new tool to define coherent upper conditional previsions arises because they cannot be obtained as extensions of linear expectations defined, by the Radon-Nikodym derivative, in the axiomatic approach ([1]); it occurs because one of the defining properties of the Radon-Nikodym derivative, that is to be measurable with respect to the σ-field of the conditioning events, contradicts a necessary condition for the coherence ([8, Theorem 1]).

The outline of the paper is the following. In Subsect. 1.2 some results proven in [10] about the disintegration property and the conglomerability for coherent upper conditional previsions defined with respect to Hausdorff outer measures are introduced. In subsection 1.3 the main results about different integral representations of coherent upper and lower conditional previsions proven in [11] are illustrated. In Sect. 2 coherent upper conditional previsions, Asymmetric and Symmetric Choquet integrals are introduced and an example is given to show that in general the Symmetric and Asymmetric Choquet integrals with respect to a coherent upper probability are not equal. In Sect. 3 we prove that the coherent upper conditional prevision defined with respect to Hausdorff outer measures is linear on the class of all continuous bounded random variables and so it can be represented by the Symmetric and Asymmetric Choquet integrals.

It is also proven that the Symmetric and Asymmetric Choquet integrals of a bounded random variable X with respect to Hausdorff outer measure coincide with the Lebesgue integral and the linearity holds for every random variable Y measurable with respect to the σ-field generated by the weak upper level sets of X.

1.1   Coherent Upper and Lower Conditional Previsions and the Disintegration Property

The vantages of using coherent upper and lower conditional previsions to represent preference orderings between random variables are that they can be extended to the class of all random variables on Ω, they are fully conglomerable and satisfy the disintegration property in many cases.

A coherent upper conditional prevision $\overline{P}(X|\mathbf{B})$ is disintegrable with respect to a partition $\mathbf{B}$ if $\overline{P}(X|\Omega) = \sum_{B \in \mathbf{B}} \overline{P}(X|B)\overline{P}(B)$

The disintegration property could be used to aggregate preferences on different conditioning events B.

For example in multi-criteria decision problem denoted by Ω the set of criteria, the elements of a partition $\mathbf{B}$ can represent clusters or macro-criteria- which are representative of the general objectives of the decision problem, as goals to pursue through the implementation of specific policies - and the elements in each B are the criteria.

To compare the random variables or the alternatives with respect to all criteria the disintegration property of the upper and lower conditional previsions can be applied to calculate $\overline{P}(X|\Omega)$ and $\underline{P}(X|\Omega)$.

For linear conditional previsions the disintegration property is equivalent to conglomerability with respect to a partition.

In [26, 6.8] full conglomerability, i.e. conglomerability with respect to all partition, is required as a rationality axiom for a coherent lower prevision since it assures that it can be coherently extended to coherent conditional previsions for any partition $\mathbf{B}$ of Ω. If the partition $\mathbf{B}$ represents an experiment that could be performed it is necessary to update the unconditional upper prevision after observing a set B of $\mathbf{B}$. Conglomerability is based on the following *conglomerability principle*: if a random variable X is B-desirable, i.e. we have a disposition to accept X for every set B in the partition $\mathbf{B}$, then X is desirable. If there is no coherent way of updating the initial prevision after learning the outcome of the experiment the lower prevision, which represents our knowledge, is unreasonable. If a coherent lower probability $\underline{P}$ is such that all the sets in the partition have zero probability, then its minimal coherent conditional prevision extension is the lower *vacuous* coherent conditional prevision. So to verify the existence of lower coherent conditional previsions coherent with $\underline{P}$ is equivalent to verify the conglomerability for all countable partition ([26, 6.8.2]). A consequence is that a fully conglomerable coherent conditional prevision may fail the disintegration property on non-countable partitions.

Disintegration property for linear prevision has been studied in [24] and for coherent lower and upper previsions has been investigated in [10, 20].

In [10] the notion of non-null partition is introduced and it has been proven that coherent upper conditional previsions defined with respect to Hausdorff outer measures satisfy the disintegration property on every countable partition and on every non-null arbitrary (non countable) partition. Coherent conditional probabilities defined by Hausdorff measure on the Borel σ-field are σ-additive disintegrations. The coherent lower prevision which represents the preference ordering in Example 1 fails the disintegration property.

1.2   Different Integral Representations for Coherent Upper and Lower Conditional Previsions

Different integrals have been introduced in literature to give an integral representation of a functional with respect to a non-additive set function: the Choquet integral [3, 6], the pan-integral [27, 28] and the concave integral [17]. They are defined by considering different sets systems: the Choquet integral is based on chains of sets, the pan-integral on finite partitions and the concave integral deals with arbitrary finite set systems. They can be used to describe different situations in practical problems. In [16] Example 4.80 these different integrals are considered to maximize the total production of a farm when the teams of workers are built according to different rules. All integrals are particular cases of decomposition integral recently introduced in [13].

Integral representations of coherent lower and upper previsions with respect to coherent lower and upper probabilities have been discussed in Doria et al. [11].

It is proven that for subadditive monotone set functions the concave integral and the pan-integral coincide.

It is also proven that, given a coherent upper conditional prevision $\overline{P}(X|B)$ and denoted by h^s its associated Hausdorff outer measure and by S the σ-field of the h^s-measurable sets, the concave integral, the pan integral and the Choquet integral with respect to Hausdorff outer measures agree with the Lebesgue integral on the class of all S-measurable random variables. It is because of the Hausdorff outer measure of a set is calculated over all δ-covers of the set and it does not depend on different set systems used to calculate the different integrals.

2   Integral Representations of a Coherent Upper Conditional Prevision by the Asymmetric Choquet Integral and the Symmetric Choquet Integral

Let (Ω, d) be a metric space and let **B** be partition of Ω. A bounded random variable is a function $X : \Omega \to \Re$ such that $|X| \le M$ for some real constant M and $L(\Omega)$ is the class of all bounded random variables defined on Ω; for every $B \in \mathbf{B}$ denote by $X|B$ the restriction of X to B and by $\sup(X|B)$ the supremum value that X assumes on B. Let $L(B)$ be the class of all bounded random variables $X|B$. Denote by I_A the indicator function of any event $A \in \wp(B)$, i.e. $I_A(\omega) = 1$ if $\omega \in A$ and $I_A(\omega) = 0$ if $\omega \in A^c$.

For every $B \in \mathbf{B}$ coherent upper conditional previsions $\overline{P}(\cdot|B)$ are functionals defined on $L(B)$ ([26]).

Definition 1. *Coherent upper conditional previsions are functionals $\overline{P}(\cdot|B)$ defined on $L(B)$, such that the following axioms of separate coherence hold for every X and Y in $L(B)$ and every strictly positive constant λ:*

(1) $\overline{P}(X|B) \le \sup(X|B)$;
(2) $\overline{P}(\lambda X|B) = \lambda \overline{P}(X|B)$ (positive homogeneity);
(3) $\overline{P}(X + Y)|B) \le \overline{P}(X|B) + \overline{P}(Y|B)$ (subadditivity);
(4) $\overline{P}(I_B|B) = 1$

Definition 2. *Given a partition **B** and a random variable $X \in L(\Omega)$ a coherent upper conditional prevision $\overline{P}(X|B)$ is a random variable on Ω equal to $\overline{P}(X|B)$ if $\omega \in B$.*

Suppose that $\overline{P}(X|B)$ is a coherent upper conditional prevision on $L(B)$ then its conjugate coherent lower conditional prevision is defined by $\underline{P}(X|B) = -\overline{P}(-X|B)$. Let K be a linear space contained in $L(B)$; if for every X belonging to K we have $P(X|B) = \underline{P}(X|B) = \overline{P}(X|B)$ then $P(X|B)$ is called a coherent *linear* conditional prevision (de Finetti [4,5], Dubins [12], Regazzini [21,22]) and it is a linear, positive and positively homogenous functional on $L(B)$.

The unconditional coherent upper prevision $\overline{P} = \overline{P}(\cdot|\Omega)$ is obtained as a particular case when the conditioning event is Ω. Coherent upper conditional probabilities are obtained when only 0–1 valued random variables are considered.

A coherent upper conditional probability μ_B on $\wp(B)$ is

(a) *submodular* or *2-alternating* if $\mu(A \cup E) + \mu(A \cap E) \le \mu(A) + \mu(E)$ for every $A, E \in \wp(B)$;
(b) *subadditive* if $\mu(A \cup E) \le \mu(A) + \mu(E)$ for every $A, E \in \wp(B)$ such that $\mu(A \cap E) = \emptyset$;
(c) *continuous from below* if $\lim_{i \to \infty} \mu(A_i) = \mu(\lim_{i \to \infty} A_i)$ for any increasing sequence of sets $\{A_i\}$, with $A_i \in \wp(B)$.

A coherent upper conditional prevision $\overline{P} : L(B) \to \Re$ can be represented as Choquet integral with respect to a coherent upper conditional probability μ on $\wp(B)$ if $\overline{P}(X) = \int X d\mu \ \forall X \in L(B)$.

A necessary and sufficient condition [9, Proposition 1] for the representation of a coherent upper conditional prevision on $L(B)$ as Choquet integral with respect to a coherent upper conditional probability μ is that μ is submodular. Then $\overline{P}(I_A) = \int I_A d\mu = \mu(A)$. For every $x \in \Re$ the *upper level sets* of a random variable X are the sets $\{X > x\} = \{\omega \in B : X(\omega) > x\}$.

The *decreasing distribution function* of X with respect to μ is the function

$$G_{\mu,X}(x) = \mu\{X > x\}.$$

It is unique except on a set with measure μ equal to zero. If μ is continuous from below then

$$G_{\mu,X}(x) = \mu\{X > x\} = \mu\{X \geq x\}.$$

Since X is a bounded random variable thus there exist a constant k such that $\tilde{X} = X + k \geq 0$ and $G_{\mu,\tilde{X}}(x) = G_{\mu,X}(x - k)$ for every real number x [6, Proposition 4.1].

The *Asymmetric Choquet integral* [6] of a bounded random variable X with respect to μ on $\wp(B)$ is defined by

$$\int^{Cho}_{\underline{B}} X d\mu = \int_0^{+\infty} \mu\left(\left\{\tilde{X} \geq x\right\}\right) dx - k\mu(\Omega).$$

Let μ be a coherent upper probability is defined on a class S containing the empty set and properly contained in $\wp(B)$ and let $\mu^*(A)$ and $\mu_*(A)$ be its outer and inner measure. To define the Choquet integral in this case the following definition is given.

Definition 3. [6, p. 49]. *Let μ_B be a coherent upper probability defined on a class S containing the empty set and properly contained in $\wp(B)$ and let $\mu_B^*(A)$ and $\underline{\mu}_B^*(A)$ be its outer and inner measures; a random variable X is upper μ-measurable if $\mu_B^*(\{X > x\}) = \underline{\mu}_B^*(\{X > x\})$ except on a a countable set. A random variable X is lower μ-measurable if $-X$ is upper μ-measurable. X is (upper, lower) S-measurable if it is (upper, lower) μ-measurable for any monotone set function μ on S. X is upper S-measurable if the class of upper level sets $\{\{X \geq x\} x \in \Re\}$ is contained in S. If S is a σ-field and the class of upper level sets of X and $-X$ belong to S then X is S-measurable, that is $X^{-1}(A) \in S$ for every Borelian set A.*

For a μ-upper measurable X the Asymmetric Choquet integral is defined by

$$\int^{Cho} X d\mu = \int X d\mu^* = \int X d\mu_*$$

If Ω is finite and μ is defined on a field S, denote by $A_1, ..., A_n$ the atoms of S, which are the minimal elements of $S - \oslash$. If the atoms A_i are enumerated so that $x_i = X(A_i)$ are in descending order, i.e. $x_1 \geq x_2 \geq ... \geq x_n$ and $x_{n+1} = 0$ the Choquet integral with respect to μ is given by

$$\int X d\mu = \sum_{i=1}^n (x_i - x_{i+1})\mu(S_i)$$

where $S_i = A_1 \cup A_2 ... \cup A_i$, and $x_{n+1} = 0$.

Any random variable X can be decomposed in its *positive part* X^+ and its *negative part* X^- given by:

$$X = X^+ - X^-; X^+ = 0 \vee X; X^- = (-X)^+$$

where $\vee$ is the maximum.

Since X^+ and X^- are comonotonic functions and X is real valued the Asymmetric Choquet integral can be written as

$$\int^{Cho} X d\mu = \int X^+ d\mu + \int (-X^-) d\mu = \int X^+ d\mu - \int X^- d\overline{\mu}$$

where $\overline{\mu}$ is the dual of μ, i.e. $\overline{\mu}(A) = \mu(\Omega) - \mu(A^c)$ for every $A \in \wp(\Omega)$.

The *Symmetric Choquet integral* of a random variable X with respect to a coherent upper probability μ is defined by

$$\int^{S-Cho} X d\mu = \int X^+ d\mu - \int X^- d\mu$$

In the following example we show that the Symmetric Choquet integral and the Asymmetric Choquet integral with respect to a coherent upper probability may not agree.

Example 2. *Let* $\Omega = \{\omega_1, \omega_2, \omega_3, \omega_4\}$ *and let* P_1 *and* P_2 *be two finitely additive probabilities defined by* $P_1(\omega_i) = \frac{1}{4}$ *for i=1,...,4 and* $P_2(\omega_1) = \frac{1}{2}, P_2(\omega_3) = \frac{1}{4}, P_2(\omega_2) = P_2(\omega_4) = \frac{1}{8}$. *Let* μ *be the set function defined on* $\wp(\Omega)$ *by the upper envelope of* P_1 *and* P_2, *i.e.* $\mu(A) = \max_{j=1,2} P_j(A)$ *for* $A \in \wp(\Omega)$. *The dual of* μ *is* $\overline{\mu} = \min_{j=1,2} P_j(A)$

Let consider the random variable X *defined by* $X(\omega_1) = -3, X(\omega_2) = -2$, $X(\omega_3) = 1, X(\omega_4) = 0$ *and* $X^+ = 0 \vee X, X^- = (-X)^+$,

We have that

$$\int^{Cho} X d\mu = \int^{Cho} X^+ d\mu - \int^{Cho} X^- d\overline{\mu} =$$

$$1\mu(\{\omega_4\}) - (3-2)\overline{\mu}(\{\omega_1\}) + (2-0)\overline{\mu}(\{\omega_1\}, \{\omega_2\}) = \frac{1}{4} - \frac{5}{4} = -1$$

$$\int^{S-Cho} X d\mu_1 = \int^{Cho} X^+ d\mu - \int^{Cho} X^- d\mu$$

$$= 1\mu(\{\omega_3\}) - [(3-2)\mu(\{\omega_1\}) + (2-0)\mu(\{\omega_1\}, \{\omega_2\})] = \frac{1}{4} - \frac{7}{4} = \frac{3}{2}.$$

so that

$$\int^{Cho} X d\mu_1 \neq \int^{S-Cho} X d\mu_1.$$

A coherent upper probability is *self-dual* if $\overline{\mu}(A) = \mu(A)$ for every $A \in \wp(\Omega)$. An additive probability is self-dual.

The Symmetric and Asymmetric Choquet integrals of a random variable X with respect to μ are equal if $X \geq 0$ or $\mu(\Omega) < +\infty$ and μ is self-dual.

The properties of the Asymmetric Choquet integral are valid, too for the Symmetric integral if the random variables are non-negative or μ is additive.

In particular

$$\int^{S-Cho} cX d\mu = c \int^{S-Cho} X d\mu \text{ for all } c \in \Re(\text{homogeneity})$$

especially

$$\int^{S-Cho} (-X)d\mu = - \int^{S-Cho} X d\mu \text{ (symmetry)}$$

3 Integral Representation of a Coherent Upper Conditional Prevision by the Symmetric Choquet Integral with Respect to Its Associated Hausdorff Outer Measure

In this section it is proven that, if the conditioning event is a set with positive and finite Hausdorff outer measure in its Hausdorff dimension, a coherent upper conditional prevision can be represented as Asymmetric Choquet integral and Symmetric Choquet integral with respect to its associated Hausdorff outer measure and the two representations agree. It occurs because the coherent upper conditional probability μ_B^* is self-dual up to a set of μ_B^*-zero probability and μ_B^* is finite.

In [7–10], a new model of coherent upper conditional probability based on Hausdorff outer measures is introduced and its properties have been studied. For the definition of Hausdorff outer measure and its basic properties see [14, 23].

Let (Ω, d) be a metric space and let **B** be partition of Ω.

Let $\delta > 0$ and let s be a non-negative number. The *diameter* of a non empty set U of Ω is defined as $|U| = \sup \{d(x, y) : x, y \in U\}$ and if a subset A of Ω is such that $A \subseteq \bigcup_i U_i$ and $0 < |U_i| \leq \delta$ for each i, the class $\{U_i\}$ is called a δ-cover of A.

The *Hausdorff s-dimensional outer measure* of A, denoted by $h^s(A)$, is defined on $\wp(\Omega)$, the class of all subsets of Ω, as

$$h^s(A) = \lim_{\delta \to 0} \inf \sum_{i=1}^{+\infty} |U_i|^s.$$

where the infimum is over all δ-covers $\{U_i\}$.

A subset A of Ω is called *measurable* with respect to the outer measure h^s if it decomposes every subset of Ω additively, that is if $h^s(E) = h^s(A \cap E) + h^s(E - A)$ for all sets $E \subseteq \Omega$.

Hausdorff s-dimensional outer measures are submodular, continuous from below and their restriction on the Borel σ-field is countably additive.

The *Hausdorff dimension* of a set A, $dim_H(A)$, is defined as the unique value, such that

$$h^s(A) = +\infty \text{ if } 0 \le s < dim_H(A),$$
$$h^s(A) = 0 \text{ if } dim_H(A) < s < +\infty.$$

Let **B** be a partition of Ω. For every $B \in \mathbf{B}$ denote by s the Hausdorff dimension of B and let h^s be the Hausdorff s-dimensional Hausdorff outer measure associated to the coherent upper conditional prevision $\overline{P}(X|B)$.

For every bounded random variable X a coherent upper conditional prevision $\overline{P}(X|B)$ is defined by the Choquet integral with respect to its associated Hausdorff outer measure if the conditioning event has positive and finite Hausdorff outer measure in its Hausdorff dimension. Otherwise if the conditioning event has Hausdorff outer measure in its Hausdorff dimension equal to zero or infinity it is defined by a 0–1 valued finitely, but not countably, additive probability.

Theorem 1. *Let (Ω, d) be a metric space and let **B** be a partition of Ω. For every $B \in \mathbf{B}$ denote by s the Hausdorff dimension of the conditioning event B and by h^s the Hausdorff s-dimensional outer measure. Let m be a 0–1 valued finitely additive, but not countably additive, probability on $\wp(B)$. Thus, for each $B \in \mathbf{B}$, the function defined on $\wp(B)$ by*

$$\overline{P}(A|B) = \frac{h^s(A \cap B)}{h^s(B)} if \ 0 < h^s(B) < +\infty$$

and by

$$\overline{P}(A|B) = m(A \cap B) if \ h^s(B) \in \{0, +\infty\}$$

is a coherent upper conditional probability.

If $B \in \mathbf{B}$ is a set with positive and finite Hausdorff outer measure in its Hausdorff dimension s the fuzzy measure μ_B^* defined for every $A \in \wp(B)$ by $\mu_B^*(A) = \frac{h^s(A \cap B)}{h^s(B)}$ is a coherent upper conditional probability, which is submodular, continuous from below and such that its restriction to the Borel σ-field is a Borel regular countably additive probability.

The coherent upper unconditional probability $\overline{P} = \mu_\Omega^*$ defined on $\wp(\Omega)$ is obtained for B equal to Ω.

In [8, Theorem 2] a new model of coherent upper conditional prevision is introduced and its properties are studied.

Theorem 2. *Let (Ω, d) be a metric space and let **B** be a partition of Ω. For every $B \in \mathbf{B}$ denote by s the Hausdorff dimension of the conditioning event B and by h^s the Hausdorff s-dimensional outer measure. Let m be a 0–1 valued finitely additive, but not countably additive, probability on $\wp(B)$. Then for each $B \in \mathbf{B}$ the functional $\overline{P}(X|B)$ defined on $L(B)$ by*

$$\overline{P}(X|B) = \frac{1}{h^s(B)} \int_B^{Cho} X dh^s \ if \ 0 < h^s(B) < +\infty$$

and by

$$\overline{P}(X|B) = m(XI_B) \ if \ h^s(B) \in \{0, +\infty\}$$

is a coherent upper conditional prevision.

When the conditioning event B has Hausdorff outer measure in its Hausdorff dimension equal to zero or infinity, an additive conditional probability is coherent if and only if it takes only $0 - 1$ values. Because linear previsions on $L(B)$ are uniquely determined by their restrictions to events, the class of linear previsions on $L(B)$ whose restrictions to events take only the values 0 and 1 can be identified with the class of $0 - 1$ valued additive probability defined on all subsets of B.

If the conditioning event B has positive and finite Hausdorff outer measure in its Hausdorff dimension in [8] the functional $\overline{P}(X|B)$ is proven to be monotone, comonotonically additive, submodular and continuous from below.

In the next theorem it is proven that on the class of continuous random variables the coherent upper conditional prevision defined in Theorem 2 is linear.

Theorem 3. *Let B be a set with positive and finite Hausdorff outer measure in its Hausdorff dimension s and let $K \subset L(B)$ be the class of continuous random variables. Then $\underline{P}(X|B) = \overline{P}(X|B)$ for every $X \in K$.*

Proof. The sets $[x; +\infty]$ for all $x \in \Re$ are Borelian sets since the class $\{[x; +\infty]x \in \Re\}$ generates the Borel σ-field so, for all $x \in \Re$ and $X \in K$, the weak upper level sets $\{X \geq x\}$ are Borelian sets since they are inverse images by a continuous function of the Borelian sets $[x, +\infty)$. Since the Borelian sets are measurable with respect to any Hausdorff outer measure we have that

$$G_{\mu_*,X}(x) = G_{\mu^*,X}(x)$$

so that

$$\overline{P}(X|B) = \frac{1}{h^s(B)} \int_B X dh^s = \frac{1}{h^s(B)} \int_0^{+\infty} G_{\mu^*,X}(x)dx =$$
$$\frac{1}{h^s(B)} \int_0^{+\infty} G_{\underline{\mu}_*,X}(x)dx = \frac{1}{h^s(B)} \int_B X dh_s = \underline{P}(X|B).\diamond$$

Corollary 1. *Let $X, Y \in K \subset L(B)$; by coherence we have*

$$\underline{P}(X + Y|B) \leq P(X + Y|B) = \frac{1}{h^s(B)} \int_B (X + Y)dh^s =$$
$$\frac{1}{h^s(B)} \int_B X dh^s + \frac{1}{h^s(B)} \int_B Y dh^s \leq \overline{P}(X + Y|B)$$

and by Theorem 3 we have $\underline{P}(X + Y|B) = P(X + Y|B) = P(X) + P(Y) = \overline{P}(X + Y|B)$.

In the next theorem it is proven that, if B is a set with positive and finite Hausdorff outer measure in its Hausdorff dimension then coherent upper and lower previsions of a bounded random variable, defined as Choquet integrals respectively with respect to the upper conditional prevision defined as in Theorem 2 and its conjugate, agree with the Lebesgue integral. For indicator functions it implies modularity except to a finite or countable set of points in B. It occurs because Hausdorff outer measures are regular and the weak upper level sets of any random variable are μ_B^*-measurable.

Theorem 4. *Let B be a set with positive and finite Hausdorff outer measure in its Hausdorff dimension s and denote by $G_{\mu_*,X}(x)$ and $G_{\mu^*,X}(x)$ respectively the decreasing distribution functions with respect to the coherent lower conditional probability $\underline{\mu}_B^*$ and the coherent upper conditional probability μ_B^* of a random variable X in $L(B)$. Then we have*

(i) $G_{\mu_*,X}(x) = G_{\mu^*,X}(x)$ except on a μ_B^* null set;
(ii) $\overline{P}(X|B) = \underline{P}(X|B)$.

Proof. (i) The coherent lower conditional probability $\underline{P} = \underline{\mu}_B^*$, which is the conjugate of $\overline{P} = \mu_B^*$ is defined in terms of Hausdorff inner measure h_s by

$$\underline{P} = \underline{\mu}_B^*(A) = \mu_B^*(B) - \mu_B^*(A^c) = \frac{h^s(B)}{h^s(B)} - \frac{h^s(A^cB)}{h^s(B)} = \frac{h^s(B)}{h_s(B)} - \frac{h^s(A^cB)}{h_s(B)} = \frac{h_s(AB)}{h_s(B)}.$$

The equalities hold since, by Theorem 4 of [10], each B is measurable with respect to $\overline{P}(\cdot|B)$, that is $h^s(B) = h_s(B)$.

We have that for every $x \in \Re$

$$G_{\mu_*,X}(x) = G_{\mu^*,X}(x)$$

except on a μ_B^*-null set.

To prove this equality, denote by μ the restriction of μ_B^* to the class of μ_B^*-measurable sets. Since Hausdorff outer measures are regular, for every weak upper level set $\{X \geq x\}$ there exists a μ_B^*-measurable set E such that

$$\mu_B^* \{X \geq x\} = \mu(E)$$

μ is σ-additive so that there is at most a countable number of sets $\{X \geq x\}$ with positive outer measure μ^*.

Let M^* be the chain of all weak upper level sets with positive outer measure μ_B^* and let ν the restriction of μ_B^* to M^*. By Theorem 2.10 and Corollary 3.2 of [6] ν is modular and there is a unique extension of ν to the field generated by M^*. This extension is σ-additive and regular and the sets belonging to M^* are μ_B^*-measurable, so that

$$G_{\mu_*,X}(x) = \underline{\mu}_B^* \{X \geq x\} = \mu_B^* \{X \geq x\} = G_{\mu^*,X}(x)$$

except on at most on a μ_B^* null set.

(ii) By (i) we have

$$\overline{P}(X|B) = \frac{1}{h^s(B)} \int_B X dh^s = \frac{1}{h^s(B)} \int_0^{+\infty} G_{\mu^*,X}(x)dx =$$

$$\frac{1}{h^s(B)} \int_0^{+\infty} G_{\mu_*,X}(x)dx = \frac{1}{h^s(B)} \int_B X dh_s = \underline{P}(X|B). \diamond$$

Corollary 2. *Let $X \in L(B)$ and let S be the field generated by the chain of the upper level sets of X and let $L^*(B)$ be the class of all upper S-measurable random variables on B. Let $Y \in L^*(B)$ by coherence we have*

$$\underline{P}(X+Y|B) \leq P(X+Y|B) = \frac{1}{h^s(B)} \int_B (X+Y)dh^s =$$

$$\frac{1}{h^s(B)} \int_B X dh^s + \frac{1}{h^s(B)} \int_B Y dh^s \leq \overline{P}(X+Y|B)$$

and by Theorem 4, since the upper level sets of Y are in S we have $\underline{P}(X+Y|B) = \overline{P}(X+Y|B)$; so we obtain that for every $X \in L(B)$ and $Y \in L^*(B)$

$$\underline{P}(X+Y|B) = \overline{P}(X+Y|B) = P(X+Y|B) = \frac{1}{h^s(B)} \int_B (X+Y)dh^s =$$

$$\frac{1}{h^s(B)} \int_B X dh^s + \frac{1}{h^s(B)} \int_B Y dh^s.$$

For indicator functions $X = I_A$ and $Y = I_C$ with $A \in \wp(B)$ and $C \in S$ the previous equality reduces to modularity except on a μ_B^* null set:

$$\mu_B^*(A \cup C) + \mu_B^*(A \cap C) = \mu_B^*(A) + \mu_B^*(C).$$

So μ_B^* is modular on S then it is additive and self-dual on S.

Remark 1. *The previous result is due to the fact that μ_B^* is regular and its restriction to the class of μ_B^*-measurable sets is countably additive.*

Theorem 5. *Let B be a set with positive and finite Hausdorff outer measure in its Hausdorff dimension then the coherent upper conditional prevision $\overline{P}(X|B)$, defined in Theorem 2 can be represented also by the Symmetric Choquet integral with respect to μ_B^*, that is*

$$\overline{P}(X|B) = \int^{Cho} X d\mu_B^* = \int^{S-Cho} X d\mu_B^*.$$

Proof. Since B is a set with positive and finite Hausdorff outer measure in its Hausdorff dimension then, by Theorem 2, the coherent upper conditional prevision $\overline{P}(X|B)$ is defined by the Asymmetric Choquet integral, which is equal to the Symmetric Choquet integral since $\mu_B^*(B) < +\infty$, μ_B^* is self-dual up to a set with μ_B^* zero probability and X^+ and X^- are comonotonic random variables, that is

$$\overline{P}(X|B) = \int^{Cho} X d\mu_B^* = \int^{Cho} X^+ d\mu_B^* + \int^{Cho} (-X^-) d\mu_B^*$$

$$= \int^{S-Cho} X^+ d\mu_B^* - \int^{S-Cho} (X^-) d\mu_B^* = \int^{S-Cho} X d\mu_B^*.\diamond$$

Example 3. *Let Ω be a finite set with cardinality equal to n then its Hausdorff dimension is 0 and the 0-dimensional Hausdorff measure is the counting measure. So the coherent probability $\mu_\Omega^* = \frac{h^0(A)}{h^0(\Omega)}$ is additive and for every bounded random variable X, such that $X(\omega_i) = x_i$ for $i = 1, ..., n$, we have that*

$$\int^{S-Cho} X d\mu_\Omega^* = \int^{Cho} X d\mu_\Omega^* = \int^{Cho} X dh^0 = \frac{1}{n} \sum_{i=1}^n x_i.$$

Example 4. *Let $\Omega = [0,1]$ and let $X(\omega) = \omega^2$. We have that*

$$\int^{Cho} X d\mu_\Omega^* = \int^{Cho} X dh^1 = \int_0^1 h^1 \{\omega \in \Omega : \omega^2 \geq x\} \, dx = \int_0^1 (1 - \sqrt{x}) \, dx = \frac{1}{3}.$$

By Theorem 5 we have that

$$\frac{1}{3} = \int^{Cho} X d\mu_\Omega^* = \int^{S-Cho} X \mu_\Omega^*.$$

Example 5. *Let $\Omega = [0,1]$ and let $X(\omega) = e^\omega$. We have that*

$$\int^{Cho} X d\mu_\Omega^* = \int^{Cho} X dh^1 = \int_0^1 h^1 \{\omega \in \Omega : e^\omega \geq x\} \, dx = e - 2.$$

By Theorem 5 we have that

$$e - 2 = \int^{Cho} X d\mu_\Omega^* = \int^{S-Cho} X \mu_\Omega^*.$$

Example 6. *Let $\Omega = [0, \frac{\pi}{2}]$ and let $X(\omega) = \sin(\omega)$. By Theorem 5 we have that*

$$\int^{S-Cho} X d\mu_\Omega^* = \int^{Cho} X \mu_\Omega^* = \int_0^1 h^1 \{\omega \in \Omega : \sin(\omega) \geq x\} \, dx = \frac{\pi}{2} + arcsin1 - 1.$$

In Example 2 the 0-dimensional Hausdorff outer measure is the counting measure and any subset of Ω is measurable with respect to it so by Corollary 1 the linearity holds for every random variable $X, Y \in L(\Omega)$.

In Examples 3, 4 and 5 the 1-dimensional Hausdorff outer measure coincide with the Lebesgue measure and by Corollary 1 we have that the linearity holds for every random variable X, Y measurable with respect to the σ-field of the Lebesgue measurable sets of $[0,1]$.

4   Conclusions

In this paper it is shown that a coherent upper conditional prevision of a bounded random variable can be represent as Asymmetric Choquet integral and Symmetric Choquet integral with respect to its associated Hausdorff outer measure and these integral representations coincide with the Lebesgue integral. This result is due to the fact that coherent upper conditional probabilities defined by Hausdorff outer measures are regular and self-dual up to a set with zero Hausdorff outer measure. Sufficient conditions for the linearity are given.

References

1. Billingsley, P.: Probability and Measure. Wiley, Hoboken (1986)
2. Brackett, M., Rivers, S., Solovey, P.: Emotional intelligence: implications for personal, social. academic, and workplace success. Soc. Pers. Psychol. Compass 5(1), 88–103 (2011)
3. Choquet, G.: Theory of capacity. Ann. Inst. Fourier **5**, 131–295 (1953)
4. de Finetti, B.: Probability, : Induction and Statistics. Wiley, New York (1970)
5. de Finetti, B.: Theory of Probability. Wiley, London (1974)
6. Denneberg, D.: Non-additive measure and integral. Kluwer Academic Publishers, Alphen aan den Rijn (1994)
7. Doria, S.: Probabilistic independence with respect to upper and lower conditional probabilities assigned by Hausdorff outer and inner measures. Int. J. Approximate Reasoning **46**, 617–635 (2007)
8. Doria, S.: Characterization of a coherent upper conditional prevision as the Choquet integral with respect to its associated Hausdorff outer measure. Ann. Oper. Res. **195**(1), 33–48 (2012)
9. Doria, S.: Symmetric coherent upper conditional prevision by the Choquet integral with respect to Hausdorff outer measure. Ann. Oper. Res. **229**(1), 377–396 (2014)
10. Doria, S.: On the disintegration property of a coherent upper conditional prevision by the Choquet integral with respect to its associated Hausdorff outer measure. Ann. Oper. Res. **216**(2), 253–269 (2017)
11. Doria, S., Dutta, B., Mesiar, R.: Integral representation of coherent upper conditional prevision with respect to its associated Hausdorff outer measure: a comparison among the Choquet integral, the pan-integral and the concave integral. Int. J. Gen Syst **216**(2), 569–592 (2018). Int. J. Gen Syst
12. Dubins, L.E.: Finitely additive conditional probabilities, conglomerability and disintegrations. Ann. Probab. **3**, 89–99 (1975)
13. Even, Y., Lehrer, E.: Decomposition-integral: unifying Choquet and concave integrals. Econ. Theory **56**, 33–58 (2014)
14. Falconer, K.J.: The Geometry of Fractal Sets. Cambridge University Press, Cambridge (1986)
15. Gabrisch, M., Labreuche, C.: The symmetric and asymmetric Choquet integrals on finite spaces for decision making. Stat. Pap. **43**(1), 37–52 (2002)
16. Grabisch, M.: Set Functions, Games and Capacities in Decision Making. Springer, Switzerland (2016). https://doi.org/10.1007/978-3-319-30690-2
17. Lehrer, E.: A new integral for capacities. Econ. Theor. **39**, 157–176 (2009)
18. Matte Blanco, I.: The Unconscious as Infinite Sets: An Essay on Bi-Logic. Gerald Duckworth, London (1975)

19. Mesiar, R., Mesiarová, A.: Fuzzy integrals and linearity. Int.J. Approx. Reason. **47**(3), 352–358 (2008)
20. Miranda, E., Zaffalon, M., de Cooman, G.: Conglomerable natural extensions. Int. J. Approximate Reasoning **53**(8), 1200–1227 (2012)
21. Regazzini, E.: Finitely additive conditional probabilities. Rend. Sem. Mat. Fis. Milano **55**, 69–89 (1985)
22. Regazzini, E.: De Finetti's coherence and statistical inference. Ann. Stat. **15**(2), 845–864 (1987)
23. Rogers, C.A.: Hausdorff Measures. Cambridge University Press, Cambridge (1970)
24. Seidenfeld, T., Schervish, M.J., Kadane, J.K.: Non-conglomerability for finite-valued, finitely additive probability. Indian J. Stat., Spec. Series A, Issue Bayesian Anal. **60**, 476–491 (1998)
25. Walley, P.: Coherent lower (and upper) probabilities, Statistics Research Report, University of Warwick (1981)
26. Walley, P.: Statistical Reasoning with Imprecise Probabilities. Chapman and Hall, London (1991)
27. Wang, Z., Klir, G.J.: Generalized Measure Theory. Springer, New York (2009). https://doi.org/10.1007/978-0-387-76852-6
28. Yang, Q.: The pan-integral on fuzzy measure space. Fuzzy Math. **3**, 107–114 (1985)

Separable Qualitative Capacities

Didier Dubois[1(✉)], Francis Faux[1], Henri Prade[1], and Agnès Rico[2]

[1] IRIT, CNRS and University of Toulouse, 31062 Toulouse cedex 9, France
dubois@irit.fr
[2] ERIC, Université Claude Bernard Lyon 1, 69100 Villeurbanne, France

Abstract. The aim of this paper is to define the counterpart of separable belief functions for capacities valued on a finite totally ordered set. Evidence theory deals with the issue of merging unreliable elementary testimonies. Separable belief functions are the results of this merging process. However not all belief functions are separable. Here, we start with a possibility distribution on the power set of a frame of discernment that plays the role of a basic probability assignment. It turns out that *any* capacity can be induced by the qualitative counterpart of the definition of a belief function (replacing sum by max). Then, we consider a qualitative counterpart of Dempster rule of combination applied to qualitative capacities, via their qualitative Möbius transforms. We study the class of capacities, called separable capacities, that can be generated by applying this combination rule to simple support capacities, each focusing on a subset of the frame of discernment. We compare this decomposition with the one of general capacities as a maximum over a finite set of necessity measures. The relevance of this framework to the problem of information fusion from unreliable sources is explained.

Keywords: Qualitative capacity · Evidence theory
Possibility theory · Information fusion

1 Introduction

Shafer's evidence theory [21] essentially deals with the fusion of information items stemming from several more or less reliable testimonies (or sources). In their most basic forms, these unreliable information items take the form of subsets of a frame of discernment (supposed to contain the value of the parameter of interest) along with weights representing the extent to which the testimonies are credible. Shafer uses as a mathematical model a positive probability distribution, called basic probability assignment, over a family of subsets said to be focal. The total quantity of belief in a particular subset is represented by the belief function and the Möbius transform allows to recover the basic probability assignment from the belief function in a univocal manner.

In this paper we focus on the representation and the management of uncertainty in information of non quantitative nature. Indeed much knowledge is qualitative, often expressed verbally or diagrammatically. We are interested in the

© Springer Nature Switzerland AG 2018
D. Ciucci et al. (Eds.): SUM 2018, LNAI 11142, pp. 124–139, 2018.
https://doi.org/10.1007/978-3-030-00461-3_9

qualitative counterpart of Shafer's evidence theory [21], where the basic probability assignment is turned into a basic possibility assignment whose weights have 1 as maximum. This idea of a basic possibility assignment dates back to a suggestion made in [12,13]. Since then, some authors have been interested by this qualitative counterpart. For example, we can mention the papers [1,2,15,25]. In [15], any monotonic set function (also called *fuzzy measure* [24]) is shown to be equal to the maximum of necessity measures and to the minimum of possibility measures. In [25], upper and lower possibilities and necessities are expressed in terms of basic possibility assignments or of necessity assignments. The interest of such possibility assignments have been already suggested in the context of information fusion [1].

We consider a set of sources informing about the value of a parameter x of interest. Each source delivers a proposition of the form $x \in A$, where A is a subset of a frame of discernment Ω. An associated weight $\sigma(A)$, belonging to a symbolic totally ordered scale, reflects the *credibility* of this elementary testimony. This is modelled by a basic possibility assignment whose weights are such that

- $\sigma(A) < 1$ (where 1 is the top of the scale) if A differs from Ω;
- $\sigma(\Omega) = 1$ (under this condition, σ is said to be *non-dogmatic*);
- $\sigma(B) = 0$ for $B \neq A, \Omega$ otherwise.

Mind that $\sigma(B) = 0$ refers to the absence of support in favor of B (not its impossibility), and the greater $\sigma(A)$, the greater the support in favor of A. In other words σ is *formally* a possibility assignment because it is normalized in the sense of a possibility distribution, but its semantics is not exactly in agreement with possibility theory. A mapping σ is also the qualitative counterpart of the basic probability assignment of a simple support belief function. In this paper we consider fuzzy measures obtained from the qualitative merging of such unreliable elementary testimonies.

In the numerical setting, simple support functions are important in Shafer's theory as they model unreliable elementary testimonies, the fusion of which (using Dempster rule of combination) yields so-called *separable belief functions*. Here we study the qualitative counterpart of separable belief functions (introduced in [6]). Namely we give necessary and sufficient conditions under which a fuzzy measure can be decomposed in terms of unreliable testimonies, and check whether this decomposition is unique. Moreover we start discussing the relevance of this framework for information fusion. In many situations such as risk analysis, experts express their knowledge about the likelihood of dreadful events verbally, and these terms are then translated into numbers on a probability scale. This paper starts an investigation on the possibility of directly expressing and merging qualitative information of this kind.

The organization of the rest of the paper is as follows. In Sect. 2, basic possibility assignments and their combination by means of a qualitative counterpart of Dempster rule of combination are recalled. In Sect. 3 we apply these notions to the combination of non-dogmatic capacities via their qualitative Möbius transforms. Section 4 lays bare the class of separable qualitative capacities that are the result of combining, by means of the qualitative Dempster rule, non-dogmatic necessity functions induced by their focal sets. This combination is compared to the decomposition of any capacity as a maximum of necessity functions. We show that any capacity has a separable approximation that dominates it. Section 5 applies these results to the fusion of unreliable testimonies. It bridges the gap with Belnap set-ups for inconsistent information handling and his 4-valued logic on the one hand, and the method of maximal consistent subsets on the other hand.

2   Framework and Notations

In evidence theory, a basic probability assignment m is used, which is a probability distribution over the power set of Ω, such that $m(\emptyset) = 0$. A belief function $Bel(A) = \sum_{\emptyset \neq B \subseteq A} m(B)$, which represents the total quantity of belief in the subset A of Ω, is associated with only one mass function m computed from Bel using the Möbius transform.

2.1   Basic Possibility Assignments

Let L be a totally ordered scale with a bottom 0 and a top 1. Given a frame of discernment Ω, information is represented by means of a mapping σ, called *basic possibility assignment*, from 2^Ω to L such that $\max_{A \subseteq \Omega} \sigma(A) = 1$ (top normalization). If we stick to Shafer's view, the σ function should reflect a set of unreliable testimonies, whereby each claim of the form $x \in A$ is weighted by $\sigma(A)$.

In this paper, in agreement with Shafer theory, we assume bottom-normal possibility assignments, i.e., $\sigma(\emptyset) = 0$. Similarly to Shafer theory, a basic possibility assignment σ defines the following set function:

$$\forall A \subseteq \Omega, \; Bel^{pos}(A) = \max_{B \subseteq A} \sigma(B). \tag{1}$$

Note that $Bel^{pos}(\emptyset) = 0$, $Bel^{pos}(\Omega) = 1$. It was very early pointed out [13] that the set function Bel^{pos} can be any fuzzy measure or capacity, that is a monotonic set function v on Ω, such that $v(A) \geq v(B)$ whenever $B \subseteq A$. Conversely, any fuzzy measure v can be put in the form $v(A) = \max_{B \subseteq A} \sigma(B)$ for a basic possibility assignment σ. In particular, the qualitative Möbius transform [9,18] of a fuzzy measure v is defined as follows:

Definition 1. *Let* $v : 2^\Omega \to L$ *be a fuzzy measure. Its qualitative Möbius transform is a set-function* v_* *defined by*

$$\forall A \subseteq \Omega, v_*(A) = \begin{cases} v(A) \text{ if } \not\exists B \subset A \text{ s.t. } v(B) = v(A), \\ 0 \text{ otherwise.} \end{cases}$$

Example 1. Consider a Boolean capacity v with range in $\{0,1\}$ and its qualitative Möbius transform v_*.

	$\{\omega_1\}$	$\{\omega_2\}$	$\{\omega_3\}$	$\{\omega_1,\omega_2\}$	$\{\omega_1,\omega_3\}$	$\{\omega_2,\omega_3\}$	Ω
v	1	0	1	1	1	1	1
v_*	1	0	1	0	0	0	0

Clearly, we have $v(A) = \max_{B \subseteq A} v_*(B)$. Like in evidence theory, the sets A such that $v_*(A) > 0$ are called focal sets. Note that for all focal sets A and B such that $B \subset A$ we have $v_*(B) < v_*(A)$. The focal set A can be viewed as the minimal set in the sense of inclusion with weight $v(A)$. Like in evidence theory, a capacity v is said to be *non-dogmatic* if Ω is a focal set of v, i.e., $v_*(\Omega) = 1$ (indeed, $0 < v_*(\Omega) < 1$ is forbidden, since then $v_*(A) < 1, \forall A \subseteq \Omega$, which would imply $v(\Omega) < 1$).

The unicity of the basic probability assignment generating a belief function is no longer satisfied in the qualitative setting as several basic possibility assignments σ may yield the same capacity v such that $v(A) = \max_{B \subseteq A} \sigma(B)$. There is actually a whole family of set functions (basic possibility assignments) generating v, namely:

$$\Sigma(v) = \{\sigma : \forall A \subseteq \Omega, v(A) = \max_{B \subseteq A} \sigma(B)\}$$

and they are such that $\sigma \in \Sigma(v)$ if and only if $\sigma \in [v_*, v]$. An equivalence relation on basic possibility assignments $\sigma_1 \sim \sigma_2$ can be defined if and only if $Bel^{pos}_{\sigma_1} = Bel^{pos}_{\sigma_2}$. The equivalence classes of relation $\sim$ are of the form $\Sigma(v)$ for all capacities v. Namely, $\sigma_1 \sim \sigma_2$ to mean there is v such that $\sigma_1 \in \Sigma(v)$ and $\sigma_2 \in \Sigma(v)$.

Remark 1. In a counterpart of Dempster's approach [7], the σ function is the result of mapping a set U equipped with a possibility distribution π to the set Ω via a multivalued mapping $\Gamma : U \to 2^\Omega$, whereby $\sigma(A) = \max_{\Gamma(u)=A} \pi(u)$[13]. The understanding of Γ is as follows: each value $u \in U$ is compatible with and only with some value $w \in \Gamma(u)$. Considering a selection function $f : U \to \Omega$ of Γ ($f(u) \in \Gamma(u), \forall u \in U$), it is clear that the possibility distribution $\pi_f(w) = \max_{f(u)=w} \pi(u)$ yields a possibility measure Π_f such that $Bel^{pos}(A) \leq \Pi_f(A) \leq Pl^{pos}(A) = \max_{\emptyset \neq A \cap B} \sigma(B)$. In other words, the imprecision due to Γ yields an imprecise possibility measure. The peculiarity of Dempster-like upper and lower possibilities is that a "lower possibility" is just any capacity, while an "upper possibility" is just a possibility measure, contrary to the quantitative case, where it is a more general set function. An upper possibility is the possibility measure whose possibility distribution is the contour function $\pi_v(w) = \max_{w \in E} v_*(E)$ of $v = Bel^{pos}$. So, Pl^{pos} is not the conjugate of Bel^{pos} (i.e., $Pl^{pos}(A) \neq 1 - Bel^{pos}(A^c)$, for the complement A^c of A). In the qualitative setting there is a disconnection between the notion of conjugate and the definition of the upper possibility. This is explained in more details in [15]. The Dempster approach to qualitative possibility can be viewed as defining imprecise possibilities, and it encompasses standard possibilistic representations as a particular case. This view of capacities as lower possibilities is at odds with the framework of our paper, where a capacity is a kind of measure of support.   □

2.2 Dempster-Like Combination of Basic Possibilistic Assignments

In Shafer's evidence theory, the well-known Dempster rule of combination [21] has a counterpart in our qualitative maxitive setting first suggested in [19]:

Definition 2. Let σ_1 and σ_2 be two basic possibility assignments. The conjunctive combination rule is defined by
$$\forall A \subseteq \Omega, \ (\sigma_1 \otimes \sigma_2)(A) = \max_{B \cap C = A} \min(\sigma_1(B), \sigma_2(C)).$$

However the conjunctive combination of equivalent basic possibility assignments (generating the same fuzzy measure) may not yield equivalent basic possibility assignments: $\sigma_1 \sim \tau_1$ and $\sigma_2 \sim \tau_2$ do not imply $\sigma_1 \otimes \sigma_2 \sim \tau_1 \otimes \tau_2$.

Example 2. Consider A, B, C with $\sigma_1(\Omega) = 1, \sigma_1(A) = \alpha > \sigma_1(A \cap B) = \beta$ where $A \cap B \neq \emptyset$, and $\sigma_1(E) = 0$ otherwise. Let $\tau_1 = \sigma_1$ but for $\tau_1(B) = \beta$. Lastly let $\sigma_2(C) = \gamma$, with $B \cap C \neq \emptyset, \sigma_2(\Omega) = 1$ and $\sigma_2(E) = 0$ otherwise. Clearly, $\sigma_1 \sim \tau_1$. Suppose $A \cap C = \emptyset$. Note that $(\sigma_1 \otimes \sigma_2)(B \cap C) = 0$ and it yields a capacity v_{12} such that $v_{12}(B \cap C) = 0$. However $(\tau_1 \otimes \sigma_2)(B \cap C) = \min(\beta, \gamma)$ yielding a capacity v'_{12} such that $v'_{12}(B \cap C) = \min(\beta, \gamma)$.   □

The set function $\sigma_1 \otimes \sigma_2$ is generally not even a basic possibility assignment because $\sigma_1 \otimes \sigma_2$ induces a monotonic set-function such that $(\sigma_1 \otimes \sigma_2)(\emptyset) \neq 0$ may occur. We may even get $(\sigma_1 \otimes \sigma_2)(\emptyset) = 1$ if the two possibility distributions σ_1 and σ_2 bear on disjoint subsets, which makes the combination ineffective. $\sigma_1 \otimes \sigma_2$ is thus a basic possibility assignment provided that there does not exist B and C such that $B \cap C = \emptyset$ and $\sigma_1(B) > 0, \sigma_2(C) > 0$.

In order to respect the closure property for this combination rule, we can, in conformity with evidence theory,

- either consider a more general class of monotonic set functions than capacities whereby $v(\emptyset) > 0$ is allowed. However it is not clear what it means.
- or modify the combination rule by bottom-renormalizing the result.

In [6], the bottom normalization condition $(\sigma_1 \otimes \sigma_2)(\emptyset) = 0$ is enforced and added to Definition 2. In the following of the paper we use this bottom-normalized conjunctive rule denoted by $\hat{\otimes}$.

Definition 3. Let σ_1 and σ_2 be two basic possibility assignments. The bottom-normalized conjunctive combination rule is defined by

$$(\sigma_1 \hat{\otimes} \sigma_2)(A) = (\sigma_1 \otimes \sigma_2)(A) \text{ if } A \neq \emptyset, \text{ and } (\sigma_1 \hat{\otimes} \sigma_2)(\emptyset) = 0.$$

The bottom-normalized combination rule is commutative; it possesses an identity: the vacuous basic possibilistic assignment σ_0, equal to 0 everywhere except on Ω ($\sigma_0 \otimes \sigma = \sigma$ for all σ); it is associative (even if not proved in [6]):

- if $A \neq \emptyset$, then $(\sigma_1 \hat{\otimes} \sigma_2) \hat{\otimes} \sigma_3(A) = (\sigma_1 \otimes \sigma_2) \otimes \sigma_3(A)$ and $\otimes$ is associative [13];
- $(\sigma_1 \hat{\otimes} \sigma_2) \hat{\otimes} \sigma_3(\emptyset) = (\sigma' \hat{\otimes} \sigma_3)(\emptyset) = 0$ and $\sigma_1 \hat{\otimes} (\sigma_2 \hat{\otimes} \sigma_3)(\emptyset) = (\sigma_1 \hat{\otimes} \sigma'')(\emptyset) = 0$.

However, the consequence of this new definition is that we may fail to preserve top-normalization via combination, when there are no B and C such that $\sigma_1(B) = \sigma_2(C) = 1$ with $B \cap C \neq \emptyset$. Moreover, if the two possibility distributions σ_1 and σ_2 bear on disjoint subsets, we may even have $(\sigma_1 \hat{\otimes} \sigma_2)(A) = 0$ for all $A \neq \emptyset$. This inconvenient does not appear if we restrict to basic possibilistic assignments such $\sigma(\Omega) = 1$, generating non-dogmatic fuzzy measures, which we shall assume in the sequel.

3   Conjunctive Combination of Qualitative Capacities

In the following, we shall apply the combination rule to non-dogmatic capacities v via their Möbius transforms. Let us start with simple support capacities. The definition of simple support functions present in the Dempster Shafer theory can be adapted to the qualitative setting.

Definition 4. *A capacity $v : 2^\Omega \to L$ is a simple support function (SSF) focused on a set S if and only if*

$$v(A) = \begin{cases} 0 & \text{if } S \not\subseteq A \\ s & \text{if } S \subseteq A \text{ but } A \neq \Omega \\ 1 & \text{if } A = \Omega \end{cases}$$

An SSF focused on S is non-dogmatic and is clearly a necessity measure we denote by N_S. The qualitative Möbius transform of such a simple support function N_S is a basic possibility assignment of the form:

$$N_{S*}(A) = \begin{cases} s & \text{if } A = S \\ 1 & \text{if } A = \Omega \\ 0 & \text{otherwise.} \end{cases}$$

Consider two simple support functions (SSFs) N_A and N_B where the focal set weights are respectively $N_{A*}(A) = \alpha \geq N_{B*}(B) = \beta$. Then the result of the combination $N_{A*} \hat{\otimes} N_{B*}$ is a basic possibility assignment σ_{AB} such that

$$\sigma_{AB}(A \cap B) = \begin{cases} \beta & \text{if } A \cap B \neq \emptyset, \\ 0 & \text{otherwise (enforced value).} \end{cases}$$

$$\sigma_{AB}(A) = \alpha, \ \sigma_{AB}(B) = \beta, \ \sigma_{AB}(\Omega) = 1.$$

The combination of Möbius transforms is not necessarily a Möbius transform as the resulting capacity v_{AB} does not have focal set B if $A \cap B \neq \emptyset$ and $\alpha > \beta$. This capacity v_{AB} thus has focal sets that depend upon A, B, α, β:

- $v_{AB*}(A \cap B) = \beta, v_{AB*}(A) = \alpha, v_{AB*}(\Omega) = 1$ if $A \cap B \neq \emptyset, A \not\subseteq B$ and $\alpha > \beta$ (since $v_{AB*}(B) = 0$);

- $v_{AB*}(A \cap B) = \alpha, v_{AB*}(\Omega) = 1$ if $A \cap B \neq \emptyset$ and $\alpha = \beta$;
- $v_{AB*}(A) = \alpha, v_{AB*}(B) = \beta, v_{AB*}(\Omega) = 1$ if $A \cap B = \emptyset$;
- When A and B are nested $(A \subset B)$ then $N_{A*} \hat{\otimes} N_{B*} = N_{A*}$ if $\beta \leq \alpha$.

The result of the conjunctive combination rules depend on the basic possibility assignments used to represent the capacity. If we combine possibilistic basic assignments σ_1, σ_2 that respectively generate capacities v_1 and v_2 and that differ from the Möbius transforms, the result of the combination will be different if we use σ_1, σ_2 and if we use Möbius transforms, as already seen in Example 2.

Example 3. We consider two SSFs. Suppose $A' \supset A$ and $B' \supset B$ and let $\sigma_1 = N_{A*}$ but for $\sigma_1(A') = \alpha$ and likewise $\sigma_2 = N_{B*}$, but for $\sigma_2(B') = \beta$. In the case when $A \cap B = \emptyset$ but $A' \cap B' \neq \emptyset$, we may have that $(N_{A*} \hat{\otimes} N_{B*})(A' \cap B') = 0$, while $(\sigma_1 \hat{\otimes} \sigma_2)(A' \cap B') = \min(\alpha, \beta)$.

If we define the conjunctive combination of two capacities via operation $\hat{\otimes}$ applied to their Möbius transforms, this lack of invariance of the combination rule for equivalent basic possibility assignments entails the following consequence: when combining v_1, v_2, v_3 as (i) first obtaining v_{12} from the basic possibilistic assignment $v_{1*} \hat{\otimes} v_{2*}$, then (ii) combining v_{12} and v_3 by computing $v_{12*} \hat{\otimes} v_{3*}$, the result may differ from the capacity derived from $\sigma_{123} = (v_{1*} \hat{\otimes} v_{2*}) \hat{\otimes} v_{3*}$, since $v_{1*} \hat{\otimes} v_{2*}$ may differ from v_{12*}.

This state of facts forces us to define the conjunctive combination of more than two (say k) capacities v_i by combining their Möbius transforms via Definition 2 in one step, avoiding the issues of associativity and lack of stability with respect to $\sim$.

Definition 5. *The conjunctive combination of any k-tuple of capacities v_i, each with focal sets A_i, consists in first computing the basic posssibilistic assignment*

$$\sigma_{\hat{\otimes}}(A) = \begin{cases} \max\{\min(v_{1*}(A_1), \ldots, v_{k*}(A_k)) : A_1, \ldots, A_k \text{ s. t. } \bigcap_{i=1}^{k} A_i = A \neq \emptyset\} \\ 0 \qquad\qquad\qquad\qquad\qquad\qquad\qquad\qquad\qquad\qquad\qquad\quad otherwise. \end{cases}$$

and the resulting capacity $v(A) = \max_{E \subseteq A} \sigma_{\hat{\otimes}}(E)$ is denoted by $v = \hat{\otimes}_{i=1}^{k} v_i$. We then call $(v_1, \ldots, v_n)$ a conjunctive decomposition of v.

If the v_i's are SSFs N_{S_i}, each focused on a subset S_i with weight α_i, the result of their conjunctive combination can be computed as follows. Note that $\sigma_{\hat{\otimes}}(A) > 0$ if and only if $A = \bigcap\{S_i \in \mathcal{T}\} \neq \emptyset$ for some family $\mathcal{T} \subseteq \{S_1, \ldots, S_k\}$ of overlapping subsets (we use $v_{i*}(S_i) = \alpha_i$ if $S_i \in \mathcal{T}$, $v_{i*}(\Omega) = 1$ and $v_{i*}(A_i) = 0$ otherwise). Then if $A \neq \Omega$,

$$\sigma_{\hat{\otimes}}(A) = \max_{\mathcal{T}: A = \bigcap\{S_i \in \mathcal{T}\}} \min_{S_i \in \mathcal{T}} \alpha_i$$

and $\sigma_{\hat{\otimes}}(A) = 0$ otherwise. Moreover $\sigma_{\hat{\otimes}}(\Omega) = 1$.

Let $\mathcal{K}_j, j = 1, \ldots, p$ be the set of maximal families of overlapping subsets of $\{S_1, \ldots, S_k\}$. The focal sets of $\hat{\otimes}_{i=1}^{k} N_{S_i}$ are thus only among the non-empty sets $\bigcap\{S_i \in \mathcal{T} \subseteq \mathcal{K}_j\}$ for some j, including sets S_i themselves. In particular, all sets of the form $\bigcap\{S_i \in \mathcal{K}_j\}, j = 1, \ldots, p$ are (disjoint) focal sets of $\hat{\otimes}_{i=1}^{k} N_{S_i}$.

4 Separable Non-dogmatic Capacities

A capacity v is said to be *separable* if and only if $v = \hat{\otimes}_{i=1}^{k} N_{S_i}$ for some SSFs. Note that in this case, v is non-dogmatic since Ω is a focal set. Each SSFs is viewed as a testimony of the form x *is* S_i whose reliability is measured by α_i.

4.1 Characterization of Separability

In the previous section, it can be seen that the capacity obtained by merging two SSFs respectively focused on sets A and B is such that it has at most three focal sets and each pair of such focal sets is either nested (e.g., $A \cap B$ and A) or disjoint (e.g., A and B if disjoint). This property holds when merging more than two SSF's and is formally expressed as follows.

Definition 6. *A family of sets $\mathcal{F}$ is said to be* disjoint-nested *if and only if the following condition holds*

$$\forall A, B \in \mathcal{F}, \ \text{either } A \cap B = \emptyset \text{ or } A \subseteq B \text{ or } B \subseteq A; \tag{2}$$

We can now formulate the main result of the paper:

Theorem 1. *Let $v : 2^{\Omega} \to [0,1]$ be a capacity and $\mathcal{F}_v$ be the set of the focal sets of v. The three following properties are equivalent:*

1. *v is separable;*
2. *$\Omega \in \mathcal{F}_v$ and $\mathcal{F}_v$ is disjoint-nested.*
3. *$v = \hat{\otimes}_{S \in \mathcal{F}_v} N_S$.*

Proof. $1 \Rightarrow 2$: We consider a separable capacity v. Since $v = \hat{\otimes}_{i=1}^{k} N_{S_i}$, suppose $A, B \in \mathcal{F}_v$ with $A \cap B \neq \emptyset$. It means that $A = \bigcap\{S_i \in \mathcal{T}_A\}$ and $B = \bigcap\{S_i \in \mathcal{T}_B\}$ for some families of overlapping subsets $\mathcal{T}_A, \mathcal{T}_B$ of $\{S_1, \ldots, S_k\}$. Suppose neither $A \subseteq B$ nor $B \subseteq A$ hold. Then the set $A \cap B$ is of the form $\bigcap\{S_i \in \mathcal{T}_A \cup \mathcal{T}_B\}$. In otherwords $\sigma_{\hat{\otimes}}(A \cap B) = \min(\sigma_{\hat{\otimes}}(A), \sigma_{\hat{\otimes}}(B))$. As a consequence either $A \notin \mathcal{F}_v$ (if $\sigma_{\hat{\otimes}}(A \cap B) = \sigma_{\hat{\otimes}}(A)$) or $B \notin \mathcal{F}_v$ (if $\sigma_{\hat{\otimes}}(A \cap B) = \sigma_{\hat{\otimes}}(B)$).

$2 \Rightarrow 3$: Consider $v' = \hat{\otimes}_{S \in \mathcal{F}_v} N_S$, with $N_{S*}(S) = v_*(S)$. Let us show that $v' = v$ under the condition (2). Suppose $A \in \mathcal{F}_v$. Then, by (2), $\forall B \in \mathcal{F}_v$, $A \cap B = \emptyset, B$, or A. So when computing $\sigma_{\hat{\otimes}}(A)$ one can only use families of sets $S_1, \ldots S_k \in \mathcal{F}_v$ such that $\cap_{i=1}^{k} S_i = A$ where each S_i contains A, and one of them is A, due to the condition (2), where $v_*(S_i) > v_*(A)$ if $S_i \neq A$. So, $\sigma_{\hat{\otimes}}(A) = v_*(A)$. If $A \notin \mathcal{F}_v$ it cannot be such that $\cap_{i=1}^{k} S_i = A$ for any $A_1, \ldots, A_k \in \mathcal{F}_v$ due to the condition (2). So, $\sigma_{\hat{\otimes}} = v_*$ and $v' = v$.

$3 \Rightarrow 1$: It is obvious. □

Remark 2. The decomposition of capacities into a combination of simple support functions is not unique. This is because if $A \subset B$ then $N_A \hat{\otimes} N_B = N_A$ if $N_B(B) \leq N_A(A)$. So we can artificially add SSFs to the decomposition of $v = \hat{\otimes}_{S \in \mathcal{F}_v} N_S$. If v is separable, then its decomposition into a combination of simple support functions based on its focal sets is minimal.

The structure of the set of focal sets of a separable capacity is very peculiar. Going top down, Ω is focal with degree 1. Then we may have disjoint focal sets, each containing a nested sequence of focal sets. In each sequence, the smallest set may also contain disjoint focal sets, and so on, recursively. In other words, for any focal set A, the set of focal sets B that contain A (if any) forms a chain of nested sets, which is another way to express the necessary and sufficient condition (2) for a capacity to be separable. Numerical belief functions with disjoint-nested focal sets have been studied in the literature. Walley [26] showed that they are the only ones whose combination is in agreement with the likelihood principle. They are also studied by Giang and Shenoy [16] in the framework of dynamic decision under uncertainty.

4.2   Non-separable Capacities

As shown in [15], each qualitative capacity v is the maximum of necessity measures: $v(A) = \max_{i=1}^{m} N_i(A)$. This decomposition is different from the one defined by the separability property using $\hat{\otimes}$. However they coincide for separable capacities.

Proposition 1. *A capacity v is separable if and only if $v = \max_{S \in \mathcal{F}_v} N_S$.*

Proof. Suppose $v = \hat{\otimes}_{S \in \mathcal{F}_v} N_S$. This is equivalent to have $v(B) = \max_{A \subseteq B} \sigma_{\hat{\otimes}}(A)$, where for $A \neq \emptyset$, $\sigma_{\hat{\otimes}}(A) = \max_{T \subseteq \mathcal{F}_v : A = \cap\{S \in T\}} \min_{S \in T} v_*(S)$. Since A is a focal set of a separable v, $\sigma_{\hat{\otimes}}(A) = v_*(A)$. Equivalently, $v(B) = \max_{S \subseteq B, S \in \mathcal{F}_v} v_*(S) = \max_{S \in \mathcal{F}_v} N_S(B)$. $\square$

Contrary to the numerical case of separable belief functions, the separability of a capacity v does not impose that the family of focal sets $\mathcal{F}_v$ is closed under intersection. For instance, the non-dogmatic capacity with focals such that $v_*(A) = \alpha > v_*(B) = \beta > v_*(A \cap B) = \gamma$ is not separable as $\hat{\otimes}(N_A, N_B, N_{A \cap B})$ since the latter is the necessity measure N with $N_*(A) = \alpha$, $N_*(A \cap B) = \beta$ (indeed $\sigma_{\hat{\otimes}}(A \cap B) = \beta$, obtained by combining A with weight α, B with weight β and Ω from $N_{A \cap B}$). Note that $N > v$. This property holds for all non-separable capacities.

Proposition 2. *Suppose v is not separable, and let $\hat{v} = \hat{\otimes}_{S \in \mathcal{F}_v} N_S$. Then $\hat{v} > v$.*

Proof. $\hat{v}$ is a separable capacity whose family of focal sets $\mathcal{F}_{\hat{v}}$ contains only some non-empty intersections of focal sets of v. We have, for $A \neq \emptyset$,

$$\sigma_{\hat{\otimes}}(A) = \max_{T \subseteq \mathcal{F}_v, \cap\{S \in T\} = A} \min_{S \in T} v_*(S) \geq v_*(A) \text{ if } A \in \mathcal{F}_v \text{ letting } T = \{A\}.$$

It is clear that $\sigma_{\hat{\otimes}}(A) \geq v_*(A)$ if $A \notin \mathcal{F}_v$. Then we have $\hat{v}(A) = \max_{E \subseteq A} \sigma_{\hat{\otimes}}(E) \geq \max_{E \subseteq A} v_*(E) = v(A)$. As v is not separable, $\hat{v}(A) > v(A)$ for some A. $\square$

So, each non-decomposable capacity possesses a separable upper approximation.

4.3 Examples of Separable Capacities

It is clear that non-dogmatic necessity measures N are separable since $\mathcal{F}_N$ is nested. Likewise, capacities whose set of focal sets contains only disjoint subsets, on top of Ω, are separable. Note that possibility measures are not separable because they are dogmatic since their focal sets are all singletons. However, non-dogmatic capacities v_π whose focal sets are singletons *but for* Ω are separable. Namely, there exists a subnormal possibility distribution π ($\max_{w \in \Omega} \pi(w) < 1$), such that v_π is defined by $v_\pi(A) = \Pi(A), A \neq \Omega$, and 1 otherwise. v_π is called a *pseudo-possibility measure* and is such that $v_\pi(A \cup B) = \max(v_\pi(A), v_\pi(B))$ if $A \cup B \neq \Omega$.

Dually, the conjugate of a pseudo-possibility measure is a capacity such that $v(A \cap B) = \min(v(A), v(B))$ whenever $A \cap B \neq \emptyset$, which can be called a *pseudo-necessity measure*.

Proposition 3. *The focal sets of a pseudo-necessity measure contain only one sequence of nested sets plus singletons with the same Möbius value.*

Proof. Let v be a pseudo-necessity measure. First define $\epsilon_v = \min_{w \in \Omega} v(\{w\})$. If $\epsilon_v = 0$ then v is a necessity measure since if $A \cap B = \emptyset$, then $\min(v(A), v(B)) \leq \min(v(A \cup \{w\}), v(B \cup \{w\})) = v(\{w\}) = 0$. Suppose $\epsilon_v > 0$. Note that $\forall w \in \Omega$, $v(\{w\}) \geq \epsilon_v$, and there can be only one w, $v(w) > \epsilon_v$. Indeed, if $v(\{w_i\}) > \epsilon_v$, for $i = 1, 2$, then we get the inequality $\min(v(\{w_1, w\}), v(\{w_2, w\})) > v(\{w\}) = \epsilon_v$.

Let N_v be the capacity defined by $N_v(A) = v(A)$ if $v(A) > \epsilon_v$, and 0 otherwise. It is clear that N_v is a necessity measure since $N_v(A \cap B) = \min(N_v(A), N_v(B))$ if $v(A \cap B) > \epsilon_v$; if $v(A \cap B) = \epsilon_v$ then $v(A \cap B) = v(A)$ or $v(B)$, so $N_v(A \cap B) = \min(N_v(A), N_v(B)) = 0$. The case when $A \cap B = \emptyset$ and $\min(v(A), v(B)) > \epsilon_v$ is impossible since $\min(v(A), v(B)) \leq \min(v(A), v(A^c)) \leq \min(v(A \cup \{w\}), v(A^c \cup \{w\})) = \epsilon_v$. Let the focal sets of N_v form the nested sequence $A_0 = \Omega \supset A_1 \supset \cdots \supset A_p$ with weights $v(\Omega) = 1 > \alpha_1 > \cdots > \alpha_p > \epsilon_v$, where $\alpha_i = v(S_i)$. The capacity v can be reconstructed for $A \neq \emptyset$ as

$$v(A) = \begin{cases} N_v(A) & \text{if } N_v(A) > 0, \\ \epsilon_v & \text{otherwise.} \end{cases}$$

Its focal sets are thus $\{\Omega, A_1 \supset \cdots \supset A_p\} \cup \{\{w\} : w \in \Omega\}$, where $v_*(\{w\}) = \epsilon_v$. They are disjoint-nested, i.e., clearly satisfy the separability condition (2).   □

Corollary 1. *A pseudo-necessity function is separable.*

But not all separable capacities take this form, since pseudo-necessities must have only one nontrivial chain of non-singleton focal sets.

Example 4. The non-dogmatic capacity v with $v(E_i) = \alpha_i > 0, i = 1, 2$ with $E_1 \cap E_2 = \emptyset$ is separable: $v = N_{E_1} \hat{\otimes} N_{E_2}$. However if there are $A_i \supset E_i, i = 1, 2$ with $A_1 \cap A_2 \neq \emptyset$, then $v(A_1 \cap A_2) = 0$, but $v(A_i) > 0, i = 1, 2$, so it is not a pseudo-necessity measure.   □

5   A Framework for Qualitative Information Fusion

Consider a set of k sources, each delivering a piece of information about an entity x in the form of a statement $x \in S_i \subset \Omega$, one can view as a testimony. The receiver may attach a weight α_i to each source i, which assesses confidence in the truth of statement $x \in S_i$. We assume no source is fully reliable (i.e., $\alpha_i < 1, i = 1, \ldots, k$), nor irrelevant (i.e., $\alpha_i > 0, i = 1, \ldots, k$).

Formally, this body of evidence can be viewed as a basic possibility assignment

$$\sigma(A) = \begin{cases} \alpha_i, & \text{if } A = S_i, i = 1, \ldots, k, \\ 1 & \text{if } A = \Omega, \\ 0 & \text{otherwise.} \end{cases}$$

5.1   Non-destructive vs. Destructive Merging

The capacity v induced by σ as $v(A) = \max_{B \subseteq A} \sigma(B)$ is non-dogmatic and achieves a representation of the set of testimonies where redundant sources are eliminated. A source i is redundant when there is $j \neq i$ such that $S_j \subseteq S_i$ and $\alpha_i \leq \alpha_j$. We can write v as $v = \max_{i=1}^k N_{S_i}$. The focal sets of v other than Ω are the set S_i corresponding to sources that are not redundant. Note that the condition $\alpha_i < 1$ is not imperative here as all capacities, including dogmatic ones take the form $v = \max_{i=1}^k N_{S_i}$ for suitable values of k and suitable sets S_i. In particular one may have $\alpha_i = 1$, $i = 1, \ldots, k$, and get a 0–1-valued capacity.

This process can be viewed as a non-destructive fusion operation, as information items from all non-redundant sources are preserved.

In contrast, if we compute possibility distributions $\pi_i(w) = \begin{cases} 1 & \text{if } w \in S_i \\ \nu(\alpha_i) & \text{otherwise.} \end{cases}$

where ν is the order-reversing map on L, the usual fusion process in possibility theory would consist in computing $\pi_\wedge = \min_{i=1}^k \pi_i$, and then obtaining the necessity measure $N(A) = \nu(\max_{w \notin A} \pi_\wedge(w)) = \min_{w \notin A} \max_{i:w \notin S_i} \alpha_i$. However, it is very likely that $\pi_\wedge$ will be subnormalized, and that a large part of the information supplied by the sources is lost. In that case, the alternative fusion method is disjunctive, and computes $\pi_\vee = \max_{i=1}^k \pi_i$, which comes down to computing the necessity measure $\min_{i=1}^k N_{S_i} < v$. Again the result is getting rid of source information, and may considerably increase ignorance. For instance, in the case of two conflicting sources S_1, S_2, such that $S_1 \cap S_2 = \emptyset$, $\pi_\wedge(w) = \nu(\max(\alpha_1, \alpha_2))$ and $\pi_\vee(w) = 1, \forall w \in S_1 \cup S_2$. The two pieces of information are destroyed by these fusion operations while they can be retrieved from v. This kind of (usual) fusion process is thus destructive.

5.2   Connection with Belnap Logic

Computing the capacity value $v(A)$ from a basic possibility assignment for all subsets A is similar to the inconsistent information management set-ups in Belnap logic [3]. In this approach, sources supply information, and each one declares each atomic proposition $p_j, j = 1, \ldots q$ of a language as true, false or unknown. Overall, each such atomic proposition is attached a value $t(p_j) \in \{T, F, N, B\}$ summarizing what sources said about p_j, and providing the epistemic status of atomic propositions:

$$t(p_j) = \begin{cases} T & \text{if at least one source declared } p_j \text{ true and none declared it false,} \\ F & \text{if at least one source declared } p_j \text{ false and none declared it true,} \\ N & \text{if no source declared } p_j \text{ true nor false,} \\ B & \text{if at least one source declared } p_j \text{ true and one declared it false,} \end{cases}$$

The values N, T, F, B respectively stand for "None, True, False, Both". The epistemic status of other propositions built from the atomic propositions p_j's is then obtained using 4-valued truth tables [4]. The set of values $\{T, F, N, B\}$ forms a bilattice, i.e., is equipped with two ordering relations: the information ordering $<_I$ such that $N <_I T <_I B$ and $N <_I F <_I B$, expressing the idea of being less informed, and the truth-ordering $<_t$ such that $F <_t N <_t T$ and $F <_t B <_t T$, expressing the idea of being less true (in fact, less credible, here).

This framework has been extended in [5] to the case where any kind of information can be supplied by sources, in the form $x \in S_i \subset \Omega$, using 0–1 capacities. Namely, we can build the Boolean (in particular, dogmatic) capacity v such that $v = \max_{i=1}^{k} N_{S_i}$ whose focal sets form the family $\mathcal{F}_v = \min_{\subset}\{S_i, i = 1, \ldots, k\}$ of sets S_i minimal for inclusion (eliminating redundant information). Note that $v(A) = 1$ means that there is a source that supports A, so that the set of Belnap truth-values can be captured by pairs $(v(A), v(A^c)) \in \{0, 1\}^2$. Namely

- T corresponds to $v(A) = 1, v(A^c) = 0$ (credibility of A);
- F corresponds to $v(A) = 0, v(A^c) = 1$ (implausibility of A);
- N corresponds to $v(A) = 0, v(A^c) = 0$ (ignorance about A);
- B corresponds to $v(A) = 1, v(A^c) = 1$ (conflicting information about A);

The framework of qualitative capacities naturally provides a valued extension of the bilattice structure, considering $(v(A), v(A^c)) \in L^2$ as a description of how much support sources of information provide to A and its negation. The information ordering is then of the form $(v(A), v(A^c)) \leq_I (v(B), v(B^c))$ if and only if $v(A) \leq v(B)$ and $v(A^c) \leq v(B^c)$, which means that sources inform less about A than about B, interpreting conflict as an excess of information. The truth-ordering $<_t$ can be extended by requiring that the opinion of sources about A be less positive than the opinion about B, which can be formalized as: $(v(A), v(A^c)) \leq_t (v(B), v(B^c)) \iff v(B) \geq v(A)$ and $v(B^c) \leq v(A^c)$.

This is a partial ordering on graded pairs. The structure $(L \otimes L, \leq_I, \leq_t)$ is a double partially ordered set that extends the bilattice structure of Belnap. It is a lattice with the disjunction $(a, a') \vee_I (b, b') = (\max(a, b), \max(a', b'))$ and the conjunction $(a, a') \wedge_I (b, b') = (\min(a, b), \min(a', b'))$. And also another lattice with operations $(a, a') \vee_t (b, b') = (\max(a, b), \min(a', b')), (a, a') \wedge_t (b, b') = (\min(a, b), \max(a', b'))$; there is also a negation that consists of switching $v(A)$ and $v(A^c)$ (which reminds of intuitionistic fuzzy sets); see [8] for algebraic considerations of such extended bilattices.

5.3   The Maximal Consistent Subsets Approach

The non-destructive approach to information fusion only collects information items supplied by sources, in the spirit of Belnap information processor, without trying to cross-fertilize them. In particular, if source 1 says $x \in S_1$ and source 2 says $x \in S_2$, we still have $v(S_1 \cap S_2) = v(S_1^c \cup S_2^c) = 0$, that is, as receivers, we do not conclude $x \in S_1 \cap S_2$. However, many fusion methods assume we can make this step unless $S_1 \cap S_2 = \emptyset$: it is a basic fusion principle in logic [20] and uncertainty theories [14]. This view can be captured if we push the previous non-destructive merging further, by constructing $\hat{v} = \hat{\otimes}_{i=1}^{k} N_{S_i}$, a non-dogmatic separable capacity, which presupposes $\alpha_i < 1, \forall i$.

This idea was briefly proposed in [1], where consistent subsets of pieces of information obtained from subsets of sources are combined, forming a lattice of arguments where the results are attached to set of sources whose credibility can be taken into account. Considering maximal subsets of consistent sources $\mathcal{K}_j, j = 1, \dots p$, each $\mathcal{K}_j$ yields a nested family of focal sets of $\hat{v}$. Namely suppose $\mathcal{K}_j = \{S_{j_1}, \dots, S_{j_{n_j}}\}$, with $\alpha_{j_1} > \cdots > \alpha_{j_{n_j}}$, then the sets $S_{j_1}, S_{j_1} \cap S_{j_2}, \dots, \cap_{\ell=1}^{n_j} S_{j_\ell}$ are focal sets of $\hat{v}$, with respective weights $\alpha_{j_1} > \cdots > \alpha_{j_{n_j}}$. There are p such chains in $\mathcal{F}_{\hat{v}}$. Again we can compare the epistemic statuses of propositions $x \in A$ from the information provided from sources after cross-checking, by applying the information and the truth orderings to pairs $(\hat{v}(A), \hat{v}(A^c)), A \subseteq \Omega$. Note that $\hat{v}$ is more informative than v since $(\hat{v}(A), \hat{v}(A^c)) \geq_I (v(A), v(A^c))$, as examplified in Example 5 and in Table 1, and already proved by Proposition 2.

Table 1. Non-destructive vs. conjunctive fusion

A	{4}	[2, 4]	{5}	[1, 3]	{2,5}
$(v(A), v(A^c))$	(0,0.4)	(0.4, 0.2)	(0, 0.8)	(0.4, 0.5)	(0, 0)
$(\hat{v}(A), \hat{v}(A^c))$	(0.5, 0.4)	(0.5, 0.4)	(0.2, 0.8)	(0.4, 0.5)	(0.2, 0)

Example 5. Suppose $\Omega = [1, 6]$ the set of integers between 1 and 6. There are 4 weighted sources and $S_1 = [1, 4], S_2 = [4, 5], S_3 = [2, 3], S_4 = [5, 6]$, with respective weights $0.8, 0.5, 0.4, 0.2$. Using the first approach defines a capacity that has these 4 focal sets plus $[1, 6]$. Using the second approach, the maximal consistent subsets are $\mathcal{K}_1 = \{[1, 4], [2, 3]\}, \mathcal{K}_2 = \{[1, 4], [4, 5]\}, \mathcal{K}_3 = \{[4, 5], [5, 6]\}$. Capacity $\hat{v}$ has focal sets with weights: $\{([2, 3], 0.4), ([1, 4], 0.8), (\{4\}, 0.5), (\{5\}, 0.2)\}$, plus $([1, 6], 1)$. Table 1 compares the two approaches via pairs $(v(A), v(A^c))$: note that precise information items, rejected by the non-destructive approach become conflicting under the conjunctive one. But regarding $\{2, 5\}$, the conjunctive combination restores consistency where the non-destructive approach is ignorant.

Example 6. Finally we can consider a qualitative counterpart of the famous Peter, Paul and Mary case after Smets [23]. A crime has been committed and

Table 2. The Peter, Paul and Mary case

	v	$\hat{v}$		v	$\hat{v}$		v	$\hat{v}$
$(\{Pa\},\{Pe,Ma\})$	$(0,\alpha)$	(α,α)	$(\{Pe\},\{Pa,Ma\})$	$(0,\beta)$	$(0,\beta)$	$(\{Ma\},\{Pe,Pa\})$	(α,α)	(α,α)

the killer is known to be among Peter, Paul and Mary. There are three pieces of evidence. One source claims the killer is a male (with weight α) and another source claims it is a female (with the same weight α). Finally, another source claims that Peter has an alibi (with stronger confidence $\beta > \alpha$). So we first define the capacity v on $\{Pe, Pa, Ma\}$ such that $v_*(\{Pe, Pa\}) = v_*(\{Ma\}) = \alpha$, and $v_*(\{Ma, Pa\}) = \beta$. Now let us combine these information items and get the separable upper approximation $\hat{v}$ of v. Its Moebius transform is the same as the one of v but for $\hat{v}_*(Pa) = \alpha = \min(v_*(\{Ma, Pa\}), v_*(\{Pe, Pa\})$. See Table 2. While, as expected, Peter is considered the least credible killer, v seems to exonerate Paul against Mary (even if information concerning her is just contradictory), while $\hat{v}$ puts Mary and Paul back on a par.

6 Conclusion

This paper has explored some formal similarities between belief functions and qualitative capacities, initiated in [15,19], by studying the counterparts of separable belief functions in relation with information fusion problems. We have focused on the merging of uncertain qualitative testimonies. We have shown that the qualitative counterpart of Dempster rule of combination applied to uncertain qualitative testimonies leads to separable capacities whose focal sets are either disjoint or nested. This notion of separability turns out to be more drastic than in the numerical setting for belief functions. Indeed, Shafer [21] has shown that the set of focal sets of a separable belief function is closed under non-empty intersections, and Smets [22] has shown that any belief function can be obtained by a kind of division between the commonalities of two separable belief functions, leading to interpret them as the fusion of both elementary testimonies and prejudices against their conjunctions [11]. In the qualitative case, the set of focal sets of separable capacities does not contain non-nested overlapping sets, so that we cannot reconstruct a non-separable capacity from two separable ones by a kind of subtraction. However, results related to symmetric minimum and maximum [17] suggest a possibility of erasing Möbius weights. Another line of study would be to replace the min and/or max in the qualitative Dempster rule by operations, inducing the leximin and leximax orderings, on multisets obtained from the concatenation of weights [9,10]. For instance, the combination of N_A and N_B with respective weights on A and B equal to α and β, would yield focal sets A with weight $(\alpha, 1)$, B with weight $(1, \beta)$ and $A \cap B$ with weight (α, β), where $(\alpha, \beta) < (\alpha, 1)$ and $(\alpha, \beta) < (1, \beta)$ in the sense of leximin. This approach may extend the range of separability for capacities.

Acknowledgements. This work is supported by ANR-11-LABX-0040-CIMI (Centre International de Mathématiques et d'Informatique) within the program ANR-11-IDEX-0002-02, project ISIPA.

References

1. Assaghir, Z., Napoli, A., Kaytoue, M., Dubois, D., Prade, H.: Numerical information fusion: lattice of answers with supporting arguments. In: International Confernce on Tools for AI, pp. 621–628. IEEE Press (2011)
2. Banon, G.J.: Constructive decomposition of fuzzy measures in terms of possibility and necessity measures. In: Proceedings of 6th International Fuzzy Systems Association Congress (IFSA 1995), Sao Paulo, pp. 217–220 (1995)
3. Belnap, N.D.: How a computer should think. In: Ryle, G. (ed.) Contemporary Aspects of Philosophy, pp. 30–56. Oriel Press, Stocksfield (1977)
4. Belnap, N.D.: A useful four-valued logic. In: Dunn, J.M., Epstein, G. (eds.) Modern Uses of Multiple-Valued Logic, pp. 8–37. D. Reidel, Dordrecht (1977)
5. Ciucci, D., Dubois, D.: A two-tiered propositional framework for handling multi-source inconsistent information. In: Antonucci, A., Cholvy, L., Papini, O. (eds.) ECSQARU 2017. LNCS (LNAI), vol. 10369, pp. 398–408. Springer, Cham (2017). https://doi.org/10.1007/978-3-319-61581-3_36
6. Chemin, M., Rico, A., Prade, H.: Decomposition of possibilistic belief functions into simple support functions. In: Melo-Pinto, P., Couto, P., Serôdio, C., Fodor, J., De Baets, B. (eds.) Eurofuse 2011. AINSC, vol. 107, pp. 31–43. Springer, Heidelberg (2011). https://doi.org/10.1007/978-3-642-24001-0_5
7. Dempster, A.P.: Upper and lower probabilities induced by a multivalued mapping. Ann. Math. Stat. **38**, 325–339 (1967)
8. Deschrijver, G., Arieli, O., Cornelis, C., Kerre, E.E.: A bilattice-based framework for handling graded truth and imprecision. Int. J. Uncertain. Fuzziness Knowl.-Based Syst. **15**(1), 13–41 (2007)
9. Dubois, D., Fargier, H.: Capacity refinements and their application to qualitative decision evaluation. In: Sossai, C., Chemello, G. (eds.) ECSQARU 2009. LNCS (LNAI), vol. 5590, pp. 311–322. Springer, Heidelberg (2009). https://doi.org/10.1007/978-3-642-02906-6_28
10. Dubois, D., Fortemps, P.: Selecting preferred solutions in the minimax approach to dynamic programming problems under flexible constraints. Eur. J. Oper. Res. **160**, 582–598 (2005)
11. Dubois, D., Faux, F., Prade, H.: Prejudiced information fusion using belief functions. In: Proceedings of SMPS/BELIEF Conference, Compiègne, France (2018)
12. Dubois, D., Prade, H.: Upper and lower possibilities induced by a multivalued mapping. In: Sanchez, E. (ed.) Fuzzy Information, Knowledge Representation and Decision Analysis, Proceedings of IFAC Symposium, pp. 174–152. Pergamon Press (1984)
13. Dubois, D., Prade, H.: Evidence measures based on fuzzy information. Automatica **21**, 547–562 (1985)
14. Dubois, D., Liu, W., Ma, J., Prade, H.: The basic principles of uncertain information fusion. An organized review of merging rules in different representation frameworks. Inf. Fusion **32**, 12–39 (2016)
15. Dubois, D., Prade, H., Rico, A.: Representing qualitative capacities as families of possibility measures. Int. J. Approx. Reason. **58**, 3–24 (2015)

16. Giang, P.H., Shenoy, P.P.: A decision theory for partially consonant belief functions. Int. J. of Approx. Reason. **52**, 375–394 (2011)
17. Grabisch, M.: The symmetric Sugeno integral. Fuzzy Sets Syst. **139**, 473–490 (2003)
18. Grabisch, M.: The Möbius transform on symmetric ordered structures and its application to capacities on finite sets. Discret. Math. **287**, 17–34 (2004)
19. Prade, H., Rico, A.: Possibilistic evidence. In: Liu, W. (ed.) ECSQARU 2011. LNCS (LNAI), vol. 6717, pp. 713–724. Springer, Heidelberg (2011). https://doi.org/10.1007/978-3-642-22152-1_60
20. Rescher, N., Manor, R.: On inference from inconsistent premises. Theory Decis. **1**, 179–219 (1970)
21. Shafer, G.: A Mathematical Theory of Evidence. Princeton University Press, Princeton (1976)
22. Smets, P.: The canonical decomposition of a weighted belief. In: Proceedings of 14th International Joint Conference on Artificial Intelligence (IJCAI), vol. 2, pp. 1896–1901. AAAI Press (1995)
23. Smets, P.: The transferable belief model for quantified belief representation. In: Gabbay, D., Smets, P. (eds.) Handbook of Defeasible Reasoning and Uncertainty Management Systems, vol. 1, pp. 267–301. Kluwer Academic Publishers, Alphen aan den Rijn (1998)
24. Sugeno, M.: Fuzzy measures and fuzzy integrals: a survey. In: Gupta, M.M. (ed.) Fuzzy Automata and Decision Processes, pp. 89–102. North-Holland, Amsterdam (1977)
25. Tsiporkova, E., De Baets, B.: A general framework for upper and lower possibilities and necessities. Int. J. Uncertain. Fuzziness Knowl.-Based Syst. **6**, 1–34 (1998)
26. Walley, P.: Belief function representation of statistical evidence. Ann. Stat. **15**(4), 1439–1465 (1987)

An Approach Based on MCDA and Fuzzy Logic to Select Joint Actions

Abdelhak Imoussaten[(✉)]

LGI2P, IMT Mines Ales, Univ Montpellier, Ales, France
abdelhak.imoussaten@mines-ales.fr

Abstract. To satisfy a fluctuating demand and achieve a high level of quality and service, companies must take into account several features when designing new products in order to become or remain market leaders. When a single company is unable to meet this objective alone, it is appropriate for it to join its actions with other companies. The product design consists of the complex task to select from various potential actions that allowing the fulfilment of several requirements: functional, technical, environmental, economic, security, etc. Furthermore, the task is even more difficult when actions are related to distinct services or companies that do not necessarily know the capacities of each others which makes complex the coordination of joint actions. Interactions between services may be affected by antagonist personal interests.

Based on a multiple criteria decision analysis (MCDA) framework and a fuzzy model that links actions to the satisfaction of objectives, this paper proposes to treat two extreme views related to the collective selection of the necessary actions to design a product: (1) The first point of view corresponds to an ideal situation where each service reveals its capacities and the unique objective is to succeed in the realization of the common goal; (2) the second point of view corresponds to a more realistic situation where only necessary information for the progress of collective action are shared and where collective and personal goals coexist and are to be taken into account. The first situation corresponds to a classical case where a single decision maker (DM) has to express his preferences then a classical optimization problem under constraints has to be solved in order to efficiently select actions. In the second situation the services do not share the same preferences and each service wants to maximize its gain, in this case we propose to build a negotiated solution between services.

Keywords: MCDA · Fuzzy logic · Constraint programming
Debate modeling

1 Introduction

When company designers/operators choose a new product or system, they have to check whether the new system satisfy company's strategic goals, customers

© Springer Nature Switzerland AG 2018
D. Ciucci et al. (Eds.): SUM 2018, LNAI 11142, pp. 140–151, 2018.
https://doi.org/10.1007/978-3-030-00461-3_10

needs and technical specifications. Furthermore, these issues are obviously not devoid of budgetary constraints.

MCDA is an interesting setting to evaluate and compare all alternatives for the new system based on their satisfaction of the previous multiple requirements. The MCDA setting provides approaches for rigorous multidimensional evaluation. Indeed a set of non correlated attributes are carefully defined to measure the characteristics of the candidate solutions. Then the measures are mixed up with the subjectivity of the decision maker (DM) to deduce the degree of satisfaction provided by each candidate solution. In MCDA we distinguish several methods depending on the number of candidate solutions, number of attributes, compensation hypothesis, etc. (see [3] for more details). In this paper we focus on Multi objective Optimisations methods. Let denote by: $S = \{s_1, s_2, ...\}$ a set of finite or infinite candidate solutions for the new system, $\{X_1, X_2, ..., X_n\}$ a finite set of attributes where X_i is the space of values taken by the ith characteristic, $N = \{1, ..., n\}$ and $X = \prod_{i \in N} X_i$. Each element of S can be represented by its results obtained on the set of attributes, then $s_k \in S$ can be represented by the vector $(x_1^k, ..., x_n^k)$ where $\forall i \in N$, x_i^k is the result of s_k on the ith characteristic. Thus, the satisfaction provided by the solution s_k can be represented by a real $v(s_k)$ where v is a bounded function ($v : X \rightarrow [0, 1]$). Finally, the problem of choosing the new system can be setted as follows:

$$\max \ v(s)$$
$$\text{s.t:}$$
$$\begin{cases} s \in S \\ c(s) \leq c_0 \end{cases} \tag{1}$$

where $c : S \rightarrow]0, +\infty[$ is the function associating to each candidate solution its cost and $c_0 > 0$ is the maximal cost allocated to the new system. When functions c and v are known we are faced to a classical optimization problem under constraints.

Usually in MCDA, a single DM holds the preferential information that allows to build the function v. Particularly, in Multi Attribute Value Theory (MAVT [1,6]) the existence of a such value function is guaranteed when DM preference relation over S denoted $\succeq$, $s_k \succeq s_{k'}$ means s_k is at least as good as $s_{k'}$, is a weak order. In this case, v represents the DM preferences over S and we have:

$$\forall s_k, s_{k'} \in S, \quad s_k \succeq s_{k'} \Leftrightarrow v(s_k) \geq v(s_{k'}). \tag{2}$$

The easiest way to find v is to build it from the DM preference relation on each attribute $i \in N$ denoted $\succeq_i$ and represented by a value function $v_i : X_i \rightarrow [0, 1]$. In [7] the conditions of a such construction are given. Furthermore, a single DM who has a perfect knowledge about implementation actions of a solution can estimate their costs. Thus, in this case of a single DM, c and v can be identified.

Unfortunately, this hypothesis of the existence of a single decision maker who knows the costs of actions and with whom we can build the utility function, is not adapted to all companies configurations. Indeed, in many cases, these actions are generally spread/distributed onto several services which are then in charge of their own actions.

Classically in MCDA, the behavioural relationship between actions related to a candidate solution and the results obtained on the system characteristics are not considered. As mentioned previously, in this case, a candidate solution is represented only by its results obtained on the set of attributes. In the case of several services, the reference to the actions composing the candidate solution is required. Indeed, it would be difficult to establish with certainty the consequences of a set of actions on the attributes since actions are related to several services and applied jointly. This is because no services has complete information about all actions.

In this situation two models are envisaged, based on distinct hypothesis. Either, services collaborate fully then they share their knowledge between each others such that all services will have complete information. The problem to resolve here is to select the subset of actions guaranteeing the "best" satisfaction of common goals at minimal cost (see [8] for more details). Or, services act as autonomous agents but are cooperating: they make their possible to reach collective objectives although they have also personal interests in the project. The more actions are carried out in a service, the greater his budget. It is a thorny problem because each service has only a partial information and don't necessarily know the capacities of other services. The problem to resolve here is to build a solution by selecting a subset of actions from all services guaranteeing the satisfaction of all services and respecting budget constraints.

The aim of this paper is to propose two different models for these two situations. In the first situation the decision process is stated in the form of an multi objective optimisation problem under constraints and in the second situation, the decision process is modelled as a debate.

The paper is structured as follows. First, Sect. 2 establishes the model of relationships between actions and objectives satisfaction. Sections 3 and 4 presents the formalization of the problem of defining joint actions composing the solution satisfying all services and respecting cost constraints respectively on the collaborative and cooperative situations.

2   Fuzzy Relationship Between Actions and the Satisfaction of Objectives

We consider a group of services denoted $\mathcal{D} = \{d_1, ..., d_l\}$ that are in charge of building a common solution for the design of a new product or system. Each service d is in charge of a set of actions $\mathcal{A}_d$. For instance, an action consists in a person performing a task, a mechanical or software component, etc. Each service proposes his actions to compose a candidate solution with other services. Thus the set of candidate solutions $S = \{s_1, s_2, ...\}$ is composed of all admissible joint actions. These actions must guarantee a good performance of the new system. We assume here that the performance of the new system induced by a set of joint actions $s \in S$ is measured by $v(s)$ where the value function v is defined in formula 2. The function v represents the overall satisfaction of the objectives set for the new system. Here, v will be built on the basis of what the services will provide as information about the effects of their actions on the satisfaction of the objectives of the system. Moreover, we consider that the value function v may be obtained through the partial value functions v_i, $i \in N$ representing the satisfaction on each single objective using an appropriate aggregation operator.

The transformation that links actions in $\mathcal{A}_d$ and the ith objective, $i \in N$, is composed of two transformations: (1) The transformations T_i $(i \in N)$ that represents the behavioral relation between inputs (actions) and each of its outputs (characteristics); (2) the value functions v_i expressing the satisfaction of the service related to the achievement of the ith objective.

Most of the time, the transformation $T = (T_1, \ldots, T_n)$ cannot be precisely known in a complex system. In some cases, the acquisition of T requires complex simulations or experiments, which are costly and time consuming (example of military architecture, [9]). Hence transformation T generally needs to be approximated. We will use a fuzzy behavioral model. Indeed, as the gathered information originates from the experts or managers perception rather than being factually measured, it is intrinsically imprecise [8].

Furthermore, each action considered must allow the satisfaction of at least one objective but it is not excluded that this action may have a negative impact on other objectives. Thus, the scale used for the evaluation of each objective must be bivariate allowing positive and negative evaluation [2]. For instance, in the automotive industry, the action of "using modular platform" reduce the weight of cars and then has a positive impact on the objective "reducing the shock with a pedestrian, animal or cyclist" but this action could impact the stability of the vehicles then it has a negative impact on the objective "safety of the driver". In our setting, we propose to use two linguistic variables to represents the impact of actions on each objective: *satisfaction* and *dissatisfaction* variables.

In [4,5,8] a degree in $]0,1]$ called the "degree of belief" for positive or negative impact of an action on an objective is introduced. The semantic of this degree is unclear and then not easy to provide by experts. In addition, the imprecision inherent to expert knowledge is not taken into account. In this paper we propose to represent the impact of actions on objectives within the setting of fuzzy logic. Thus several linguistic terms can be associated to each of the two previous linguistic variables. For example, terms *low*, *medium* and *strong* can be used.

Definition 1. *Consider the ith objective, denoted Obj_i where $i \in N$, and an action $a \in \mathcal{A}$. Either, action a induce a satisfaction on Obj_i in a such case S_i^a denote the associated linguistic term describing the degree of a such satisfaction. Or, action a induce a dissatisfaction on Obj_i in a such case D_i^a denote the associated linguistic term describing the degree of a such dissatisfaction.*

The bivariate fuzzy model described in Definition 1 can be represented through a digraph between $\mathcal{A}$ and N, denoted $Dig(\mathcal{A}, N)$, such that (Fig. 1

shows an example): the arc between a and Obj_i defined

$$\text{Arc}(a, Obj_i) = \begin{cases} +S_i^a & \text{in case of satisfaction} \\ -D_i^a & \text{in case of dissatisfaction} \end{cases} \tag{3}$$

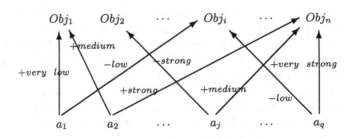

Fig. 1. Digraph of actions-objectives relationships

As we can remark in the example given above of automotive industry, a single action can not satisfy all the objectives. Thus several actions are required. Let consider a subset of actions ap and an objective Obj_i. Several actions in ap may induce satisfaction on Obj_i and several other actions in ap may induce dissatisfaction on Obj_i. We denote by $A_i^S(ap)$ (resp. $A_i^D(ap)$) the crisp subset of actions in ap that induce satisfaction (resp. dissatisfaction) on Obj_i.

To obtain the overall satisfaction and dissatisfaction of the joint implementation of actions in ap, it is important to consider all the interactions and results emerging from the joint implementation of some actions. The simplest solution is to evaluate the impact of all possible combinations of actions. However, when the set of actions is very large, this solution seems unrealistic. To simplify, we consider again a soft model based on mathematical operator applied to fuzzy subsets. In order to reduce the number of actions that do the same thing, we consider that the satisfaction and dissatisfaction caused by the joint actions can not be additive. Thus, based on cautious behaviour, we keep all the maximum possibility degrees of satisfaction and dissatisfaction in the aggregated result by using a conjunctive operator. Finally, we propose to synthesize the satisfaction and dissatisfaction of ap on Obj_i as follows:

$$\mu_{S_i^{ap}} = \bigvee_{a \in A_i^S(ap)} \mu_{S_i^a}; \tag{4}$$

$$\mu_{D_i^{ap}} = \bigvee_{a \in A_i^D(ap)} \mu_{D_i^a}. \tag{5}$$

where $\mu_{S_i^a}$ (resp. $\mu_{D_i^a}$) is the fuzzy subset associated to linguistic term S_i^a (resp. D_i^a).

The next step is to verify if service is satisfied by the impact of actions in ap on Obj_i. This is done here by comparing the center of area of the resulting two fuzzy subsets $\mu_{S_i^{ap}}$ and $\mu_{D_i^{ap}}$ to two thresholds $\alpha_i, \beta_i \in]0,1]$ fixing the limits of satisfaction and dissatisfaction degrees.

Definition 2. *Let consider an action plan ap, an objective Obj_i and two thresholds $\alpha_i, \beta_i \in]0,1]$. We consider that ap satisfies Obj_i if:*

1. $\dfrac{\int y \times \mu_{S_i^{ap}}(y)dy}{\int \mu_{S_i^{ap}}(y)dy} \geq \alpha_i$ *and*

2. $\dfrac{\int y \times \mu_{D_i^{ap}}(y)dy}{\int \mu_{D_i^{ap}}(y)dy} \leq \beta_i.$

Equations (4), (5) and Definition 2 concerns a single objective but can be extended to any subset of objectives.

The principle extension of Zadeh allows aggregating satisfactions and dissatisfactions as follows:

$$\mu_{S_N^{ap}}(y) = \bigvee_{\substack{(y_1,\ldots,y_n) \\ \phi_S(y_1,\ldots,y_n)=y}} \bigwedge_{i \in N} \mu_{S_i^{ap}}(y_i). \tag{6}$$

where ϕ_S is an aggregation operator representing the DM trade-off on the satisfaction of objectives.

$$\mu_{D_N^{ap}}(y) = \bigvee_{\substack{(y_1,\ldots,y_n) \\ \phi_D(y_1,\ldots,y_n)=y}} \bigwedge_{i \in N} \mu_{D_i^{ap}}(y_i). \tag{7}$$

where ϕ_D is an aggregation operator representing the DM trade-off on the dissatisfaction of objectives.

Note that ϕ_S and ϕ_D are two operators applied to a vector of real values from $[0,1]^n$ to $[0,1]$. They can be a classical operators like *min, max* or *weighted average* operator or more sophisticated operators like Choquet integral or Sugeno integral.

We can also define the satisfaction and dissatisfactions of ap on a subset of objectives $I \subset N$ by fixing the results of ap on the remaining objectives, *i.e.*, $N \backslash I$, at given values:

$$\mu_{S_I^{ap}}(y) = \bigvee_{\substack{y_I \\ \phi_S(y_I, y_{N \backslash I}^0)=y}} \bigwedge_{i \in I} \mu_{S_i^{ap}}(y_i). \tag{8}$$

$$\mu_{D_I^{ap}}(y) = \bigvee_{\substack{y_I \\ \phi_D(y_I, z_{N \backslash I}^0)=y}} \bigwedge_{i \in I} \mu_{D_i^{ap}}(y_i). \tag{9}$$

where the notation y_I represents the components of the vector $(y_1, \ldots, y_n)$ for the indices in I and the notation $(y_I, y_{N \backslash I}^0)$ represents the vector z where $z_i = y_i$ if $i \in I$ and $z_i = y_i^0$ if $i \in N \setminus I$.

In the similar way as in Definition 2, we verify if ap satisfy the subset of objectives I.

Definition 3. *Let consider an action plan ap, a subset of objectives $I \subseteq N$ and two thresholds $\alpha_I, \beta_I \in]0, 1]$. We consider that ap satisfies objectives in I if:*

1. $s_I(ap) = \dfrac{\int y \times \mu_{S_I^{ap}}(y) dy}{\int \mu_{S_I^{ap}}(y) dy} \geq \alpha_I$ and

2. $d_I(ap) = \dfrac{\int y \times \mu_{D_I^{ap}}(y) dy}{\int \mu_{D_I^{ap}}(y) dy} \leq \beta_I$.

If the two previous conditions are satisfied, $s_I(ap)$ is considered as the degree of satisfaction of objectives on I by ap, otherwise $s_I(ap) = 0$.

Note that the dissatisfaction degree will be less mentioned than the satisfaction degree in the following. Indeed, the satisfaction the degree is the aggregated $s_I(ap)$ result of ap, including dissatisfaction degree, and if $s_I(ap) = 0$ it means that $d_I(ap) \leq \beta_I$.

3   Selecting Joint Actions from Collaborative Services

An operational cost $c(ap)$ is associated with each action plan ap. We assume for any action a, $c(\{a\}) > 0$ and: $c(ap) = \displaystyle\sum_{a \in ap} c(\{a\})$.

In most instances, a cost constraint is applicable to the new system such that the cost of an action plan cannot exceed a predetermined budget denoted $b_0 \in]0, \infty]$. In addition, the new system may require a degree of satisfaction higher than a predetermined threshold $\alpha_0 \in]0, 1]$. Nevertheless, services of course prefer the solution with the highest satisfaction degree when the cost is the same; inversely, they prefer the cheapest solutions when the satisfaction degree is the same.

Let us define the Pareto order $\prec_{\text{Pareto}}$ over action plans:

$$
\begin{aligned}
&(s_N(ap), -c(ap)) \prec_{\text{Pareto}} (s_N(ap'), -c(ap')) \\
&\Leftrightarrow \quad [(s_N(ap) < s_N(ap')) \text{ and } (c(ap) \geq c(ap'))] \\
&\qquad \text{or } [(s_N(ap) \leq s_N(ap')) \text{ and } (c(ap) > c(ap'))]
\end{aligned}
\tag{10}
$$

Action plan ap is dominated by action plan ap' if $(s_N(ap), -c(ap)) \prec_{\text{Pareto}} (s_N(ap'), -c(ap'))$.

In collaborative approach where service share all their knowledge of the system before collective action, the complete digraph $Dig(\mathcal{A}, N)$ is known and the multi-objective optimization problem of formula (1) is adapted for our fuzzy model as follows:

$$
\begin{aligned}
&\max \ (s_N(ap), -c(ap)) \\
&\text{s.t:} \\
&\begin{cases}
ap \subseteq \mathcal{A} = \displaystyle\bigcup_{d \in \mathcal{D}} \mathcal{A}_d \\
s_N(ap) \geq \alpha_0 \\
c(ap) \leq b_0
\end{cases}
\end{aligned}
\tag{11}
$$

When the number of actions is very large the resolution of this optimization problem clearly raises a combinatorial problem.

The hypothesis of collaboration leading to the optimisation problem of formula (11) means that the services involved in the process of designing the new system share their partial knowledge on the fuzzy model of relationship between action and objectives so that all services have full knowledge of the model before committing collective action. This assumption is not always valid; service heads are not always willing to share knowledge of their own design service. However, they are willing to cooperate to build a common solution. Thus collaborative approach is not often feasible in practice. It seems more natural to consider that the teams will cooperate more than collaborate. In the following section, we address the problem of selecting joint actions from cooperative services.

4   Selecting Joint Actions from Cooperative Services to Achieve Common Goals

In the case of cooperating services, the model is constructed gradually: each service contributes when it is required, there is no a priori planning. The collective choice of action plan is thus modelled as a debate. Services exchange knowledge and negotiate the way actions will be distributed. The more actions are carried out in a service the greater his budget.

We consider that no service is able to possess the action plan satisfying all the objectives. Thus, each service needs actions from others to build joint action plan.

4.1   General Principle for Debate Structure

Let consider a service $d \in \mathcal{D}$. Actions in $\mathcal{A}_d$ induce impacts only on a subset $I_d = \{i_1^d, \ldots, i_{|I_d|}^d\} \subseteq N$. The partial digraph $Dig(\mathcal{A}_d, I_d)$ of service d is presented in Fig. 2.

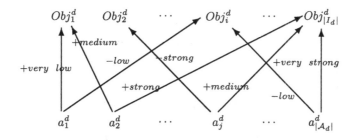

Fig. 2. Partial digraph of actions-objectives relationships

For simplification purpose and to limit the information required from the experts, we consider that the thresholds of satisfaction and dissatisfaction for subsets of objectives can be deduced from those of single objective as follows: $\forall I \subseteq N$, $\alpha_I = \min_{i \in I} \alpha_i$, and $\beta_I = \max_{i \in I} \beta_i$. It comes $\alpha_0 = \min_{i \in N} \alpha_i$. This is an optimistic way of considering aggregated satisfaction and dissatisfaction thresholds.

Services would propose their actions in order to increase their budget, then we need to manage the turn of propositions. This is done by selecting the service having the best offer for the common objective at the current stage. Then services have to learn from the previous propositions in order to make adequate next proposition.

Indeed, at the beginning, the services do not know the impacts induced by the actions of others services. At this stage, the information shared between the services are: the set of objectives to be satisfied N, the constraints α_i, β_i, $\forall i \in N$ and the constraint concerning the cost of the proposed actions that should not exceed the budget b_0. However, each service d start with a fictive complete digraph by introducing fictive actions that are supposed to be owned by other services and satisfying objectives that are not satisfied by d.

When a proposal is made, the service reveals required information on the proposed actions. Each service has to take advantage from this information and complete his partial digraph to make successful propositions.

Let us introduce time variable t explicitly in the notations. Let ap_t be the common action plan build at time t and I_t the subset of objectives satisfied, i.e., $s_{I_t}(ap_t) \geq \alpha_0$, by ap_t. The debate organization is broken down into the following steps:

- At $t = 0$, $ap_0 = \emptyset$ and $I_0 = \emptyset$.
- At $t \geq 1$, the current action plan is ap_{t-1} and the satisfied objectives are in I_{t-1}.
  1. Each service propose an action plan to complete ap_{t-1}. The new proposal should satisfy new objectives in $N \setminus I_{t-1}$ and should guarantee $s_{I(t)} > 0$. He adopts a suitable strategy to make a proposal that maximizes his earnings (see Subsect. 4.2). The algorithm solving the optimisation problem (11) is used locally by the service to select actions forming his proposal;
  2. A selection is made for the next action plan to add to ap_{t-1} (see Subsect. 4.3);
  3. Action plan ap_t satisfying I_t is built by adding new actions to ap_t;
  4. Each service updates his information about the proposed actions (see Subsect. 4.4). Thus the partial digraph of a service is time depending, denoted $Dig(\mathcal{A}_d(t), I_d(t))$.
- When at $t \geq 1$ no proposal can be made by services because of cost constraint, the debate starts from the beginning, i.e. the resulting common action plan is emptied, and services update their information.
- The debate ends when $I_{t+1} = N$ and cost constraint is respected.

Note that when a joint action plan satisfying all the objectives is found all other remaining actions are no longer receivable.

4.2 Individual Action Plan Choice Strategies

At each step t, all the services with available actions must adopt an appropriate strategy to propose the most relevant actions. The stake is twofold: (1) the actions proposed by a service must be chosen at time t and, (2) the common action plan must be successful.

Let $AP_d(t)$ the Pareto front of solutions for the optimisation problem (11) for the digraph $Dig(\mathcal{A}_d(t-1), I_d(t-1))$. The following criteria may be introduced to select a solution $sap_d(t)$ from $AP_d(t)$ depending on the debate stage:

- $sap_d(t)$ may maximize the number of satisfied objectives, $i.e.$, there is no cost conflict;
- $sap_d(t)$ may maximize the satisfaction degree;
- $sap_d(t)$ has minimal cost.

4.3 Fair Resources Sharing

Let denote $props(t)$ the set containing the next propositions of the services (each service propose a single sub action plan).

Let consider the following notations:

- $G_d^{max} = \sum\limits_{a \in \mathcal{A}_d} c(a)$ the maximal expected gain for service d,
- $G_d(ap_t) = \sum\limits_{a \in ap_t \cap \mathcal{A}_d} c(a)$ the individual gain of service d from the common action plan ap_t.

To quantify the loss of earnings for service d with respect to ap_t we can use the following formula: $G_d^{max} - G_d(ap_t)$. However, for homogeneity reasons, the formula (12) is preferred.

$$\rho(d, ap_t) = (G_d^{max} - G_d(ap_t))/G_d^{max}. \tag{12}$$

We assume that each service have at least an action with strictly positive cost.

We consider in our approach that the group tries to avoid an unfair sharing of the allocated budget by minimizing the loss of earnings of the worst paid service. The worst paid service with respect to to common action plan ap_t is defined as follow:

$$d(ap_t) = arg \max_{d \in \mathcal{D}}[\rho(d, ap_t)]. \tag{13}$$

It comes that the new proposition $sap(t)$ should verify:

$$sap(t) = arg \min_{sap \in props(t)} \max_{d \in \mathcal{D}}[\rho(d, sap \cup ap_{t-1})]. \tag{14}$$

The new common action is $ap_t = sap(t) \cup ap_{t-1}$.

4.4 Digraphs Updating

Let $d(t)$ the service owner of the proposal $sap(t) \subseteq \mathcal{A}_{d(t)}$. Service $d \in \mathcal{D} \setminus \{d(t)\}$ a priori don't know satisfactions and dissatisfactions induced by actions in $sap(t)$. The service $d(t)$ must give the necessary information about his actions so that the other services can complete their digraph and make the next proposals. $d(t)$ should reveal the following information:

- the new satisfied objectives $I^s_{d(t)}$;
- the new dissatisfied objectives $I^d_{d(t)}$;
- the maximum dissatisfaction degree induced on $I^d_{d(t)}$;
- the cost of actions in $sap(t)$;
- the new satisfaction degree.

From those information and from the fact that new actions should guarantee $s_{I(t)} > 0$, services in $\mathcal{D} \setminus \{d(t)\}$ have a quantity of helpful information to continue the construction of their complete digraph.

5 Conclusion

This article offers an aid to the evaluation of the degree of satisfaction of a new product or system to a set of expected objectives when various candidate solution are possible and several teams are involved in the project. The evaluation focuses on the actions leading to the implementation of new product or system. We have shown the link between this problem and a classical decision-making process based on MCDA approach and we have distinguished two configurations of services in a company: collaborative services and cooperative services. In the first configuration a multi-objective optimisation problem is stated to resolve the problem of selecting the optimal set of actions with regard to cost and satisfaction degree constraints. In the second configuration, the problem is stated as a debate between services where the objective is to build a common action plan satisfying a set of objectives under the same previous constraints. The debate modelling is better suited to practical situations where each service controls his own know-how, and only shares the part of his knowledge which is required achieving the global objective, defending his own interests and not necessarily revealing his weaknesses. The debate model can be also seen as a decision-support system used by a service (or a group of services) during the real debate to optimize his own interest. In the two cases fuzzy logic and MCDA tools are used to represent the relationship between actions and objectives satisfaction.

This is a preliminary work. In order to complete our debate modelling, several points in our approach deserve to be deepened in the future works. For instance, we are currently working on refining and finalizing the knowledge updating phase. In addition, we are working to provide an illustrative example of the approach. Furthermore, we will work to establish links between this work and existing approach dealing with debate modelling as the fields of game theory and theory of argumentation.

References

1. Dyer, J.S.: Multiattribute utility theory (MAUT). In: Greco, S., Ehrgott, M., Figueira, J. (eds.) Multiple Criteria Decision Analysis. ISOR, vol. 233, pp. 285–314. Springer, New York (2016). https://doi.org/10.1007/978-1-4939-3094-4_8
2. Grabisch, M., Greco, S., Pirlot, M.: Bipolar and bivariate models in multicriteria decision analysis: descriptive and constructive approaches. Int. J. Intell. Syst. **23**(9), 930–969 (2008)
3. Greco, S., Figueira, J., Ehrgott, M.: Multiple Criteria Decision Analysis. Springer, Heidelberg (2005). https://doi.org/10.1007/b100605
4. Imoussaten, A., Montmain, J., Trousset, F., Labreuche, C.: Multi-criteria improvement of options. In: European Society for Fuzzy Logic and Technology, p. 1 (2011)
5. Imoussaten, A., Trousset, F., Montmain, J.: Improving performances in a company when collective strategy comes up against individual interests. In: EUSFLAT Conference, pp. 904–911 (2011)
6. Keeney, R.L., Raiffa, H.: Decision analysis with multiple conflicting objectives. In: Decision with Multiple Objectives. Wiley, New York (1976)
7. Krantz, D.H., Luce, R.D., Suppes, P., Tversky, A.: Foundations of Measurement (Additive And Polynomial Representations), vol. 1. Academic Press, New York (1971)
8. Montmain, J., Labreuche, C., Imoussaten, A., Trousset, F.: Multi-criteria improvement of complex systems. Inf. Sci. **291**, 61–84 (2015)
9. Pignon, J.P., Labreuche, Ch.: A methodological approach for operational and technical experimentation based evaluation of systems of systems architectures. In: International Conference on Software & Systems Engineering and their Applications (ICSSEA), Paris, France, 4–6 December 2007 (2007)

Discovering Ordinal Attributes Through Gradual Patterns, Morphological Filters and Rank Discrimination Measures

Christophe Marsala[1]([✉]), Anne Laurent[2], Marie-Jeanne Lesot[1], Maria Rifqi[3], and Arnaud Castelltort[2]

[1] Sorbonne Université, CNRS, Laboratoire d'Informatique de Paris 6, LIP6, 75005 Paris, France
{christophe.marsala,marie-jeanne.lesot}@lip6.fr
[2] LIRMM, Université de Montpellier, CNRS, Montpellier, France
{anne.laurent,arnaud.castelltort}@umontpellier.fr
[3] LEMMA, Université Panthéon-Assas, Paris, France
maria.rifqi@u-paris2.fr

Abstract. This paper proposes to exploit heterogeneous data, i.e. data described by both numerical and categorical features, so as to gain knowledge about the categorical attributes from the numerical ones. More precisely, it aims at discovering whether, according to a given data set, based on information provided by the numerical attributes, some categorical attributes actually are ordinal ones and, additionally, at establishing ranking relations between the category values. To that aim, the paper proposes the 3-step methodology OSACA, standing for Order Seeking Algorithm for Categorical Attributes: it first consists in extracting gradual patterns from the numerical attributes, to identify rich ranking information about the data; it then applies mathematical morphology tools, more precisely alternated filters, to induce an associated order on the categorical attributes. The third step evaluates the quality of the candidate rankings through an original measure derived from the rank entropy discrimination.

Keywords: Heterogeneous data · Ordinal attributes
Gradual patterns · Rank discrimination measure
Mathematical morphology

1 Introduction

By definition, heterogeneous data are described using several types of features, including numerical and categorical attributes. Research issues about such data usually aim at simultaneously exploiting all the attributes, raising questions about how to combine the information they respectively provide. For instance, clustering tasks can then be addressed using relational approaches and appropriate distance measures [3,13], classification can be performed using decision trees, that can efficiently process both types of attributes [19].

© Springer Nature Switzerland AG 2018
D. Ciucci et al. (Eds.): SUM 2018, LNAI 11142, pp. 152–163, 2018.
https://doi.org/10.1007/978-3-030-00461-3_11

Table 1. Illustrative toy data set.

Id	X	Y	Colour
o_1	0	1	Blue
o_2	1.2	1.5	Blue
o_3	1.8	1.6	Red
o_4	2.3	9.3	Yellow
o_5	2.5	9.8	Red
o_6	3.0	2.1	Blue
o_7	4.8	3.2	Yellow
o_8	5.0	8.5	Yellow

This paper proposes another point of view, aiming at exploiting the information provided by the numerical attributes so as to gain knowledge about the categorical ones. More precisely, the goal is to determine whether some of the categorical features actually are ordinal, and additionally identify the associated order.

For an illustration, consider the toy data set given in Table 1 where the data points are described by two numerical attributes and a categorical ones. It can be considered that, for this data set, the numerical attributes lead to the knowledge that the categorical attribute is ordinal, with partial order blue $\prec$ yellow.

To that aim, the paper proposes an original method called OSACA, standing for Order Seeking Algorithm for Categorical Attributes, that combines three tools, as detailed in Sect. 2, namely gradual patterns, mathematical morphology and rank entropy discrimination: gradual patterns are patterns of the form *the higher/lower a_1, ..., the higher/lower a_k* where a_i are numerical attributes. In the proposed approach, they are extracted to discover how the objects of the data base can be ranked. These rankings are then processed by mathematical morphology tools, more precisely alternated filters, to induce candidate orders on the categorical attributes. The candidates are finally evaluated through measures derived from the rank entropy discrimination.

The paper is organized as follows: Sect. 2 recalls some preliminaries, for the three types of tools, Sect. 3 describes the proposed approach and Sect. 4 presents illustrative results showing the relevance of OSACA. Section 5 concludes the paper and gives some directions for future works.

2   Preliminaries

This section provides some reminder about the three major tools used in the proposed method, successively gradual patterns, rank entropy measures and mathematical morphology tools.

Throughout the paper, $\Omega = \{o_1, \ldots, o_n\}$ denotes a set of n objects, or data points, described by a set $m + p$ attributes made of the union of the set $\mathcal{N}$, containing m numerical attributes, and the set $\mathcal{C}$, containing p categorical attributes. The value of attribute a for object o is denoted $o[a]$.

2.1   Gradual Patterns

Gradual patterns extract linguistic knowledge from data described by numerical attributes that can be expressed by patterns of the form *the higher/lower a_1, ..., the higher/lower a_k* where $a_i \in \mathcal{N}$, e.g. *the higher the budget, the higher the number of Champion's cup wins.* Initially introduced in the fuzzy implication formalism [6,7,10], gradual itemsets have then been interpreted as expressing constraints on attribute covariations. Within this framework, several interpretations of the constraints have in turn been proposed, as regression [11], correlation of induced order [2,12] or identification of compatible object subsets [4,5]. Each interpretation is associated with the definition of a support measure to quantify the validity of gradual itemsets and with methods for the identification of the itemsets that are frequent according to these support definitions.

This section focuses on the approach based on the identification of compatible object subsets [4,5], which is the one exploited in OSACA.

Definition 1 (Gradual Item). *A gradual item i is a pair $(a, *)$ where $a \in \mathcal{N}$ is a numerical attribute and $* \in \{\uparrow, \downarrow\}$ is the variation direction.*

Definition 2 (Gradual Pattern). *A gradual pattern P of size k is a set of k gradual items $\{(a_1, *_1), \ldots, (a_k, *_k)\}$, interpreted as their conjunction.*

It should be noted that there is no causality in such a pattern.

For instance $(budget, \uparrow)$ is a gradual item meaning *the higher the budget*, and $\{(X, \uparrow), (Y, \uparrow)\}$ is a gradual pattern meaning *the higher X* and *the higher Y*.

The main question is then the evaluation of the quality of candidate gradual patterns, with respect to the considered data set. One approach [4,5] proposes to consider that a pattern is all the truer as it occurs frequently in the data, measuring the truth degree by its *support*. In regular frequent patterns and association rules, computing the support amounts to counting the number of objects containing the pattern [1]. For gradual patterns, counting the support requires to rank the objects with respect to the pattern. This relies on the definition of the pattern-induced precedence relation, which defines a partial order on the objects:

Definition 3 (Object Precedence w.r.t. a Gradual Pattern). *Given a data set Ω, two objects $o, o' \in \Omega$ and a gradual pattern $P = \{(a_1, *_1), \ldots, (a_k, *_k)\}$, the precedence relation $o \prec_P o'$ in Ω holds if and only if $\forall j \in [1, k]$:*

- *if $*_j = \uparrow$ then $o[a_j] < o'[a_j]$*
- *if $*_j = \downarrow$ then $o[a_j] > o'[a_j]$*

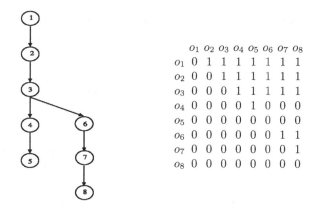

	o_1	o_2	o_3	o_4	o_5	o_6	o_7	o_8
o_1	0	1	1	1	1	1	1	1
o_2	0	0	1	1	1	1	1	1
o_3	0	0	0	1	1	1	1	1
o_4	0	0	0	0	1	0	0	0
o_5	0	0	0	0	0	0	0	0
o_6	0	0	0	0	0	0	1	1
o_7	0	0	0	0	0	0	0	1
o_8	0	0	0	0	0	0	0	0

Fig. 1. Precedence graph and matrix for pattern $\{(X,\uparrow),(Y,\uparrow)\}$ and data from Table 1. To make the graph simple, transitive edges are omitted.

For instance, considering the data set given in Table 1 and for the pattern $P = \{(X,\uparrow),(Y,\uparrow)\}$, it holds that $o_3 \prec_P o_4$, as $o_3[X] = 1.8 < 2.3 = o_4[X]$ and $o_3[Y] = 1.6 < 9.3 = o_4[Y]$.

The precedence relation leads to the definition of a precedence graph, where nodes represent objects and the directed edges the existence of a precedence relation, as illustrated on Fig. 1. The graph can equivalently be represented by its adjacency matrix, also shown on Fig. 1.

The precedence relation then leads to the definition of a support path:

Definition 4 (Support Path). *Given a data set Ω and a gradual pattern $P = \{(a_1, *_1), \ldots, (a_k, *_k)\}$, a support path p of length l is an ordered subset $\langle o_{\rho_1}, \ldots, o_{\rho_l} \rangle$ containing l objects from Ω such that $\forall j \in [1, l-1]$, $o_{\rho_j} \prec_P o_{\rho_{j+1}}$.*

We denote by $\mathcal{P}$ the set of all the paths supporting P.

Definition 5 (Support by Longest Path). *Given a data set Ω and a gradual pattern P, the support of P in Ω is defined by the length of the longest support path, relative to the total number of objects:* $supp(P) = \dfrac{1}{|\Omega|} \max_{p \in \mathcal{P}} (length(p))$.

For instance, the existing support paths for the pattern $P = (X \uparrow, Y \uparrow)$ are $\langle o_1, o_2, o_3, o_4, o_5 \rangle$ and $\langle o_1, o_2, o_3, o_6, o_7, o_8 \rangle$ (and all sub-chains from these chains). The support of P is thus $supp(P) = \frac{6}{8} = 0.75$.

Based on this quality criterion, one can then define the gradual patterns of interest, which must be both frequent and maximal:

Definition 6 (Frequent Gradual Pattern). *Given a data set Ω and a minimal support value minsup, a gradual pattern P is a* frequent *gradual pattern if $supp(P) \geq minsup$. In this case, P is said to be* frequent.

Definition 7 (Maximal Frequent Gradual Pattern). *Given a data set Ω and a minimal support value minsup, a frequent gradual pattern P is said to be maximal if there does not exist any frequent pattern P' such that $P \subset P'$.*

The GARE algorithm [5] proposes an efficient method to mine the maximal frequent gradual patterns according to these definitions.

2.2   Rank Discrimination Measures

Rank discrimination measures [16,17] have been studied in the setting of a specific classification task, when the class (or label) to be predicted is ordinal, and not categorical as it is usually the case. We propose to describe these data as *ordinally labelled*. The aim of such a task is to preserve a possible order on the class in the trained classifier and to highlight a gradual relation between the numerical attributes and the class.

For instance, monotonic decision trees [17] allow to address this task. Such decision trees are built from the attributes that are supposed to be the most gradually related to the class. They rely on the notion of rank discrimination measures: the latter are an extension of discrimination measures used in the attribute selection step of classic decision tree building, so as to favour attributes whose behaviour satisfies the monotonicity aim.

Formally, classically labelled (numerical) datasets first, only contain numerical attributes, i.e. $C = \emptyset$. Second, each data point is associated with a class by a labelling function $\lambda : \Omega \to C$ where $C = \{c_1, \ldots c_k\}$ is an ordered set of class, *i.e.* associated with a ranking $\prec_C$.

Rank discrimination measures are based on the concept of *dominance* [8]: any object $o \in \Omega$ is associated to the sets $[o]_\lambda^\leq$ and $[o]_a^\leq$ for any attribute a, respectively defined as

$$[o]_\lambda^\leq = \{o' \in \Omega \; : \; \lambda(o) \preceq_C \lambda(o')\}$$
$$[o]_a^\leq = \{o' \in \Omega \; : \; o[a] \leq o'[a]\}$$

In this paper, we focus on the rank Shannon entropy that corresponds to a rank version of the classical Shannon entropy. It is obtained by substituting the conditional probabilities with the corresponding measures obtained from the dominant sets, as shown in the next definition. The rank Shannon entropy can also be considered as a measure comparing the orders of two ordered sets: Ω_λ the set of objects from Ω ordered according to λ, and Ω_a the set of objects from Ω ordered according to a. We thus denote $H_S^*(\Omega_\lambda | \Omega_a)$ the rank Shannon entropy of attribute a with respect to the class labelling λ.

Definition 8 (Rank Shannon Entropy [9]). *For an ordinally labelled (numerical) dataset Ω, the rank Shannon entropy of attribute a with respect to the class labelling λ is computed as*

$$H_S^*(\Omega_\lambda | \Omega_a) = -\frac{1}{|\Omega|} \sum_{o \in \Omega} \log_2 \left(\frac{|[o]_\lambda^\leq \cap [o]_a^\leq|}{|[o]_a^\leq|} \right)$$

2.3   Mathematical Morphology to Induce Order

Mathematical morphology [20] defines a set of tools for the identification of spatial structures as the shape and size of objects, it has been extensively used for image processing and functional analysis. One-dimensional mathematical morphology [14,15] applies to symbolic words, obtained as data transcriptions on a set of symbols. In particular, it has been proposed as an efficient approach to perform an automatic fuzzy partition method, to discretise a numerical attribute: it makes it possible to identify class homogeneous intervals, tolerating some noise in the intervals, i.e. to highlight homogeneous class kernels. One-dimensional mathematical morphology has also been applied to enrich gradual patterns through characterising clauses exploiting categorical attributes [18].

Considering a numerical attribute a whose universe X_a has to be partitioned, first, a "word" is built, made by the values of the class for all $o \in \Omega$ ordered by the corresponding values of a for o. Then, a smoothing of this word is done to highlight consistent kernels of class values. During this smoothing, maximally homogeneous sequences of letters are searched for to define kernels for the fuzzy sets. To obtain such a sequence, a *morphological filter* is applied to the word using a particular class as structuring element.

A morphological filter is defined as the composition of *opening* and *closure* operators, themselves defined as composition of *dilation* and *erosion* operators: according to the class c used as structuring element, the dilation operator enables the merging of two sequences of letters c separated by a "small" number of other classes; the erosion operator enables the deletion of very small sequences of letters c [15].

3   Proposed Method

This section describes the proposed OSACA approach which consists in three steps, described in turn in the following subsections: the first one exploits the numerical attributes to highlight rich ordering information, inducing an object ranking, based on gradual patterns extraction and their supporting paths. The second step considers these paths from the point of view of the categorical attributes and processes the induced class words using a morphological alternated filter, so as to smooth them and identify kernels of homogeneous values and lead to a candidate (partial) ranking on the categorical attribute. The third step evaluates the candidate rankings and outputs the relevant ones, it is based on quality measures regarding both the ranking compatibility to the order induced by the gradual pattern and the number of involved objects.

3.1   Identification of Compatible Object Subsets

The first step exploits the numerical attributes to extract rich knowledge through gradual patterns using the covariation interpretation: they allow to maximally combine the attributes so as to derive rich ordering information, inducing an object ranking.

We propose to apply the GARE algorithm [5], whose theoretical principles are recalled in Sect. 2.1, as it offers the additional advantage of providing a subset of objects satisfying the induced ranking.

Given the data set Ω, this step outputs a set of gradual patterns from the numerical attributes $\mathcal{N}$, each of them being associated to a set of supporting paths $\mathcal{P}$.

3.2 Construction of Candidate Rankings

In the second step, for each gradual pattern P extracted by GARE in the previous step, each associated supporting path p is considered in turn to build a tentative ranking on each categorical attribute $c \in C$. In the following, we consider the categorical attribute as a labelling function: for any $o \in \Omega$, $\lambda(o) = o[c]$.

Given a path $p \in \mathcal{P}$, i.e. an ordered set of object $p = \langle o_1, \ldots, o_{|p|} \rangle$, and λ, the construction of a candidate ranking on C first builds the word made of the corresponding sequence of categorical values $w_p = \langle \lambda(o_1), \ldots, \lambda(o_{|p|}) \rangle$, denoted $w = \langle w_1, \ldots w_p \rangle$ below.

The method then consists in smoothing w in order to highlight kernels for each class. As recalled in Sect. 2.3, morphological alternated filters are appropriate tools to perform such an operation. In the particular setting considered here, a specific dilation operator needs to be defined: it must indeed enable the deletion of any class distinct from the structuring element, which is not possible in the classical approaches [14,15]. We thus introduce a *strong dilation* operator to enable the growth of a sequence of a class even if it induces the deletion of other classes. This strong dilation is thus used to define the corresponding alternate filter F.

The obtained filtered word $w'_p = F(w_p)$ then allows to define a partial order of the values taken by the considered attribute c: for each value c_i, the largest kernel associated to c_i in w'_p is retained (the first one in case of ties), and the order of these kernels provides the ranking of the associated values.

It can occur that no kernel for c_i survives after the filtering. In this case, the class is not involved in the ranking and the result is a partial order $\prec_p$ of the values of C induced by p. We denote by $\lambda^* \subseteq \lambda$ the subsets of labels that are comparable according to $\prec_p$. In the extreme case where $\lambda^* = \emptyset$, no order can be induced for this class: the corresponding attribute is only categorical and not, even partially, ordinal.

In the case of the toy data given in Table 1, considering the gradual pattern $P = \{(X, \uparrow), (Y, \uparrow)\}$, and the maximal support path $p = \langle o_1, o_2, o_3, o_6, o_7, o_8 \rangle$, the obtained word is $w_p = \langle$ *blue, blue, red, blue, yellow, yellow* $\rangle$. The strong dilatation by *blue* delete the *red* label and the forthcoming erosion leads to $\langle$ *blue, blue, ·, yellow, yellow* $\rangle$, suggesting the ranking *blue* $\prec_p$ *yellow*. The dot value corresponds to a value transformed in "no class" during the filtering [15].

3.3   Evaluation of a Candidate Ranking

In this step, any order $\prec_p$ identified in the previous step is evaluated by combining several quality criteria.

First, a notion of support is considered so as to quantify the proportion of data points concerned by the candidate ranking. This support is defined as the sum of the sizes of the kernels considered for building the ranking, divided by the path length. For the considered example, the support of the candidate partial ranking *blue* $\prec_p$ *yellow* is therefore $4/6 = 0.666$.

Second, the quality of the compatibility between the candidate ranking of categorical values and the pattern ranking is considered. It corresponds to a rank Shannon entropy applied to the subset of all data points Ω that are present in the considered path p with the orders $\prec_p$ on the categorical attribute and $\prec_P$ on the numerical attributes involved in the considered pattern.

It must be underlined that the induced ranking $\prec_p$ may not apply to all points in p, as it can be only partial, as it is for instance the case for the considered toy example. Therefore, we introduce, as a second quality criterion of the candidate rankings, the partial rank Shannon entropy defined as follows.

Definition 9 (Partial Rank Shannon Entropy). *Given P a gradual pattern, $\prec_P$ the induced data ranking, p a support path associated to P, c a categorical attribute, $\prec_p$ the categorical value ranking induced by p on c, and λ^* the subset of c values ordered according to $\prec_p$, the partial rank Shannon entropy is defined as:*

$$H_S^*(\Omega_{\lambda^*}|p) = -\frac{1}{|p|} \sum_{o \in p} \log_2 \left(\frac{|[o]_{\lambda^*}^{\prec_p} \cap [o]^{\prec_P}|}{|[o]^{\prec_P}|} \right)$$

It can be noted that, contrary to the rank Shannon entropy, the intersection $[o]_{\lambda^*}^{\prec_p} \cap [o]^{\prec_P}$ can be empty. In this case, the corresponding term in the sum is set to 0.

For instance, for the considered example, the partial rank Shannon entropy equals $-\frac{1}{6}\left(\log_2(\frac{5}{6}) + \log_2(\frac{5}{6}) + 0 + \log_2(\frac{3}{3}) + \log_2(\frac{2}{2}) + \log_2(\frac{1}{1}) \right) = 0.0659$. This small value highlights the fact that the ranking *blue* $\prec_p$ *yellow* should thus be considered.

4   Experiments

This section describes the experiments conducted to illustrate the OSACA approach on a set with 100 randomly generated data. Each data point is described by two numerical attributes, X and Y, and is associated with a categorical attribute, also called class, among the unordered set {blue, red, yellow, green}.

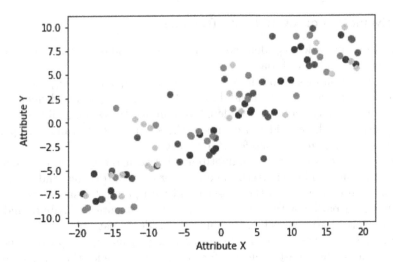

Fig. 2. Gradual relation on X and Y but no order on the class. (Color figure online)

4.1 Gradual Relation on X and Y but No Order on the Class

In the first experiment, represented on Fig. 2, the class values are not correlated with the numerical attributes: it is expected that no ranking is identified.

As there is a gradual relation between the two attributes X and Y, the gradual pattern extraction provides several paths for the gradual pattern $P = \{(X, \uparrow), (Y, \uparrow)\}$. There are 16 longest paths, each containing 34 objects. Paths are rather long as they contain one third of the objects of the dataset, which highlights the gradual relation.

After the morphological filtering on each of these paths, no kernel can be found and thus, as expected, no order on the class can be identified.

4.2 Gradual Relation on X and Y and Partial Order on the Class

In the second experiment, three of the four class values are generated with an underlying ranking (see Fig. 3): smaller values of the numerical attributes are associated with class value blue, middle values with yellow, and higher values with green. The red class is uniformly spread among the universes.

As in the previous experiment, the gradual relation between the two numerical attributes X and Y is highlighted by the fact that several paths for the gradual pattern $\{(X, \uparrow), (Y, \uparrow)\}$ are found: 2160 longest paths, each containing 24 objects are found. They are still rather long as they contain one quarter of the objects of the dataset, which highlights the existing gradual relation between X and Y.

In this case, the ranking of the class values is successfully identified: after the morphological filtering on each of these paths, 1800 paths among the 2160 highlight the order blue $\prec$ yellow $\prec$ green, with a support ranging from 0.75 to

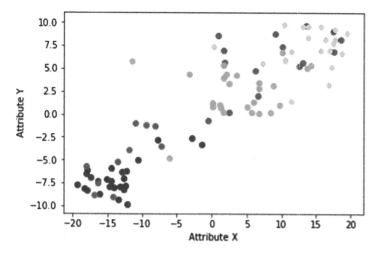

Fig. 3. Gradual relation on X and Y and partial order on the class. (Color figure online)

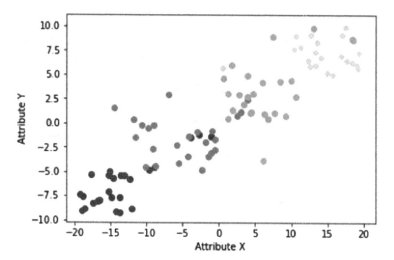

Fig. 4. Gradual relation on X and Y and order on the class (Color figure online)

0.875 and a very small partial rank Shannon entropy, ranging from 0 to 0.008. The other paths highlight the order blue $\prec$ green with a support ranging from 0.583 to 0.625 (Fig. 4).

4.3   Gradual Relation on X and Y and Order on the Class

In the third experiment, all 4 class values are ordered (see Fig. 3): the expected ranking should indicate blue before red, before yellow, before green.

As in the previous experiment, the gradual relation between the two attributes X and Y is highlighted by the fact that several paths for the gradual pattern $\{(X, \uparrow), (Y, \uparrow)\}$ are found: 16 longest paths, each containing 34 objects.

As expected, after the morphological filtering on each of these paths, all paths induce the ranking blue $\prec$ red $\prec$ yellow $\prec$ green. In all cases, the support is 1 and the partial rank Shannon entropy is very small, ranging from 0.02 to 0.03, highlighting the existence of the order on the class.

5   Conclusion and Future Works

The OSACA method proposed in this paper makes it possible to identify, among categorical attributes describing a set of data, the ones that actually are ordinal. Moreover, when such an attribute is pointed out, OSACA proposes a tentative ranking over its values and assesses the quality if this candidate by numerical measures, including an original ranking entropy measure. When applied to toy data sets, the method appears to provide relevant results, and opens many directions for future works, beside experiments on real data sets.

One question is to determine which gradual patterns should be considered: OSACA considers all gradual patterns whose support is greater than the minimum support. It may be relevant to consider only patterns with maximum number of attributes or only patterns with maximum support value. Indeed, due to the anti-monotonicity property, it is known that the longer a pattern, the lower the support. Besides, it may be relevant to choose a subset of categorical attributes to be studied, or even only one. Indeed, OSACA scans all categorical attributes from the database while some of them may not be interesting for end-users. Finally, it may be considered to evaluate the quality of the method by proposing an adaptation of the classical notions of precision and recall, respectively describing the extent to which the identified candidates, as well as the associated orders, are relevant and the extent to which the expected ordinal attributes have been retrieved, together with their rankings.

References

1. Agrawal, R., Imieliński, T., Swami, A.: Mining association rules between sets of items in large databases. In: Proceedings of the SIGMOD International Conference on Management of Data, SIGMOD, pp. 207–216. ACM (1993)
2. Berzal, F., Cubero, J.C., Sanchez, D., Vila, M.A., Serrano, J.M.: An alternative approach to discover gradual dependencies. Int. J. Uncertainty Fuzziness Knowl. Based Syst. **15**(5), 559–570 (2007)
3. Cheung, Y.M., Jia, H.: Categorical-and-numerical-attribute data clustering based on a unified similarity metric without knowing cluster number. Pattern Recognit. **46**(8), 2228–2238 (2013)
4. Di Jorio, L., Laurent, A., Teisseire, M.: Fast extraction of gradual association rules: a heuristic based method. In: Proceedings of the 5th International Conference on Soft computing as Transdisciplinary Science and Technology (CSTST), pp. 205–210 (2008)

5. Di-Jorio, L., Laurent, A., Teisseire, M.: Mining frequent gradual itemsets from large databases. In: Adams, N.M., Robardet, C., Siebes, A., Boulicaut, J.-F. (eds.) IDA 2009. LNCS, vol. 5772, pp. 297–308. Springer, Heidelberg (2009). https://doi.org/10.1007/978-3-642-03915-7_26

6. Dubois, D., Prade, H.: Gradual inference rules in approximate reasoning. Inf. Sci. **61**(1–2), 103–122 (1992)

7. Galichet, S., Dubois, D., Prade, H.: Imprecise specification of ill-known functions using gradual rules. Int. J. Approx. Reason. **35**(3), 205–222 (2004)

8. Greco, S., Matarazzo, B., Slowinski, R.: Rough approximation by dominance relations. Int. J. Intell. Syst. **17**(2), 153–171 (2002)

9. Hu, Q.H., Guo, M.Z., Yu, D.R., Liu, J.F.: Information entropy for ordinal classification. Sci. China Inf. Sci. **53**, 1188–1200 (2010)

10. Hüllermeier, E.: Implication-based fuzzy association rules. In: De Raedt, L., Siebes, A. (eds.) PKDD 2001. LNCS (LNAI), vol. 2168, pp. 241–252. Springer, Heidelberg (2001). https://doi.org/10.1007/3-540-44794-6_20

11. Hüllermeier, E.: Association rules for expressing gradual dependencies. In: Elomaa, T., Mannila, H., Toivonen, H. (eds.) PKDD 2002. LNCS, vol. 2431, pp. 200–211. Springer, Heidelberg (2002). https://doi.org/10.1007/3-540-45681-3_17

12. Laurent, A., Lesot, M.-J., Rifqi, M.: GRAANK: exploiting rank correlations for extracting gradual itemsets. In: Andreasen, T., Yager, R.R., Bulskov, H., Christiansen, H., Larsen, H.L. (eds.) FQAS 2009. LNCS (LNAI), vol. 5822, pp. 382–393. Springer, Heidelberg (2009). https://doi.org/10.1007/978-3-642-04957-6_33

13. Li, C., Biswas, G.: Unsupervised learning with mixed numeric and nominal data. IEEE Trans. Knowl. Data Eng. **14**(4), 673–690 (2002)

14. Marsala, C.: Fuzzy partitioning methods. In: Pedrycz, W. (ed.) Granular Computing: An Emerging Paradigm. STUDFUZZ, vol. 70, pp. 163–186. Springer, Heidelberg (2001). https://doi.org/10.1007/978-3-7908-1823-9_8

15. Marsala, C., Bouchon-Meunier, B.: Fuzzy partioning using mathematical morphology in a learning scheme. In: Proceedings of the 5th IEEE International Conference on Fuzzy Systems, vol. 2, pp. 1512–1517 (1996)

16. Marsala, C., Petturiti, D.: Hierarchical model for rank discrimination measures. In: van der Gaag, L.C. (ed.) ECSQARU 2013. LNCS (LNAI), vol. 7958, pp. 412–423. Springer, Heidelberg (2013). https://doi.org/10.1007/978-3-642-39091-3_35

17. Marsala, C., Petturiti, D.: Rank discrimination measures for enforcing monotonicity in decision tree induction. Inf. Sci. **291**, 143–171 (2015)

18. Oudni, A., Lesot, M.J., Rifqi, M.: Characterisation of gradual itemsets through" especially if" clauses based on mathematical morphology tools. In: Proceedings of the conference of the European Society for Fuzzy Logic and Technology (EUSFLAT), pp. 826–833 (2013)

19. Quinlan, J.R.: Induction of decision trees. Mach. Learn. **1**(1), 86–106 (1986)

20. Serra, J.: Introduction to mathematical morphology. Comput. Vis. Graph. Image Process. **35**(3), 283–305 (1986)

Distribution-Aware Sampling of Answer Sets

Matthias Nickles[(✉)]

School of Engineering and Informatics, National University of Ireland, Galway,
Galway, Ireland
matthias.nickles@nuigalway.ie

Abstract. Distribution-aware answer set sampling has a wide range of
potential applications, for example in the area of probabilistic logic programming or for the computation of approximate solutions of combinatorial or search problems under uncertainty. This paper introduces
algorithms for the sampling of answer sets under given probabilistic constraints. Our approaches allow for the specification of probability distributions over stable models using probabilistically weighted facts and
rules as constraints for an approximate sampling task with specifiable
accuracy. At this, we do not impose any independence requirements on
random variables. An experimental evaluation investigates the performance characteristics of the presented algorithms.

Keywords: Nonmonotonic Probabilistic Logic Programming
Answer Set Programming (ASP) · Relational Artificial Intelligence
Probabilistic Programming

1 Introduction

This work addresses the problem of *answer set* [9] sampling, defined as the computation of a multiset of answer sets approximately sampled from a probability distribution constrained by probabilistically weighted atoms (with weighted
models as an instance of this scenario). We construe distribution-aware sampling as an optimization problem where users specify targeted probability distributions over answer sets by assigning probabilities to arbitrary normal rules
(translated into plain answer set programs and weighted atoms using syntactic
transformations) - that is, basically in the same way in which many probabilistic
logic programming frameworks specify uncertainty. Search algorithms are then
employed to find a sample which *approximately* reflects the given distribution by
minimizing an associated specific cost (loss) function. At this, we do not impose
restrictions (such as a requirement of independence of random variables) on the
logic programs or the probabilistic rules or weights, except consistency. One of
our approaches is an enhancement and modification of the current state-of-the-
art Answer Set Programming (ASP) solving algorithm. Experiments indicate

© Springer Nature Switzerland AG 2018
D. Ciucci et al. (Eds.): SUM 2018, LNAI 11142, pp. 164–180, 2018.
https://doi.org/10.1007/978-3-030-00461-3_12

that this leads to performance gains over solutions which employ an external conventional ASP solver.

An envisaged application area (which is itself outside the scope of this introductory paper) is query answering in probabilistic logic programming (by summing over a sample of weighted possible worlds to obtain an approximate result). Generally, potential applications include all cases where the solution sets of combinatorial or search problems with informative priors over rules and facts can be approximated using samples of models. To approach our objective technically, we present and compare several algorithms: our main contribution is a new algorithm (Sect. 5.2) which extends the state-of-the-art ASP solving approach (Conflict-Driven Nogood Learning, CDNL-ASP) with probabilistic capabilities. Since our approach is an optimization method, we also introduce new algorithms which instantiate or enhance suitable existing optimization methods (SGD (Sect. 4.1) and the Lagrange multipliers method (Sect. 4.2)) for answer set sampling, to provide a basis for benchmarking. We also compare (in Sect. 5.1) our new algorithms with an improved variant of the (to our best knowledge) only existing approach to a (somewhat different) form of distribution-aware sampling in ASP [15], namely simulated annealing with answer sets.

2 Related Work

Despite its expected usefulness, sampling of answer sets has not gained much attention so far. To our best knowledge, the main algorithm introduced in this paper is the first realistic and scalable distribution-aware sampling approach in ASP which supports arbitrary distributions over atoms and rules.

Existing approaches which use sampling in ASP employ near-uniform sampling [11,15] using parity (XOR) constraints [10], a family of approaches devised for Boolean assignment sampling with SAT solvers (made efficient using specialized solvers such as CryptoMiniSat [19]). While (near-)uniform sampling might be required as a subroutine in certain other forms of sampling (we will use it this way in some of the algorithms presented here), it is a different task compared to distribution-aware sampling - the former can be expressed in terms of the latter by assigning identical weights to all possible worlds (written as conjunctions), but this is might not be a practicable approach. [2] proposes a form of distribution-aware sampling for SAT, but its scenario and prerequisites are different from ours: it addresses SAT instead of ASP and it does not consider sampling based on probabilistic rules and facts (utilizing ASP's nature as a declarative programming language). Its algorithms are also not significantly related to ours (building on concepts from weighted model counting and parity constraints instead of our mostly machine learning-based approaches).

A difference to existing approaches to optimization in ASP, SAT and related paradigms such as constraint solving is that we aim for optimal multisets of models, not optimal individual models, preference relations among individual models, or individual variables.

PSAT (probabilistic Boolean Satisfiability) [7] tackles a closely related but different problem than ours. While we are interested in finding multisets of models under probabilistic constraints which are very similar to those in PSAT, PSAT determines whether or not there exists a probability distribution over all models which fit the given PSAT instance. The related SSAT problem [17] assigns probabilities to random variables to be true, and a solution is an assignment of truth values to existential variables which yields the maximum formula probability of satisfaction. A form of distribution-aware sampling was proposed in [15], however, with certain limitations (we examine a variant of this approach later in this paper). Our work is also related to Probabilistic Databases [4] (in the sense of relational databases with explicit support for probabilistic uncertainty), in particular those which allow for attribute-level uncertainty. With this area, we share the challenge that the number of possible worlds induced by uncertainty can be very large (exponential in the number of random attributes) - hence the need for approximation methods. However, their semantics and representation formats are different from ASP. An envisaged use case for our approach is as an implementation tool for Probabilistic Logic Programming (PLP) languages, in particular PLP under stable model semantics, with existing such frameworks including [1,12,14,15].

3   Preliminaries and Notation

Recall that a (normal) *logic program* is a finite set of *normal rules* of the form $h :- b_1, ..., b_m,$ *not* $b_{m+1}, ..., not$ b_n (with $0 \leq m \leq n$). h and the b_i are atoms. *not* represents default negation. Rules without body literals are called *facts*. The set of literals in the body of a rule r is denoted as $body(r)$, its head atom is denoted as $head(r)$. Most other syntactic constructs supported by contemporary ASP solvers (like choice rules or classical negation) can be translated into (sets of) normal rules. We only consider ground rules and ground atoms in this work. The *answer sets* (*stable models*) of a normal logic program are as defined in [9]. Throughout the paper, we use the term "model" or "possible world" in the sense of answer set. Ψ_{asp} denotes the set of all answer sets of answer set program *asp*. We often write s_i to denote the i-th element of a set s under an arbitrary but fixed order. We enhance logic programs to *weighted programs* by allowing for probabilities (called *target weights* or just *weights*) attached to arbitrary rules:

Definition 1. *Weighted program*
A weighted program is a finite set of tuples (called weighted rules*) of the following form:* $(w, \ h :- b_1, ..., b_m, \ not \ b_{m+1}, ..., not \ b_n)$, *with* $0 \leq m \leq n, w \in (0; 1]$ *and atoms* b_i. *The* w *are called* weights *of the rules.* $w(r)$ *denotes the weight* w *of a rule* r.

Specifying probability distributions directly over entire possible worlds is a straightforward instance of the above.

Do avoid the unnecessary dealing with weighted rules with nonempty bodies, we de-sugar all programs into (1) a set of weighted atoms (called *weight associations*) and a plain answer set program called *spanning program* consisting of non-weighted rules (similar to the normalization step commonly applied in PSAT [7]):

Definition 2. *Spanning program*
The spanning program $span(wp)$ *obtained from weighted program* wp *is the normal logic program* $\{h :- not\ aux_i \mid \exists(w_i, r_i) \in wp : h = head(r_i), body(r_i) = \emptyset, w_i < 1\}$
$\cup\{aux_i :- not\ h \mid \exists(w_i, r_i) \in wp : h = head(r_i), body(r_i) = \emptyset, w_i < 1\}$
$\cup\{h :- l_1, ..., l_n,\ not\ aux_i \mid n > 0, \exists(w_i, r_i) \in wp : h = head(r_i), body(r_i) = \{l_1, ..., l_n\}, w_i < 1\}$
$\cup\{aux_i :- l_1, ..., l_n,\ not\ h \mid n > 0, \exists(w_i, r_i) \in wp : h = head(r_i), body(r_i) = \{l_1, ..., l_n\}, w_i < 1\}$
$\cup\{r_i \mid \exists(w_i, r_i) \in wp, w_i = 1\}.$
The aux_i *are fresh symbols per each rule* r_i *in* wp. *The rules which contain* aux_i *are called* spanning rules *or* spanning formulas. *Remark: in extended ASP syntax, spanning rules for facts can be written as choice rules* $\{h\}$, *and further equivalent forms exist.*

The concept of spanning programs and rules is borrowed from [12,15]. Spanning rules allow our algorithms to make a choice between including or omitting uncertain atoms in a model, however, the user can alternatively provide some other "spanning constructs" to this end. Our algorithms do *not* require answer set programs to have the form of spanning programs, they work with arbitrary plain answer set programs together with *atom weight associations*, consistency assumed. We nevertheless assume henceforth that all plain answer set programs are spanning programs. The spanning program induces a simplified weighted program where only atoms can carry weights, which we call *associated* with the respective uncertain atoms:

Definition 3. *Atom weight associations*
The atom weight associations *(or atom target weights) for the spanning program* $asp = span(wp)$ *obtained from weighted program* wp *is the set* $\phi^{asp} = \{head(r_i) \mapsto w_i \mid \exists(w_i, r_i) \in wp : w_i < 1, body(r_i) = \emptyset\}$
$\cup\{aux_i \mapsto 1-w_i \mid \exists r_i \in span(wp):r_i = aux_i :- b_1, ..., b_n,\ not\ h, w_i = w(h :- l_1, ..., l_n)\}$

We denote the set of all atoms in the weight associations set (i.e., the uncertain atoms) as θ^{asp} and individual uncertain atoms as θ_i^{asp}. We write $\phi^{asp}(\theta_i^{asp})$ or ϕ_i^{asp} to denote the weight of atom θ_i^{asp}. We omit the ^{asp} where it is clear from the context which program is meant. To distinguish the user-defined atom weights ϕ from actual atom frequencies (normalized counts of models where the atom holds within some possibly incomplete sample of models), we write φ for those frequencies instead of ϕ.

3.1 Sample Distributions

Informally, the sampling process iteratively adds models to the sample under construction until either for each uncertain atom its frequency in the sample (number of answer sets in the sample multiset where the atom holds, normalized by the sample's multiset cardinality) is equal to its given weight, or close to the given weight up to some specified *error threshold* (also called *accuracy*), as detailed later. If, for example, the given weight of atom a is 0.4, the sample should contain ~40% not necessarily different answer sets where a holds (the frequencies of the individual answer sets in the sample where a holds add up to ~40%). It should be observed that atom weights do, in general, not directly tell us anything about the probabilities of models or formulas, in contrast to *Weighted Model Counting* (WMC).

To get an idea of the worst-case complexity involved, consider that deciding whether a propositional normal logic program has some answer set is an NP-complete task [5]. Obtaining a distribution over given answer sets which observes the weight-induced probabilistic constraints is a linear programming problem (polynomial time complexity) and sampling from that distribution can also done in polynomial time. However, these considerations do not necessarily tell us much about the actual feasibility of real-world tasks; e.g., the linear programming task can become infeasible if the number of possible worlds is large, and this number blows up in space exponentially in the number of uncertain atoms. It should also be considered that a major strength of ASP is its ability to encode complex problems and to be efficient for *some* instances of these.

We define a (distribution-aware) *sample* of answer sets of answer set program asp as a non-empty multiset $[m_i]$, $Supp([m_i]) = \Psi_{asp}$, such that the probability that a specific answer set m_i is included in this sample (with replacement) is defined by a discrete probability distribution over answer sets with probability mass function $pmf^{asp}(m_i) = Pr(m_i)$, where $m_i \in \Psi_{asp}$ and $Pr(m_i)$ is the probability of m_i. A sample of a weighted program is defined as above with asp being the spanning program.

A weighted answer set program wp induces a set of probability distributions as follows:

Definition 4. *Target probability distributions*
We call set $pds = \{pmf_i^{asp}\}$ *the set of weight-induced target probability distributions (over the answer sets* $m_i \in \Psi_{asp}$ *of spanning program* $asp = span(wp)$*)*
iff $\forall pmf_i^{asp} \in pds \ \forall \theta_j \in \theta : \sum_{m_i \in \Psi_{asp}, \theta_j \in m_i} pmf_i^{asp}(m_i) = \phi_j.$

The condition for pds above defines a system of linear equations which is the typical way of defining possible world probabilities in Nilsson-/PSAT-style [16] probabilistic logics: per each uncertain atom θ_k, we get an equation $v(k, m_1)pmf_i^{asp}(m_1) + ... + v(k, m_n)pmf_i^{asp}(m_n) = \phi(k)$, with $\Psi_{asp} = \{m_i\}$, $n = |\Psi_{asp}|$, and $v(k, m_j) = 1$ if $\theta_k \in m_j$ and $v(k, m_j) = 0$ otherwise. The solution sets pmf_i^{asp} of this equation systems with the two extra constraints $pmf_i^{asp}(m_i) \in [0, 1]$ and $\sum_i pmf_i^{asp}(m_i) = 1$ are precisely the elements of pds.

We say that together with the spanning program, the user-specified atom weights ϕ "define" the set *pds* of target distributions. Observe that the probabilities of possible worlds (answer sets of the spanning program) may be zero.

As usual in Nilsson-style PLP, the weighted program typically under-specifies the equation system, so typically multiple (even infinitely many) correct solutions exist. However, only one of the algorithms presented in this paper guarantees a maximum entropy target distribution, whereas the others only guarantee (assuming sampling is precise) that per each run, sampling takes place from *one* of the target distributions.

In this work, we only consider weighted programs for which the above system of equations and extra conditions is solvable, which requires that the spanning program is logically consistent and that the given weights do not cause probabilistic inconsistency.

To allow the user to trade off accuracy against speed, our algorithms proceed until a given *error threshold (accuracy)* ψ is reached for the Mean Squared Error (MSE) *cost*:

Definition 5. *Cost function*
The cost function *used in this work is defined as* $cost(\phi, \varphi) = \frac{1}{n}\sum_{i=1}^{n}(\phi_i - \varphi_i)^2$ *(Mean Squared Error (MSE); squared Euclidean distance normalized with the number n of uncertain atoms) over given atom target weights ϕ and a list φ of frequencies of these atoms.*

For convenience, we also define a variant which operates on sample multisets:
$$cost(\phi, sample) = cost(\phi, \varphi) \text{ where } \varphi_k = \frac{|[pws_j| \ pws_j \in sample, \theta_k \in pws_j]|}{|sample|} \text{ for}$$
each $k, \theta_k \in \theta$

MSE is a common choice for the cost function in Gradient Descent and other Monte Carlo-style iterative algorithms such as ours. Although in principle other cost functions could be used too, MSE is used in this paper for all algorithms (and with the same threshold in all experiments, see Sect. 6), thus making the accuracies achieved with the different algorithms comparable with each other. MSE expresses the acceptable minimum accuracy directly in terms of the maximum acceptable difference between an uncertain atom's given weight and its actual weight in the sample (sum of the frequencies of its possible worlds in the sample).

To show that, in the precise case (sampling until $cost(\phi, \varphi) = 0$) the resulting sample s matches a target distribution from *pds*, we need to show that the vector of model frequencies $(\frac{multi(m_1)}{|s|}, ..., \frac{multi(m_n)}{|s|})$ in sample s (where $multi(m_i)$ denotes the multiplicity of answer set m_i within s) approaches solution of the equation system above (the two extra constraints are trivially fulfilled). This can be seen from $cost(\phi, \varphi) = 0 \Rightarrow \forall k \in 1..|\theta| : \phi_k = \varphi_k$ and rewriting, for every uncertain atom θ_k, $\phi_k = \varphi_k = \frac{|[m_j \ | \ m_j \in s, \theta_k \in m_j]|}{|s|} =$

$$\sum_{m_i \in s, \theta_k \in m_i} \frac{multi(m_i)}{|s|} = v(k, m_1)\frac{multi(m_1)}{|s|} + ... + v(k, m_n)\frac{multi(m_n)}{|s|}, \text{ with}$$
$v(k, m) = 1$ if $\theta_k \in m$ and $v(k, m) = 0$ otherwise.

If error threshold ψ for $cost(\phi, \varphi)$ is larger zero, the equation system above becomes a system of inequalities of the form $\phi(k) - \sqrt{n\psi} \leq ... \leq \phi(k) + \sqrt{n\psi}$ per each uncertain atom θ_k, with n being the number of uncertain atoms (that is, setting ψ allows for a controlled amount of inaccuracy with which the sample must reflect the given weights). However, $\pm\sqrt{n\psi}$ is just the worst case.

4   Two-Step Approaches

We distinguish two categories of sampling approaches: (1) *Two-step approaches* which compute an estimate of one target probability distribution over possible worlds (the answer sets of the spanning program, or a subset thereof) and then sample models from that distribution (or omit the first step and sample from a given target distribution), and (2) *direct approaches* which compute a sample directly from a program and atom weights. The two-step approaches presented here have the advantage of being easy to implement on top of some off-the-shelf ASP solver, and their two steps can be used separately from each other (e.g., if only the distribution is required but not the sampling), or (with a trivial tweak) to fix one target distribution (step one) and then to apply step two multiple times to sample multiple times from that distribution. However, as we will see later, they perform much worse than the direct approaches in our experiments. A special case of (1) are approaches where a distribution is already given (perhaps obtained using some tool like LP^{MLN} [12]). We have also experimented with Linear Programming (LP) using the Simplex algorithm for the determination of a target distribution, but found it infeasible for our application for anything but small systems, due to the typically large number of answer sets (corresponding to the number of unknowns in the LP objective function, which is over possible worlds, not atoms).

4.1   Stochastic Gradient Descent with MCMC

The first approach we present uses Stochastic Gradient Descent (SGD) for the target probability distribution search, followed by an application of a Markov Chain Monte Carlo (MCMC) method. The overall approach does not seem to be practical (see Sect. 6), and is reported only for its simplicity. In the instance shown below (Algorithm 1) we use Metropolis-Hastings (MH), but other forms of MCMC would certainly be usable too. The purpose of the MCMC step is to sample a specified number of models in accordance with the previously approximated distribution.

   Algorithm 1 (SGD-ASP-MH) starts with calling DISTRISGD with arguments n (requested number of sampled models), *maxTrials*, *asp* (the spanning program of the weighted program), γ (SGD learning rate. We used 1e−4, a common default, in our experiments) and ψ (error threshold (accuracy), see Sect. 6 for value used). θ and ϕ are the program's uncertain atoms and their given probabilities, as described in the previous section. Procedure UNISAMPLE first fetches a set of approximately-uniformly "sufficiently large" (see further below) sampled set of answer sets with the help of some external existing answer set solver using some additional near-uniform sampling procedure. In the subsequent SGD loop, the parameters whose values we want to learn are the initially unknown probabilities of these answer sets which the subsequent MH part should use for the actual sampling. In each trial (loop iteration), we select randomly one of the uncertain atoms θ_i, and compute a gradient ∇ whose subtraction from the current estimate of possible world probabilities decrements the cost function (actually, we use Momentum updates in Algorithm 1 which are slightly more complicated, see below). The cost function calculates how "close" the uncertain atom weights as predicted under the current estimated possible world distribution are to the given target uncertain atom weights. Using MSE (see Sect. 3) as cost function $cost(\phi, \varphi) = \frac{1}{n} \sum_{i=1}^{n} cost_i$ with $cost_i = (\phi_i - \varphi_i)^2$ and the estimates of possible world probabilities as parameters, we get the following negative gradient vector (per each uncertain atom with index i - in each SGD iteration, our algorithm chooses i randomly) $\nabla_i = -\nabla cost_i = -2(\phi_i - \varphi_i) \left(\frac{\partial}{\partial pws_j} \sum_{k:\theta_i \in pws_k} prPws_k \right)$ where $prPws$ is a tuple consisting of the current estimates of the probabilities of possible worlds $\{pws_k\}$. This simplifies to 0 for possible worlds pws_j in which atom θ_i does not hold, and to $-2(\phi_i - \varphi_i)$ for all other possible worlds.

Function $updws(pws, prPws)$ computes the current estimates φ_i of the probabilities of the uncertain atoms, given the current model probability estimates $prPws$, as the sums of the estimated probabilities of those answer sets where the respective atom is true. After having estimated a probability distribution (the time critical part of this approach), we use a simple Metropolis-Hastings (MH) sampler (MH) to sample from that distribution, using the uniform distribution as symmetric proposal distribution (so we actually utilize the Metropolis variant of MH). It should be observed that sampling from the distribution outputted by the SGD part using naive sampling would only be possible for small systems, due to the typically very large number of models and minuscule probabilities of individual answer sets in a target distribution with larger systems.

The SGD part of Algorithm 1 is fairly standard, but the fact that we are working directly with probabilities (which need to be normalized) instead of some unrestricted kinds of parameters somewhat complicates the basic SGD algorithm: To ensure that the sum of all probabilities of all possible worlds remains (about) 1 after each gradient descent step, we add a normalization step by adding a (virtual) fact **true** to every answer set and assign atom **true** the weight 1 - this way, Algorithm 1 aims at keeping the sum of all possible world probabilities, i.e., all possible worlds where **true** holds, close to 1. Another - purely technical - modification is the use of *Momentum* in our updates (this is why we use *addons* instead of directly incrementing or decrementing the possible world probabilities). Experiments with other GD learning rate adaptation heuristics (Adagrad and Adam) did not improve the performance in our experiments.

Algorithm 1 SGD with subsequent Metropolis-Hastings Sampler (SGD-ASP-MH)

1: **procedure** SAMPLESGDMH(n (requested number of sampled models), $maxTrials$, asp (spanning program), θ (uncertain atoms), ϕ (weights per atom), γ (SGD learn rate), ψ (error threshold)) (see beginning of Sect. 4.1)
2: $\quad (pws, prPws) \leftarrow$ DISTRISGD($maxTrials, asp, \theta, \phi, \gamma, \psi$)
3: $\quad$ **return** MH($n, pws, prPws$)
4: **procedure** DISTRISGD($maxTrials, asp, \theta, \phi, \gamma, \psi$)
5: $\quad pws \leftarrow$ UNISAMPLE($asp, *$) $\qquad \triangleright$ possible worlds
6: $\quad$ Initialize $prPws_i \leftarrow 1/|pws|$, for all $i \in \{1..|pws|\}$
7: $\quad$ Initialize $addons_i \leftarrow 0$, for all $i \in \{1..|pws|\}$
8: $\quad \varphi \leftarrow updws(pws, prPws)$ $\qquad \triangleright$ sets each φ_j to $\sum_{i, \theta_j \in pws_i} prPu$
9: $\quad t \leftarrow 1$
10: $\quad$ **repeat**
11: $\quad\quad i \leftarrow rand_1^{|\theta|}$
12: $\quad\quad \nabla_i \leftarrow -2(\phi_i - \varphi_i)$
13: $\quad\quad$ **for each** $j \in \{k : \theta_i \in pws_k\}$ **do**
14: $\quad\quad\quad addons_j \leftarrow addons_j + \gamma * \nabla_i$
15: $\quad\quad\quad prPws_j \leftarrow prPws_j - addons_j$
16: $\quad\quad \varphi \leftarrow updws(pws, prPws)$
17: $\quad\quad k \leftarrow t + 1$
18: $\quad$ **until** $cost(\phi, \varphi) \leq \psi \vee t > maxTrials$
19: $\quad$ **return** $(pws, prPws)$ $\qquad \triangleright$ poss. worlds pws and their probabilities pr
20: **procedure** MH($n, pws, prPws$)
21: $\quad sample \leftarrow [\,], lmi \leftarrow rand_1^{|pws|}$ $\qquad \triangleright$ index of last sampled model, in
22: $\quad$ **repeat**
23: $\quad\quad candModelIndex \leftarrow rand_1^{|pws|}$
24: $\quad\quad \alpha \leftarrow prPws_{candModelIndex} / prPws_{lmi}$
25: $\quad\quad$ **if** $rand_0^1 \leq \alpha$ **then** $lmi \leftarrow candModelIndex$
26: $\quad\quad sample \leftarrow sample \uplus \{pws_{lmi}\}$
27: $\quad\quad \varphi_k \leftarrow \frac{|[pws_j \mid pws_j \in sample, \theta_k \in pws_j]|}{|sample|}$ for each φ_k
28: $\quad$ **until** $|sample| = n$
29: $\quad$ **return** $sample$

The major bottleneck of Algorithm 1 is the fact that it first needs to obtain a sufficiently extensive set of answer sets to work on. What "sufficiently" means needs to be experimentally determined - our implementation starts with a small set and requests further models from the external ASP solver if the SGD part does not converge within the specified number of trials (Algorithms 1 and 2 are somewhat simplified in the sense that they assume that helper procedure UNISAMPLE always delivers a sufficient set of answer sets from the external solver). Technically, in our implementation of UNISAMPLE, we fetch bulks of models from an external solver (Clingo 5.2/clasp) with call arguments which cause a high degree of randomness in its decision making. This does not achieve real uniformity, but is much faster than the usual parity (XOR) constraint-based method, and was sufficient to converge to a target distribution in our experiments. Note that line 23 ensures that the MH proposal distribution is truly uniform (and thus symmetric).

4.2 Lagrange Multiplier Method with MCMC

The second algorithm (LMM-ASP-MH) we consider uses the well-known method of Lagrange multipliers (LMM) for determining a probability distribution over models using optimization, followed by the same MH-based sampling step as in SGD-ASP-MH. The only difference to existing implementations of LMM (except for the sampling step) is the fact that we work with answer sets, which does not require an adaptation of the LMM algorithm. Using Lagrange multipliers to solve for possible world weights is one of the oldest techniques used in probabilistic logic, dating back to [3,16].

The benefit over the other algorithms in this paper is that it results in a maximum-entropy distribution if the set of possible worlds is complete. The overall LMM-ASP-MH algorithm is largely identical with Algorithm 1, with the only significant difference that instead of SGD we now use the Lagrange multipliers method to approximate a maximum entropy probability distribution over answer sets. The actual LMM algorithm (replacing procedure DISTRISGD) and its theoretical analysis are omitted here for lack of space, details can be found in [16,18]. As we will see in Sect. 6, this approach also performs quite badly in our experiments - however, its results are less prone to information bias.

5   Direct Approaches

The following distribution-aware sampling approaches generate the sample directly from the given answer set program and list of atom probabilities, in the (justified) hope that this is more efficient than two-step approaches. It should be kept in mind though (see Sect. 3) that while the multiset they produce reflect one target distribution, results from different runs of the algorithm can be from different target distributions. The two-step approaches in the previous section can overcome this issue by fixing a certain target distribution with step one and then applying step two multiple times. With the direct approaches, we can obtain one sample multiset in the sense of Sect. 3 and then sample from it uniformly multiple times until the desired number of models is reached.

5.1   Simulated Annealing for Answer Set Sampling

Algorithm 2 shows how to use Simulated Annealing (SA) for sampling answer sets (SimA-ASP), based on the SA variant proposed in [15], with the main difference that the variant presented here uses a threshold of the MSE cost function as convergence criterion (like all algorithms in this paper). Without this improvement, the algorithm proposed in [15] would not be guaranteed to converge, with the specified accuracy, to a target distribution in the sense of this paper. Note that [15] also provides a much more efficient sampling algorithm for the special case that all uncertain atoms are mutually independent, a restriction which we do not make in this paper (the algorithms in this work allow for arbitrary dependencies among uncertain atoms).

Simulated Annealing can be seen as an adaption of Metropolis-Hastings to hard global optimization problems, so the core of this algorithm is similar to procedure MH in Algorithm 1. Basically, it explores the space of answer sets, iteratively evaluating uniformly-sampled candidate models regarding whether or not accepting them would decrease the cost function. A cooling scheme with slowly decreasing temperature decreases the probability of accepting worse answer sets (wrt. cost function change) into the sample, trading-off search space exploration against greediness.

nhSize specifies the neighborhood size, that is, the number of uniformly sampled candidate models from which the top-ranked is selected in each iteration (15 in our experiments) and ψ is the error threshold. We use 1e−5 as initial temperature *initialTemp*.

Algorithm 2 Answer Set Sampling with Simulated Annealing (SimA-ASP)

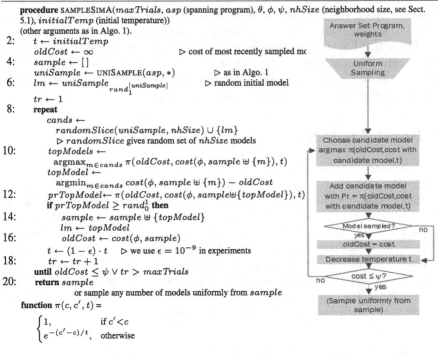

procedure SAMPLESIMA(*maxTrials*, *asp* (spanning program), θ, ϕ, ψ, *nhSize* (neighborhood size, see Sect. 5.1), *initialTemp* (initial temperature))
(other arguments as in Algo. 1).
2:     $t \leftarrow initialTemp$
       $oldCost \leftarrow \infty$ ▷ cost of most recently sampled mc
4:     $sample \leftarrow []$
       $uniSample \leftarrow$ UNISAMPLE(*asp*, *)   ▷ as in Algo. 1
6:     $lm \leftarrow uniSample_{rand_1^{|uniSample|}}$ ▷ random initial model

       $tr \leftarrow 1$
8:     **repeat**
          $cands \leftarrow$
             $randomSlice(uniSample, nhSize) \cup \{lm\}$
             ▷ *randomSlice* gives random set of *nhSize* models
10:       $topModels \leftarrow$
             $\text{argmax}_{m \in cands} \pi(oldCost, cost(\phi, sample \uplus \{m\}), t)$
          $topModel \leftarrow$
             $\text{argmin}_{m \in cands} cost(\phi, sample \uplus \{m\}) - oldCost$
12:       $prTopModel \leftarrow \pi(oldCost, cost(\phi, sample \uplus \{topModel\}), t)$
          if $prTopModel \geq rand_0^1$ **then**
14:          $sample \leftarrow sample \uplus \{topModel\}$
             $lm \leftarrow topModel$
16:          $oldCost \leftarrow cost(\phi, sample)$
          $t \leftarrow (1 - \epsilon) \cdot t$ ▷ we use $\epsilon = 10^{-9}$ in experiments
18:       $tr \leftarrow tr + 1$
       until $oldCost \leq \psi \vee tr > maxTrials$
20:    **return** *sample*
          or sample any number of models uniformly from *sample*
    function $\pi(c, c', t) =$

$$\begin{cases} 1, & \text{if } c' < c \\ e^{-(c'-c)/t}, & \text{otherwise} \end{cases}$$

22: **endFunction**

5.2   Gradient-Guided Sampling with CDNL-Style ASP Solving

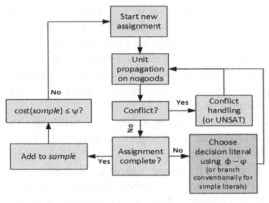

Fig. 1. Outline CDNL-PROP-ASP

To overcome the issues of the previously presented algorithms (mainly the required large amount of candidate models), we introduce an approach (called *Probabilistic Optimization in Conflict-Driven Nogood Learning* (CDNL-PROP-ASP)) which does not require any external source of answer sets (in fact, it does not require any existing answer set or SAT solver at all) and copes much better with large search spaces.

Our approach, whose core part is outlined in Fig. 1, equips the state-of-the-art approach to answer set solving CDNL-ASP [8] with Gradient Descent steps.

Algorithm 3 (which uses a simpler approach to loop handling compared to [8], to simplify experimental implementation) iteratively adds literals to a partial Boolean assignment (a sequence of literals) until all literals are covered and no *nogood* is violated (nogoods correspond to clauses with all literals negated). The procedure for this is guided by three factors: a set of nogoods initially obtained from Clark's completion of the answer set program and enhanced during conflict resolution, further nogoods added when a discovered supported model of a non-tight program is not a stable model, and an approach to select so-called decision literals such that the set of all sampled stable models adheres to our probabilistic constraints. The first two points are very briefly recalled at the end of this section, but for a complete description we refer to [8,13] (for our rather elementary loop handling approach), due to lack of space. The main difference to existing variants of CDNL is our decision literal choice policy: In line 32 in procedure COMPMODELCAND, $da \leftarrow \text{argmax}_{a \in \theta \wedge a \notin as \wedge \neg a \notin as} |\phi(a) - \varphi(a)|$ gives the not yet assigned uncertain atom $\in \theta$ where the absolute difference between target (given) probability and the atom's actual frequency in the current incomplete sample is largest, thus greedily selecting the uncertain atom which is currently most "wrong" (in terms of its weight in the sample) as the next decision literal. Then, we add this atom to the partial assignment with probability ≈ 1 if $\varphi(da) - \phi(da) < 0$. Otherwise we add its negation (so intuitively, we add the uncertain atom depending on whether the current sample is "lacking" that atom or not). If no uncertain atom is uncovered in the partial assignment, we select the decision literal using some conventional branching heuristics such as VSIDS.

Theorem 1. *Algorithm 3 terminates (with maxTrials $= \infty$ and $\epsilon > 0$) and $\forall m \in sample : m \in \Psi_{asp}$, where asp is the spanning program, if $\exists s, \text{Supp}(s) = \Psi_{asp} : cost(\phi, s) \leq \psi$.*

As for the not directly sampling-related aspects of Algorithm 3, we just recall some essentials: If a conflict (a nogood violation) is encountered, the conflict is analyzed in CONFLICTANALYSIS and the algorithm jumps back to one of the previous decision levels, and adds a new nogood (conflict analysis and nogood learning sets CDNL apart from the older DPLL-style approaches to SAT/ASP solving). Every literal that is added to the partial assignment is added at a certain decision level; each time we make a decision (i.e., when we add a literal which is not enforced to hold deductively given the nogoods), we increase the decision level. Sub-procedure UNITPROP adds non-decision literals which must hold deductively (forced by so-called unit-resulting nogoods). CONFLICTANALYSIS and UNITPROP are as in [8] except for loop-handling (see below), and are omitted here. $atoms(asp)$ and $bodies(asp)$ denote the atom respectively body literals for answer set program asp (body literals are literals which represent entire rule bodies). Argument Ξ represents the set of all nogoods for program asp. Nogoods are derived from and sufficient to represent all rules and facts, st. for an answer set m, $\forall \xi \in \Xi : \xi \not\subseteq m$. Loop nogood generation differs from [8] by using the simpler ASSAT approach [13], just to simplify our initial implementation.

Algorithm 3 Probabilistic Optimization in Conflict-Driven Nogood Learning (CDNL-PROP-ASP)

---

    procedure SAMPLECDNL($maxTrials, asp, \Xi, \theta, \phi, \psi$)   ▷ Arguments as in Algorithm 1, for Ξ see Sect. 5.2
        $sample \leftarrow [\,], \; t \leftarrow 1$
3:    **repeat**
        $bounced \leftarrow true$
        repeat
6:         $cand \leftarrow$ COMPMODELCAND($asp, \Xi, \theta, \phi, \psi$)
           if $cand = $ UNSAT **then return** UNSAT
           if $SM^{asp}(cand)$ **then**   ▷ Stable model check (omit if program is known to be tight)
9:          $sample \leftarrow sample \uplus \{cand\}$
         $\varphi_k \leftarrow \dfrac{|[pws_j \mid pws_j \in sample, \theta_k \in pws_j]|}{|sample|}$ for each $k, \phi_k \in \phi$
         $bounced \leftarrow false$
12:      **else**
         add loop nogoods to Ξ as described in [13]
        until $\neg bounced$
15:     $t \leftarrow t + 1$
      until $cost(\phi, \varphi) \leq \psi \vee t > maxTrials$ ▷ Function $cost$ as defined in Sect. 3
      return $sample$ or sample any number of models uniformly from $sample$
18: **procedure** COMPMODELCAND($asp, \Xi, \theta, \phi, \psi$)
      $as \leftarrow \emptyset$ ▷ The partial assignment (incomplete model candidate)
      $level \leftarrow 0$ ▷ initial decision level
21:   **while** $true$ **do**
      $(as, \varepsilon) \leftarrow$ UNITPROP(as, Ξ)
      if $\varepsilon \neq \emptyset$ **then**   ▷ Conflict (violated nogood ε)
24:      **if** $level = 0$ **then return** UNSAT
        $(newNogood, unitLit, newLevel) \leftarrow$ CONFLICTANALYSIS($\varepsilon_1, asp, \Xi, as$)
        $\Xi \leftarrow \Xi \cup \{newNogood\}$
27:      $as \leftarrow as \backslash \{lit \in as : newLevel < level(lit)\}$ ▷ $level(lit)$ gives the decision level of lit
        $level \leftarrow newLevel$
        $as \leftarrow as \cup \neg unitLit$
30:     **else if** $|as| = |atoms(asp) \cup bodies(asp)|$ **then**
       return $as^T \cap atoms(asp)$ ▷ Model found; as^T represents the positive literals in as
      else
        $da \leftarrow \operatorname*{argmax}_{a \in \theta \wedge a \notin as \wedge \neg a \notin as} |\phi(a) - \varphi(a)|$
33:      **if** $da = \emptyset$ **then**
        $decLit \leftarrow$ SELECT(as, asp, Ξ)   ▷ Choose decision literal using
                               some branching heuristics (e.g., VSIDS) from literals not yet assigned in as
        $dlPr \leftarrow 1$
36:      **else**
        $decLit \leftarrow$ random element from da
        $dlPr \leftarrow \begin{cases} 1 - \epsilon, & \text{if } \varphi(decLit) - \phi(decLit) < 0 \\ \epsilon, & \text{otherwise} \end{cases}$ (ϵ some positive value $\ll 1$)

39:     $level \leftarrow level + 1$
      if $rand_0^1 < dlPr$ **then** $as \leftarrow as \cup decLit$ **else** $as \leftarrow as \cup \neg decLit$

---

6   Experiments

In the following, SGD-ASP and LMM-ASP denote the first steps of algorithms SGD-ASP-MH (Sect. 4.1) and LMM-ASP-MH (Sect. 4.2), that is, without the Metropolis step. To make results comparable with each other, all experiments (except one where we plot durations in dependency of accuracies) use 0.01 as MSE error threshold for the cost checks in the optimization loops, and all distribution-aware sampling tasks sample 100000 answer sets.

In our first experiment, the random variables are independent biased coins. We generate a number of answer set programs with an increasing number of coins, each with a given random probability for "heads". Despite its simplicity, this experiment is suitable for showing how well each of the approaches scales with the number of random variables if these are mutually independent (while our approach is not restricted to this, it is a typical use case in PLP). CDNL-PROP-ASP scales much better here than the other approaches (up to thousands of coins) - however, it should be kept in mind that, with the exception of Sect. 4.2, in this paper we only consider approaches which sample from a target distribution which is not necessarily a maximum entropy distribution.

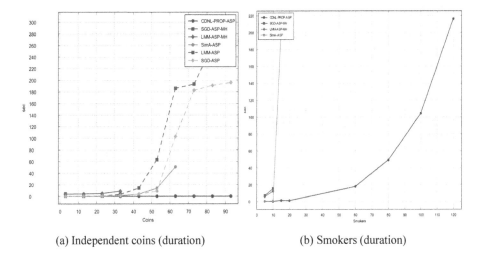

(a) Independent coins (duration)   (b) Smokers (duration)

Fig. 2. Performance (1)

Figure 2a shows the durations of this task. Per each point on the x-axis, a random program was generated and for each sampling approach, the respective task run over five trials. Our next experiment is the "Smokers"-scenario, a classic benchmarking scenario for probabilistic logics. However, variants exist - ours is different from [6,15] in that it does not make independence assumptions. A number of people smoke with probability 0.8. With probability 0.2, smoking causes cancer. With probability 0.6, the fact that two different people influence each other is added to the program. With probability 0.8, a smoker who influences another person causes the other one to smoke too. Observe that with 60 smokers, the probabilistic answer set program already comprises ca. 2200 random variables (from probabilistic rules/facts) and ca. 4500 rules and facts.

Figure 2b shows the durations (in seconds) of this task for all approaches. The figure shows the expected exponential increase in time, but also that even with 100 smokers (a probabilistic answer set program with ca. 12000 rules and facts, including almost 6000 uncertain rules and facts) this task can still be solved under 4 min on an average laptop. In a third experiment, we generate and sample from random graphs. Results for the random graph task are shown in

Fig. 3a. This task has a significantly larger number of plain rules than Smokers or Coins, and much steeper duration increase in the number of items; e.g., with 50 edges, the resulting weighted program has more than 82000 rules (and about 1600 uncertain atoms). The graphs have been generated as follows: One third of the all possible edge candidates are part of the graph, one third of the candidates are not part of the graph, and for the remaining third, the edge candidate is part of the graph with random probability (the edges' weight, e.g., 0.7: `edge(16,22)`). For each possible pairing of different probabilistic edges, this pair is disallowed with probability 0.0001 (`:- edge(a,b), edge(c,d)`), in order to introduce some dependencies among edges. Paths exist according to the following "hard" (unweighted) rules (which are grounded using Scala code, to avoid expensive external grounder calls):

```
path(X,Y) :- edge(X,Y).
path(X,Y) :- edge(X,Z),Y != Z, path(Z,Y).
```

It should be mentioned that calling Clingo adds an operating system-dependent time overhead of around 30–120 ms per each run which requires an external ASP solver. A minor random overhead was also introduced by JVM garbage collection events.

Figure 3b shows how the duration of Smoker and Random Graph sampling tasks depend from the specified accuracy (error thresholds), for CDNL-PROP-ASP.

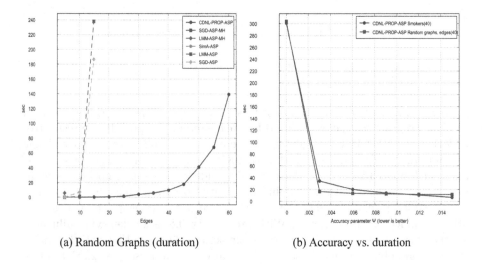

(a) Random Graphs (duration)   (b) Accuracy vs. duration

Fig. 3. Performance (2)

We have used Scala 2.12.4 for all implementations. Everywhere, cut-off graphs indicate JVM out of memory or timeout after 6 min (that is why, e.g., the SGD-ASP-MH curve is barely visible in Fig. 2a, timing out shortly after 40 coins). All experiments have been conducted on a i7-4810MQ machine (2.8 GHz).

Overall, CDNL-PROP-ASP is the clear winner with our suite of experiments. All two-step approaches in this paper as well as Simulated Annealing, which nevertheless performs significantly better than the two-step approaches, suffer from the fact that they need to examine (reject or keep) a potentially very large number of candidate answer sets, whereas CDNL-PROP-ASP computes answer sets by adding only those atoms to the current partial assignment which are deemed necessary to reach a target distribution (instead of accepting or rejecting entire externally computed answer sets). The results for LMM-ASP and SGD-ASP indicate that for the LMM and SGD algorithms, indeed the first step is the most costly part.

7   Conclusion

This paper has introduced new approaches to the distribution-aware sampling of answer sets. Samples act as approximate representatives of solution sets of answer set programs with probabilistic constraints, which opens up a wide range of potential use cases, particularly in PLP and other areas of Relational AI. To our best knowledge, we have introduced the first realistic and general approach to distribution-aware answer set sampling, and in particular the first approach which enhances the state-of-the-art CDNL-style ASP solving algorithm with cost-driven sampling capabilities. Planned work comprises the equipment of existing modular ASP solvers with this feature and the introduction of guarantees for entropy maximization to this algorithm.

References

1. Baral, C., Gelfond, M., Rushton, N.: Probabilistic reasoning with answer sets. Theory Practice Logic Program. **9**(1), 57–144 (2009)
2. Chakraborty, S., Fremont, D.J., Meel, K.S., Seshia, S.A., Vardi, M.Y.: Distribution-aware sampling and weighted model counting for SAT. In: Proceedings of the 28th AAAI Conference on Artificial Intelligence (AAAI), pp. 1722–1730, July 2014
3. Cheeseman, P.: A method of computing generalized Bayesian probability values for expert systems. In: Proceedings of the Eighth International Joint Conference on Artificial Intelligence, IJCAI 1983, vol. 1, pp. 198–202. Morgan Kaufmann Publishers Inc., Burlington (1983)
4. Dalvi, N., Suciu, D.: Efficient query evaluation on probabilistic databases. VLDB J. **16**(4), 523–544 (2007)
5. Dantsin, E., Eiter, T., Gottlob, G., Voronkov, A.: Complexity and expressive power of logic programming. ACM Comput. Surv. **33**(3), 374–425 (2001)
6. Dries, A., et al.: ProbLog2: probabilistic logic programming. In: Bifet, A., May, M., Zadrozny, B., Gavalda, R., Pedreschi, D., Bonchi, F., Cardoso, J., Spiliopoulou, M. (eds.) ECML PKDD 2015. LNCS (LNAI), vol. 9286, pp. 312–315. Springer, Cham (2015). https://doi.org/10.1007/978-3-319-23461-8_37
7. Finger, M., Bona, G.D.: Probabilistic satisfiability: logic-based algorithms and phase transition. In: Proceedings of the 22nd International Joint Conference on Artificial Intelligence (IJCAI 2011) (2011)

8. Gebser, M., Kaufmann, B., Neumann, A., Schaub, T.: Conflict-driven answer set solving. In: Proceedings of the 20th International Joint Conference on Artificial Intelligence (IJCAI 2007) (2007)

9. Gelfond, M., Lifschitz, V.: The stable model semantics for logic programming. In: Proceedings of the 5th International Conference on Logic Programming, vol. 161 (1988)

10. Gomes, C.P., Sabharwal, A., Selman, B.: Near-uniform sampling of combinatorial spaces using XOR constraints. In: NIPS, pp. 481–488 (2006)

11. Greßler, A., Oetsch, J., Tompits, H.: Harvey: a system for random testing in ASP. In: Balduccini, M., Janhunen, T. (eds.) LPNMR 2017. LNCS (LNAI), vol. 10377, pp. 229–235. Springer, Cham (2017). https://doi.org/10.1007/978-3-319-61660-5_21

12. Lee, J., Wang, Y.: A probabilistic extension of the stable model semantics. In: 2015 AAAI Spring Symposium on Logical Formalizations of Commonsense Reasoning (2015)

13. Lin, F., Zhao, Y.: ASSAT: computing answer sets of a logic program by SAT solvers. Artif. Intell. **157**(1), 115–137 (2004)

14. Ng, R.T., Subrahmanian, V.S.: Stable semantics for probabilistic deductive databases. Inf. Comput. **110**(1), 42–83 (1994)

15. Nickles, M., Mileo, A.: A hybrid approach to inference in probabilistic non-monotonic logic programming. In: 2nd International Workshop on Probabilistic Logic Programming (PLP 2015) (2015)

16. Nilsson, N.J.: Probabilistic logic. Artif. Intell. **28**(1), 71–87 (1986)

17. Papadimitriou, C.H.: Games against nature. J. Comput. Syst. Sci. **31**(2), 288–301 (1985)

18. Rödder, W., Kern-Isberner, G.: Representation and extraction of information by probabilistic logic. Inf. Syst. **21**(8), 637–652 (1996)

19. Soos, M.: CryptoMiniSat – a SAT solver for cryptographic problems (2009). http://planete.inrialpes.fr/~soos/CryptoMiniSat/index.html

Consequence-Based Axiom Pinpointing

Ana Ozaki and Rafael Peñaloza[✉]

KRDB Research Centre, Free University of Bozen-Bolzano, Bolzano, Italy
penaloza@inf.unibz.it

Abstract. Axiom pinpointing refers to the problem of finding the axioms in an ontology that are relevant for understanding a given entailment or consequence. One approach for axiom pinpointing, known as *glass-box*, is to modify a classical decision procedure for the entailments into a method that computes the solutions for the pinpointing problem. Recently, consequence-based decision procedures have been proposed as a promising alternative for tableaux-based reasoners for standard ontology languages. In this work, we present a general framework to extend consequence-based algorithms with axiom pinpointing.

Keywords: Consequence-based reasoning · Non-standard reasoning Consequence management

1 Introduction

Ontologies are now widely used in various domains such as medicine [22–24], biology [26], chemistry [10], geography [17,18] and many others [29], to represent conceptual knowledge in a formal and easy to understand manner. It is a multi-task effort to construct and maintain such ontologies, often containing thousands of concepts. As these ontologies increase in size and complexity, it becomes more and more challenging for an ontology engineer to understand which parts of the ontology cause a certain consequence to be entailed. If, for example, this consequence is an error, the ontology engineer would want to understand its precise causes, and correct it with minimal disturbances to the rest of the ontology.

To support this task, a technique known as *axiom pinpointing* was introduced in [25]. The goal of axiom pinpointing is to identify the minimal subontologies (w.r.t. set inclusion) that entail a given consequence; we call these sets *MinAs*. There are two basic approaches to axiom pinpointing. The *black-box approach* [20] uses repeated calls to an unmodified decision procedure to find these MinAs. The *glass-box approach* [4,5], on the other hand, modifies the decision algorithm to generate the MinAs during one execution. In reality, glass-box methods do not explicitly compute the MinAs, but rather a compact representation of them known as the *pinpointing formula*. In this setting, each axiom of the ontology is labelled with a unique propositional symbol. The pinpointing formula is a (monotone) Boolean formula, satisfied exactly by those valuations

© Springer Nature Switzerland AG 2018
D. Ciucci et al. (Eds.): SUM 2018, LNAI 11142, pp. 181–195, 2018.
https://doi.org/10.1007/978-3-030-00461-3_13

which evaluate to true the labels of the axioms in the ontology which cause the entailment of the consequence. Thus, the formula points out to the user the relevant parts of the ontology for the entailment of a certain consequence, where disjunction means alternative use of the axioms and conjunction means that the axioms are jointly used.

Axiom pinpointing can be used to enrich a decision procedure for entailment checking by further presenting to the user the axioms which cause a certain consequence. Since glass-box methods modify an existing decision procedure, they require a specification of the decision method to be studied. Previously, general methods for extending tableaux-based and automata-based decision procedures to axiom pinpointing have been studied in detail [4,5]. Classically, automata-based decision procedures often exhibit optimal worst-case complexity, but the most efficient reasoners for standard ontology languages are tableaux-based. When dealing with pinpointing extensions one observes a similar behaviour: the automata-based axiom pinpointing approach preserves the complexity of the original method, while tableau-based axiom pinpointing is not even guaranteed to terminate in general. However, the latter are more goal-directed and lead to a better run-time in practice.

A different kind of reasoning procedure that is gaining interest is known as the consequence-based method. In this setting, rules are applied to derive explicit consequences from previously derived knowledge. Consequence-based decision procedures often enjoy optimal worst-case complexity and, more recently, they have been presented as a promising alternative for tableaux-based reasoners for standard ontology languages [8,9,14,15,27,28,30]. Consequence-based algorithms have been previously described as simple variants of tableau algorithms [6], and as syntactic variants of automata-based methods [12]. They share the positive complexity bounds of automata, and the goal-directed nature of tableaux.

In this work, we present a general approach to produce axiom pinpointing extensions of consequence-based algorithms. Our driving example and use case is the extension of the consequence-based algorithm for entailment checking for the prototypical ontology language $\mathcal{ALC}$ [15]. We show that the pinpointing extension does not change the ExpTime complexity of the consequence-based algorithm for $\mathcal{ALC}$.

2 Preliminaries

We briefly introduce the notions needed for this paper. We are interested in the problem of understanding the causes for a consequence to follow from an ontology. We consider an abstract notion of ontology and consequence relation. For the sake of clarity, however, we instantiate these notions to the description logic $\mathcal{ALC}$.

2.1   Axiom Pinpointing

To keep the discourse as general as possible, we consider an *ontology language* to define a class $\mathfrak{A}$ of *axioms*. An *ontology* is then a finite set of axioms; that is, a finite subset of $\mathfrak{A}$. We denote the set of all ontologies as $\mathfrak{O}$. A *consequence property* (or *c-property* for short) is a binary relation $\mathcal{P} \subseteq \mathfrak{O} \times \mathfrak{A}$ that relates ontologies to axioms. If $(\mathcal{O}, \alpha) \in \mathcal{P}$, we say that α is a *consequence* of $\mathcal{O}$ or alternatively, that $\mathcal{O}$ *entails* α.

We are only interested in relations that are monotonic in the sense that for any two ontologies $\mathcal{O}, \mathcal{O}' \in \mathfrak{O}$ and axiom $\alpha \in \mathfrak{A}$ such that $\mathcal{O} \subseteq \mathcal{O}'$, if $(\mathcal{O}, \alpha) \in \mathcal{P}$ then $(\mathcal{O}', \alpha) \in \mathcal{P}$. In other words, adding more axioms to an ontology will only increase the set of axioms that are entailed from it. For the rest of this paper whenever we speak about a c-property, we implicitly assume that it is monotonic in this sense.

Notice that our notions of ontology and consequence property differ from previous work. In [4,5], c-properties are defined using two different types of statements and ontologies are allowed to require additional structural constraints. The former difference is just syntactic and does not change the generality of our approach. In the latter case, our setting becomes slightly less expressive, but at the benefit of simplifying the overall notation and explanation of our methods. Our results can be easily extended to the more general setting from [4,5].

When dealing with ontology languages, one is usually interested in deciding whether an ontology $\mathcal{O}$ entails an axiom α; that is, whether $(\mathcal{O}, \alpha) \in \mathcal{P}$. In axiom pinpointing, we are more interested in the more detailed question of *why* it is a consequence. More precisely, we want to find the minimal (w.r.t. set inclusion) sub-ontologies $\mathcal{O}' \subseteq \mathcal{O}$ such that $(\mathcal{O}', \alpha) \in \mathcal{P}$ still holds. These subsets are known as *MinAs* [4,5], *justifications* [13], or *MUPS* [25]—among many other names—in the literature. Rather than enumerating all these sub-ontologies explicitly, one approach is to compute a formula, known as the pinpointing formula, that encodes them.

Formally, suppose that every axiom $\alpha \in \mathfrak{A}$ is associated with a unique propositional variable $\mathsf{lab}(\alpha)$, and let $\mathsf{lab}(\mathcal{O})$ be the set of all the propositional variables corresponding to axioms in the ontology $\mathcal{O}$. A *monotone Boolean formula* ϕ over $\mathsf{lab}(\mathcal{O})$ is a Boolean formula using only variables in $\mathsf{lab}(\mathcal{O})$ and the connectives for conjunction ($\wedge$) and disjunction ($\vee$). The constants $\top$ and $\bot$, always evaluated to true and false, respectively, are also monotone Boolean formulae. We identify a propositional valuation with the set of variables which are true in it. For a valuation $\mathcal{V}$ and a set of axioms $\mathcal{O}$, the $\mathcal{V}$-*projection of* $\mathcal{O}$ is the set $\mathcal{O}_{\mathcal{V}} := \{\alpha \in \mathcal{O} \mid \mathsf{lab}(\alpha) \in \mathcal{V}\}$. Given a c-property $\mathcal{P}$ and an axiom $\alpha \in \mathfrak{A}$, a monotone Boolean formula ϕ over $\mathsf{lab}(\mathcal{O})$ is called a *pinpointing formula* for $(\mathcal{O}, \alpha)$ w.r.t $\mathcal{P}$ if for every valuation $\mathcal{V} \subseteq \mathsf{lab}(\mathcal{O})$:

$$(\mathcal{O}_{\mathcal{V}}, \alpha) \in \mathcal{P} \text{ iff } \mathcal{V} \text{ satisfies } \phi.$$

2.2   Description Logics

Description logics (DLs) [3] are a family of knowledge representation formalisms that have been successfully applied to represent the knowledge of many application domains, in particular from the life sciences [29]. We briefly introduce, as a prototypical example, $\mathcal{ALC}$, which is the smallest propositionally closed description logic.

Given two disjoint sets N_C and N_R of *concept names* and *role names*, respectively, $\mathcal{ALC}$ *concepts* are defined through the grammar rule:

$$C ::= A \mid \neg C \mid C \sqcap C \mid \exists r.C,$$

where $A \in N_C$ and $r \in N_R$. A *general concept inclusion* (GCI) is an expression of the form $C \sqsubseteq D$, where C, D are $\mathcal{ALC}$ concepts. A *TBox* is a finite set of GCIs.

The semantics of this logic is given in terms of *interpretations* which are pairs of the form $\mathcal{I} = (\Delta^{\mathcal{I}}, \cdot^{\mathcal{I}})$ where $\Delta^{\mathcal{I}}$ is a finite set called the *domain*, and $\cdot^{\mathcal{I}}$ is the *interpretation function* that maps every concept name $A \in N_C$ to a set $A^{\mathcal{I}} \subseteq \Delta^{\mathcal{I}}$ and every role name $r \in N_R$ to a binary relation $r^{\mathcal{I}} \subseteq \Delta^{\mathcal{I}} \times \Delta^{\mathcal{I}}$. The interpretation function is extended to arbitrary $\mathcal{ALC}$ concepts inductively as shown in Fig. 1. Following this semantics, we introduce the usual abbreviations $C \sqcup D := \neg(\neg C \sqcap \neg D)$, $\forall r.C := \neg(\exists r.\neg C)$, $\bot := A \sqcap \neg A$, and $\top := \neg\bot$. That is, $\top$ stands for a (DL) tautology, and $\bot$ for a contradiction. The interpretation $\mathcal{I}$ *satisfies* the GCI $C \sqsubseteq D$ iff $C^{\mathcal{I}} \subseteq D^{\mathcal{I}}$. It is a *model* of the TBox $\mathcal{T}$ iff it satisfies all the GCIs in $\mathcal{T}$.

$$(\neg C)^{\mathcal{I}} := \Delta^{\mathcal{I}} \setminus C^{\mathcal{I}}$$
$$(C \sqcap D)^{\mathcal{I}} := C^{\mathcal{I}} \cap D^{\mathcal{I}}$$
$$(\exists r.C)^{\mathcal{I}} := \{d \in \Delta^{\mathcal{I}} \mid \exists e \in C^{\mathcal{I}}.(d, e) \in r^{\mathcal{I}}\}$$

Fig. 1. Semantics of $\mathcal{ALC}$

One of the main reasoning problems in DLs is to decide *subsumption* between two concepts C, D w.r.t. a TBox $\mathcal{T}$; that is, to verify that every model of the TBox $\mathcal{T}$ also satisfies the GCI $C \sqsubseteq D$. If this is the case, we denote it as $\mathcal{T} \models C \sqsubseteq D$. It is easy to see that the relation $\models$ defines a c-property over the class $\mathfrak{A}$ of axioms containing all possible GCIs; in this case, an ontology is a TBox.[1]

The following example instantiates the basic ideas presented in this section.

---

[1] $\mathcal{ALC}$ ontologies often include also an *ABox* with facts about individuals. We disregard that part, as it is irrelevant for our example setting.

Example 1. Consider for example the $\mathcal{ALC}$ TBox $\mathcal{T}_{\mathsf{exa}}$ containing the axioms

$$A \sqsubseteq \exists r.A : \mathsf{ax}_1, \qquad\qquad \exists r.A \sqsubseteq B : \mathsf{ax}_2,$$

$$A \sqsubseteq \forall r.B : \mathsf{ax}_3, \qquad\qquad A \sqcap B \sqsubseteq \bot : \mathsf{ax}_4,$$

where $\mathsf{ax}_i, 1 \leq i \leq 4$ are the propositional variables labelling the axiom. It is easy to see that $\mathcal{T}_{\mathsf{exa}} \models A \sqsubseteq \bot$, and there are two justifications for this fact; namely, the TBoxes $\{\mathsf{ax}_1, \mathsf{ax}_2, \mathsf{ax}_4\}$ and $\{\mathsf{ax}_1, \mathsf{ax}_3, \mathsf{ax}_4\}$. From this, it follows that $\mathsf{ax}_1 \wedge \mathsf{ax}_4 \wedge (\mathsf{ax}_2 \vee \mathsf{ax}_3)$ is a pinpointing formula for $A \sqsubseteq \bot$ w.r.t. $\mathcal{T}_{\mathsf{exa}}$.

3   Consequence-Based Algorithms

Abstracting from particularities, a *consequence-based algorithm* works on a set $\mathcal{A}$ of *consequences*, which is expanded through rule applications. Algorithms of this kind have two phases. The *normalization* phase first transforms all the axioms in an ontology into a suitable normal form. The *saturation* phase initializes the set $\mathcal{A}$ of *derived consequences* with the normalized ontology and applies the rules to expand it. The set $\mathcal{A}$ is often called a *state*. As mentioned, the initial state $\mathcal{A}_0$ contains the normalization of the input ontology $\mathcal{O}$. A *rule* is of the form $\mathcal{B}_0 \rightarrow \mathcal{B}_1$, where $\mathcal{B}_0, \mathcal{B}_1$ are finite sets of consequences. This rule is *applicable* to the state $\mathcal{A}$ if $\mathcal{B}_0 \subseteq \mathcal{A}$ and $\mathcal{B}_1 \not\subseteq \mathcal{A}$. Its *application* extends $\mathcal{A}$ to $\mathcal{A} \cup \mathcal{B}_1$. $\mathcal{A}$ is *saturated* if no rule is applicable to it. The method *terminates* if $\mathcal{A}$ is saturated after finitely many rule applications, independently of the rule application order chosen. For the rest of this section and most of the following, we assume that the input ontology is already in this normal form, and focus only on the second phase.

Given a rule $R = \mathcal{B}_0 \rightarrow \mathcal{B}_1$, we use $\mathsf{pre}(R)$ and $\mathsf{res}(R)$ to denote the sets $\mathcal{B}_0$ of premises that trigger R and $\mathcal{B}_1$ of consequences resulting of its applicability, respectively. If the state $\mathcal{A}'$ is obtained from $\mathcal{A}$ through the application of the rule R, we write $\mathcal{A} \rightarrow_R \mathcal{A}'$, and denote $\mathcal{A} \rightarrow \mathcal{A}'$ if the precise rule used is not relevant. As usual, $\rightarrow^*$ denotes the transitive and reflexive closure of $\rightarrow$.

Consequence-based algorithms derive, in a single execution, several axioms that are entailed from the input ontology. Obviously, in general they cannot generate *all* possible entailed axioms, as such a set may be infinite (e.g., in the case of $\mathcal{ALC}$). Thus, to define correctness, we need to specify for every ontology $\mathcal{O}$, a finite set $\delta(\mathcal{O})$ of *derivable consequences* of $\mathcal{O}$. This set is assumed to be provided as part of the consequence-based algorithm.

Definition 2 (Correctness). *A consequence-based algorithm is* correct *for the consequence property $\mathcal{P}$ if for every ontology $\mathcal{O}$, the following two conditions hold: (i) it terminates, and (ii) if $\mathcal{O} \rightarrow^* \mathcal{A}$ and $\mathcal{A}$ is saturated, then for every derivable consequence $\alpha \in \delta(\mathcal{O})$ it follows that $(\mathcal{O}, \alpha) \in \mathcal{P}$ iff $\alpha \in \mathcal{A}$.*

That is, the algorithm is correct for a property if it terminates and is sound and complete w.r.t. the finite set of derivable consequences $\delta(\mathcal{O})$.

Notice that the definition of correctness requires that the resulting set of consequences obtained from the application of the rules is always the same, independently of the order in which the rules are applied. In other words, if $\mathcal{O} \rightarrow^* \mathcal{A}$, $\mathcal{O} \rightarrow^* \mathcal{A}'$, and $\mathcal{A}, \mathcal{A}'$ are both saturated, then $\mathcal{A} = \mathcal{A}'$. This is a fundamental property that will be helpful for showing correctness of the pinpointing extensions in the next section.

Table 1. $\mathcal{ALC}$ consequence-based algorithm rules $\mathcal{B}_0 \rightarrow \mathcal{B}_1$.

$\mathcal{B}_0$		$\mathcal{B}_1$
1: $\emptyset$		$H \sqcap A \sqsubseteq A$
2: $H \sqcap \neg A \sqsubseteq N \sqcup A$		$H \sqcap \neg A \sqsubseteq N$
3: $H \sqsubseteq N_1 \sqcup A_1, \ldots, H \sqsubseteq N_n \sqcup A_n$	$A_1 \sqcap \cdots \sqcap A_n \sqsubseteq N$	$H \sqsubseteq \bigsqcup_{i=1}^{n} N_i \sqcup N$
4: $H \sqsubseteq N \sqcup A$	$A \sqsubseteq \exists r.B$	$H \sqsubseteq N \sqcup \exists r.B$
5: $H \sqsubseteq M \sqcup \exists r.K, K \sqsubseteq N \sqcup A$	$\exists r.A \sqsubseteq B$	$H \sqsubseteq M \sqcup B \sqcup \exists r.(K \sqcap \neg A)$
6: $H \sqsubseteq M \sqcup \exists r.K, K \sqsubseteq \bot$		$H \sqsubseteq M$
7: $H \sqsubseteq M \sqcup \exists r.K, H \sqsubseteq N \sqcup A$	$A \sqsubseteq \forall r.B$	$H \sqsubseteq M \sqcup N \sqcup \exists r.(K \sqcap B)$

A well-known example of a consequence-based algorithm is the $\mathcal{ALC}$ reasoning method from [27]. To describe this algorithm we need some notation. A *literal* is either a concept name or a negated concept name. Let H, K denote (possibly empty) conjunctions of literals, and M, N are (possibly empty) disjunctions of concept names. For simplicity, we treat these conjunctions and disjunctions as sets. The normalization phase transforms all GCIs to be of the form:

$$\prod_{i=1}^{n} A_i \sqsubseteq \bigsqcup_{j=1}^{m} B_j, \qquad A \sqsubseteq \exists r.B, \qquad A \sqsubseteq \forall r.B, \qquad \exists r.A \sqsubseteq B.$$

For a given $\mathcal{ALC}$ TBox $\mathcal{T}$, the set $\delta(\mathcal{T})$ of derivable consequences contains all GCIs of the form $H \sqsubseteq M$ and $H \sqsubseteq N \sqcup \exists r.K$. The saturation phase initializes $\mathcal{A}$ to contain the axioms in the (normalized) TBox, and applies the rules from Table 1 until a saturated state is found. After termination, one can check that for every derivable consequence $C \sqsubseteq D$ it holds that $\mathcal{T} \models C \sqsubseteq D$ iff $C \sqsubseteq D \in \mathcal{A}$; that is, this algorithm is correct for the property [27].

Example 3. Recall the $\mathcal{ALC}$ TBox $\mathcal{T}_{\text{exa}}$ from Example 1. Notice that all axioms in this TBox are already in normal form; hence the normalization step does not modify it. The consequence-based algorithm starts with $\mathcal{A} := \mathcal{T}_{\text{exa}}$ and applies the rules until saturation. One possible execution of the algorithm is

$$\mathcal{A}_0 \rightarrow_5 A \sqsubseteq B \rightarrow_3 A \sqsubseteq \bot \rightarrow_7 A \sqsubseteq \exists r.(A \sqcap B) \rightarrow^* \ldots,$$

where $\mathcal{A}_0$ contains $\mathcal{T}_{\text{exa}}$ and the result of adding all the tautologies generated by the application of Rule 1 over it (see Fig. 2). Since rule applications only extend the set of consequences, we depict exclusively the newly added consequence; e.g., the first rule application $\mathcal{A}_0 \rightarrow_1 A \sqsubseteq A$ is in fact representing $\mathcal{A}_0 \rightarrow_1 \mathcal{A}_0 \cup \{A \sqsubseteq A\}$. When the execution of the method terminates, the set of consequences $\mathcal{A}$ contains $A \sqsubseteq \bot$; hence we can conclude that this subsumption follows from $\mathcal{T}_{\mathsf{exa}}$. Notice that other consequences (e.g., $A \sqsubseteq B$) are also derived from the same execution.

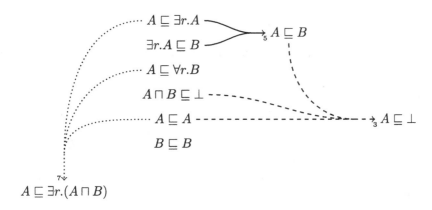

Fig. 2. An execution of the $\mathcal{ALC}$ consequence-based algorithm over $\mathcal{T}_{\mathsf{exa}}$ from Example 3. Arrows point from the premises to the consequences generated by the application of the rule denoted in the subindex.

For the rest of this paper, we consider an arbitrary, but fixed, consequence-based algorithm, that is correct for a given c-property $\mathcal{P}$.

4   The Pinpointing Extension

Our goal is to extend consequence-based algorithms from the previous section to methods that compute pinpointing formulae for their consequences. We achieve this by modifying the notion of states, and the rule applications on them. Recall that every axiom α in the class $\mathfrak{A}$ (in hence, also every axiom in the ontology $\mathcal{O}$) is labelled with a unique propositional variable $\mathsf{lab}(\alpha)$. In a similar manner, we consider sets of consequences $\mathcal{A}$ that are labelled with a monotone Boolean formula. We use the notation $\mathcal{A}^{pin}$ to indicate that the elements in a the set $\mathcal{A}$ are labelled in this way, and use $\alpha : \varphi_\alpha \in \mathcal{A}^{pin}$ to express that the consequence α, labelled with the formula φ_α, belongs to $\mathcal{A}^{pin}$. A *pinpointing state* is a set of labelled consequences. We assume that each consequence in this set is labelled with only one formula. For a set of labelled consequences $\mathcal{A}^{pin}$ and a set of (unlabelled) consequences X, we define $\mathsf{fm}(X, \mathcal{A}^{pin}) := \bigwedge_{\alpha \in X} \varphi_\alpha$, where $\varphi_\alpha = \bot$ if $\alpha \notin \mathcal{A}$.

A consequence-based algorithm A induces a pinpointing consequence-based algorithm A^{pin} by modifying the notion of rule application, and dealing with pinpointing states, instead of classical states, through a modification of the formulae labelling the derived consequences.

Definition 4 (Pinpointing Application). *The rule* $R = \mathcal{B}_0 \to \mathcal{B}_1$ *is pinpointing applicable* to the pinpointing state A^{pin} *if* $\mathsf{fm}(\mathcal{B}_0, A^{pin}) \not\models \mathsf{fm}(\mathcal{B}_1, A^{pin})$. *The* pinpointing application *of this rule modifies* A^{pin} *to:*

$$\{\alpha : \varphi_\alpha \vee \mathsf{fm}(\mathcal{B}_0, A^{pin}) \mid \alpha \in \mathcal{B}_1, \alpha : \varphi_\alpha \in A^{pin}\} \cup (A^{pin} \setminus \{\alpha : \varphi_\alpha \mid \alpha \in \mathcal{B}_1\}).$$

The pinpointing state A^{pin} *is* pinpointing saturated *if no rule is pinpointing applicable to it.*

Example 5. Consider again the TBox $\mathcal{T}_{\mathsf{exa}}$ from Example 1. At the beginning of the execution of the pinpointing algorithm, the set of consequences is the TBox, with each axiom labelled by the unique propositional variable representing it; that is $\mathcal{T}^{pin} = \{A \sqsubseteq \exists r.A : \mathsf{ax}_1, \exists r.A \sqsubseteq B : \mathsf{ax}_2, A \sqsubseteq \forall r.B : \mathsf{ax}_3, A \sqcap B \sqsubseteq \bot : \mathsf{ax}_4\}$. A pinpointing application of Rule 1 adds the new consequence $A \sqsubseteq A : \top$, where the tautology $\top$ labelling this consequence arises from the fact that rule 1 has no premises. At this point, one can pinpointing apply Rule 5 with

$$\mathcal{B}_0 = \{A \sqsubseteq \exists r.A, \ A \sqsubseteq A, \ \exists r.A \sqsubseteq B\}, \qquad \mathcal{B}_1 = \{A \sqsubseteq B\}$$

(see the solid arrow in Fig. 2). In this case, $\mathsf{fm}(\mathcal{B}_0, A^{pin}) = \mathsf{ax}_1 \wedge \top \wedge \mathsf{ax}_2$ and $\mathsf{fm}(\mathcal{B}_1, A^{pin}) = \bot$ because the consequence $A \sqsubseteq B$ does not belong to A^{pin} yet. Hence $\mathsf{fm}(\mathcal{B}_0, A^{pin}) \not\models \mathsf{fm}(\mathcal{B}_1, A^{pin})$, and the rule is indeed pinpointing applicable. The pinpointing application of this rule adds the new labelled consequence $A \sqsubseteq B : \mathsf{ax}_1 \wedge \mathsf{ax}_2$ to A^{pin}. Then, Rule 3 becomes pinpointing applicable with

$$\mathcal{B}_0 = \{A \sqsubseteq B, \ A \sqsubseteq A, \ A \sqcap B \sqsubseteq \bot\},$$

which adds $A \sqsubseteq \bot : \mathsf{ax}_1 \wedge \mathsf{ax}_2 \wedge \mathsf{ax}_4$ to the set of consequences. Then, Rule 7 over the set of premises

$$\mathcal{B}_0 = \{A \sqsubseteq \exists r.A, \ A \sqsubseteq A, \ A \sqsubseteq \forall r.B\},$$

yields the new consequence $A \sqsubseteq \exists r.(A \sqcap B) : \mathsf{ax}_1 \wedge \mathsf{ax}_3$.

Notice that, at this point Rule 6 is not applicable in the classical case over the set of premises $\mathcal{B}_0 = \{A \sqsubseteq \exists r.(A \sqcap B), A \sqcap B \sqsubseteq \bot\}$ because its (regular) application would add the consequence $A \sqsubseteq \bot$ that was already derived. However,

$$\mathsf{fm}(\mathcal{B}_0, A^{pin}) = \mathsf{ax}_1 \wedge \mathsf{ax}_3 \wedge \mathsf{ax}_4 \not\models \mathsf{ax}_1 \wedge \mathsf{ax}_2 \wedge \mathsf{ax}_4 = \varphi_{A \sqsubseteq \bot} = \mathsf{fm}(\mathcal{B}_1, A^{pin});$$

hence, the rule is in fact pinpointing applicable. The pinpointing application of this Rule 6 substitutes the labelled consequence $A \sqsubseteq \bot : \mathsf{ax}_1 \wedge \mathsf{ax}_2 \wedge \mathsf{ax}_4$ with the consequence $A \sqsubseteq \bot : (\mathsf{ax}_1 \wedge \mathsf{ax}_2 \wedge \mathsf{ax}_4) \vee (\mathsf{ax}_1 \wedge \mathsf{ax}_3 \wedge \mathsf{ax}_4)$. The pinpointing extension will then continue applying rules until a saturated state is reached. This execution is summarized in Fig. 3. At that point, the set of labelled consequences will contain, among others, $A \sqsubseteq \bot : (\mathsf{ax}_1 \wedge \mathsf{ax}_2 \wedge \mathsf{ax}_4) \vee (\mathsf{ax}_1 \wedge \mathsf{ax}_3 \wedge \mathsf{ax}_4)$. The label of this consequence corresponds to the pinpointing formula that was computed in Example 1.

We denote as $\mathcal{A}^{pin} \rightarrow_R \mathcal{B}^{pin}$ the fact that $\mathcal{B}^{pin}$ is obtained from the pinpointing application of the rule R to $\mathcal{A}^{pin}$. As before, we drop the subscript R if the name of the rule is irrelevant and write simply $\mathcal{A}^{pin} \rightarrow \mathcal{B}^{pin}$. The pinpointing extension starts, as the classical one, with the set of all normalized axioms. For the rest of this section, we assume that the input ontology is already normalized, and hence each axiom in the initial pinpointing state is labelled with its corresponding propositional variable. In the next section we show how to deal with normalization.

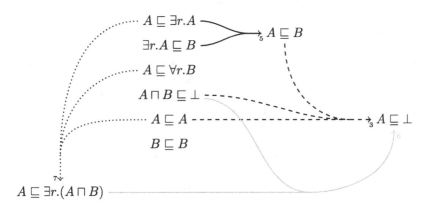

Fig. 3. Pinpointing application of rules over $\mathcal{T}_{\mathsf{exa}}$ in Example 5. Arrows point from the premises to the consequences generated by the pinpointing application of the rule denoted in the subindex.

Notice that if a rule R is applicable to some state $\mathcal{A}$, then it is also pinpointing applicable to it. This holds because the regular applicability condition requires that at least one consequence α in $\mathsf{res}(R)$ should not exist already in the state $\mathcal{A}$, which is equivalent to having the consequence $\alpha : \bot \in \mathcal{A}^{pin}$. Indeed, we used this fact in the first pinpointing rule applications of Example 5. If the consequence-based algorithm is correct, then it follows by definition that for any saturated state $\mathcal{A}$ obtained by a sequence of rule applications from $\mathcal{O}$, $\mathcal{O} \models \alpha$ iff $\alpha \in \mathcal{A}$. Conversely, as shown next, every consequence created by a pinpointing rule application is also generated by a regular rule application. First, we extend the notion of a $\mathcal{V}$-projection to sets of consequences (i.e., states) in the obvious manner: $\mathcal{A}_\mathcal{V} := \{\alpha \mid \alpha : \varphi_\alpha \in \mathcal{A}^{pin}, \mathcal{V} \models \varphi\}$.

Lemma 6. *Let $\mathcal{A}^{pin}, \mathcal{B}^{pin}$ be pinpointing states and let $\mathcal{V}$ be a valuation. If $\mathcal{A}^{pin} \rightarrow^* \mathcal{B}^{pin}$ then $\mathcal{A}_\mathcal{V} \rightarrow^* \mathcal{B}_\mathcal{V}$.*

Proof. We show that if $\mathcal{A}^{pin} \rightarrow_R \mathcal{B}^{pin}$ then $\mathcal{A}_\mathcal{V} \rightarrow_R \mathcal{B}_\mathcal{V}$ or $\mathcal{A}_\mathcal{V} = \mathcal{B}_\mathcal{V}$, where $R = \mathcal{B}_0 \rightarrow \mathcal{B}_1$ is a rule. If $\mathcal{V}$ does not satisfy $\mathsf{fm}(\mathcal{B}_0, \mathcal{A}^{pin})$ then $\mathcal{A}_\mathcal{V} = \mathcal{B}_\mathcal{V}$ since the labels of the newly added assertions are not satisfied by $\mathcal{V}$, and the disjunction with $\mathsf{fm}(\mathcal{B}_0, \mathcal{A}^{pin})$ does not change the evaluation of the modified labels under $\mathcal{V}$. On the other hand, if $\mathcal{V}$ satisfies $\mathsf{fm}(\mathcal{B}_0, \mathcal{A}^{pin})$ then $\mathcal{B}_0 \subseteq \mathcal{A}_\mathcal{V}$. If $\mathcal{B}_1 \not\subseteq \mathcal{A}_\mathcal{V}$ then $\mathcal{A}_\mathcal{V} \rightarrow_R \mathcal{B}_\mathcal{V}$. Otherwise, again we have $\mathcal{A}_\mathcal{V} = \mathcal{B}_\mathcal{V}$. □

Since all the labels are monotone Boolean formulae, it follows that the valuation $\mathcal{V}_\top = \mathsf{lab}(\mathcal{O})$ that makes every propositional variable true satisfies all labels, and hence for every pinpointing state $\mathcal{A}^{pin}$, $\mathcal{A}_{\mathcal{V}_\top} = \mathcal{A}$. Lemma 6 hence entails that the pinpointing extension of the consequence-based algorithm A does not create new consequences, but only labels these consequences. Termination of the pinpointing extension then follows from the termination of the consequence-based algorithm and the condition for pinpointing rule application that entails that, whenever a rule is pinpointing applied, the set of labelled consequences is necessarily modified either by adding a new consequence, or by modifying the label of at least one existing consequence to a weaker (i.e., more general) monotone Boolean formula. Since there are only finitely many monotone Boolean formulas over $\mathsf{lab}(\mathcal{O})$, every label can be changed finitely many times only.

It is in fact possible to get a better understanding of the running time of the pinpointing extension of a consequence-based algorithm. Suppose that, on input $\mathcal{O}$, the consequence-based algorithm A stops after at most $f(\mathcal{O})$ rule applications. Since every rule application must add at least one consequence to the state, the saturated state reached by this algorithm will have at most $f(\mathcal{O})$ consequences. Consider now the pinpointing extension of A. We know, from the previous discussion, that this pinpointing extension generates the same set of consequences. Moreover, since there are $2^{|\mathcal{O}|}$ possible valuations over $\mathsf{lab}(\mathcal{O})$, and every pinpointing rule application that does not add a new consequence must generalize at least one formula, the labels of each consequence can be modified at most $2^{|\mathcal{O}|}$ times. Overall, this means that the pinpointing extension of A stops after at most $2^{|\mathcal{O}|} f(\mathcal{O})$ rule applications. We now formalize this result.

Theorem 7. *If a consequence-based algorithm A stops after at most $f(\mathcal{O})$ rule applications, then A^{pin} stops after at most $2^{|\mathcal{O}|} f(\mathcal{O})$ rule applications.*

Another important property of the pinpointing extension is that saturatedness of a state is preserved under projections.

Lemma 8. *Let $\mathcal{A}^{pin}$ be a pinpointing state and $\mathcal{V}$ a valuation. If $\mathcal{A}^{pin}$ is pinpointing saturated then $\mathcal{A}_\mathcal{V}$ is saturated.*

Proof. Suppose there is a rule R such that R is applicable to $\mathcal{A}_\mathcal{V}$. This means that $\mathcal{B}_0 \subseteq \mathcal{A}_\mathcal{V}$ and $\mathcal{B}_1 \not\subseteq \mathcal{A}_\mathcal{V}$. We show that R is pinpointing applicable to $\mathcal{A}^{pin}$. Since $\mathcal{B}_0 \subseteq \mathcal{A}_\mathcal{V}$, $\mathcal{V}$ satisfies $\mathsf{fm}(\mathcal{B}_0, \mathcal{A}^{pin})$. As $\mathcal{B}_1 \not\subseteq \mathcal{A}_\mathcal{V}$, there is $\alpha \in \mathcal{B}_1$ such that either $\alpha \notin \mathcal{A}$ or $\alpha : \varphi_\alpha \in \mathcal{A}^{pin}$ but $\mathcal{V}$ does not satisfy φ_α. In the former case, R is clearly pinpointing applicable to $\mathcal{A}^{pin}$. In the latter, $\mathsf{fm}(\mathcal{B}_0, \mathcal{A}^{pin}) \not\models \mathsf{fm}(\mathcal{B}_1, \mathcal{A}^{pin})$ since $\mathcal{V}$ satisfies $\mathsf{fm}(\mathcal{B}_0, \mathcal{A}^{pin})$ but not $\mathsf{fm}(\mathcal{B}_1, \mathcal{A}^{pin})$. □

We can now show that the pinpointing extension of a consequence-based algorithm is indeed a pinpointing algorithm; that is, that when a saturated pinpointing state $\mathcal{A}^{pin}$ is reached from rule applications starting from $\mathcal{O}^{pin}$, then for every $\alpha : \varphi_\alpha \in \mathcal{A}^{pin}$, φ_α is a pinpointing formula for α w.r.t. $\mathcal{O}^{pin}$.

Theorem 9 (Correctness of Pinpointing). *Let $\mathcal{P}$ be a c-property on axiomatized inputs for $\mathfrak{I}$ and $\mathfrak{A}$. Given a correct consequence-based algorithm A for $\mathcal{P}$, for every axiomatized input $\Gamma = (\mathcal{O}, \alpha) \in \mathcal{P}$, where $\mathcal{O}$ is normalized, then*

if $\mathcal{O}^{pin} \rightsquigarrow^* \mathcal{A}^{pin}$, $\alpha : \varphi_\alpha \in \mathcal{A}^{pin}$, and $\mathcal{A}^{pin}$ is pinpointing saturated, then φ_α is a pinpointing formula for Γ w.r.t. $\mathcal{P}$.

Proof. We want to show that φ_α is a pinpointing formula for $\mathcal{P}$ and $(\alpha, \mathcal{O})$. That is, for every valuation $\mathcal{V} \subseteq \mathsf{lab}(\mathcal{O})$: $(\alpha, \mathcal{O}_\mathcal{V}) \in \mathcal{P}$ iff $\mathcal{V}$ satisfies φ_α.

Assume that $(\alpha, \mathcal{O}_\mathcal{V}) \in \mathcal{P}$, i.e., $\mathcal{O}_\mathcal{V} \models \alpha$, and let $\mathcal{A}_0 = \mathcal{O}_\mathcal{V}$. Since A terminates on every input, there is a saturated state $\mathcal{B}$ such that $\mathcal{A}_0 \rightarrow^* \mathcal{B}$. Completeness of A then implies that $\alpha \in \mathcal{B}$. By assumption, $\mathcal{A}_0^{pin} \rightsquigarrow^* \mathcal{A}^{pin}$ and $\mathcal{A}^{pin}$ is pinpointing saturated. By Lemma 6 it follows that $\mathcal{O}_\mathcal{V} \rightarrow^* \mathcal{A}_\mathcal{V}$, and by Lemma 8, $\mathcal{A}_\mathcal{V}$ is saturated. Hence, since A is correct, $\mathcal{A}_\mathcal{V} = \mathcal{B}$. This implies that $\mathcal{V} \models \varphi_\alpha$ because $\alpha \in \mathcal{A}_\mathcal{V}$.

Conversely, suppose that $\mathcal{V}$ satisfies φ_α. By assumption, $\alpha : \varphi_\alpha \in \mathcal{A}^{pin}$, $\mathcal{O}^{pin} \rightsquigarrow^* \mathcal{A}^{pin}$, and $\mathcal{A}^{pin}$ is saturated. By Lemma 6, $\mathcal{O}_\mathcal{V} \rightarrow^* \mathcal{A}_\mathcal{V}$. Since $\mathcal{V}$ satisfies φ_α, $\alpha \in \mathcal{A}_\mathcal{V}$. Then, by soundness of A, $\mathcal{O}_\mathcal{V} \models \alpha$. □

As it was the case for classical consequence-based algorithms, their pinpointing extensions can apply the rules in any desired order. The notion of correctness of consequence-based algorithms guarantees that a saturated state will always be found, and the result will be the same, regardless of the order in which the rules are applied. We have previously seen that termination transfers also the pinpointing extensions. Theorem 9 also shows that the formula associated to the consequences derived is always equivalent.

Corollary 10. *Let $\mathcal{A}^{pin}, \mathcal{B}^{pin}$ two pinpointing saturated states, $\mathcal{O}$ an ontology, and α a consequence such that $\alpha : \varphi_\alpha \in \mathcal{A}^{pin}$ and $\alpha : \psi_\alpha \in \mathcal{B}^{pin}$. If $\mathcal{O}^{pin} \rightsquigarrow^* \mathcal{A}^{pin}$ and $\mathcal{O}^{pin} \rightsquigarrow^* \mathcal{B}^{pin}$, then $\varphi_\alpha \equiv \psi_\alpha$.*

To finalize this section, we consider again our running example of deciding subsumption in $\mathcal{ALC}$ described in Sect. 3. It terminates after an exponential number of rule applications on the size of the input TBox $\mathcal{T}$. Notice that every pinpointing rule application requires an entailment test between two monotone Boolean formulas, which can be decided in non-deterministic polynomial time on $|\mathcal{O}|$. Thus, it follows from Theorem 7 that the pinpointing extension of the consequence-based algorithm for $\mathcal{ALC}$ runs in exponential time.

Corollary 11. *Let $\mathcal{T}$ be an $\mathcal{ALC}$ TBox, and C, D two $\mathcal{ALC}$ concepts. A pinpointing formula for $C \sqsubseteq D$ w.r.t. $\mathcal{T}$ is computable in exponential time.*

5   Dealing with Normalization

Throughout the last two sections, we have disregarded the first phase of the consequence-based algorithms in which the axioms in the input ontology are transformed into a suitable normal form. In a nutshell, the normalization phase takes every axiom in the ontology and substitutes it by a set of simpler axioms that are, in combination, equivalent to the original one w.r.t. the set of derivable consequences. For example, in $\mathcal{ALC}$ the axiom $A \sqsubseteq B \sqcap C$ is *not* in normal form. During the normalization phase, it would then be substituted by the two axioms $A \sqsubseteq B$, $A \sqsubseteq C$, which in combination provide the exact same constraints as the original axiom.

Obviously, in the context of pinpointing, we are interested in finding the set of *original* axioms that cause the consequence of interest, and not those in normal form; in fact, normalization is an internal process of the algorithm, and the user should be agnostic to the internal structures used. Hence, we need to find a way to track the original axioms.

To solve this, we slightly modify the initialization of the pinpointing extension. Recall from the previous section that, if the input ontology is already in normal form, then we initialize the algorithm with the state that contains exactly that ontology, where every axiom is labelled with the unique propositional variable that represents it. If the ontology is not originally in normal form, then it is first normalized. In this case, we set as the initial state the newly normalized ontology, but every axiom is labelled with the disjunction of the variables representing the axioms that generated it. This disjunction is used to represent the fact that the normalized axiom may be generated in more than one way. The following example explains this idea.

Example 12. Consider a variant $\mathcal{T}'_{\text{exa}}$ of the $\mathcal{ALC}$ TBox from Example 1 that is now formed by the three axioms

$$A \sqsubseteq \exists r.A \sqcap \forall r.B : \mathsf{ax}'_1, \qquad\qquad \exists r.A \sqsubseteq B : \mathsf{ax}'_2,$$
$$A \sqsubseteq \forall r.B : \mathsf{ax}'_3 \qquad\qquad A \sqcap B \sqsubseteq \bot : \mathsf{ax}'_4.$$

Obviously, the first axiom ax'_1 is not in normal form, but can be normalized by substituting it with the two axioms $A \sqsubseteq \exists r.A$, $A \sqsubseteq \forall r.B$. Thus, the normalization step yields the same TBox $\mathcal{T}_{\text{exa}}$ from Example 1. However, instead of using different propositional variables to label these two axioms, they just inherit the label from the axiom that generated them; in this case ax'_1. However, the axiom $A \sqsubseteq \forall r.B$ is also caused by axiom ax'_3. Thus, the pinpointing algorithm is initialized with

$$\mathcal{A}^{pin} = \{A \sqsubseteq \exists r.A : \mathsf{ax}'_1, \exists r.A \sqsubseteq B : \mathsf{ax}'_2, A \sqsubseteq \forall r.B : \mathsf{ax}'_1 \vee \mathsf{ax}'_3, A \sqcap B \sqsubseteq \bot : \mathsf{ax}'_4\}.$$

Following the same process as in Example 5, we see that we can derive the consequence $A \sqsubseteq \bot : \mathsf{ax}'_1 \wedge \mathsf{ax}'_4$. Hence $\mathsf{ax}'_1 \wedge \mathsf{ax}'_4$ is a pinpointing formula for $A \sqsubseteq \bot$ w.r.t. $\mathcal{T}'_{\text{exa}}$. It can be easily verified that this is in fact the case.

Thus, the normalization phase does not affect the correctness, nor the complexity of the pinpointing extension of a consequence-based algorithm.

6 Conclusions

We presented a general framework to extend consequence-based algorithms with axiom pinpointing. These algorithms often enjoy optimal upper bound complexity and can be efficiently implemented in practice. Our focus in this paper and use case is for the prototypical ontology language $\mathcal{ALC}$. We emphasize that this is only one of many consequence-based algorithms available. The completion-based algorithm for $\mathcal{EL}^+$ [2] is obtained by restricting the assertions to be of the form $A \sqsubseteq B$ and $A \sqsubseteq \exists r.B$ with $A, B \in N_C \cup \{\top\}$ and $r \in N_R$, and adding one rule to handle role constructors. Other examples of consequence-based methods include LTUR approach for Horn clauses [19], and methods for more expressive and Horn DLs [8,14,16].

Understanding the axiomatic causes for a consequence, and in particular the pinpointing formula, has importance beyond MinA enumeration. For example, the pinpointing formula also encodes all the ways to *repair* an ontology [1]. Depending on the application in hand, a simpler version of the formula can be computed, potentially more efficiently. This idea has already been employed to find good approximations for MinAs [7] and lean kernels [21] efficiently.

As future work, it would be interesting to investigate how algorithms for query answering in an ontology-based data access setting can be extended with the pinpointing technique. The pinpointing formula in this case could also be seen as a provenance polynomial, as introduced by Green et al. [11], in database theory. Another direction is to investigate axiom pinpointing in decision procedures for non-monotonic reasoning, where one would also expect the presence of negations in the pinpointing formula.

References

1. Arif, M.F., Mencía, C., Marques-Silva, J.: Efficient axiom pinpointing with EL2MCS. In: KI, pp. 225–233 (2015)
2. Baader, F., Brandt, S., Lutz, C.: Pushing the $\mathcal{EL}$ envelope. In: IJCAI, pp. 364–369 (2005)
3. Baader, F., Calvanese, D., McGuinness, D., Nardi, D., Patel-Schneider, P.F. (eds.): The Description Logic Handbook: Theory, Implementation, and Applications. Cambridge University Press, Cambridge (2003)
4. Baader, F., Peñaloza, R.: Automata-based axiom pinpointing. J. Autom. Reason. **45**(2), 91–129 (2010)
5. Baader, F., Peñaloza, R.: Axiom pinpointing in general tableaux. J. Log. Comput. **20**(1), 5–34 (2010)
6. Baader, F., Peñaloza, R., Suntisrivaraporn, B.: Pinpointing in the description logic $\mathcal{EL}^+$. In: Hertzberg, J., Beetz, M., Englert, R. (eds.) KI 2007. LNCS (LNAI), vol. 4667, pp. 52–67. Springer, Heidelberg (2007). https://doi.org/10.1007/978-3-540-74565-5_7
7. Baader, F., Suntisrivaraporn, B.: Debugging SNOMED CT using axiom pinpointing in the description logic $\mathcal{EL}^+$. In: Proceedings of the 3rd Knowledge Representation in Medicine (KR-MED 2008): Representing and Sharing Knowledge Using SNOMED. CEUR-WS, vol. 410 (2008)

8. Bate, A., Motik, B., Grau, B.C., Simancik, F., Horrocks, I.: Extending consequence-based reasoning to SRIQ. In: KR, pp. 187–196 (2016)

9. Cucala, D.T., Grau, B.C., Horrocks, I.: Consequence-based reasoning for description logics with disjunction, inverse roles, and nominals. In: Proceedings of the 30th International Workshop on Description Logics, Montpellier, France, 18–21 July 2017 (2017)

10. Degtyarenko, K., et al.: ChEBI: a database and ontology for chemical entities of biological interest. Nucleic Acids Res. **36**(suppl 1), D344–D350 (2008)

11. Green, T.J., Karvounarakis, G., Tannen, V.: Provenance semirings. In: Proceedings of the Twenty-Sixth ACM SIGACT-SIGMOD-SIGART Symposium on Principles of Database Systems, Beijing, China, 11–13 June 2007, pp. 31–40 (2007)

12. Hutschenreiter, L., Peñaloza, R.: An automata view to goal-directed methods. In: Drewes, F., Martín-Vide, C., Truthe, B. (eds.) LATA 2017. LNCS, vol. 10168, pp. 103–114. Springer, Cham (2017). https://doi.org/10.1007/978-3-319-53733-7_7

13. Kalyanpur, A., Parsia, B., Horridge, M., Sirin, E.: Finding all justifications of OWL DL entailments. In: Aberer, K. (ed.) ASWC/ISWC -2007. LNCS, vol. 4825, pp. 267–280. Springer, Heidelberg (2007). https://doi.org/10.1007/978-3-540-76298-0_20

14. Kazakov, Y.: Consequence-driven reasoning for Horn SHIQ ontologies. In: Boutilier, C. (ed.) IJCAI 2009, pp. 2040–2045 (2009)

15. Kazakov, Y., Klinov, P.: Bridging the gap between tableau and consequence-based reasoning. In: Informal Proceedings of the 27th International Workshop on Description Logics, Vienna, Austria, 17–20 July 2014, pp. 579–590 (2014)

16. Kazakov, Y., Krötzsch, M., Simancik, F.: The incredible ELK - from polynomial procedures to efficient reasoning with $\mathcal{EL}$ ontologies. JAR **53**(1), 1–61 (2014)

17. Kuipers, B.J.: An ontological hierarchy for spatial knowledge. In: Proceedings of the 10th International Workshop on Qualitative Reasoning About Physical Systems, Fallen Leaf Lake, California, USA (1996)

18. McMaster, R.B., Usery, E.L.: A Research Agenda for Geographic Information Science, vol. 3. CRC Press, Boca Raton (2004)

19. Minoux, M.: LTUR: a simplified linear-time unit resolution algorithm for Horn formulae and computer implementation. Inf. Process. Lett. **29**(1), 1–12 (1988)

20. Parsia, B., Sirin, E., Kalyanpur, A.: Debugging OWL ontologies. In: WWW, pp. 633–640 (2005)

21. Peñaloza, R., Mencía, C., Ignatiev, A., Marques-Silva, J.: Lean kernels in description logics. In: Blomqvist, E., Maynard, D., Gangemi, A., Hoekstra, R., Hitzler, P., Hartig, O. (eds.) ESWC 2017. LNCS, vol. 10249, pp. 518–533. Springer, Cham (2017). https://doi.org/10.1007/978-3-319-58068-5_32

22. Price, C., Spackman, K.: SNOMED clinical terms. Br. J. Healthc. Comput. Inf. Manag. **17**(3), 27–31 (2000)

23. Rector, A.L., Solomon, W.D., Nowlan, W.A., Rush, T.W., Zanstra, P.E., Claassen, W.M.: A terminology server for medical language and medical information systems. Methods Inf. Med. **34**(1–2), 147–157 (1995)

24. Ruch, P., Gobeill, J., Lovis, C., Geissbühler, A.: Automatic medical encoding with SNOMED categories. BMC Med. Inform. Decis. Mak. **8**(Suppl. 1), S6 (2008)

25. Schlobach, S., Cornet, R.: Non-standard reasoning services for the debugging of description logic terminologies. In: Proceedings of the 18th International Joint Conference on Artificial Intelligence, IJCAI 2003, pp. 355–360. Morgan Kaufmann Publishers Inc. (2003)

26. Sidhu, E.S., Dillon, T.S., Chang, E., Sidhu, B.S.: Protein ontology development using OWL. In: OWL Experiences and Directions Workshop, OWLED, p. 188 (2005)

27. Simancik, F., Kazakov, Y., Horrocks, I.: Consequence-based reasoning beyond Horn ontologies. In: IJCAI 2011, pp. 1093–1098. IJCAI/AAAI (2011)

28. Simancik, F., Motik, B., Horrocks, I.: Consequence-based and fixed-parameter tractable reasoning in description logics. Artif. Intell. **209**, 29–77 (2014)

29. Smith, B.: The OBO foundry: coordinated evolution of ontologies to support biomedical data integration. Nat. Biotechnol. **25**, 1251–1255 (2007)

30. Wang, C., Hitzler, P.: Consequence-based procedure for description logics with self-restriction. In: Semantic Web and Web Science - 6th Chinese Semantic Web Symposium and 1st Chinese Web Science Conference, CSWS 2012, Shenzhen, China, 28–30 November 2012, pp. 169–180 (2012)

Probabilistic Semantics for Categorical Syllogisms of Figure II

Niki Pfeifer[1]($\boxtimes$) and Giuseppe Sanfilippo[2]($\boxtimes$)

[1] Munich Center for Mathematical Philosophy, LMU Munich,
Geschwister-Scholl-Platz 1, 80539 Munich, Germany
niki.pfeifer@lmu.de
[2] Department of Mathematics and Computer Science,
University of Palermo, Via Archirafi 34, 90123 Palermo, Italy
giuseppe.sanfilippo@unipa.it

Abstract. A coherence-based probability semantics for categorical syllogisms of Figure I, which have transitive structures, has been proposed recently (Gilio, Pfeifer, & Sanfilippo [15]). We extend this work by studying Figure II under coherence. Camestres is an example of a Figure II syllogism: from *Every P is M* and *No S is M* infer *No S is P*. We interpret these sentences by suitable conditional probability assessments. Since the probabilistic inference of $\bar{P}|S$ from the premise set $\{M|P, \overline{M}|S\}$ is not informative, we add $p(S|(S \vee P)) > 0$ as a probabilistic constraint (i.e., an "existential import assumption") to obtain probabilistic informativeness. We show how to propagate the assigned (precise or interval-valued) probabilities to the sequence of conditional events $(M|P, \overline{M}|S, S|(S \vee P))$ to the conclusion $\bar{P}|S$. Thereby, we give a probabilistic meaning to the other syllogisms of Figure II. Moreover, our semantics also allows for generalizing the traditional syllogisms to new ones involving generalized quantifiers (like *Most S are P*) and syllogisms in terms of defaults and negated defaults.

Keywords: Categorical syllogisms · Coherence · Conditional events Defaults · Generalized quantifiers · Imprecise probability

1 Motivation and Outline

There is a long tradition in logic to investigate categorical syllogisms that goes back to Aristotle's *Analytica Priora*. However, not many authors proposed *probabilistic* semantics for categorical syllogisms (see, e.g., [9,10,13,15,23,38]) to overcome formal restrictions imposed by logic, like its *monotonicity* (i.e., the

N. Pfeifer and G. Sanfilippo—Shared first authorship (both authors contributed equally to this work).

N. Pfeifer—Supported by his DFG grant PF 740/2-2 (within the SPP1516).

G. Sanfilippo—Partially supported by the "National Group for Mathematical Analysis, Probability and their Applications" (GNAMPA – INdAM).

© Springer Nature Switzerland AG 2018
D. Ciucci et al. (Eds.): SUM 2018, LNAI 11142, pp. 196–211, 2018.
https://doi.org/10.1007/978-3-030-00461-3_14

inability to retract conclusions in the light of new evidence) or its qualitative nature (i.e., the inability to express *degrees of belief*). The main goal of building a probabilistic semantics is therefore to manage nonmonotonicity and degrees of belief, which are necessary for the formalization of commonsense reasoning. Although this paper is about probabilistic reasoning, applications of our results may include (i) relating ancient syllogisms to nonmonotonic reasoning by proposing a new nonmonotonic rule of inference and (ii) proposing a new rationality framework for the psychology of reasoning, specifically, for reasoning about a particular set of quantified statements (see, e.g., [29–32,35]). Moreover, (iii) our results are also applicable in formal semantics: specifically, our probabilistic approach is scalable in the sense that the proposed semantics allows for managing not only traditional logical quantifiers but also the much bigger superset of generalized quantifiers.

What are classical categorical syllogisms? They are valid argument forms consisting of two premises and a conclusion, which are composed of basic syllogistic sentence types (see, e.g., [28]): (A) *Every a is b*, (E) *No a is b*, (I) *Some a is b*, and (O) *Some a is not b*, where "a" and "b" denote two of the three categorical terms M ("middle term"), P ("predicate term"), or S ("subject term"). The M term appears only in the premises and are combined with P (in the first premise) and S (in the second premise). The predicates contained in the conclusion appear only in the order (S, P). By all possible permutations of the predicate order, four syllogistic *figures* result under the given restrictions. Syllogisms of Figure I, for example, have a transitive structure, i.e., M *is* P, S *is* M, therefore S *is* P. Consider *(Modus) Barbara* as an instance of a syllogism of Figure I: *Every M is P, Every S is M, therefore Every S is P*. The syllogism's name traditionally encodes logical properties. For the present purpose, we only recall that vocals refer to the syllogistic sentence type: for instance, B̲a̲r̲b̲a̲r̲a̲ involves only sentences of type (A) (see, e.g., [28] for details). Our paper is based on [15,19], where a coherence-based probability semantics for categorical syllogisms of Figure I was studied. We extend this work to Figure II, which has the following structure: P *is* M, S *is* M, therefore S *is* P. *Camestres* is an instance of a Figure II syllogism: *Every P is M, No S is M, therefore No S is P*. Camestres involves the sentence types (A) and (E).

While the S, M, and P terms are interpreted as predicate terms in first order logic, we interpret them as events as follows. Imagine a random experiment where the (random) outcome is denoted by X. Consider, for example, the predicate S. Depending on the outcome of the experiment, X may satisfy or not satisfy the predicate S. Then, we denote by E_S the event "X satisfies S" (the event E_S is true if X satisfies the predicate S and E_S is false if X does not satisfy S). We conceive the predicate S as the event E_S, which will be true or false. Thus, we simply identify E_S by S (in this sense S is both a predicate and an event). The same reasoning applies to the P and M terms, which are in our context both predicates and events. On the level of events, we associate pairs of predicates (S, P) with the corresponding conditional event $P|S$. On the level of probability assessments, we interpret the degree of belief in syllogistic sentence

(A) by $p(P|S) = 1$, (E) by $p(\bar{P}|S) = 1$, (I) by $P(P|S) > 0$, and we interpret (O) by $p(\bar{P}|S) > 0$ (see also [9,15] for similar interpretations and [33,34] for basic relations among these probabilistic sentence types). Thus, (A) and (E) are interpreted as precise probability assessments and (I) and (O) by imprecise probability assessments.

We note that, like the probabilistic Modus Barbara [15], the probabilistic Camestres is not probabilistically informative without existential import assumption: indeed, $p(M|P) = 1, p(\bar{M}|S) = 1 \implies 0 \leqslant p(\bar{P}|S) \leqslant 1$. We propose to add the conditional event existential import (i.e., $p(S|(S \vee P)) > 0$, which was originally proposed in the context of Weak Transitivity, see [15]) to the premise set to make Camestres probabilistically informative:

(Camestres) $p(M|P) = 1, p(\bar{M}|S) = 1$, and $p(S|(S \vee P)) > 0 \implies p(\bar{P}|S) = 1$.

After recalling some preliminary notions and results in Sect. 2, we show how to propagate the assigned probabilities to the sequence of conditional events $(M|P, \bar{M}|S, S|(S \vee P))$ to the conclusion $\bar{P}|S$ in Sect. 3. This result is applied in Sect. 4, where we give a probabilistic meaning to the other syllogisms of Figure II. Section 5 concludes by remarks on further applications (generalized quantifiers and nonmonotonic reasoning) and future work.

2   Preliminary Notions and Results

In this section we recall selected key features of coherence (for more details see, e.g., [5,7,11,12,20,21,27,37]). Given two events E and H, with $H \neq \perp$, the *conditional event* $E|H$ is defined as a three-valued logical entity which is *true* if EH (i.e., $E \wedge H$) is true, *false* if $\bar{E}H$ is true, and *void* if H is false. In betting terms, assessing $p(E|H) = x$ means that, for every real number s, you are willing to pay an amount $s \cdot x$ and to receive s, or 0, or $s \cdot x$, according to whether EH is true, or $\bar{E}H$ is true, or $\bar{H}$ is true (i.e., the bet is called off), respectively. In these cases the random gain is $\mathcal{G} = sH(E - x)$. More generally speaking, consider a real-valued function $p : \mathcal{K} \to \mathbb{R}$, where $\mathcal{K}$ is an arbitrary (possibly not finite) family of conditional events. Let $\mathcal{F} = (E_1|H_1, \ldots, E_n|H_n)$ be a sequence of conditional events, where $E_i|H_i \in \mathcal{K}$, $i = 1, \ldots, n$, and let $\mathcal{P} = (p_1, \ldots, p_n)$ be the vector of values $p_i = p(E_i|H_i)$, where $i = 1, \ldots, n$. We denote by $\mathcal{H}_0$ the disjunction $H_1 \vee \cdots \vee H_n$. With the pair $(\mathcal{F}, \mathcal{P})$ we associate the random gain $\mathcal{G} = \sum_{i=1}^{n} s_i H_i (E_i - p_i)$, where $s_1, \ldots, s_n$ are n arbitrary real numbers. $\mathcal{G}$ represents the net gain of n transactions. Let $\mathcal{G}_{\mathcal{H}_0}$ denote the set of possible values of $\mathcal{G}$ restricted to $\mathcal{H}_0$, that is, the values of $\mathcal{G}$ when at least one conditioning event is true.

Definition 1. *Function p defined on $\mathcal{K}$ is* coherent *if and only if, for every integer n, for every sequence $\mathcal{F}$ of n conditional events in $\mathcal{K}$ and for every $s_1, \ldots, s_n$, it holds that:* $\min \mathcal{G}_{\mathcal{H}_0} \leqslant 0 \leqslant \max \mathcal{G}_{\mathcal{H}_0}$.

Intuitively, Definition 1, means in betting terms that a probability assessment is coherent if and only if, in any finite combination of n bets, it cannot happen that the values in $G_{\mathcal{H}_0}$ are all positive, or all negative (*no Dutch Book*).

We recall the fundamental theorem of de Finetti for conditional events, which states that a coherent assessment of premises can always be coherently extended to a conclusion:

Theorem 1. *Let a coherent probability assessment* $\mathcal{P} = (p_1, \ldots, p_n)$ *on a sequence* $\mathcal{F} = (E_1|H_1, \ldots, E_n|H_n)$ *be given. Moreover, given a further conditional event* $E_{n+1}|H_{n+1}$. *Then, there exists a suitable closed interval* $[z', z''] \subseteq [0, 1]$ *such that the extension* $(\mathcal{P}, z)$ *of* $\mathcal{P}$ *to* $(\mathcal{F}, E_{n+1}|H_{n+1})$ *is coherent if and only if* $z \in [z', z'']$.

For applying Theorem 1, we now recall an algorithm which allows for computing the interval of coherent extensions $[z', z'']$ on $E_{n+1}|H_{n+1}$ from a coherent probability assessment $\mathcal{P}$ on a finite sequence of conditional events $\mathcal{F}$ (see [15, Algorithm 1], which is originally based on [5, Algorithm 2]).

Algorithm 1. Let $\mathcal{F} = (E_1|H_1, \ldots, E_n|H_n)$ be a sequence of conditional events and $\mathcal{P} = (p_1, \ldots, p_n)$ be a coherent precise probability assessment on $\mathcal{F}$, where $p_j = p(E_j|H_j) \in [0, 1]$, $j = 1, \ldots, n$. Moreover, let $E_{n+1}|H_{n+1}$ be a further conditional event and denote by J_{n+1} the set $\{1, \ldots, n+1\}$. The steps below describe the computation of the lower bound z' (resp., the upper bound z'') for the coherent extensions $z = p(E_{n+1}|H_{n+1})$.

- **Step 0.** Expand the expression $\bigwedge_{j \in J_{n+1}} (E_j H_j \vee \bar{E}_j H_j \vee \bar{H}_j)$ and denote by $C_1, \ldots, C_m$ the constituents contained in $\mathcal{H}_0 = \bigvee_{j \in J_{n+1}} H_j$ associated with $(\mathcal{F}, E_{n+1}|H_{n+1})$. Then, construct the following system in the unknowns $\lambda_1, \ldots, \lambda_m, z$

$$
\begin{cases}
\sum_{r:C_r \subseteq E_{n+1}H_{n+1}} \lambda_r = z \sum_{r:C_r \subseteq H_{n+1}} \lambda_r \, ; \\
\sum_{r:C_r \subseteq E_j H_j} \lambda_r = p_j \sum_{r:C_r \subseteq H_j} \lambda_r, \quad j \in J_n \, ; \\
\sum_{r \in J_m} \lambda_r = 1; \quad \lambda_r \geq 0, \, r \in J_m \, .
\end{cases}
\tag{1}
$$

- **Step 1.** Check the solvability of system (1) under the condition $z = 0$ (resp., $z = 1$). If the system (1) is not solvable go to Step 2, otherwise go to Step 3.
- **Step 2.** Solve the following linear programming problem

$$
\text{Compute:} \qquad \gamma' = \min \sum_{r:C_r \subseteq E_{n+1}H_{n+1}} \lambda_r
$$

$$
\text{(respectively:} \qquad \gamma'' = \max \sum_{r:C_r \subseteq E_{n+1}H_{n+1}} \lambda_r \text{)}
$$

subject to:

$$
\begin{cases}
\sum_{r:C_r \subseteq E_j H_j} \lambda_r = p_j \sum_{r:C_r \subseteq H_j} \lambda_r, \, j \in J_n \, ; \\
\sum_{r:C_r \subseteq H_{n+1}} \lambda_r = 1; \, \lambda_r \geq 0, \, r \in J_m.
\end{cases}
$$

The minimum γ' (respectively the maximum γ'') of the *objective function* coincides with z' (respectively with z'') and the procedure stops.

– **Step 3.** For each subscript $j \in J_{n+1}$, compute the maximum M_j of the function $\Phi_j = \sum_{r:C_r \subseteq H_j} \lambda_r$, subject to the constraints given by the system (1) with $z = 0$ (respectively $z = 1$). We have the following three cases:
 1. $M_{n+1} > 0$;
 2. $M_{n+1} = 0$, $M_j > 0$ for every $j \neq n+1$;
 3. $M_j = 0$ for $j \in I_0 = J \cup \{n+1\}$, with $J \neq \varnothing$.
 In the first two cases $z' = 0$ (respectively $z'' = 1$) and the procedure stops. In the third case, defining $I_0 = J \cup \{n+1\}$, set $J_{n+1} = I_0$ and $(\mathcal{F}, \mathcal{P}) = (\mathcal{F}_J, \mathcal{P}_J)$, where $\mathcal{F}_J = (E_i | H_i : i \in J)$ and $\mathcal{P}_J = (p_i : i \in J)$. Then, go to Step 0.

The procedure ends in a finite number of cycles by computing the value z' (respectively z'').

Remark 1. Assuming $(\mathcal{P}, z)$ on $(\mathcal{F}, E_{n+1} | H_{n+1})$ coherent, each solution $\Lambda = (\lambda_1, \ldots, \lambda_m)$ of System (1) is a coherent extension of the assessment $(\mathcal{P}, z)$ to the sequence $(C_1 | \mathcal{H}_0, \ldots, C_m | \mathcal{H}_0)$.

Definition 2. *An imprecise, or set-valued, assessment $\mathcal{I}$ on a finite sequence of n conditional events $\mathcal{F}$ is a (possibly empty) set of precise assessments $\mathcal{P}$ on $\mathcal{F}$.*

Definition 2, introduced in [14], states that an *imprecise (probability) assessment* $\mathcal{I}$ on a finite sequence $\mathcal{F}$ of n conditional events is just a (possibly empty) subset of $[0, 1]^n$. We recall the notions of g-coherence and total-coherence for imprecise (in the sense of set-valued) probability assessments [15].

Definition 3. *Let a sequence of n conditional events $\mathcal{F}$ be given. An imprecise assessment $\mathcal{I} \subseteq [0, 1]^n$ on $\mathcal{F}$ is g-coherent if and only if there exists a coherent precise assessment $\mathcal{P}$ on $\mathcal{F}$ such that $\mathcal{P} \in \mathcal{I}$.*

Definition 4. *An imprecise assessment $\mathcal{I}$ on $\mathcal{F}$ is totally coherent (t-coherent) if and only if the following two conditions are satisfied: (i) $\mathcal{I}$ is non-empty; (ii) if $\mathcal{P} \in \mathcal{I}$, then $\mathcal{P}$ is a coherent precise assessment on $\mathcal{F}$.*

We denote by Π the set of *all coherent precise* assessments on $\mathcal{F}$. We recall that if there are no logical relations among the events $E_1, H_1, \ldots, E_n, H_n$ involved in $\mathcal{F}$, that is $E_1, H_1, \ldots, E_n, H_n$ are logically independent, then the set Π associated with $\mathcal{F}$ is the whole unit hypercube $[0, 1]^n$. If there are logical relations, then the set Π *could be* a strict subset of $[0, 1]^n$. As it is well known $\Pi \neq \varnothing$; therefore, $\varnothing \neq \Pi \subseteq [0, 1]^n$.

Remark 2. We observe that:

$$\mathcal{I} \text{ is g-coherent} \iff \Pi \cap \mathcal{I} \neq \varnothing$$
$$\mathcal{I} \text{ is t-coherent} \iff \varnothing \neq \Pi \cap \mathcal{I} = \mathcal{I}.$$

Then: $\mathcal{I}$ is t-coherent $\Rightarrow \mathcal{I}$ is g-coherent.

Given a g-coherent assessment $\mathcal{I}$ on a sequence of n conditional events $\mathcal{F}$, for each coherent precise assessment $\mathcal{P}$ on $\mathcal{F}$, with $\mathcal{P} \in \mathcal{I}$, we denote by $[\alpha_{\mathcal{P}}, \beta_{\mathcal{P}}]$ the interval of coherent extensions of $\mathcal{P}$ to $E_{n+1}|H_{n+1}$; that is, the assessment $(\mathcal{P}, z)$ on $(\mathcal{F}, E_{n+1}|H_{n+1})$ is coherent if and only if $z \in [z'_{\mathcal{P}}, z''_{\mathcal{P}}]$. Then, defining the set

$$\Sigma = \bigcup_{\mathcal{P} \in \Pi \cap \mathcal{I}} [z'_{\mathcal{P}}, z''_{\mathcal{P}}], \tag{2}$$

for every $z \in \Sigma$, the assessment $\mathcal{I} \times \{z\}$ is a g-coherent extension of $\mathcal{I}$ to $(\mathcal{F}, E_{n+1}|H_{n+1})$; moreover, for every $z \in [0,1] \setminus \Sigma$, the extension $\mathcal{I} \times \{z\}$ of $\mathcal{I}$ to $(\mathcal{F}, E_{n+1}|H_{n+1})$ is not g-coherent. We say that Σ is the *set of coherent extensions* of the imprecise assessment $\mathcal{I}$ on $\mathcal{F}$ to the conditional event $E_{n+1}|H_{n+1}$.

3   Figure II: Propagation of Probability Bounds

In this section, we prove the precise and imprecise probability propagation rules for the inference from $(B|C, \bar{B}|A, A|A \vee C)$ to $\bar{C}|A$. We apply our results in Sect. 4, where we give a probabilistically informative interpretation of categorical syllogisms of Figure II.

Remark 3. Let A, B, C be logically independent events. It can be proved that the assessment (x, y, z) on $(B|C, \bar{B}|A, \bar{C}|A)$ is coherent for every $(x, y, z) \in [0,1]^3$, that is, the imprecise assessment $\mathcal{I} = [0,1]^3$ on $(B|C, \bar{B}|A, \bar{C}|A)$ is totally coherent. For this it is sufficient to check that each of the eight vertices of the unit cube is coherent. Coherence can be checked, for example, by applying Algorithm 1 of [14] or by the CkC-package [3]. Moreover, it can also be proved that the assessment (x, y, t) on $(B|C, \bar{B}|A, A|A \vee C)$ is coherent for every $(x, y, t) \in [0,1]^3$, that is, the imprecise assessment $\mathcal{I} = [0,1]^3$ on $(B|C, \bar{B}|A, A|A \vee C)$ is totally coherent.

Given a coherent probability assessment (x, y, t) on the sequence of conditional events $(B|C, \bar{B}|A, A|A \vee C)$. The next result allows for computing the lower and upper bounds, z' and z'' respectively, for the coherent extension $z = p(\bar{C}|A)$.

Theorem 2. *Let A, B, C be three logically independent events and $(x, y, t) \in [0,1]^3$ be a (coherent) assessment on the family $(B|C, \bar{B}|A, A|A \vee C)$. Then, the extension $z = P(\bar{C}|A)$ is coherent if and only if $z \in [z', z'']$, where*

$$[z', z''] = \begin{cases} [0,1], & \text{if } t \leqslant x + yt \leqslant 1, \\ [\dfrac{x + yt - 1}{t\,x}, 1], & \text{if } x + yt > 1, \\ [\dfrac{t - x - yt}{t\,(1 - x)}, 1], & \text{if } x + yt < t. \end{cases}$$

Proof. We now apply Algorithm 1 in a symbolic way.
Computation of the lower probability bound z' on $\bar{C}|A$.
Input. $\mathcal{F} = (B|C, \bar{B}|A, A|A \vee C)$, $E_{n+1}|H_{n+1} = \bar{C}|A$.

Step 0. The constituents associated with $(B|C, \bar{B}|A, A|A \vee C, \bar{C}|A)$ and contained in $\mathcal{H}_0 = A \vee C$ are $C_1 = ABC, C_2 = AB\bar{C}, C_3 = A\bar{B}C, C_4 = A\bar{B}\bar{C}, C_5 = \bar{A}BC,$ and $C_6 = \bar{A}\bar{B}C$. We construct the following starting system with unknowns $\lambda_1, \ldots, \lambda_6, z$ (see Remark 1):

$$\begin{cases} \lambda_2 + \lambda_4 = z(\lambda_1 + \lambda_2 + \lambda_3 + \lambda_4), \ \lambda_1 + \lambda_5 = x(\lambda_1 + \lambda_3 + \lambda_5 + \lambda_6), \\ \lambda_3 + \lambda_4 = y(\lambda_1 + \lambda_2 + \lambda_3 + \lambda_4), \\ \lambda_1 + \lambda_2 + \lambda_3 + \lambda_4 = t(\lambda_1 + \lambda_2 + \lambda_3 + \lambda_4 + \lambda_5 + \lambda_6), \\ \lambda_1 + \lambda_2 + \lambda_3 + \lambda_4 + \lambda_5 + \lambda_6 = 1, \ \lambda_i \geq 0, \ i = 1, \ldots, 6. \end{cases} \quad (3)$$

Step 1. By setting $z = 0$ in System (3), we obtain

$$\begin{cases} \lambda_2 + \lambda_4 = 0, \ \lambda_1 + \lambda_5 = x, \\ \lambda_3 = y(\lambda_1 + \lambda_3), \ \lambda_1 + \lambda_3 = t, \\ \lambda_1 + \lambda_3 + \lambda_5 + \lambda_6 = 1, \\ \lambda_i \geq 0, \ i = 1, \ldots, 6. \end{cases} \Longleftrightarrow \begin{cases} \lambda_1 = t(1 - y), \ \lambda_2 = 0, \ \lambda_3 = yt, \\ \lambda_4 = 0, \ \lambda_5 = x - t(1 - y), \\ \lambda_6 = 1 - x - yt, \\ \lambda_i \geq 0, \ i = 1, \ldots, 6. \end{cases} \quad (4)$$

The solvability of System (4) is a necessary condition for the coherence of the assessment $(x, y, t, 0)$ on $(B|C, \bar{B}|A, A|A \vee C, \bar{C}|A)$. As $(x, y, t) \in [0, 1]^3$, it holds that: $\lambda_1 = t(1 - y) \geq 0, \lambda_3 = yt \geq 0$. Thus, System (4) is solvable if and only if $\lambda_5 \geq 0$ and $\lambda_6 \geq 0$, that is

$$t - yt \leq x \leq 1 - yt \Longleftrightarrow t \leq x + yt \leq 1.$$

We distinguish two cases: (i) $x + yt > 1 \vee x + yt < t$; (ii) $t \leq x + yt \leq 1$. In Case (i), System (4) is not solvable (which implies that the coherent extension z of (x, y, t) must be positive). Then, we go to Step 2 of the algorithm where the (positive) lower bound z' is obtained by optimization. In Case (ii), System (4) is solvable and in order to check whether $z = 0$ is a coherent extension, we go to Step 3.

    Case (i). We observe that in this case t cannot be 0. By Step 2 we have the following linear programming problem:
Compute $z' = \min(\lambda_2 + \lambda_4)$ subject to:

$$\begin{cases} \lambda_1 + \lambda_5 = x(\lambda_1 + \lambda_3 + \lambda_5 + \lambda_6), \ \lambda_3 + \lambda_4 = y(\lambda_1 + \lambda_2 + \lambda_3 + \lambda_4), \\ \lambda_1 + \lambda_2 + \lambda_3 + \lambda_4 = t(\lambda_1 + \lambda_2 + \lambda_3 + \lambda_4 + \lambda_5 + \lambda_6), \\ \lambda_1 + \lambda_2 + \lambda_3 + \lambda_4 = 1, \ \lambda_i \geq 0, \ i = 1, \ldots, 6. \end{cases} \quad (5)$$

In this case, the constraints in (5) can be rewritten in the following way

$$\begin{cases} \lambda_1 + \lambda_5 = x(\lambda_1 + \lambda_3 + \lambda_5 + \lambda_6), \\ \lambda_3 + \lambda_4 = y, \ \lambda_5 + \lambda_6 = \frac{1-t}{t}, \\ \lambda_1 + \lambda_2 + \lambda_3 + \lambda_4 = 1, \\ \lambda_i \geq 0, \ i = 1, \ldots, 6, \end{cases} \Leftrightarrow \begin{cases} 1 - y - \lambda_2 + \lambda_5 = x(1 - \lambda_2 - \lambda_4 + \frac{1-t}{t}), \\ \lambda_3 = y - \lambda_4, \ \lambda_6 = \frac{1-t}{t} - \lambda_5, \\ \lambda_1 = 1 - y - \lambda_2, \\ \lambda_i \geq 0, \ i = 1, \ldots, 6, \end{cases}$$

or equivalently

$$\begin{cases} x\lambda_4 + (1 - y) + \lambda_5 = \lambda_2(1 - x) + \frac{x}{t}, \ \lambda_3 = y - \lambda_4, \\ \lambda_5 = \frac{1-t}{t} - \lambda_6, \ \lambda_1 = 1 - y - \lambda_2, \ \lambda_i \geq 0, \ i = 1, \ldots, 6. \end{cases}$$

We distinguish two (alternative) cases: $(i.1)$ $x + yt > 1$; $(i.2)$ $x + yt < t$.
Case $(i.1)$. The constraints in (5) can be rewritten in the following way

$$\begin{cases} x(\lambda_2 + \lambda_4) = \frac{x}{t} - (1 - y) - \frac{1-t}{t} + \lambda_2 + \lambda_6, \ \lambda_3 = y - \lambda_4, \\ \lambda_5 = \frac{1-t}{t} - \lambda_6, \ \lambda_1 = 1 - y - \lambda_2, \ \lambda_i \geqslant 0, \ i = 1, \ldots, 6. \end{cases}$$

As $x > 1 - ty$, we observe that $x > 0$. Then, the minimum of $z = \lambda_2 + \lambda_4$, obtained when $\lambda_2 = \lambda_6 = 0$, is

$$z' = \frac{1}{x}\left(\frac{x}{t} - (1 - y) - \frac{1 - t}{t}\right) = \frac{x - t + yt - 1 + t}{xt} = \frac{x + yt - 1}{xt}. \tag{6}$$

By choosing $\lambda_2 = \lambda_6 = 0$ the constraints in (5) are satisfied with

$$\begin{cases} \lambda_1 = 1 - y, \ \lambda_2 = 0, \ \lambda_3 = y - \frac{x+yt-1}{xt}, \ \lambda_4 = \frac{x+yt-1}{xt}, \\ \lambda_5 = \frac{1-t}{t}, \ \lambda_6 = 0, \ \lambda_i \geqslant 0, \ i = 1, \ldots, 6. \end{cases}$$

In particular $\lambda_3 \geqslant 0$ is satisfied because the condition $\frac{x+yt-1}{xt} \leqslant y$, which in this case amounts to $yt(1 - x) \leq 1 - x$, is always satisfied. Then, the *procedure stops* yielding as *output* $z' = \frac{x+yt-1}{xt}$.
Case $(i.2)$. The constraints in (5) can be rewritten in the following way

$$\begin{cases} (1 - y) - \frac{x}{t} + \lambda_5 + \lambda_4 = \lambda_2(1 - x) - x\lambda_4 + \lambda_4, \ \lambda_3 = y - \lambda_4, \\ \lambda_6 = \frac{1-t}{t} - \lambda_5, \ \lambda_1 = 1 - y - \lambda_2, \ \lambda_i \geqslant 0, \ i = 1, \ldots, 6, \end{cases}$$

or equivalently

$$\begin{cases} (\lambda_2 + \lambda_4)(1 - x) = (1 - y) - \frac{x}{t} + \lambda_4 + \lambda_5, \ \lambda_3 = y - \lambda_4, \\ \lambda_6 = \frac{1-t}{t} - \lambda_5, \ \lambda_1 = 1 - y - \lambda_2, \ \lambda_i \geqslant 0, \ i = 1, \ldots, 6. \end{cases}$$

As $t - yt - x > 0$, that is $x < t(1 - y)$, it holds that $x < 1$. Then, the minimum of $z = \lambda_2 + \lambda_4$, obtained when $\lambda_4 = \lambda_5 = 0$, is

$$z' = \frac{1}{1 - x}\left(1 - y - \frac{x}{t}\right) = \frac{t - yt - x}{(1 - x)t} \geqslant 0.$$

We observe that by choosing $\lambda_4 = \lambda_5 = 0$ the constraints in (5) are satisfied, indeed they are

$$\begin{cases} \lambda_1 = 1 - y, \ \lambda_2 = \frac{t-yt-x}{(1-x)t}, \ \lambda_3 = y, \ \lambda_4 = 0, \\ \lambda_5 = 0, \ \lambda_6 = \frac{1-t}{t}, \ \lambda_i \geqslant 0, \ i = 1, \ldots, 6. \end{cases}$$

Then, the *procedure stops* yielding as *output* $z' = \frac{t-yt-x}{(1-x)t}$.
    Case (ii). We take Step 3 of the algorithm. We denote by Λ and $\mathcal{S}$ the vector of unknowns $(\lambda_1, \ldots, \lambda_6)$ and the set of solution of System (4), respectively. We consider the following linear functions (associated with the conditioning events $H_1 = C, H_2 = H_4 = A, H_3 = A \vee C$) and their maxima in $\mathcal{S}$:

$$\begin{aligned} \Phi_1(\Lambda) &= \sum_{r:C_r \subseteq C} \lambda_r = \lambda_1 + \lambda_3 + \lambda_5 + \lambda_6, \\ \Phi_2(\Lambda) &= \Phi_4(\Lambda) = \sum_{r:C_r \subseteq A} \lambda_r = \lambda_1 + \lambda_2 + \lambda_3 + \lambda_4, \\ \Phi_3(\Lambda) &= \sum_{r:C_r \subseteq A \vee C} \lambda_r = \lambda_1 + \lambda_2 + \lambda_3 + \lambda_4 + \lambda_5 + \lambda_6, \\ M_i &= \max_{\Lambda \in \mathcal{S}} \Phi_i(\Lambda), \quad i = 1, 2, 3, 4. \end{aligned} \tag{7}$$

By (4) we obtain: $\Phi_1(\Lambda) = 1$, $\Phi_2(\Lambda) = \Phi_4(\Lambda) = t$, $\Phi_3(\Lambda) = 1$, $\forall \Lambda \in \mathcal{S}$. Then, $M_1 = 1$, $M_2 = M_4 = t$, and $M_3 = 1$. We consider two subcases: $t > 0$; $t = 0$. If $t > 0$, then $M_4 > 0$ and we are in the first case of Step 3. Thus, the *procedure stops* and yields $z' = 0$ as *output*.

If $t = 0$, then $M_1 > 0, M_3 > 0$ and $M_2 = M_4 = 0$. Hence, we are in third case of Step 3 with $J = \{2\}, I_0 = \{2, 4\}$ and the procedure restarts with Step 0, with $\mathcal{F}$ replaced by $\mathcal{F}_J = (\bar{B}|A)$.

(2nd cycle) Step 0. The constituents associated with $(\bar{B}|A, \bar{C}|A)$, contained in A, are $C_1 = ABC, C_2 = AB\bar{C}, C_3 = A\bar{B}C, C_4 = A\bar{B}\bar{C}$. The starting system is

$$\begin{cases} \lambda_3 + \lambda_4 = y(\lambda_1 + \lambda_2 + \lambda_3 + \lambda_4), \ \lambda_2 + \lambda_4 = z(\lambda_1 + \lambda_2 + \lambda_3 + \lambda_4), \\ \lambda_1 + \lambda_2 + \lambda_3 + \lambda_4 = 1, \ \lambda_i \geqslant 0, \ i = 1, \dots, 4. \end{cases} \quad (8)$$

(2nd cycle) Step 1. By setting $z = 0$ in System (8), we obtain

$$\{\lambda_1 = 1 - y, \ \lambda_2 = \lambda_4 = 0, \ \lambda_3 = y, \ \lambda_i \geqslant 0, \ i = 1, \dots, 4. \quad (9)$$

As $y \in [0, 1]$, System (9) is always solvable; thus, we go to Step 3.

(2nd cycle) Step 3. We denote by Λ and $\mathcal{S}$ the vector of unknowns $(\lambda_1, \dots, \lambda_4)$ and the set of solution of System (9), respectively. The conditioning events are $H_2 = A$ and $H_4 = A$; then the associated linear functions are: $\Phi_2(\Lambda) = \Phi_4(\Lambda) = \sum_{r:C_r \subseteq A} \lambda_r = \lambda_1 + \lambda_2 + \lambda_3 + \lambda_4$. From System (9), we obtain: $\Phi_2(\Lambda) = \Phi_4(\Lambda) = 1$, $\forall \Lambda \in \bar{\mathcal{S}}$; so that $M_2 = M_4 = 1$. We are in the first case of Step 3 of the algorithm; then the *procedure stops* and yields $z' = 0$ as *output*.

To summarize, for any $(x, y, t) \in [0, 1]^3$ on $(B|C, \bar{B}|A, A|A \vee C)$, we have computed the coherent lower bound z' on $\bar{C}|A$. In particular, if $t = 0$, then $z' = 0$. We also have $z' = 0$, when $t > 0$ and $t \leqslant x + yt \leqslant 1$, that is when $0 < t \leqslant x + yt \leqslant 1$. Then, we can write that $z' = 0$, when $t \leqslant x + yt \leqslant 1$. Otherwise, we have two cases: $(i.1)$ $z' = \frac{x+yt-1}{xt}$, if $x + yt > 1$; $(i.2)$ $z' = \frac{t-yt-x}{(1-x)t}$, if $x + yt < t$.

Computation of the Upper Probability Bound z'' on $\bar{C}|A$
Input and *Step 0* are the same as in the proof of z'.
Step 1. By setting $z = 1$ in System (3), we obtain

$$\begin{cases} \lambda_1 + \lambda_3 = 0, \ \lambda_5 = x(\lambda_5 + \lambda_6), \\ \lambda_4 = y(\lambda_2 + \lambda_4), \ \lambda_2 + \lambda_4 = t, \\ \lambda_2 + \lambda_4 + \lambda_5 + \lambda_6 = 1, \\ \lambda_i \geqslant 0, \ i = 1, \dots, 6. \end{cases} \Longleftrightarrow \begin{cases} \lambda_1 = \lambda_3 = 0, \ \lambda_2 = t(1 - y), \\ \lambda_4 = yt, \ \lambda_5 = x(1 - t), \\ \lambda_6 = (1 - x)(1 - t), \\ \lambda_i \geqslant 0, \ i = 1, \dots, 6. \end{cases} \quad (10)$$

As $(x, y, t) \in [0, 1]^3$, System (10) is solvable and we go to Step 3.
Step 3. We denote by Λ and $\mathcal{S}$ the vector of unknowns $(\lambda_1, \dots, \lambda_6)$ and the set of solution of System (10), respectively. We consider the functions given in (7). From System (10), we obtain $M_1 = x(1 - t) + (1 - x)(1 - t) = 1 - t$, $M_2 = M_4 = t$, and $M_3 = 1$. If $t > 0$, then $M_4 > 0$ and we are in the first case of Step 3. Thus, the *procedure stops* and yields $z'' = 1$ as *output*. If $t = 0$,

then $M_1 > 0, M_3 > 0$ and $M_2 = M_4 = 0$. Hence, we are in the third case of Step 3 with $J = \{2\}, I_0 = \{2, 4\}$ and the procedure restarts with Step 0, with $\mathcal{F}$ replaced by $\mathcal{F}_J = (E_2|H_2) = (\bar{B}|A)$ and $\mathcal{P}$ replaced by $\mathcal{P}_J = y$.

(2nd cycle) Step 0. This is the same as the (2nd cycle) Step 0 in the proof of z'.

(2nd cycle) Step 1. By setting $z = 1$ in System (3), we obtain

$$\{\lambda_1 + \lambda_3 = 0, \;\; \lambda_4 = y, \;\; \lambda_2 = 1 - y, \;\; \lambda_i \geqslant 0, \;\; i = 1, \ldots, 4. \tag{11}$$

As $y \in [0, 1]$, System (11) is always solvable; thus, we go to Step 3.

(2nd cycle) Step 3. Like in the (2nd cycle) Step 3 of the proof of z', we obtain $M_4 = 1$. Thus, the procedure stops and yields $z'' = 1$ as output. To summarize, for any assessment $(x, y, t) \in [0, 1]^3$ on $(B|C, \bar{B}|A, A|A \vee C)$, we have computed the coherent upper probability bound z'' on $\bar{C}|A$, which is always $z'' = 1$. $\square$

Remark 4. We observe that in Theorem 2 we do not presuppose, differently from the classical approach, positive probability for the conditioning events (A and C). For example, even if we assume $p(A|A \vee C) = t > 0$ we do not require positive probability for the conditioning event A, and $p(A)$ could be zero (indeed, since $p(A) = p(A \wedge (A \vee C)) = p(A|A \vee C)p(A \vee C)$, $p(A) > 0$ implies $p(A|A \vee C) > 0$, but not vice versa). Moreover, we used a general and global approach for obtaining the inference rule in Theorem 2 (see [1,2,6,8,24,25] on local versus global approaches).

The next result is based on Theorem 2 and presents the set of the coherent extensions of a given interval-valued probability assessment $\mathcal{I} = ([x_1, x_2] \times [y_1, y_2] \times [t_1, t_2]) \subseteq [0, 1]^3$ on the sequence on $(B|C, \bar{B}|A, A|A \vee C)$ to the further conditional event $\bar{C}|A$.

Theorem 3. *Let A, B, C be three logically independent events and $\mathcal{I} = ([x_1, x_2] \times [y_1, y_2] \times [t_1, t_2]) \subseteq [0, 1]^3$ be an imprecise assessment on $(B|C, \bar{B}|A, A|A \vee C)$. Then, the set Σ of the coherent extensions of $\mathcal{I}$ on $\bar{C}|A$ is the interval $[z^*, z^{**}]$, where*

$$[z^*, z^{**}] = \begin{cases} [0, 1], & \text{if } (x_2 + y_2 t_1 \geqslant t_1) \wedge (x_1 + y_1 t_1 \leqslant 1), \\ [\frac{x_1 + y_1 t_1 - 1}{t_1 x_1}, 1], & \text{if } x_1 + y_1 t_1 > 1, \\ [\frac{t_1 - x_2 - y_2 t_1}{t_1 (1 - x_2)}, 1], & \text{if } x_2 + y_2 t_1 < t_1. \end{cases}$$

Proof. As from Remark 3 the set $[0, 1]^3$ on $(B|C, \bar{B}|A, A|A \vee C)$ is totally coherent, then $\mathcal{I}$ is totally coherent too. Then, $\Sigma = \bigcup_{\mathcal{P} \in \mathcal{I}} [z'_\mathcal{P}, z''_\mathcal{P}] = [z^*, z^{**}]$, where $z^* = \inf_{\mathcal{P} \in \mathcal{I}} z'_\mathcal{P}$ and $z^{**} = \sup_{\mathcal{P} \in \mathcal{I}} z''_\mathcal{P}$. We distinguish three alternative cases: (i) $x_1 + y_1 t_1 > 1$; (ii) $x_2 + y_2 t_1 < t_1$; (iii) $(x_2 + y_2 t_1 \geqslant t_1) \wedge (x_1 + y_1 t_1 \leqslant 1)$. Of course, for all three cases $z^{**} = \sup_{\mathcal{P} \in \mathcal{I}} z''_\mathcal{P} = 1$.

Case (i). We observe that the function $x + yt : [0, 1]^3$ is non-decreasing in the arguments x, y, t. Then, in this case, $x + yt \geqslant x_1 + y_1 t_1 > 1$ for every $\mathcal{P} = (x, y, t) \in \mathcal{I}$ and hence by Theorem 2 $z'_\mathcal{P} = f(x, y, t) = \frac{x + yt - 1}{tx}$ for every $\mathcal{P} \in \mathcal{I}$.

Moreover, $f(x, y, t) : [0, 1]^3$ is non-decreasing in the arguments x, y, t, thus $z^* = \frac{x_1 + y_1 t_1 - 1}{t_1 x_1}$.

Case (ii). We observe that the function $x + yt - t : [0, 1]^3$ is non-decreasing in the arguments x, y and non-increasing in the argument t. Then, in this case, $x + yt - t \leqslant x_2 + y_2 t_1 - t_1 < 0$ for every $\mathcal{P} = (x, y, t) \in \mathcal{I}$ and hence by Theorem 2 $z'_{\mathcal{P}} = g(x, y, t) = \frac{t - x - yt}{t(1-x)}$ for every $\mathcal{P} \in \mathcal{I}$. Moreover, $g(x, y, t) : [0, 1]^3$ is non-increasing in the arguments x, y and non-decreasing in the argument t. Thus, $z^* = \frac{t_1 - x_2 - y_2 t_1}{t_1(1-x_2)}$. Case (iii). In this case there exists a vector $(x, y, t) \in \mathcal{I}$ such that $t \leqslant x + yt \leqslant 1$ and hence by Theorem 2 $z'_{\mathcal{P}} = 0$. Thus, $z^* = 0$.   □

Remark 5. By instantiating Theorem 3 with the imprecise assessment $\mathcal{I} = \{1\} \times [y_1, 1] \times [t_1, 1]$, where $t_1 > 0$, we obtain the following lower and upper bounds for the conclusion $[z^*, z^{**}] = [y_1, 1]$. Thus, for every $t_1 > 0$: z^* depends only on the value of y_1.

4   Some Categorical Syllogisms of Figure II

In this section we consider examples of probabilistic categorical syllogisms of Figure II (Camestres, Camestrop, Baroco, Cesare, Cesaro, Festino) by suitable instantiations in Theorem 2. We consider three events P, M, S corresponding to the predicate, middle, and the subject term, respectively.

Camestres. The direct probabilistic interpretation of the categorical syllogism *"Every P is M, No S is M, therefore No S is P"* would correspond to infer $p(\bar{P}|S) = 1$ from the premises $p(M|P) = 1$ and $p(\bar{M}|S) = 1$; however, this inference is not justified. Indeed, by Remark 3, a probability assessment $(1, 1, z)$ on $(M|P, \bar{M}|S, \bar{P}|S)$ is coherent for every $z \in [0, 1]$. In order to construct a probabilistically informative version of Camestres, a further constraint of the premise set is needed. Like in [15], we use the *conditional event existential import* for further constraining the premise set: this is defined by the conditional probability of the conditioning event of the conclusion given the disjunction of all conditioning events. For categorical syllogisms of Figure II the conditional event existential import is $p(S|(S \vee P)) > 0$. Then, by instantiating S, M, P in Theorem 2 for A, B, C with $x = y = 1$ and $t > 0$ it follows that $z' = \frac{x + yt - 1}{t x} = 1$. Then,

$$p(M|P) = 1, \ p(\bar{M}|S) = 1, \ \text{and} \ p(S|(S \vee P)) > 0 \Longrightarrow p(\bar{P}|S) = 1. \qquad (12)$$

Therefore, inference (12) is a probabilistically informative version of Camestres.

By instantiating S, M, P in Theorem 3 for A, B, C with $x_1 = x_2 = 1$, $y_1 = .6$, $y_2 = .9$, $t_1 > 0$, and $t_2 = 1$, we obtain $z^* = y_1 = .6$ and $z^{**} = 1$, i.e.,

$$p(M|P) = 1, \ .6 \leqslant p(\bar{M}|S) \leqslant .9, \ \text{and} \ p(S|(S \vee P)) \geqslant t_1 > 0 \Longrightarrow p(\bar{P}|S) \geqslant .6. \qquad (13)$$

This can be seen as an extension of Camestres to generalized quantifiers. Specifically, the second premise can be used to represent a generalized quantified statement like *At least most but not all S are not-M* and the conclusion can represent

At least most S are not-P. Of course, the specific values involved in the premises are context dependent (see also [4, 26]).

We observe that, by Remark 3, every direct probabilistic interpretation of the other categorical syllogisms of Figure II are probabilistically non-informative without the further probabilistic constraint $p(S|(S \vee P)) > 0$. In what follows, we show how to construct probabilistically informative versions of other categorical syllogisms of Figure II by suitable instantiations of Theorem 2.

Camestrop. From (12) it also follows that

$$p(M|P) = 1, \ p(\overline{M}|S) = 1, \ \text{and} \ p(S|(S \vee P)) > 0 \Longrightarrow p(\overline{P}|S) > 0, \qquad (14)$$

which is a probabilistic informative interpretation of Camestrop (*Every P is M, No S is M, therefore Some S is not P*) under the existential import assumption $(p(S|(S \vee P)) > 0)$.

Baroco. By instantiating S, M, P in Theorem 2 for A, B, C with $x = 1, y > 0$ and $t > 0$ it follows that $z' = \frac{x+yt-1}{tx} = \frac{1+yt-1}{t} = y > 0$. Then,

$$p(M|P) = 1, \ p(\overline{M}|S) > 0, \ \text{and} \ p(S|(S \vee P)) > 0 \Longrightarrow p(\overline{P}|S) > 0. \qquad (15)$$

Therefore, inference (15) is a probabilistically informative version of Baroco (*Every P is M, Some S is not M, therefore Some S is not P*) under the existential import.

Cesare. The direct probabilistic interpretation of the categorical syllogism "*No P is M, Every S is M, therefore No S is P*" would correspond to infer $p(\overline{P}|S) = 1$ from the premises $p(\overline{M}|P) = 1$ and $p(M|S) = 1$; However, this inference is not probabilistically informative because it is obtained from Camestres when M is replaced by $\overline{M}$. By instantiating S, M, P in Theorem 2 for A, B, C with $x = y = 0$ and $t > 0$ it follows that $z' = \frac{t-x-yt}{t(1-x)} = 1$. Then,

$$p(M|P) = 0, \ p(\overline{M}|S) = 0, \ \text{and} \ p(S|(S \vee P)) > 0 \Longrightarrow p(\overline{P}|S) = 1.$$

or equivalently,

$$p(\overline{M}|P) = 1, \ p(M|S) = 1, \ \text{and} \ p(S|(S \vee P)) > 0 \Longrightarrow p(\overline{P}|S) = 1. \qquad (16)$$

Therefore, inference (16) is a probabilistically informative version of Cesare under the existential import assumption.

By instantiating S, M, P in Theorem 3 for A, B, C with $x_1 = x_2 = 0, y_1 = .1, y_2 = .4, t_1 > 0$, and $t_2 = 1$, we obtain $z^* = 1 - y_2 = .6$ and $z^{**} = 1$; i.e.,

$$p(M|P) = 0, \ .1 \leqslant p(\overline{M}|S) \leqslant .4, \ \text{and} \ p(S|(S \vee P)) \geqslant t_1 > 0 \Longrightarrow p(\overline{P}|S) \geqslant .6,$$

which is equivalent to

$$p(\overline{M}|P) = 1, \ .6 \leqslant p(M|S) \leqslant .9, \ \text{and} \ p(S|(S \vee P)) \geqslant t_1 > 0 \Longrightarrow p(\overline{P}|S) \geqslant .6. \qquad (17)$$

Equation (17) is a generalized version of Cesare, where the second premise can represent a generalized quantified statement like *At least most but not all S are M* and the conclusion can represent *At least many S are not-P*.

Cesaro. From (16) it also follows that

$$p(\overline{M}|P) = 1,\ p(M|S) = 1,\ \text{and}\ p(S|(S \vee P)) > 0 \implies p(\bar{P}|S) > 0, \qquad (18)$$

which is a probabilistically informative interpretation of Cesaro (*No P is M, Every S is M, therefore Some S is not P*) under the existential import.

Festino. By instantiating S, M, P in Theorem 2 for A, B, C with $x = 0, y < 1$ and $t > 0$, as $x + yt < t$, it follows that $z' = \frac{t-x-yt}{t(1-x)} = \frac{t-yt}{t} > 1 - y > 0$. Then,

$$p(\overline{M}|P) = 1,\ p(M|S) > 0,\ \text{and}\ p(S|(S \vee P)) > 0 \implies p(\bar{P}|S) > 0. \qquad (19)$$

Therefore, inference (19) is a probabilistically informative version of Festino (*No P is M, Some S is M, therefore Some S is not P*) under the existential import.

Remark 6. We observe that, traditionally, the conclusions of logically valid categorical syllogisms of Figure II are neither in the form of sentence type I (*some*) nor of A (*every*). In terms of our probability semantics, indeed, this must be the case even if the existential import assumption $p(S|(S \vee P)) > 0$ is made: according to Theorem 2, the upper the bound for the conclusion $p(\bar{P}|S)$ is always 1; thus, neither sentence type I ($p(P|S) > 0$, i.e. $p(\bar{P}|S) < 1$) nor sentence type A ($p(P|S) = 1$, i.e. $p(\bar{P}|S) = 0$) can be validated.

Remark 7. We recall that $p(S) = p(S \wedge (S \vee P)) = p(S|(S \vee P))P(S \vee P)$. Hence, if we assume that $p(S)$ is positive, then $p(S|(S \vee P))$ must be positive too (the converse, however, does not hold). Therefore, the inferences (12)–(19) hold if $p(S|(S \vee P)) > 0$ is replaced $p(S) > 0$. The constraint $p(S) > 0$ can be seen as a stronger version of an existential import assumption compared to the conditional event existential import.

5   Concluding Remarks

In this paper we observed that an existential import assumption is required for the probabilistic validity of syllogisms of Figure II, which we expressed in terms of a probability constraint. Then, we proved probability propagation rules for categorical syllogisms of Figure II. We applied the probability propagation rules to show the validity of the probabilistic versions of the traditional categorical syllogisms of Figure II (i.e., Camestres, Camestrop, Baroco, Cesare, Cesaro, Festino).

We note that, by setting appropriate thresholds, our semantics also allows for generalizing the traditional syllogisms to new ones involving generalized quantifiers (like interpreting *Most S are P* by $p(P|S) \geqslant t$, where t is a given— usually context dependent—threshold like >.5). Probabilistic syllogisms are a much more plausible rationality framework for studying categorical syllogisms compared to the traditional syllogisms, as "truly" *all-* and *existentially quantified* statements are hardly ever used in commonsense contexts (even if people mention words like "all", they usually don't mean *all* in a strictly universal sense). Indeed, quantified statements are usually not falsified by one exception (while the universal quantifier $\forall$ does not allow for exceptions) and quantify over more than just *at least one thing* (while the existential quantifier $\exists$ is weak in the sense that it is true when it holds for at least one thing).

Furthermore, as proposed in [15], the basic syllogistic sentence types can also be interpreted as instances of defaults (i.e., (A) by $S \mathrel{|\!\sim} P$ and (E) by $S \mathrel{|\!\sim} \bar{P}$) and negated defaults (i.e., (I) by $S \mathrel{|\!\not\sim} \bar{P}$ and (O) by $S \mathrel{|\!\not\sim} P$): thus, we can also build a bridge from categorical syllogisms of Figure II to default reasoning. Camestres, for example, has the following form in terms of defaults: (A) $P \mathrel{|\!\sim} M$, (E) $S \mathrel{|\!\sim} \bar{M}$, and the existential import $(S \vee P) \mathrel{|\!\not\sim} \bar{S}$ implies (E) $S \mathrel{|\!\sim} \bar{P}$. This version of Camestres can serve as a valid inference rule for nonmonotonic reasoning.

We will devote future work to deepen our results and to extend our coherence-based probability semantics to other categorical syllogisms. In particular, we plan to further generalize categorical syllogisms by applying the theory of compounds of conditionals under coherence (see, e.g. [16,17,22]): as shown in the context of conditional syllogisms [18,36,37], this theory allows for managing logical operations on conditionals and iterated conditionals. Iterated conditionals can be used for interpreting categorical syllogisms with statements like *If S_1 are P_1, then S_2 are P_2* (i.e., $(P_2|S_2)|(P_1|S_1)$).

Acknowledgments. We thank three anonymous reviewers for their useful comments and suggestions.

References

1. Amarger, S., Dubois, D., Prade, H.: Constraint propagation with imprecise conditional probabilities. In: Proceedings of UAI 1991, pp. 26–34. Morgan Kaufmann, Burlington (1991)
2. Amarger, S., Dubois, D., Prade, H.: Handling imprecisely-known conditional probabilities. In: Hand, D. (ed.) AI and Computer Power: the Impact on Statistics, pp. 63–97. Chapman & Hall, London (1994)
3. Baioletti, M., Capotorti, A., Galli, L., Tognoloni, S., Rossi, F., Vantaggi, B.: CkC (Check Coherence package; version e6, November 2016). http://www.dmi.unipg.it/~upkd/paid/software.html
4. Barwise, J., Cooper, R.: Generalized quantifier and natural language. Linguist. Philos. **4**, 159–219 (1981)
5. Biazzo, V., Gilio, A.: A generalization of the fundamental theorem of de Finetti for imprecise conditional probability assessments. Int. J. Approximate Reason. **24**(2–3), 251–272 (2000)
6. Biazzo, V., Gilio, A., Lukasiewicz, T., Sanfilippo, G.: Probabilistic logic under coherence: complexity and algorithms. Ann. Math. Artif. Intell. **45**(1–2), 35–81 (2005)

7. Biazzo, V., Gilio, A., Sanfilippo, G.: Coherent conditional previsions and proper scoring rules. In: Greco, S., Bouchon-Meunier, B., Coletti, G., Fedrizzi, M., Matarazzo, B., Yager, R.R. (eds.) IPMU 2012. CCIS, vol. 300, pp. 146–156. Springer, Heidelberg (2012). https://doi.org/10.1007/978-3-642-31724-8_16

8. Capotorti, A., Galli, L., Vantaggi, B.: Locally strong coherence and inference with lower-upper probabilities. Soft Comput. **7**(5), 280–287 (2003)

9. Chater, N., Oaksford, M.: The probability heuristics model of syllogistic reasoning. Cogn. Psychol. **38**, 191–258 (1999)

10. Cohen, A.: Generics, frequency adverbs, and probability. Linguist. Philos. **22**, 221–253 (1999)

11. Coletti, G., Scozzafava, R.: Probabilistic Logic in a Coherent Setting. Kluwer, Dordrecht (2002)

12. Coletti, G., Scozzafava, R., Vantaggi, B.: Possibilistic and probabilistic logic under coherence: default reasoning and system P. Math. Slovaca **65**(4), 863–890 (2015)

13. Dubois, D., Godo, L., López De Màntaras, R., Prade, H.: Qualitative reasoning with imprecise probabilities. J. Intell. Inf. Syst. **2**(4), 319–363 (1993)

14. Gilio, A., Ingrassia, S.: Totally coherent set-valued probability assessments. Kybernetika **34**(1), 3–15 (1998)

15. Gilio, A., Pfeifer, N., Sanfilippo, G.: Transitivity in coherence-based probability logic. J. Appl. Logic **14**, 46–64 (2016)

16. Gilio, A., Sanfilippo, G.: Conditional random quantities and iterated conditioning in the setting of coherence. In: van der Gaag, L.C. (ed.) ECSQARU 2013. LNCS (LNAI), vol. 7958, pp. 218–229. Springer, Heidelberg (2013). https://doi.org/10.1007/978-3-642-39091-3_19

17. Gilio, A., Sanfilippo, G.: Generalized logical operations among conditional events. Appl. Intell. (in press). https://doi.org/10.1007/s10489-018-1229-8

18. Gilio, A., Over, D.E., Pfeifer, N., Sanfilippo, G.: Centering and compound conditionals under coherence. In: Ferraro, M.B., et al. (eds.) Soft Methods for Data Science. AISC, vol. 456, pp. 253–260. Springer, Cham (2017). https://doi.org/10.1007/978-3-319-42972-4_32

19. Gilio, A., Pfeifer, N., Sanfilippo, G.: Transitive reasoning with imprecise probabilities. In: Destercke, S., Denoeux, T. (eds.) ECSQARU 2015. LNCS (LNAI), vol. 9161, pp. 95–105. Springer, Cham (2015). https://doi.org/10.1007/978-3-319-20807-7_9

20. Gilio, A., Sanfilippo, G.: Probabilistic entailment in the setting of coherence: the role of quasi conjunction and inclusion relation. Int. J. Approximate Reason. **54**(4), 513–525 (2013)

21. Gilio, A., Sanfilippo, G.: Quasi conjunction, quasi disjunction, t-norms and t-conorms: probabilistic aspects. Inf. Sci. **245**, 146–167 (2013)

22. Gilio, A., Sanfilippo, G.: Conditional random quantities and compounds of conditionals. Studia Logica **102**(4), 709–729 (2014)

23. Lambert, J.H.: Neues Organon oder Gedanken über die Erforschung und Bezeichnung des Wahren und dessen Unterscheidung vom Irrthum und Schein. Wendler, Leipzig (1764)

24. Lukasiewicz, T.: Local probabilistic deduction from taxonomic and probabilistic knowledge-bases over conjunctive events. Int. J. Approximate Reason. **21**, 23–61 (1999)

25. Lukasiewicz, T.: Probabilistic deduction with conditional constraints over basic events. J. Artif. Intell. Res. **10**, 199–241 (1999)

26. Peters, S., Westerståhl, D.: Quantifiers in Language and Logic. Oxford University Press, Oxford (2006)

27. Petturiti, D., Vantaggi, B.: Envelopes of conditional probabilities extending a strategy and a prior probability. Int. J. Approximate Reason. **81**, 160–182 (2017)
28. Pfeifer, N.: Contemporary syllogistics: comparative and quantitative syllogisms. In: Kreuzbauer, G., Dorn, G.J.W. (eds.) Argumentation in Theorie und Praxis: Philosophie und Didaktik des Argumentierens, pp. 57–71. Lit Verlag, Wien (2006)
29. Pfeifer, N.: The new psychology of reasoning: a mental probability logical perspective. Thinking Reason. **19**(3–4), 329–345 (2013)
30. Pfeifer, N.: Reasoning about uncertain conditionals. Studia Logica **102**(4), 849–866 (2014)
31. Pfeifer, N., Douven, I.: Formal epistemology and the new paradigm psychology of reasoning. Rev. Philos. Psychol. **5**(2), 199–221 (2014)
32. Pfeifer, N., Kleiter, G.D.: Towards a mental probability logic. Psychol. Belgica **45**(1), 71–99 (2005)
33. Pfeifer, N., Sanfilippo, G.: Probabilistic squares and hexagons of opposition under coherence. Int. J. Approximate Reason. **88**, 282–294 (2017)
34. Pfeifer, N., Sanfilippo, G.: Square of opposition under coherence. In: Ferraro, M.B. (ed.) Soft Methods for Data Science. AISC, vol. 456, pp. 407–414. Springer, Cham (2017). https://doi.org/10.1007/978-3-319-42972-4_50
35. Pfeifer, N., Tulkki, L.: Conditionals, counterfactuals, and rational reasoning. An experimental study on basic principles. Minds Mach. **27**(1), 119–165 (2017)
36. Sanfilippo, G., Pfeifer, N., Gilio, A.: Generalized probabilistic modus ponens. In: Antonucci, A., Cholvy, L., Papini, O. (eds.) ECSQARU 2017. LNCS (LNAI), vol. 10369, pp. 480–490. Springer, Cham (2017). https://doi.org/10.1007/978-3-319-61581-3_43
37. Sanfilippo, G., Pfeifer, N., Over, D.E., Gilio, A.: Probabilistic inferences from conjoined to iterated conditionals. Int. J. Approximate Reason. **93**, 103–118 (2018)
38. Martin, T.: J.-H. Lambert's theory of probable syllogisms. Int. J. Approximate Reason. **52**(2), 144–152 (2011)

Measuring Disagreement Among Knowledge Bases

Nico Potyka[(⊠)]

University of Osnabrück, Osnabrück, Germany
npotyka@uos.de

Abstract. When combining beliefs from different sources, often not only new knowledge but also conflicts arise. In this paper, we investigate how we can measure the disagreement among sources. We start our investigation with disagreement measures that can be induced from inconsistency measures in an automated way. After discussing some problems with this approach, we propose a new measure that is inspired by the η-inconsistency measure. Roughly speaking, it measures how well we can satisfy all sources simultaneously. We show that the new measure satisfies desirable properties, scales well with respect to the number of sources and illustrate its applicability in inconsistency-tolerant reasoning.

1 Introduction

One challenge in logical reasoning are conflicts between given pieces of information. Therefore, a considerable amount of work has been devoted to repairing inconsistent knowledge bases [1,2] or performing paraconsistent reasoning [3–5]. Inconsistency measures [6,7] quantify the degree of inconsistency and help analyzing and resolving conflicts. While work on measuring inconsistency was initially inspired by ideas from repairing knowledge bases and paraconsistent reasoning [8], inconsistency measures also inspired new repair [9,10] and paraconsistent reasoning mechanisms [11,12].

Here, we are interested in belief profiles $(\kappa_1, \ldots, \kappa_n)$ rather than single knowledge bases κ. Intuitively, we can think of each κ_i as the set of beliefs of an agent. Our goal is then to measure the disagreement among the agents. A natural idea is to reduce measuring disagreement to measuring inconsistency by transforming multiple knowledge bases to a single base using multiset union or conjunction. However, both approaches have some flaws as we will discuss in the following. This observation is similar to the insight that merging belief profiles should be guided by other principles than repairing single knowledge bases [13]. We will therefore propose some new principles for measuring disagreement and introduce a new measure that complies with them.

After explaining the necessary basics in Sect. 2, we will discuss the relationship between inconsistency measures and disagreement measures in Sect. 3. To begin with, we will define disagreement measures as functions with two basic

© Springer Nature Switzerland AG 2018
D. Ciucci et al. (Eds.): SUM 2018, LNAI 11142, pp. 212–227, 2018.
https://doi.org/10.1007/978-3-030-00461-3_15

properties that seem quite indisputable. We will then show that disagreement measures induced from inconsistency measures by taking the multiset union or conjunction satisfy these basic desiderata and give us some additional guarantees. In Sect. 4, we will propose some stronger principles for measuring disagreement. One key idea is to allow resolving conflicts by majority decisions. We will show that many measures that are induced from inconsistency measures must necessarily violate some of these principles. In Sect. 5, we will then introduce a new disagreement measure that is inspired by the η-inconsistency measure from [6]. Intuitively, it attempts to satisfy all agents' beliefs as well as possible and then measures the average dissatisfaction. We will show that the measure satisfies the principles proposed in Sect. 4 and some other properties that correspond to principles for measuring inconsistency. To give additional motivation for this work, we will sketch how the measure can be used for belief merging and inconsistency-tolerant reasoning at the end of Sect. 5.

2   Basics

We consider a propositional logical language $\mathcal{L}$ built up over a finite set $\mathcal{A}$ of propositional atoms using the usual connectives. Satisfaction of formulas $F \in \mathcal{L}$ by valuations $v : \mathcal{A} \to \{0, 1\}$ is defined as usual. A *knowledge base* κ is a nonempty finite multiset over $\mathcal{L}$. $\mathcal{K}$ denotes the set of all knowledge bases. An n-tuple $\mathcal{B} = (\kappa_1, \dots, \kappa_n) \in \mathcal{K}^n$ is called a *belief profile*. We let $\bigsqcup \mathcal{B} = \bigsqcup_{i=1}^{n} \kappa_i$, where $\sqcup$ denotes multiset union. Note that using multisets is crucial to avoid information loss when several sources contain syntactically equal beliefs. For instance, $\{\neg a\} \sqcup \{a\} \sqcup \{a\} = \{\neg a, a, a\}$. We let $\mathcal{B} \circ \kappa = (\kappa_1, \dots, \kappa_n, \kappa)$, that is, $\mathcal{B} \circ \kappa$ is obtained from $\mathcal{B}$ by adding κ at the end of the profile. Furthermore, we let $\mathcal{B} \circ^1 \kappa = \mathcal{B} \circ \kappa$ and $\mathcal{B} \circ^k \kappa = (\mathcal{B} \circ^{k-1} \kappa) \circ \kappa$ for $k > 1$. That is, $\mathcal{B} \circ^k \kappa$ is obtained from $\mathcal{B}$ by adding k copies of κ. We call a non-contradictory formula f *safe in* κ iff f and κ are built up over distinct variables from $\mathcal{A}$. Intuitively, adding a safe formula to κ cannot introduce any conflicts.

A model of κ is a valuation v that satisfies all $f \in \kappa$. We denote the set of all *models* of κ by $\text{Mod}(\kappa)$. If $\text{Mod}(\kappa) \neq \emptyset$, we call κ consistent and inconsistent otherwise. A *minimal inconsistent (maximal consistent) subset* of κ is a subset of κ that is inconsistent (consistent) and minimal (maximal) with this property. If $\text{Mod}(\kappa) \subseteq \text{Mod}(\kappa')$, we say that κ entails κ' and write $\kappa \models \kappa'$. If $\kappa \models \kappa'$ and $\kappa' \models \kappa$, we call κ and κ' equivalent and write $\kappa \equiv \kappa'$. If $\kappa = \{f\}$ and $\kappa' = \{g\}$ are singletons, we just write $f \models g$ or $f \equiv g$.

An inconsistency measure $\mathcal{I} : \mathcal{K}^n \to \mathbb{R}_0^+$ maps knowledge bases to nonnegative degrees of inconsistency. The most basic example is the *drastic measure* that yields 0 if the knowledge base is consistent and 1 otherwise [14]. Hence, it basically performs a satisfiability test. There exist various other measures, see [15] for a recent overview. While there is an ongoing debate about what properties an inconsistency measure should satisfy, there is general agreement that it should be *consistent* in the sense that $\mathcal{I}(\kappa) = 0$ if and only if κ is consistent. Hence, the inconsistency value is greater than zero if and only if κ is inconsistent.

Various other properties of inconsistency measures have been discussed [14–16]. We will present some of these later, when talking about corresponding properties of disagreement measures.

3   Induced Disagreement Measures

To begin with, we define disagreement measures as functions over the set of all belief profiles $\bigcup_{n=1}^{\infty} \mathcal{K}^n$ that satisfy two basic desiderata.

Definition 1 (Disagreement Measure). *A disagreement measure is a function* $\mathcal{D} : \bigcup_{n=1}^{\infty} \mathcal{K}^n \to \mathbb{R}_0^+$ *such that for all belief profiles* $\mathcal{B} = (\kappa_1, \ldots, \kappa_n)$, *we have*

1. *Consistency:* $\mathcal{D}(\mathcal{B}) = 0$ *iff* $\bigsqcup_{i=1}^{n} \kappa_i$ *is consistent.*
2. *Symmetry:* $\mathcal{D}(\mathcal{B}) = \mathcal{D}(\kappa_{\sigma(1)}, \ldots, \kappa_{\sigma(n)})$ *for each permutation* σ *of* $\{1, \ldots, n\}$.

Consistency generalizes the corresponding property for inconsistency measures. Symmetry assures that the disagreement value is independent of the order in which the knowledge bases are presented. It is similar to *Anonymity* in social choice theory [17] and guarantees equal treatment of different sources.

Note that each disagreement measure $\mathcal{D}$ induces a *corresponding inconsistency measure* $\mathcal{I}_\mathcal{D} : \mathcal{K} \to \mathbb{R}_0^+$ defined by $\mathcal{I}_\mathcal{D}(\kappa) = \mathcal{D}(\kappa)$. Conversely, we can induce disagreement measures from inconsistency measures as we discuss next.

3.1   ⊔-Induced Disagreement Measures

It is easy to see that each inconsistency measure induces a corresponding disagreement measure by taking the multiset union of knowledge bases in the profile.

Proposition 1 (⊔-induced Measure). *If* $\mathcal{I}$ *is an inconsistency measure, then the function* $\mathcal{D}_\mathcal{I}^\sqcup : \bigcup_{n=1}^{\infty} \mathcal{K}^n \to \mathbb{R}_0^+$ *defined by* $\mathcal{D}_\mathcal{I}^\sqcup(\mathcal{B}) = \mathcal{I}(\bigsqcup \mathcal{B})$ *for all* $\mathcal{B} \in \mathcal{K}^n$ *is a disagreement measure. We call* $\mathcal{D}_\mathcal{I}^\sqcup$ *the measure* ⊔-*induced by* $\mathcal{I}$.

What can we say about the properties of ⊔-induced measures? As we explain first, many properties for inconsistency measures have a natural generalization to disagreement measures that is compatible with ⊔-induced measures in the following sense.

Definition 2 (Corresponding Properties). *Let P be a property for inconsistency measures and let P' be a property for disagreement measures. We call* (P, P') *a pair of corresponding properties iff*

1. *if an inconsistency measure $\mathcal{I}$ satisfies P, then the ⊔-induced measure* $\mathcal{D}_\mathcal{I}^\sqcup$ *satisfies P',*
2. *if a disagreement measure $\mathcal{D}$ satisfies P', then the corresponding inconsistency measure $\mathcal{I}_\mathcal{D}$ satisfies P.*

One big class of properties for inconsistency measures gives guarantees about the relationship between inconsistency values when we extend the knowledge bases by particular formulas. We start with a general lemma and give some examples in the subsequent proposition.

Lemma 1 (Transfer Lemma). *Let R be a binary relation on $\mathbb{R}$ and let $C \subseteq \mathcal{K}^3$ be a ternary constraint on knowledge bases. Given a property for inconsistency measures*

$$\text{For all } \kappa, S, T \in \mathcal{K}, \text{ if } C(\kappa, S, T) \text{ then } \mathcal{I}(\kappa \sqcup S) \, R \, \mathcal{I}(\kappa \sqcup T), \tag{1}$$

define a property for disagreement measures as follows:

$$\text{For all } \kappa_1, \ldots, \kappa_n, S, T \in \mathcal{K}, \text{ if } C(\bigsqcup_{i=1}^{n} \kappa_i, S, T) \text{ then}$$

$$\mathcal{D}(\kappa_1 \sqcup S, \kappa_2, \ldots, \kappa_n) \, R \, \mathcal{D}(\kappa_1 \sqcup T, \kappa_2, \ldots, \kappa_n). \tag{2}$$

Then $((1), (2))$ is a pair of corresponding properties.

Remark 1. The reader may wonder why the corresponding property looks only at the first argument. Note that by symmetry of disagreement measures, the same is true for all other arguments. For instance, we have $\text{Inc}^*(\kappa_1, \kappa_2 \sqcup S) = \text{Inc}^*(\kappa_2 \sqcup S, \kappa_1) \, R \, \text{Inc}^*(\kappa_2 \sqcup T, \kappa_1) = \text{Inc}^*(\kappa_1, \kappa_2 \sqcup T)$.

We now apply Lemma 1 to some basic properties for inconsistency measures from [14] and adjunction invariance from [16] that will play an important role later.

Proposition 2. *The following are pairs of corresponding properties for inconsistency and disagreement measures:*

- Monotony:
  $\mathcal{I}(\kappa) \leq \mathcal{I}(\kappa \sqcup \kappa')$
  $\mathcal{D}(\kappa_1, \kappa_2, \ldots, \kappa_n) \leq \mathcal{D}(\kappa_1 \sqcup \kappa', \kappa_2, \ldots, \kappa_n)$
- Dominance: *For $f, g \in \mathcal{L}$ such that $f \models g$ and $f \not\models \bot$,*
  $\mathcal{I}(\kappa \sqcup \{f\}) \geq \mathcal{I}(\kappa \sqcup \{g\})$
  $\mathcal{D}(\kappa \sqcup \{f\}, \kappa_2, \ldots, \kappa_n) \geq \mathcal{D}(\kappa \sqcup \{g\}, \kappa_2, \ldots, \kappa_n)$
- Safe Formula Independence: *If $f \in \mathcal{L}$ is safe in κ, then*
  $\mathcal{I}(\kappa \sqcup \{f\}) = \mathcal{I}(\kappa)$
  If $f \in \mathcal{L}$ is safe in $\bigsqcup_{i=1}^{n} \kappa_i$, then
  $\mathcal{D}(\kappa_1 \sqcup \{f\}, \kappa_2, \ldots, \kappa_n) = \mathcal{D}(\kappa_1, \kappa_2, \ldots, \kappa_n)$
- Adjunction Invariance: *For all $f, g \in \mathcal{L}$,*
  $\mathcal{I}(\kappa \cup \{f, g\}) = \mathcal{I}(\kappa \cup \{f \wedge g\})$
  $\mathcal{D}(\kappa_1 \cup \{f, g\}, \kappa_2, \ldots) = \mathcal{D}(\kappa_1 \cup \{f \wedge g\}, \kappa_2, \ldots)$

Monotony demands that adding knowledge can never decrease the disagreement value. Dominance says that replacing a claim with a (possibly weaker) implication of the original claim can never increase the disagreement value. Safe Formula Independence demands that a safe formula does not affect the disagreement value. Adjunction invariance says that it makes no difference whether two pieces of information are presented independently or as a single formula.

Example 1. The inconsistency measure $\mathcal{I}_{LP_m}$ that was discussed in [18] satisfies Monotony, Dominance, Safe Formula Independence and Adjunction Invariance. From Proposition 2, we can conclude that the $\sqcup$-induced disagreement measure $\mathrm{Inc}_{LP_m}^{\sqcup}$ satisfies the corresponding properties for disagreement measures.

What we can take from our discussion so far is that each inconsistency measure induces a disagreement measure with similar properties. As it turns out, each $\sqcup$-induced disagreement measure satisfies an additional property and, in fact, only the $\sqcup$-induced measures do. We call this property *partition invariance*. Intuitively, partition invariance means that the disagreement value depends only on the pieces of information in the belief profile and is independent of the distribution of these pieces. In the following proposition, a partition of a multiset M is a sequence of non-empty multisets $M_1, \ldots, M_k$ such that $\bigsqcup_{i=1}^{k} M_i = M$.

Proposition 3 (Characterizations of Induced Families). *The following statements are equivalent:*

1. $\mathcal{D}$ is $\sqcup$-*induced by an inconsistency measure.*
2. $\mathcal{D}$ is $\sqcup$-*induced by* $\mathcal{I}_{\mathcal{D}}$.
3. $\mathcal{D}$ *is* partition invariant, *that is, for all $\kappa \in \mathcal{K}$ and for all partitions $\bigsqcup_{i=1}^{n_1} P_i = \bigsqcup_{i=1}^{n_2} P_i' = \kappa$ of κ, we have that $\mathcal{D}(P_1, \ldots, P_{n_1}) = \mathcal{D}(P_1', \ldots, P_{n_2}')$.*

So the $\sqcup$-induced disagreement measures are exactly the partition invariant measures. However, partition variance can be undesirable in some scenarios.

Example 2. Consider the political goals 'increase wealth of households' (h), 'increase wealth of firms' (f), 'increase wages' (w). Suppose there are three political parties whose positions we represent in the profile

$$\mathcal{B} = (\{f, w, f \to w\}, \{w, h, w \to h\}, \{f, \neg w, w \to \neg f\}).$$

In this scenario, the parties only disagree about w. We modify $\mathcal{B}$ by moving $w \to \neg f$ from the third to the second party:

$$\mathcal{B}' = (\{f, w, f \to w\}, \{w, h, w \to h, w \to \neg f\}, \{f, \neg w\}).$$

The conflict with respect to w remains, but party 2's positions now imply $\neg f$. Since we now have an additional conflict with respect to f, we would expect $\mathcal{D}(\mathcal{B}) < \mathcal{D}(\mathcal{B}')$.

Partition invariant measures are unable to detect the difference in Example 2. Since partition invariance is an inherent property of $\sqcup$-induced measures, we should also investigate non-$\sqcup$-induced measures.

3.2 $\wedge$-Induced Disagreement Measures

Instead of taking the multiset union of all knowledge bases in the profile, we can also just replace each knowledge base with the conjunction of the formulas that it contains in order to induce a disagreement measure.

Proposition 4 ($\wedge$-induced Measure). *If $\mathcal{I}$ is an inconsistency measure, then $\mathcal{D}_{\mathcal{I}}^{\wedge} : \bigcup_{n=1}^{\infty} \mathcal{K}^n \to \mathbb{R}_0^+$ defined by $\mathcal{D}_{\mathcal{I}}^{\wedge}(\mathcal{B}) = \mathcal{I}(\bigsqcup_{\kappa \in \mathcal{B}} \{\bigwedge_{F \in \kappa} F\})$ for $\mathcal{B} \in \mathcal{K}^n$ is a disagreement measure. We call $\mathcal{D}_{\mathcal{I}}^{\wedge}$ the measure $\wedge$-induced by $\mathcal{I}$.*

By repeated application of adjunction invariance (c.f. Proposition 2), one can show that each adjunction invariant inconsistency measure satisfies $\mathcal{I}(\kappa) = \mathcal{I}(\{\bigwedge_{f \in \kappa} f\})$, see [16], Proposition 9. We can use this result to show that for adjunction invariant inconsistency measures, the $\wedge$-induced and the $\sqcup$-induced measures are equal.

Corollary 1. *If $\mathcal{I}$ is an adjunction invariant inconsistency measure, then $\mathcal{D}_{\mathcal{I}}^{\wedge} = \mathcal{D}_{\mathcal{I}}^{\sqcup}$.*

This is actually the only case in which the $\wedge$-induced measure can be $\sqcup$-induced.

Proposition 5. *Let $\mathcal{I}$ be an inconsistency measure. $\mathcal{D}_{\mathcal{I}}^{\wedge}$ is $\sqcup$-induced if and only if $\mathcal{I}$ is adjunction invariant.*

The $\sqcup$-induced disagreement measures are characterized by partition invariance. Adjunction invariance plays a similar role for $\wedge$-induced measures.

Proposition 6. *For each inconsistency measure $\mathcal{I}$, $\mathcal{D}_{\mathcal{I}}^{\wedge}$ satisfies adjunction invariance.*

Note that the inconsistency measure $\mathcal{I}_{\mathcal{D}_{\mathcal{I}}^{\wedge}}$ induced by $\mathcal{D}_{\mathcal{I}}^{\wedge}$ will also be adjunction invariant. Therefore, $\mathcal{I}_{\mathcal{D}_{\mathcal{I}}^{\wedge}} \neq \mathcal{I}$ if $\mathcal{I}$ is not adjunction invariant. In particular, $\mathcal{D}_{\mathcal{I}}^{\wedge}$ can be a rather coarse measure if $\mathcal{I}$ is not adjunction invariant.

Example 3. The inconsistency measure $\mathcal{I}_{MI}$ from [18] counts the number of minimal inconsistent sets of a knowledge base. $\mathcal{I}_{MI}$ is not adjunction invariant. For instance, $\mathcal{I}_{MI}(\{a, \neg a, a \wedge b\}) = 2$ because $\{a, \neg a\}$ and $\{\neg a, a \wedge b\}$ are the only minimal inconsistent sets. However, $\mathcal{I}_{MI}(\{a \wedge \neg a \wedge a \wedge b\}) = 1$ because the only minimal inconsistent set is the knowledge base itself. Furthermore, we will have $\mathcal{D}_{\mathcal{I}_{MI}}^{\wedge}(\kappa) = 1$ whenever $\bigwedge_{f \in \kappa} f$ is inconsistent and $\mathcal{D}_{\mathcal{I}_{MI}}^{\wedge}(\kappa) = 0$ otherwise. Hence, the inconsistency measure corresponding to $\mathcal{D}_{\mathcal{I}_{MI}}^{\wedge}$ is the drastic measure.

Proposition 6 tells us that $\wedge$-induced measures are necessarily adjunction invariant. Whether or not each adjunction invariant disagreement measure is $\wedge$-induced is currently an open question. However, we have the following result.

Proposition 7. *If $\mathcal{D}$ satisfies adjunction invariance and*

$$\mathcal{D}(\{f_1\}, \ldots, \{f_n\}) = \mathcal{D}(\bigsqcup_{i=1}^{n} \{f_i\}), \tag{3}$$

then $\mathcal{D}$ is $\wedge$-induced by an inconsistency measure.

We call property (3) *singleton union invariance* in the following. While adjunction invariance and singleton union invariance are sufficient for being $\wedge$-induced, they are no longer necessary as the following example illustrates.

Example 4. Consider again the inconsistency measure $\mathcal{I}_{MI}$ from [18] that was explained in Example 3. We have $\mathcal{D}^{\wedge}_{\mathcal{I}_{MI}}(\{a \wedge b\}, \{\neg a \wedge b\}, \{a \wedge \neg b\}) = \mathcal{I}_{MI}(\{a \wedge b, \neg a \wedge b, a \wedge \neg b\}) = 3$ by definition of the $\wedge$-induced measure. However, $\mathcal{D}^{\wedge}_{\mathcal{I}_{MI}}(\{a \wedge b, \neg a \wedge b, a \wedge \neg b\}) = \mathcal{I}_{MI}(\{a \wedge b \wedge \neg a \wedge b \wedge a \wedge \neg b\}) = 1$. Hence, $\mathcal{D}^{\wedge}_{\mathcal{I}_{MI}}$ is not singleton union invariant.

We close this section by showing that the set of disagreement measures $\sqcup$-induced and $\wedge$-induced from inconsistency measures are neither equal nor disjoint.

To begin with, the $\mathcal{I}_{LP_m}$ inconsistency measure that was discussed in [18] is adjunction invariant. Therefore, $\mathcal{D}^{\sqcup}_{\mathcal{I}_{LP_m}} = \mathcal{D}^{\wedge}_{\mathcal{I}_{LP_m}}$ according to Corollary 1. Hence, the intersection of $\sqcup$-induced and $\wedge$-induced disagreement measures is non-empty.

In order to show that there are partition invariant measures that are not adjunction invariant and vice versa, we use the minimal inconsistent set measure $\mathcal{I}_{MI}$ from [18]. As demonstrated in Example 3, $\mathcal{I}_{MI}$ is not adjunction invariant. Therefore, the Transfer Lemma implies that $\mathcal{D}^{\sqcup}_{\mathcal{I}_{MI}}$ is not adjunction invariant either. Hence, $\mathcal{D}^{\sqcup}_{\mathcal{I}_{MI}}$ cannot be $\wedge$-induced according to Proposition 6.

On the other hand, $\mathcal{D}^{\wedge}_{\mathcal{I}_{MI}}$ is adjunction invariant because each $\wedge$-induced measure is. However, since $\mathcal{I}_{MI}$ is not adjunction invariant, we know from Proposition 5 that $\mathcal{D}^{\wedge}_{\mathcal{I}_{MI}}$ is not $\sqcup$-induced. Hence, $\mathcal{D}^{\wedge}_{\mathcal{I}_{MI}}$ is an example of a disagreement measure that is $\wedge$-induced, but not $\sqcup$-induced.

We illustrate our findings in Fig. 1. The $\wedge$-induced incompatibility measures are a subset of the adjunction invariant measures (Proposition 6). The fact that all measures in the intersection of partition invariant and adjunction invariant measures are $\wedge$-induced follows from observing that partition invariance implies singleton union invariance (3) and Proposition 7.

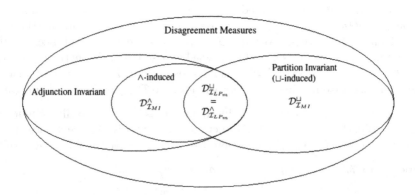

Fig. 1. Venn diagram illustrating induced measures in the space of all disagreement measures.

4   Principles for Measuring Disagreement

As illustrated in Fig. 1, induced measures correspond to disagreement measures with very specific properties. $\sqcup$-induced measures are necessarily partition invariant. This may be undesirable in certain applications as illustrated in Example 2. If an inconsistency measure is adjunction invariant, the $\wedge$-induced measure will also be partition invariant. If it is not adjunction invariant, the $\wedge$-induced measure will not be partition invariant, but the measure may become rather coarse as illustrated in Example 3. This is some evidence that it is worth investigating non-induced measures. To further distinguish inconsistency from disagreement measures, we will now propose some stronger principles that go beyond our basic desiderata from Definition 1.

To guide our intuition, we think of each knowledge base as the belief set of an agent. We say that κ_i *contradicts* κ_j if $\kappa_i \cup \kappa_j$ is inconsistent. To begin with, let us consider an agent whose beliefs do not contradict any consistent position (its knowledge base is tautological). When adding such an agent to a belief profile, the disagreement value should not increase. Dually, if we add an agent that contradicts every position (its knowledge base is inconsistent), the disagreement value should not decrease. This intuition is captured by the following principles.

Tautology. Let $\mathcal{B} \in \mathcal{K}^n$ and let $\kappa_\top \in \mathcal{K}$ be tautological. Then $\mathcal{D}(\mathcal{B} \circ \kappa_\top) \leq \mathcal{D}(\mathcal{B})$.

Contradiction. Let $\mathcal{B} \in \mathcal{K}^n$ and let $\kappa_\bot \in \mathcal{K}$ be contradictory. Then $\mathcal{D}(\mathcal{B} \circ \kappa_\bot) \geq \mathcal{D}(\mathcal{B})$.

Inconsistency measures focus mainly on the existence of conflicts. However, in a multiagent setting, conflicts can often be resolved by majority decisions. Given a belief profile $\mathcal{B} = (\kappa_1, \ldots, \kappa_n) \in \mathcal{K}^n$, we call a subset $C \subseteq \{1, \ldots, n\}$ a *consistent coalition* iff $\bigcup_{i \in C} \kappa_i$ is consistent. We say that κ_j is *involved in a conflict in* $\mathcal{B}$ iff there is a consistent coalition C such that $\kappa_j \cup \bigcup_{i \in C} \kappa_i$ is inconsistent. Our next principle demands that conflicts can be eased by majority decisions.

Majority. Let $\mathcal{B} = (\kappa_1, \ldots, \kappa_n) \in \mathcal{K}^n$. If κ_j is consistent and involved in a conflict, then there is a $k \in \mathbb{N}$ such that $\mathcal{D}(\mathcal{B} \circ^k \kappa_j) < \mathcal{D}(\mathcal{B})$.

Intuitively, Majority says that we can decrease the severity of a conflict by giving sufficient support for one of the conflicting positions. It does not matter what position we choose as long as this position is consistent. In future work, one may look at alternative principles based on other methods to make group decisions [17], but Majority seems to be a natural starting point.

Majority implies that we can strictly decrease the disagreement value by adding copies of one consistent position. However, this does not imply that the disagreement value will vanish. If we keep adding copies, the disagreement value will necessarily decrease but it may converge to a value strictly greater than 0. While one may argue that the limit should be 0 if almost all agents agree, one may also argue that the limit should be bounded from below by a positive constant if an unresolved conflict remains. We therefore do not strengthen majority. Instead, we consider an additional principle that demands that the limit is indeed 0 if the majority agrees on all non-contradictory positions. This intuition is captured by the next principle.

Majority Agreement in the Limit. Let $\mathcal{B} \in \mathcal{K}^n$. If M is a $\subseteq$-maximal consistent subset of $\bigsqcup \mathcal{B}$, then $\lim_{k \to \infty} \mathcal{D}(\mathcal{B} \circ^k M) = 0$.

We close this section with an impossibility result: Monotony and Partition Invariance cannot be satisfied jointly with our majority principles. The reason is that such measures can never decrease when receiving new information as explained in the following proposition.

Proposition 8. *If $\mathcal{D}$ satisfies Monotony and Partition Invariance, then $\mathcal{D}(\mathcal{B} \circ^k \kappa) \geq \mathcal{D}(\mathcal{B})$ for all $\mathcal{B} \in \mathcal{K}^n, \kappa \in \mathcal{K}, k \in \mathbb{N}$.*

The conditions of Proposition 8 are in particular met by several induced measures.

Corollary 2. *Every disagreement measure that is*

- *partition invariant and monotone or*
- *$\sqcup$-induced from a monotone inconsistency measure or*
- *$\wedge$-induced from a monotone and adjunction invariant inconsistency measure*

violates Majority *and* Majority Agreement in the Limit.

5   The η-Disagreement Measure

We now consider a novel disagreement measures inspired by the η-inconsistency measure from [6]. Roughly speaking, the *η-inconsistency measure* attempts to maximize the probability of all formulas within a knowledge base. By subtracting this probability from 1, we get an inconsistency value. In order to assign probabilities to formulas, we consider probability distributions over the set of all valuations $\Omega = \{v \mid v : \mathcal{A} \to \{0,1\}\}$ of our language. Given a probability distribution $\pi : \Omega \to [0,1]$ ($\sum_{v \in \Omega} \pi(v) = 1$) and a formula $F \in \mathcal{L}$, we let

$$P_\pi(F) = \sum_{v \models F} \pi(v).$$

Intuitively, $P_\pi(F)$ is the probability that F is true with respect to π. The η-inconsistency measure from [6] is defined by

$$\mathcal{I}_\eta(\kappa) = 1 - \max\{p \mid \exists \pi : \forall F \in \kappa : P_\pi(F) \geq p\}.$$

This formula describes the intuition that we explained in the beginning. $p^* = \max\{p \mid \exists \pi : \forall F \in \kappa : P_\pi(F) \geq \kappa\}$ is the maximum probability that all formulas in κ can simultaneously take. We will have $p^* = 1$ if and only if κ is consistent [6].

Let us first look at the disagreement measures induced by $\mathcal{I}_\eta$. $\mathcal{I}_\eta$ satisfies Monotony [15]. Therefore, $\mathcal{D}^\sqcup_{\mathcal{I}_\eta}$ will violate our majority principles as explained in Corollary 2. However, $\mathcal{I}_\eta$ is not adjunction invariant [15]. Therefore, Proposition 5 implies that $\mathcal{D}^\sqcup_{\mathcal{I}_\eta} \neq \mathcal{D}^\wedge_{\mathcal{I}_\eta}$. Still, $\mathcal{D}^\wedge_{\mathcal{I}_\eta}$ does not satisfy our majority principles either.

Example 5. Let $\mathcal{B} = (\{a\}, \{\neg a\})$. Since $P_\pi(a) = 1 - P_\pi(\neg a)$, we have for all $n \in \mathbb{N}$

$$\mathcal{D}^\wedge_{\mathcal{I}_\eta}(\{a\}, \{\neg a\}) = \mathcal{I}_\eta(\{a, \neg a\}) = 0.5$$

$$= \mathcal{I}_\eta(\{a, \neg a\} \sqcup \bigsqcup_{i=1}^{n} \{a\}) = \mathcal{D}^\wedge_{\mathcal{I}_\eta}((\{a\}, \{\neg a\}) \circ^n \{a\}).$$

However, we can modify the definition of the η-inconsistency measure in order to get a disagreement measure that satisfies our desiderata. If we think of $P_\pi(F)$ as the degree of belief in F, then we should try to find a π such that the beliefs of all agents are satisfied as well as possible. To do so, we can first look at how well π satisfies the beliefs of each agent and then look at how well π satisfies the agents' beliefs overall. To measure satisfaction of one agent's beliefs, we take the minimum of all probabilities assigned to the formulas in the agent's knowledge base. Formally, for all probability distributions π and knowledge bases κ over our language, we let

$$s_\pi(\kappa) = \min\{P_\pi(F) \mid F \in \kappa\}.$$

and call $s_\pi(\kappa)$ the *degree of satisfaction* of κ. In order to measure satisfaction of a belief profile, we take the average degree of satisfaction of the knowledge bases in the profile. Formally, we let for all probability distributions π and belief profiles $\mathcal{B}$

$$S_\pi(\mathcal{B}) = \frac{1}{|\mathcal{B}|} \sum_{\kappa \in \mathcal{B}} s_\pi(\kappa)$$

and call $S(\mathcal{B})$ the *degree of satisfaction* of $\mathcal{B}$. We now define a new disagreement measure. Intuitively, it attempts to maximize the degree of satisfaction of the profile. By subtracting the maximum degree of satisfaction from 1, we get a disagreement value.

Definition 3 (η-Disagreement Measure). *The η-Disagreement Measure is defined by*

$$\mathcal{D}_\eta(\mathcal{B}) = 1 - \max\{p \mid \exists \pi : S_\pi(\mathcal{B}) = p\}.$$

To begin with, we note that $\mathcal{D}_\eta$ is a disagreement measures as defined in Definition 1 and can be computed by linear programming techniques.

Proposition 9. *$\mathcal{D}_\eta$ is a disagreement measures and can be computed by solving a linear optimization problem.*

As we show next, $\mathcal{D}_\eta$ is neither $\sqcup$- nor $\wedge$-induced from any inconsistency measure. According to Propositions 3 and 6, it suffices to show that it is neither partition invariant nor adjunction invariant.

Example 6. Consider again the belief profiles $\mathcal{B}$ and $\mathcal{B}'$ from Example 2. We have $\mathcal{D}_\eta(\mathcal{B}) \approx 0.33$ and $\mathcal{D}_\eta(\mathcal{B}) \approx 0.44$. As desired, $\mathcal{D}_\eta$ recognizes the increased disagreement in the profile. In particular, $\mathcal{D}_\eta$ is not partition invariant.

Example 7. To see that $\mathcal{D}_\eta$ is not adjunction invariant, note that $\mathcal{D}_\eta(\{a, \neg a\}) = 0.5$, whereas $\mathcal{D}_\eta(\{a \wedge \neg a\}) = 1$ (contradictory formulas have probability 0 with respect to each π). Hence, $\mathcal{D}_\eta$ is also not adjunction invariant.

$\mathcal{D}_\eta$ satisfies our four principles for measuring disagreement as we show next. To begin with, we note that the disagreement value necessarily decreases as the proportion of agreeing agents increases.

Proposition 10. *Let $\mathcal{B} \in \mathcal{K}^n$. If $\mathcal{B}$ contains a consistent coalition of size k, then $\mathcal{D}_\eta(\mathcal{B}) \leq 1 - \frac{k}{n}$.*

Proposition 10 implies, in particular, that the disagreement value goes to 0 as the proportion of agreeing agents $\frac{k}{n}$ goes to 1. Therefore, $\mathcal{D}_\eta$ satisfies our majority principles.

Corollary 3. $\mathcal{D}_\eta$ *satisfies Majority and Majority Agreement in the Limit.*

Tautology and Contradiction are also satisfied and can be strengthened slightly.

Proposition 11. $\mathcal{D}_\eta$ *satisfies Tautology and Contradiction. Furthermore,*

- *If $\mathcal{D}_\eta(\mathcal{B}) > 0$, then $\mathcal{D}_\eta(\mathcal{B} \circ \kappa_\top) < \mathcal{D}_\eta(\mathcal{B})$.*
- *If $\mathcal{D}_\eta(\mathcal{B}) < 1$, then $\mathcal{D}_\eta(\mathcal{B} \circ \kappa_\perp) > \mathcal{D}_\eta(\mathcal{B})$.*

Regarding the properties corresponding to principles for measuring inconsistency from Proposition 2, $\mathcal{D}_\eta$ satisfies all of them except Adjunction Invariance (Example 7).

Proposition 12. $\mathcal{D}_\eta$ *satisfies Monotony, Dominance and Safe Formula Independence.*

We already know that $\mathcal{D}_\eta$ yields 0 if and only if all knowledge bases in the profile are consistent with each other. In the following proposition, we explain in what cases it takes the maximum value 1.

Proposition 13. *Let $\mathcal{B} \in \mathcal{K}^n$. We have $\mathcal{D}_\eta(\mathcal{B}) = 1$ iff all κ_i contain at least one contradictory formula.*

Intuitively, if there is a knowledge base that does not contain any contradictory formulas, then all beliefs of one agent can be partially satisfied and the disagreement value with respect to $\mathcal{D}_\eta$ cannot be 1. So the degree of disagreement can only be maximal if each agent has contradictory beliefs.

In some applications, we may want to restrict to belief profiles with consistent knowledge bases. We can rescale $\mathcal{D}_\eta$ for this purpose. Proposition 10 gives us the following upper bounds on the disagreement value.

Corollary 4. *Let* $\mathcal{B} = (\kappa_1, \ldots, \kappa_n)$. *If some* κ_i *is consistent, then* $\mathcal{D}_\eta(\mathcal{B}) \leq 1 - \frac{1}{n}$.

The bound in Corollary 4 is actually tight even if all knowledge bases in the profile are individually consistent as we explain in the following example.

Example 8. For $n = 2$ agents, we have $\mathcal{D}_\eta(\{a\}, \{\neg a\}) = \frac{1}{2}$. For $n = 3$, we have $\mathcal{D}_\eta(\{a \wedge b\}, \{\neg a \wedge b\}, \{\neg b\}) = \frac{2}{3}$. In general, if we have n satisfiable but pairwise inconsistent $(F_i \wedge F_j \equiv \bot)$ formulas $F_1, \ldots, F_n$, then $\mathcal{D}_\eta(\{F_1\}, \ldots, \{F_n\}) = 1 - \frac{1}{n}$.

Hence, if we want to restrict to consistent knowledge bases, we can renormalize $\mathcal{D}_\eta$ by multiplying by $\frac{n}{n-1}$. The disagreement value will then be maximal whenever all agents have pairwise inconsistent beliefs.

As explained in Proposition 9, computing $\mathcal{D}_\eta(\mathcal{B})$ is a linear optimization problem. Interior-point methods can solve these problems in polynomial time in the number of optimization variables and constraints [19]. While the number of optimization variables is exponential in the number of atoms $|\mathcal{A}|$ of our language, the number of constraints is linear in the number of formulas in all knowledge bases in the profile. Roughly speaking, computing $\mathcal{D}_\eta(\mathcal{B})$ is very sensitive to the number of atoms, but scales well with respect to the number of agents. In the language of parameterized complexity theory [20], computing $\mathcal{D}_\eta(\mathcal{B})$ is fixed-parameter tractable (that is, polynomial if we fix the number of atoms).

Proposition 14. *Computing* $\mathcal{D}_\eta(\mathcal{B})$ *is fixed-parameter tractable with parameter* $|\mathcal{A}|$.

While interior-point methods give us a polynomial worst-case guarantee, they are often outperformed in practice by the simplex algorithm. The simplex algorithm has exponential runtime for some artificial examples, but empirically runs in time linear in the number of optimization variables (exponential in $|\mathcal{A}|$) and quadratic in the number of constraints (quadratic in the overall number of formulas in the belief profile) [19].

In the long-term, our goal is to reason over belief profiles that contain conflicts among agents. While we must leave a detailed discussion for future work, we will now sketch how the η-disagreement measure can be used for this purpose. The optimal solutions of the linear optimization problem corresponding to $\mathcal{D}_\eta$ form a topologically closed and convex set of probability distributions. This allows us to compute lower and upper bound on the probability (or more intuitively, the degree of belief) of formulas with respect to the optimal solutions that minimize disagreement. This is similar to the probabilistic entailment problem [21], where we compute lower and upper bounds with respect to probability distributions that satisfy probabilistic knowledge bases. If, for a belief profile $\mathcal{B}$, the lower bound of the formula F is l and the upper bound is u, we write $P_\mathcal{B}(F) = [l, u]$. If $l = u$, we just write $P_\mathcal{B}(F) = l$. We call $P_\mathcal{B}$ the *aggregated group belief*.

Example 9. Suppose we have 100 reviews about a restaurant. While most reviewers agree that the food (f) and the service (s) are good, two reviewers disagree about the interior design (d) of the restaurant. Let us assume that

$\mathcal{B} = (((\{d, f, s\}, \{\neg d, f, s\}) \circ^{95} \{f, s\}) \circ^3 \{\neg f, \neg s\})$. We have $\mathcal{D}_\eta(\mathcal{B}) \approx 0.03$. Intuitively, the degree of disagreement among agents is low because the majority of agents seem not to care about the interior design. The aggregated group beliefs for the atoms in this example are $P_\mathcal{B}(d) = 0.5, P_\mathcal{B}(f) = 1, P_\mathcal{B}(s) = 1$.

We can use $P_\mathcal{B}$ to define an entailment relation. For instance, we could say that $\mathcal{B}$ entails F iff the lower bound is strictly greater than 0.5. Then, in Example 9, $P_\mathcal{B}$ entails f and s, but neither d nor $\neg d$.

6 Related Work

The authors in [22] considered the problem of measuring disagreement in limited choice problems, where each agent can choose from a finite set of alternatives. The measures are basically defined by counting the decisions and relating the counts. The authors give intuitive justification for their measures, but do not consider general principles. In order to transfer their approach to our setting, one may identify atomic formulas with alternatives in their framework, but it is not clear how this approach could be extended to knowledge bases that contain complex formulas.

Some other conflict measures have been considered in non-classical frameworks. These measures are often closer to distance measures because they mainly compare how close two quantitative belief representations like probability functions, belief functions or fuzzy membership functions are [23–25]. In [26], some compatibility measures for Markov logic networks have been proposed. The measures are normalized and the maximum degree of compatibility can be related to a notion of coherence of Markov logic networks. However, this notion cannot be transferred to classical knowledge bases easily.

As we discussed, measuring disagreement is closely related to measuring inconsistency [6,14,27] and merging knowledge bases [28–30]. The principles *Majority* and *Majority Agreement in the Limit* from Sect. 4 are inspired by *Majority merging operators* that allow that a sufficiently large interest group can determine the merging outcome. The η-disagreement measure is perhaps most closely related to *model-based operators* and DA^2 *operators*, which attempt to minimize some notion of distance between interpretations and the models of the knowledge bases in the profile. In contrast, the η-disagreement measure minimizes a probabilistic degree of dissatisfaction of the belief profile.

[31] introduced some entailment relations based on consensus in belief profiles. We will investigate relationships to entailment relations derived from the η-disagreement measure in future work.

7 Conclusions and Future Work

In this paper, we investigated approaches to measuring disagreement among knowledge bases. In principle, inconsistency measures can be applied for this purpose by transforming belief profiles to single knowledge bases. However, we noticed some problems with this approach. For instance, many measures that are naively induced from inconsistency measures violate Majority and Agreement in the Limit as explained in Corollary 2. Even though this problem does not apply to measures $\wedge$-induced from inconsistency measures that violate adjunction invariance, these induced measures show another problem: they may be unable to notice that a conflict can be resolved by giving up parts of agents' beliefs. For instance, the measures $\mathcal{D}^\wedge_{\mathcal{I}_{MI}}$ and $\mathcal{D}^\wedge_{\mathcal{I}_\eta}$ cannot distinguish the profiles $(\{a, b\}, \{\neg a, b\})$ and $(\{a\}, \{\neg a\})$ because $\mathcal{I}_{MI}$ and $\mathcal{I}_\eta$ cannot distinguish the knowledge bases $\{a \wedge b, \neg a \wedge b\}$ and $\{a, \neg a\}$.

The η-inconsistency measure $\mathcal{D}_\eta$ satisfies our principles for measuring disagreement and some other basic properties that correspond to principles for measuring inconsistency. Since $\mathcal{D}_\eta$ can perform satisfiability tests, we cannot expect to compute disagreement values in polynomial time with respect to the number of atoms. However, if our agents argue only about a moderate number of statements (we fix the number of atoms), the worst-case runtime is polynomial with respect to the number of agents.

In the long-term, we are in particular interested in reasoning over belief profiles that contain conflicts. We can use the η-inconsistency measure for this purpose as we sketched at the end of Sect. 5. However, the aggregated group belief $P_\mathcal{B}$ does not behave continuously. For instance, if we gradually increase the support for $\neg s$ in Example 9, $P_\mathcal{B}(s)$ will not gradually go to 0, but will jump to an undecided state like 0.5 or will jump to 0 at some point. This is not a principal problem for defining an entailment relation that either says that a formula is entailed or not entailed by a profile. However, a continuous notion of group beliefs would allow us to shift the focus from measuring disagreement among agents to measuring disagreement about statements (logical formulas). We could do so by measuring how well we can bound the aggregated beliefs about the formulas in the profile away from 0.5. However, if $P_\mathcal{B}$ does not behave continuously, this approach will give us a rather coarse measure (basically three-valued). Therefore, an interesting question for future research is whether we can modify $\mathcal{D}_\eta$ or design other measures that give us an aggregated group belief with a more continuous behavior.

References

1. Kalyanpur, A., Parsia, B., Sirin, E., Cuenca-Grau, B.: Repairing unsatisfiable concepts in OWL ontologies. In: Sure, Y., Domingue, J. (eds.) ESWC 2006. LNCS, vol. 4011, pp. 170–184. Springer, Heidelberg (2006). https://doi.org/10.1007/11762256_15
2. Lehmann, J., Bühmann, L.: ORE - a tool for repairing and enriching knowledge bases. In: Patel-Schneider, P.F., et al. (eds.) ISWC 2010. LNCS, vol. 6497, pp. 177–193. Springer, Heidelberg (2010). https://doi.org/10.1007/978-3-642-17749-1_12
3. Benferhat, S., Dubois, D., Prade, H.: Some syntactic approaches to the handling of inconsistent knowledge bases: a comparative study part 1: the flat case. Studia Logica **58**(1), 17–45 (1997)

4. Arieli, O., Avron, A., Zamansky, A.: What is an ideal logic for reasoning with inconsistency? In: IJCAI 2011, pp. 706–711 (2011)
5. Priest, G.: Paraconsistent logic. In: Gabbay, D.M., Guenthner, F. (eds.) Handbook of Philosophical Logic, pp. 287–393. Springer, Dordrecht (2002). https://doi.org/10.1007/978-94-017-0460-1_4
6. Knight, K.: Measuring inconsistency. J. Philos. Log. **31**(1), 77–98 (2002)
7. Grant, J., Hunter, A.: Measuring inconsistency in knowledgebases. J. Intell. Inf. Syst. **27**(2), 159–184 (2006)
8. Hunter, A., Konieczny, S.: Approaches to measuring inconsistent information. In: Bertossi, L., Hunter, A., Schaub, T. (eds.) Inconsistency Tolerance. LNCS, vol. 3300, pp. 191–236. Springer, Heidelberg (2005). https://doi.org/10.1007/978-3-540-30597-2_7
9. Thimm, M.: Measuring inconsistency in probabilistic knowledge bases. In: UAI 2009, pp. 530–537. AUAI Press (2009)
10. Muiño, D.P.: Measuring and repairing inconsistency in probabilistic knowledge bases. Int. J. Approx. Reason. **52**(6), 828–840 (2011)
11. Potyka, N.: Linear programs for measuring inconsistency in probabilistic logics. In: KR 2014, pp. 568–577 (2014)
12. Potyka, N., Thimm, M.: Probabilistic reasoning with inconsistent beliefs using inconsistency measures. In: IJCAI 2015, pp. 3156–3163 (2015)
13. Konieczny, S.: On the difference between merging knowledge bases and combining them. In: KR, pp. 135–144 (2000)
14. Hunter, A., Konieczny, S.: Measuring inconsistency through minimal inconsistent sets. In: KR 2008, pp. 358–366 (2008)
15. Thimm, M.: On the compliance of rationality postulates for inconsistency measures: a more or less complete picture. KI-Künstliche Intelligenz **31**, 1–9 (2016)
16. Besnard, P.: Revisiting postulates for inconsistency measures. In: Fermé, E., Leite, J. (eds.) JELIA 2014. LNCS (LNAI), vol. 8761, pp. 383–396. Springer, Cham (2014). https://doi.org/10.1007/978-3-319-11558-0_27
17. Zwicker, W.: Introduction to the theory of voting. In: Handbook of Computational Social Choice, pp. 23–56 (2016)
18. Hunter, A., Konieczny, S.: On the measure of conflicts: shapley inconsistency values. Artif. Intell. **174**(14), 1007–1026 (2010)
19. Matousek, J., Gärtner, B.: Understanding and Using Linear Programming. Universitext. Springer, Heidelberg (2007). https://doi.org/10.1007/978-3-540-30717-4
20. Flum, J., Grohe, M.: Parameterized Complexity Theory. Springer, Heidelberg (2006). https://doi.org/10.1007/3-540-29953-X
21. Hansen, P., Jaumard, B.: Probabilistic satisfiability. In: Kohlas, J., Moral, S. (eds.) Handbook of Defeasible Reasoning and Uncertainty Management Systems, pp. 321–367. Springer, Dordrecht (2000). https://doi.org/10.1007/978-94-017-1737-3_8
22. Whitworth, B., Felton, R.: Measuring disagreement in groups facing limited-choice problems. ACM SIGMIS Database **30**(3–4), 22–33 (1999)
23. Liu, W.: Analyzing the degree of conflict among belief functions. Artif. Intell. **170**(11), 909–924 (2006)
24. Castiñeira, E.E., Cubillo, S., Montilla, W.: Measuring incompatibility between atanassovs intuitionistic fuzzy sets. Inf. Sci. **180**(6), 820–833 (2010)
25. Jousselme, A.L., Maupin, P.: Distances in evidence theory: comprehensive survey and generalizations. Int. J. Approx. Reason. **53**(2), 118–145 (2012)

26. Thimm, M.: Coherence and compatibility of markov logic networks. In: ECAI 2014, pp. 891–896. IOS Press (2014)
27. Grant, J., Hunter, A.: Distance-based measures of inconsistency. In: van der Gaag, L.C. (ed.) ECSQARU 2013. LNCS (LNAI), vol. 7958, pp. 230–241. Springer, Heidelberg (2013). https://doi.org/10.1007/978-3-642-39091-3_20
28. Baral, C., Kraus, S., Minker, J.: Combining multiple knowledge bases. IEEE Trans. Knowl. Data Eng. **3**(2), 208–220 (1991)
29. Liberatore, P., Schaerf, M.: Arbitration (or how to merge knowledge bases). IEEE Trans. Knowl. Data Eng. **10**(1), 76–90 (1998)
30. Konieczny, S., Pérez, R.P.: Logic based merging. J. Philos. Log. **40**(2), 239–270 (2011)
31. Grégoire, É., Konieczny, S., Lagniez, J.: On consensus extraction. In: IJCAI 2016, pp. 1095–1101. AAAI Press, New York (2016)

On Enumerating Models for the Logic of Paradox Using Tableau

Pilar Pozos-Parra[1]([⊠]), Laurent Perrussel[2], and Jean Marc Thévenin[2]

[1] ALTEAP, Instituto Tecnológico y de Estudios Superiores de Occidente, Tlaquepaque, Jalisco, Mexico
mariapozos@iteso.mx
[2] IRIT, Université Toulouse 1 Capitole, Toulouse, France
{laurent.perrussel,thevenin}@ut-capitole.fr

Abstract. We extend the classic propositional tableau method in order to compute the models given by the semantics of the Priest's paraconsistent logic of paradox. Without loss of generality, we assume that the knowledge base is represented through propositional statements in NNF, which leads to use only two rules from the classical propositional tableau calculus for computing the paraconsistent models. We consider multisets to represent branches of the tableau tree and we extend the classical closed branches in order to compute the paradoxical models of formulas of the knowledge base. A sound and complete algorithm is provided.

Keywords: Inconsistent information · 3-valued logic · Tableau

1 Introduction

Two main approaches have been explored to deal with inconsistencies. The first approach consists in providing paraconsistent logics with a theory of reasoning for handling inconsistent information. Among these works two widely used are: three valued logic $\mathcal{LP}$ [14] and its variant $\mathcal{LP}_m$ [15] focusing on minimally inconsistent models of $\mathcal{LP}$. The great interest of $\mathcal{LP}_m$ is to allow classical reasoning on inconsistent information, under some constraints. The second approach consists in measuring the degree of inconsistency of the knowledge base [10]. Inconsistency measures can be used to automatically recover consistent information [8], to determine the reliability assessment of the sources [5] and then help to handle inconsistencies. Numerous measures of inconsistency have been proposed [7,9,17]. The contension inconsistency measure I_c and its normalized version I_{LP_m} [10,17] are two popular syntax independent inconsistency measures. I_c is based on the minimum number of inconsistent atoms in $\mathcal{LP}_m$ models of the knowledge base K and I_{LP_m} is the ratio between I_c and the total number of atoms in K.

Under these considerations, having a decision procedure to produce the $\mathcal{LP}_m$ models of a set of formulas is of prime interest since these models allow to check the satisfiability of an inconsistent set of formulas using $\mathcal{LP}_m$ and all

© Springer Nature Switzerland AG 2018
D. Ciucci et al. (Eds.): SUM 2018, LNAI 11142, pp. 228–242, 2018.
https://doi.org/10.1007/978-3-030-00461-3_16

the minimal inconsistent models are required to compute syntax independent inconsistency measures such as I_c and I_{LP_m}. To the best of our knowledge, few existing research addresses enumeration of models of some paraconsistent logic. This paper proposes a tableau-based decision procedure to compute all the $\mathcal{LP}$ and the $\mathcal{LP}_m$ models of a set of formulas K, which is sound and complete.

The tableau method, originally designed for classical propositional logic [2, 6,16], is very intuitive and brings together the proof-theory and the semantic approaches of a logical system. It tries to build models of an input set of formulas by sequentially decomposing formulas according to decomposition rules based on the logical structure of formulas. The decomposition is organized in a tree called *tableau*. As soon as an inconsistency is found on a branch of this tree, the decomposition of the branch is stopped and the branch is considered as closed. A branch which is decomposed until no more decomposition rule can be applied outputs a set of models[3] of the initial set of formulas. At the end of the process if all the branches are closed, then the set of input formulas is considered as unsatisfiable.

The tableau method has also been used for proof systems using paraconsistent logics [1,4]. Some work is focussed on paraconsistent logics $\mathcal{LP}$ and $\mathcal{LP}_m$ [3,12]. The main concern is to show good reasoning properties of the underlying paraconsistent logics and/or that proof systems are sound and complete, decidable and can be implemented. Like tableau methods designed for classical propositional logic, theses tableau methods define conditions to close branches of the tableau tree, which are specific to the underlying paraconsistent logic and which are necessary for reasoning. For instance to prove that $S \models_{\mathcal{LP}_m} A$ one have to show that the tableau of $S \wedge \neg A$ is closed [12]. Due to these closing conditions, these tableau methods can not produce all the paraconsistent models of an input set of formulas.

We propose an adaptation of the tableau method to enumerate the 3-valued models of a set of formulas according to the Logic of Paradox $\mathcal{LP}$ and the Minimally Inconsistent Logic of Paradox $\mathcal{LP}_m$. When an inconsistency is found in a branch, the corresponding inconsistent symbols are set to a specific valued called *Both* instead of closing the branch so that the decomposition process can go further. We provide an algorithm of this adaptation of the tableau method and show that it is sound and complete. We also show that, as a result of this algorithm, we can get a set of *succinct models* which summarizes all the $\mathcal{LP}$ (respectively $\mathcal{LP}_m$) models of the input formulas. For the sake of conciseness and without loss of generality, we assume that the initial set of formulas is represented through propositional statements in NNF, which leads to only use two rules from the classical propositional tableau calculus for computing the paraconsistent models. The adaptation we propose can be extended to other tableau method with more than two rules and which does not require propositional statement to be in NNF.

---

[3] Strictly speaking, the method does not exactly provide a model considering that some symbols of the language may not get a truth value, however, the symbols that get a truth value are sufficient to satisfy the input formulas.

The rest of the paper is organized as follows. After providing some technical preliminaries and a short introduction to $\mathcal{LP}$ and $\mathcal{LP}_m$ in Sect. 2, the proposal and its corresponding algorithms are presented in Sect. 3 while complexity and validity are provided in Sect. 4. Conclusion is given in Sect. 5.

2 Preliminaries

We consider a propositional language $\mathcal{L}$ using a finite set of atoms $P = \{p_1, p_2, ..., p_n\}$ where the formulas are in Negation Normal Form (NNF). A formula ϕ is in NNF iff all operators are negation ($\neg$), conjunction ($\wedge$) and disjunction ($\vee$), and negation operator only appears in front of atomic symbols. A *literal* is an atom or its negation.

A *knowledge base* K is a finite multiset of propositional formulas of $\mathcal{L}$. $At(K)$ denotes the set of atoms appearing in K. A literal is *pure* in K iff it appears in K and its negation does not.

In classical propositional logic an interpretation w is a total function from P to $\{F, T\}$, where F and T respectively represent *False* and *True* such that $F < T$. The set of interpretations is denoted by $\mathcal{W}$, its elements will be denoted by sets of valued propositional symbols

2.1 Logic of Paradox

The logic of paradox $\mathcal{LP}$ introduced in [14] is one of the simplest paraconsistent logics. While classical logic assumes that no formula can be both *True* and *False*, $\mathcal{LP}$ interprets such formulas as paradoxical formulas and can reason with them. $\mathcal{LP}$ is a 3-valued logic such that an interpretation w is a total function from P to $\{F, B, T\}$ where B represents *Both* true and false, and F and T stand as usual for the truth values *False* and *True*, such that $F < B < T$. B value characterises inconsistency and allows to find models for classically inconsistent set of formulas. An interpretation w is a $\mathcal{LP}$ model of a formula ϕ, denoted $w \models_{\mathcal{LP}} \phi$ iff w assigns to ϕ the truth value T or B in the $\mathcal{LP}$ truth-functional way as in Tables 1 and 2. These truth tables are the same as the truth tables proposed by Kleene 3-valued logic [11]. However, in Kleene's logic the only designated truth value is T while T and B are the designated truth values in $\mathcal{LP}$.

Implication and equivalence are defined in terms of conjunction and disjunction as usual: $\phi \rightarrow \psi \equiv \neg\phi \vee \psi$ and $\phi \leftrightarrow \psi \equiv (\neg\phi \vee \psi) \wedge (\phi \vee \neg\psi)$.

If w is a $\mathcal{LP}$ model for every formula in a knowledge base K then we write $w \models_{\mathcal{LP}} K$. The set of $\mathcal{LP}$ models of a knowledge base K is denoted by $[\![K]\!]_{\mathcal{LP}}$. There is at least one $\mathcal{LP}$ model for every knowledge base K, i.e. if $w(p_i) = B$ for $i = 1, \ldots, n$, then $w(K) = B$. The set of $\mathcal{LP}$ models of the language is denoted by $\mathcal{W}_{\mathcal{LP}}$ and its elements will be denoted by sets of the form $\{l_1, ..., l_n\}$ where $l_i = p_i$ iff $w(p_i) = T$, $l_i = \pm p_i$ iff $w(p_i) = B$ and $l_i = \neg p_i$ iff $w(p_i) = F$.

Example 1. Consider the set of formulas $K_1 = \{a, b \vee c, \neg a \wedge \neg b\}$; in classical and Kleene logics K_1 is inconsistent ($K_1 \models \perp$). However, in $\mathcal{LP}$ we have the following models: $[\![K_1]\!]_{\mathcal{LP}} = \{\{\pm a, \pm b, c\}, \{\pm a, \pm b, \pm c\}, \{\pm a, \pm b, \neg c\}, \{\pm a, \neg b, c\}, \{\pm a, \neg b, \pm c\}\}$.

Table 1. Truth table for negation in the logic of paradox

$w(\phi)$	$w(\neg\phi)$
T	F
B	B
F	T

Table 2. Truth table for conjunction and disjunction in the logic of paradox

$w(\phi)$	$w(\psi)$	$w(\phi \wedge \psi)$	$w(\phi \vee \psi)$
T	T	T	T
T	B	B	T
T	F	F	T
B	T	B	T
B	B	B	B
B	F	F	B
F	T	F	T
F	B	F	B
F	F	F	F

2.2 Minimally Inconsistent $\mathcal{LP}$ models

Models which minimize the number of inconsistent symbols are the most "informative" ones. We can see in Example 1 that the most informative model is $\{\pm a, \neg b, c\}$ which states that only a is conflicting and recovering consistency for K_1 entails to delete some formulas involving a. Notice that if a pure literal is only involved in conjunctions of K, then half of the models set value B to the symbol associated to this literal. These models are of low interest as there is no conflict (no contradiction can arise) on pure literals.

In order to avoid unnecessary B valuations Priest [15] introduced the Minimally Inconsistent Logic of Paradox denoted $\mathcal{LP}_m$, that reasons only with $\mathcal{LP}$ models which are minimally inconsistent. Logic $\mathcal{LP}_m$ allows to rank $\mathcal{LP}$ models of K in order to find minimally inconsistent $\mathcal{LP}$ models. To this end, Priest first introduced the notion of *inconsistent part of an interpretation* w, denoted $w!$, as the subset of atomic symbols set to value B. Then the $\mathcal{LP}$ models of K w and w' can be ranked such that $w < w'$ iff $w! \subset w'!$. Finally, the $\mathcal{LP}$ model w of K is a minimally inconsistent $\mathcal{LP}$ model of K iff if $w' < w$ then $w' \not\models_{\mathcal{LP}} K$.

Oller in [13] provided a slightly different definition of minimally inconsistent model, which is the same as the definition used for the inconsistency measure I_{LP_m} [10]: ranking is no longer based on set inclusion but on counting. A model will then be minimal if it minimizes the number of inconsistent symbols. In the following, we commit to Oller's definition of minimal inconsistent model:

Definition 1 (Minimally inconsistent $\mathcal{LP}$-models). *The set of $\mathcal{LP}_m$ models of K denoted $[\![K]\!]_{\mathcal{LP}_m}$ are defined as follows:*
$$[\![K]\!]_{\mathcal{LP}_m} = \{w | w \models_{\mathcal{LP}} K \text{ and } |w!| \leq |w'!| \text{ for all } w' \models_{\mathcal{LP}} K\}.$$

Example 2. Consider the set of formulas $K_1 = \{a, b \vee c, \neg a \wedge \neg b\}$, in the $\mathcal{LP}_m$ we have the following model: $[\![K_1]\!]_{\mathcal{LP}_m} = \{\{\pm a, \neg b, c\}\}$.

2.3   Succinct Models

A *Partial interpretation* $\ddot{w}$ is a partial function from P to $\{F, B, T\}$ for $\mathcal{LP}$ logic (or to $\{F, T\}$ for classical logic). Partial interpretations are denoted by sets of the form $\{l_{i_1}, ..., l_{i_m}\}$ where $i_m \leq |P|$, $l_i = p_i$ iff $\ddot{w}(p_i) = T$, $l_i = \pm p_i$ iff $\ddot{w}(p_i) = B$ and $l_i = \neg p_i$ iff $\ddot{w}(p_i) = F$. A *succinct model* $\ddot{w}$ of the knowledge base K is a partial interpretation such that for each interpretation w superset of $\ddot{w}$, $w \models_{\mathcal{LP}} K$:

$$\ddot{w} \models_{\mathcal{LP}} K \iff w \models_{\mathcal{LP}} K \text{ for all } w \supseteq \ddot{w}$$

A similar relation is also defined for classical logic (value B is no longer possible):

$$\ddot{w} \models K \iff w \models K \text{ for all } w \supseteq \ddot{w}$$

A succinct model gives the subset of propositional symbols that have to be set to a specific value to satisfy K while the rest of the propositional symbols can be set to any value among *True*, *Both* or *False* without changing the satisfaction of K. For example, for $P = \{a, b\}$ and $K_2 = \{\neg b, a \vee \neg b\}$, we have the succinct model $\ddot{w} = \{\neg b\}$ which summarises the three models $\{\{a, \neg b\}, \{\pm a, \neg b\}, \{\neg a, \neg b\}\}$ (or the two models $\{\{a, \neg b\}, \{\neg a, \neg b\}\}$ in classical logic). A succinct model is actually the semantic representation of an implicant. Notice that the previous notion of minimal $\mathcal{LP}$ models can be extended to succinct model

Definition 2. *A succinct model $\ddot{w}$ of a knowledge base K is said to be* minimal *iff there is no succinct model $\ddot{w}'$ of K such that $|\ddot{w}'!| < |\ddot{w}!|$.*

It is clear that minimal succinct models represent in a compact way $\mathcal{LP}_m$ models: the minimal inconsistent part of a $\mathcal{LP}_m$ model must belong to a succinct model which is also minimal and vice versa.

Proposition 1. *Let $\ddot{w}$ be a minimal succinct model and w a $\mathcal{LP}_m$ model. The number of inconsistent symbols is the same in both models: $|\ddot{w}!| = |w!|$.*

Proof. ($\Leftarrow$) as $\ddot{w}$ is a succinct model, then there exists a model w such that for all $p \notin \ddot{w}$, $w(p) \neq B$. Hence, the number of inconsistent symbols in w is the same as in $\ddot{w}$ otherwise $\ddot{w}$ is not minimal. ($\Rightarrow$) If w is a minimally inconsistent model, any succinct model $\ddot{w} \subseteq w$ have the same symbols set to value B. If they have more symbols set to values B they are not minimal. If they have less symbols set to value B, $\exists w' \supseteq \ddot{w}$ s.t. $w' \nmodels_{\mathcal{LP}} K$.

Proposition 2. $\forall w \in [\![K]\!]_{\mathcal{LP}_m}$, $\exists \ddot{w}$ a minimal succinct model s.t. $w \supseteq \ddot{w}$.

It means that computing the minimal paradoxical models of some knowledge base K boils down to computing its succinct models.

3   Tableau and Succinct Models for $\mathcal{LP}$ and $\mathcal{LP}_m$

3.1   Classical Tableau

The principle behind classical semantic tableaux is very simple: searching models for a set of formulas by decomposing the formulas into sets of literals. The corresponding decomposition is often represented as a tree whose nodes are labelled by sets composed of formulas or atoms. The initial set of formulas labels the root of the tree; each node has one or two child nodes depending on how a formula included in the node label is decomposed according to its principal operator [2]. For a formula in NNF, we only need two tableau decomposition rules, one for conjunctions and one for disjunctions as shown in Fig. 1.

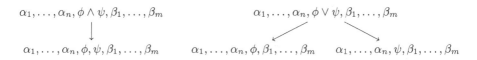

Fig. 1. Tableau decomposition rules for conjunctions and disjunctions

A node labelled by a set containing at least a pair of opposed literals is marked *closed* so that the decomposition process is stopped and the node becomes a leaf. A node labelled by a set only composed of literals is a leaf. If the label of this leaf does not contain any pair of opposed literals it can be used to produce a set of models. It is actually a *succinct model* of the input set of formulas.

3.2   Extending Tableau for Finding Succinct $\mathcal{LP}$-Models

Extending the tableau method to produce succinct $\mathcal{LP}$ models is straightforward. Nodes labelled by a set containing pairs of opposed literals are no longer considered as closed so that any branch of the tree is decomposed until reaching a node labelled by a set only composed of literals. Indeed, each pair of opposed literals represents a paradoxical symbol so that value B will be assigned to the corresponding symbol in the succinct $\mathcal{LP}$ model derived from the corresponding leaf of the tree. Figure 2 illustrates this process for formulas in $K_3 = \{a, \neg a \wedge \neg b, b \vee c, \neg d\}$: after decomposing formula $\neg a \wedge \neg b$ we get a pair of opposed literals for symbol a, however, the corresponding node is not closed and is further decomposed into two branches according to the decomposition of formula $b \vee c$. The two nodes obtained after this decomposition correspond to two leafs since all formulas of K_3 have been decomposed. The first leaf contains opposite literals for atomic symbols a and b resulting in succinct model $\ddot{w}_1 = \{\pm a, \pm b, \neg d\}$. Second leaf contains one pair of opposite literals for a representing the succinct model $\ddot{w}_2 = \{\pm a, \neg b, c, \neg d\}$. This extension of the tableau method is detailed by Algorithms 1 and 2, where all the set and multiset are ordered.

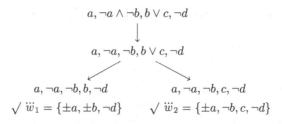

Fig. 2. Extended tableau for set of formulas $\{a, \neg a \wedge \neg b, b \vee c, \neg d\}$

Algorithm 1: *succinct_models(K)*

Data:
    K : multiset of formulas
Result:
    $\ddot{W}$ set of succinct $\mathcal{LP}$ models of K
1 **begin**
2     $\ddot{W} \leftarrow \{\{\}\}$;
3   **for** $i \leftarrow 1 \ldots |K|$ **do**
4         $\ddot{W} \leftarrow add\_literals(\ddot{W}, K(i))$
5   **end**
6 **end**

Algorithm 1 computes the set of succinct $\mathcal{LP}$ models $\ddot{W}$ from a set of formulas K by fully decomposing one formula at each iteration. *Full decomposition* means to consider all literals of the formula until no decomposition rule can be applied. More precisely, at iteration i, we expand each partial succinct model with the literals of the i-th formula.

Algorithm 2 details Function $add\_literals(\ddot{W}, formula)$ which consists of the following two alternatives:

$alternative_1$ (Lines 2–9) handles the base case of a formula only composed of one literal by adding this literal to each succinct model under construction.
$alternative_2$ (Lines 10–20) decomposes the formula by recursive calls along the edges of the tableau tree.

Example 3 illustrates how every succinct model under construction is expanded at each iteration.

Example 3. Consider again set of formulas $K_3 = \{a, b \vee c, \neg a \wedge \neg b, \neg d\}$. The value of $\ddot{W}$ after each call is the following:

1. call of $add\_literals(\{\{\}\}, a)$ gives $\ddot{W} = \{\{a\}\}$
2. call of $add\_literals(\{\{a\}\}, \neg a \wedge \neg b)$ gives $\ddot{W} = \{\{\pm a, \neg b\}\}$
3. call of $add\_literals(\{\{\pm a, \neg b\}\}, b \vee c)$ gives $\ddot{W} = \{\{\pm a, \pm b\}, \{\pm a, \neg b, c\}\}$
4. call of $add\_literals(\{\{\pm a, \pm b\}, \{\pm a, \neg b, c\}\}, \neg d)$ gives $\ddot{W} = \{\{\pm a, \pm b, \neg d\}, \{\pm a, \neg b, c, \neg d\}\}$, which represent the succinct $\mathcal{LP}$ models of K.

Algorithm 2: $add\_literals(\ddot{W}, formula)$

Data:

$\ddot{W}$: Set of succinct models to be completed

$formula$: Formula to decompose

Result:

$Temp$: Set of succinct models satisfying $formula$

1 **begin**

2 $\quad$ **if** $is\_literal(formula)$ **then**

3 $\quad\quad$ $p \leftarrow At(formula)$;

4 $\quad\quad$ **for** $i \leftarrow 1 \ldots |\ddot{W}|$ **do**

5 $\quad\quad\quad$ **if** $\pm p \notin \ddot{W}(i)$ **then**

6 $\quad\quad\quad\quad$ **if** $\ddot{W}(i) \not\models \neg formula$ **then** $Temp(i) \leftarrow \ddot{W}(i) \cup \{formula\}$ **else**
$\quad\quad\quad\quad$ $Temp(i) \leftarrow \ddot{W}(i) \setminus \{p, \neg p\} \cup \{\pm p\}$

7 $\quad\quad\quad$ **end**

8 $\quad\quad$ **end**

9 $\quad$ **else**

10 $\quad\quad$ $(operator, operand_1, operand_2) \leftarrow decompose(formula)$;

11 $\quad\quad$ **if** $operator = `\wedge`$ **then**

12 $\quad\quad\quad$ $Temp \leftarrow add\_literals(\ddot{W}, operand_1)$;

13 $\quad\quad\quad$ $Temp \leftarrow add\_literals(Temp, operand_2)$;

14 $\quad\quad$ **else**

15 $\quad\quad\quad$ $Temp_1 \leftarrow add\_literals(\ddot{W}, operand_1)$;

16 $\quad\quad\quad$ $Temp_2 \leftarrow add\_literals(\ddot{W}, operand_2)$;

17 $\quad\quad\quad$ $Temp \leftarrow Temp_1 \cup Temp_2$;

18 $\quad\quad$ **end**

19 $\quad$ **end**

20 $\quad$ $return(Temp)$;

21 **end**

From the succinct models we can then compute $\mathcal{LP}$ and $\mathcal{LP}_m$ models. In the following section we propose the algorithm that extends the succinct $\mathcal{LP}$ models of K produced by Algorithm 1, in order to get $[\![K]\!]_{\mathcal{LP}}$ and $[\![K]\!]_{\mathcal{LP}_m}$.

3.3 Computing $[\![K]\!]_{\mathcal{LP}}$ and $[\![K]\!]_{\mathcal{LP}_m}$

Algorithm 3 implements the transformation of succinct models into $\mathcal{LP}$ models.

For each succinct model (Line 3), the algorithm creates a new set of temporary and partial interpretations W_{Temp} (Line 4) and extends these interpretations one symbol at a time for each symbol of the language (Lines 5–16). For each symbol $P(j)$ the current set of partial interpretations is replicated in three new sets: one where previous partial interpretations are extended with the current symbol set to value B (Line 7), one where previous partial interpretations are extended with the current symbol set to value T if $\neg P(j)$ is not in the current succinct model (Lines 8–10) and one where previous partial interpretations are extended with the current symbol set to value F if $P(j)$ is not in the current

---

Algorithm 3: $extend\_succinct\_models(\ddot{W}, P)$

    Data:
       $\ddot{W}$: Set of succinct models
       P : Set of atoms of language
    Result:
       $Temp$: Set of extended models

1 **begin**
2     $Temp \leftarrow \{\{\}\}$;
3   **for** $i \leftarrow 1 \ldots |\ddot{W}|$ **do**
4         $W_{Temp} \leftarrow \{\{\}\}$; $n \leftarrow 1$;
5   **for** $j \leftarrow 1 \ldots |P|$ **do**
6             $W^+ \leftarrow \{\}$; $W^{\pm} \leftarrow \{\}$; $W^- \leftarrow \{\}$;
7   **for** $h \leftarrow 1 \ldots n$ **do** $W^{\pm}(h) \leftarrow W_{Temp}(h) \cup \{\pm P(j)\}$
8   **if** $\neg P(j) \notin \ddot{W}(i)$ and $\pm P(j) \notin \ddot{W}(i)$ **then**
9   **for** $h \leftarrow 1 \ldots n$ **do** $W^+(h) \leftarrow W_{Temp}(h) \cup \{P(j)\}$
10   **end**
11   **if** $P(j) \notin \ddot{W}(i)$ and $\pm P(j) \notin \ddot{W}(i)$ **then**
12   **for** $h \leftarrow 1 \ldots n$ **do** $W^-(h) \leftarrow W_{Temp}(h) \cup \{\neg P(j)\}$
13   **end**
14            $W_{Temp} \leftarrow W^+ \cup W^{\pm} \cup W^-$;
15            $n \leftarrow size(W_{Temp})$;
16   **end**
17         $Temp \leftarrow Temp \cup W_{Temp}$;
18   **end**
19     $return(Temp)$;
20 **end**

---

succinct model (Lines 11–13). Consequently, a symbol which is not in the current succinct model is added to the three copies since it can get any value: a symbol which occurs as $\pm P(j)$ is added only to the first copy, while the other two types of symbol occurrences are added to two copies one with value B and one with one of the values T or F. Finally, the three copies of partial interpretations are assigned as the new current set of partial interpretations (Line 14). Suppose that the Algorithm 3 receives as input a set of succinct models $\ddot{W}$ of some knowledge base K, then every $w \in Temp$ is also a $\mathcal{LP}$ model of K:

Lemma 1. *Let K be a knowledge base represented w.r.t. set P. Let $\ddot{W}$ be a set of succinct models of K: if $W = extend\_succinct\_models(\ddot{W}, P)$, then for every $w \in W$, $w \models_{\mathcal{LP}} K$.*

Proof. Each iteration on symbols (Line 5) extends the current succinct model $\ddot{w}_i \in \ddot{W}$ in an incremental way. If the current symbol $P(j)$ has no valuation in $\ddot{w}_i$ then $P(j)$ can get any of the three values. Thus three copies of $\ddot{w}_i$ are created. Otherwise, at most two copies of $\ddot{w}_i$ are created, one with the value $\pm P(j)$ and one with the value of $P(j)$ in $\ddot{w}_i$ if it is different of $\pm P(j)$. At the end, copies may or may not be superset of $\ddot{w}_i$. First case, assume that $\ddot{w}_i \subseteq w$:

according to the definition of succinct model, all interpretations $w \supseteq \ddot{w}_i$ are models of K. Second case if $\ddot{w}_i \not\subseteq w$, then it occurs that for each symbol p such that $\ddot{w}_i(p) \neq w(p)$, $\ddot{w}_i(p) \neq B$ and $w(p) = B$ thanks to Lines 7, 8 and 11. Hence, according to Tables 1 and 2, either K has a truth value equal to B or it is still equal to T. In any case, the semantics of the logic $\mathcal{LP}$ leads to conclude that $w \models_{\mathcal{LP}} K$. Then in both cases w is a model of K.

Algorithm 3 can easily be adapted to compute the set of models $[\![K]\!]_{\mathcal{LP}_m}$ if it takes as input the set of minimal succinct models instead of all succinct models. First, thanks to Proposition 2, all the $[\![K]\!]_{\mathcal{LP}_m}$ models can be generated from the set of minimal succinct models. Next, thanks to Proposition 1, we only need to consider T and F values for the symbols not yet set in the succinct models.

4   Complexity and Validity

4.1   Complexity

Let us now discuss the complexity of Algorithms 1 and 2. First, recall that we only allow negation in front of a variable as we only consider the restricted language NNF. Transformation of any arbitrary formula to NNF is polynomial.

Assume a set of formulas K. The complexity of Algorithm 1 is in $O(|K|)$ without considering the complexity of function *add_literals*. Let us now consider some formula ϕ to be decomposed. Function *add_literals*$(\ddot{W}, \phi)$ applies either *alternative*$_1$ or *alternative*$_2$. *Alternative*$_1$ adds the current literal to each succinct model of $\ddot{W}$ and is in $O(|\ddot{W}|)$. *Alternative*$_2$ decomposes ϕ by recursive calls. If the total number of conjunctions and disjunctions in ϕ is m, *alternative*$_2$ is in $O(2 \times m)$. The complexity of the process for computing the succinct models is mostly due to *alternative*$_1$ considering that the number $|\ddot{W}|$ of succinct models produced grows exponentially with respect to the number of disjunctions appearing in formulas. Indeed, the number of succinct models is duplicated at each disjunction (Lines 16–18 of function *add_literals*). Let $\widehat{\phi}$ be the formula of K with the largest number of disjunctions and let $|\widehat{\phi}|$ denote the number of disjuncts in $\widehat{\phi}$. If the knowledge base K contains only conjunctions (there is only one alternative and $|\widehat{\phi}| = 1$), Algorithm 1 produces exactly one succinct model.

Proposition 3. *Let K be a knowledge base with no disjunction ($|\widehat{\phi}| = 1$). Let $\ddot{W} = succinct\_models(K)$, then $|\ddot{W}| = 1$.*

Proof. Line 2 of function *succinct_models* initializes $\ddot{W}$ with an empty succinct model. Then Lines 6, 7, 13 and 14 of function *add_literals* always extend this initial succinct model which is never duplicated.

We can give an upper bound of the number of succinct models produced for the case in which K contains disjunctions.

Proposition 4. *Let K be a knowledge base, $\widehat{\phi}$ be the formula of K with the largest number of disjunctions and let $|\widehat{\phi}|$ denote the number of disjuncts in $\widehat{\phi}$. Let $\ddot{W} = succinct\_models(K)$. An upper bound of $|\ddot{W}|$ is $|\widehat{\phi}|^{|K|}$ and the complexity of function $succinct\_models(K)$ is in $O(|\widehat{\phi}|^{|K|})$.*

Proof. Let us focus on the case where K contains $n > 1$ formulas with disjunctions and suppose that after $n - 1$ iterations of the main loop of Algorithm 1, the size of the ongoing set of succinct models is equal to S_{n-1}. Let us focus on the case where n-th formula contains disjunctions. Assume that $\phi_n = \widehat{\phi}$ then according to Lines 16 and 17 of Algorithm 2, the size of $Temp$ returned by Line 21 is equal to $S_n = |\phi_n| \times S_{n-1}$. Hence, by assuming that each formula in K contains $|\widehat{\phi}|$ disjuncts, the upper bound of $|\ddot{W}|$ is $|\widehat{\phi}|^{|K|}$ and the complexity of function $succinct\_models(K)$ is in $O(|\widehat{\phi}|^{|K|})$.

Algorithm 3 is also exponential as shown by the following proposition.

Proposition 5. *Let S be the number of succinct models of K and $\ddot{w}_m$ be a minimal succinct model of K: let k be the number of symbols in $\ddot{w}_m$ and l the number of symbols with values F or T. The complexity for computing paradoxical models as given by Algorithm 3 is in $O(S \times 2^l \times 3^{|P|-k})$.*

Proof. Suppose there is only one succinct model assigning values to k symbols and $l \leq k$ symbols are assigned value F or T. $Temp$ contains $2^l \times 3^{|P|-k}$ succinct models, hence the complexity is in $O(2^l \times 3^{|P|-k})$. Suppose now there are n succinct models considered as input. Suppose that for iteration of Line 3, i is equal to $n - 1$ and the current size of $Temp$ is S_{n-1}. Again n succinct model is composed of k symbols and the number of new paradoxical models is bounded by $S_n = S_{n-1} + 2^l \times 3^{|P|-k}$. Hence, the overall complexity is in $O(n \times 2^l \times 3^{|P|-k})$.

It is worth noting that the number of succinct $\mathcal{LP}$ models of K is actually much smaller than the number of $\mathcal{LP}$ models of K. For instance, as mentioned previously, if K contains only conjunctions we get one succinct $\mathcal{LP}$ model while the number of $\mathcal{LP}$ models is $2^l \times 3^{|P|-|At(K)|}$.

4.2 Soundness and Completeness

The combination of the three algorithms is sound and complete, that is it produces exactly $[\![K]\!]_{\mathcal{LP}}$. We only need to show that succinct models output by Algorithm 1 are actually models of the knowledge base K considered as input. Next, Lemma 1 guarantees that these succinct models can be transformed into paradoxical models.

Theorem 1 (Soundness). *If $\ddot{w} \in succinct\_models(K)$ then $\ddot{w} \models_{\mathcal{LP}} K$.*

Proof. (sketch) We prove that Algorithms 1 and 2 are sound by induction over the cardinality of K:

(Base case) Suppose that K contains one formula ϕ and let us consider the three subcases about the formula structure: ϕ is either a literal, a conjunction or a disjunction. We prove by induction on the structure of ϕ that Algorithms 1 and 2 are sound.

Base case. If ϕ is a literal l associated to atom p_i then, output $Temp$ contains one unique partial interpretation $\{l\}$ which is a succinct model of ϕ.

Induction. Suppose that for any subformula ϕ' of ϕ, function $add\_literals$ $(\ddot{W}, \phi')$ returns the set of succinct models of ϕ'. We only focus on the *conjunction* case (as disjunction case is similar): Line 13 calls $add\_literals(\{\},$ $operand_1)$ and returns a set $Temp$ of succinct models of $operand_1$ (by induction). Next step Line 14 calls $add\_literals(Temp, operand_2)$. Let us focus on the case where $operand_2$ is a literal l; then, $add\_literals$ applies $alternative_1$ (Lines 2–9) which returns a set $Temp$ of extended partial interpretations $\ddot{w}'$ such that for all $\ddot{w} \in Temp$, $\ddot{w}' = \ddot{w} \ddot{\cup} \{l\}$ such that:

$$\ddot{w} \ddot{\cup} \{l\} \begin{cases} \ddot{w}' = \ddot{w} \cup \{l\} & \text{if } \ddot{w} \not\models \neg l \\ \ddot{w}' = \ddot{w} \setminus \{l, \neg l\} \cup \pm At(l) \text{ otherwise.} \end{cases}$$

Each extended interpretation $\ddot{w}' \in Temp$ is clearly a succinct model of $operand_1 \wedge operand_2$ which sets the interpretation value to B if a contradiction has been found or to T otherwise. For the general case "$operand_2$ is not a literal", Line 14 extends each model $\ddot{w}$ of $operand_1$ with each succinct models $\ddot{w}'$ of $operand_2$. Thus function $add\_literals(Temp, operand_2)$ returns a set of succinct interpretations $\ddot{w} \ddot{\cup} \ddot{w}'$.

(Induction) Let us now consider that K contains n formulas. Suppose that up to formula $i - 1$, Algorithm 1 outputs the set $\ddot{W}_{i-1}$ of succinct models of $\cup_{j=1...i-1} K(j)$. Function $add\_literals(\ddot{W}_{i-1}, \phi_n)$ is then called (Line 4 of Algorithm 1). Again, we have to consider the three subcases with respect to the structure of ϕ_i: ϕ_i is either a literal, a conjunction or a disjunction. Proof is again by induction on the structure and similar to the Base case previously described except for literals.

The immediate consequence is the overall soundness of Algorithms 1–3:

Corollary 1. *If $w \in extend\_succinct\_models(succinct\_models(K), P)$ then $w \models_{\mathcal{LP}} K$.*

Proof. It is a direct consequence of Lemma 1 and Theorem 1.

Let us now show that the algorithms are also *complete*:

Theorem 2 (completeness). *Let w_0 be an $\mathcal{LP}$ interpretation. If $w_0 \models_{\mathcal{LP}} K$ then $w_0 \in extend\_succinct\_models(succinct\_models(K), P)$.*

Proof. (sketch) Let us consider the different cases with respect to the size of K and the structure of the formulas.

(Base Case) Suppose that K contains only one formula ϕ and let us focus the literal and conjunction subcases:

If $\phi = l$, then $succinct\_models(K) = \{l\}$. First, set $\{l\}$ is produced by Algorithm 2. Next, if the symbol P associated to l has B value in w_0 then Algorithm 3 will create a new model with this value; otherwise if P has value T or F in w_0 then Algorithm 3 creates a model w where P's value is identical to the one in w_0. Next, for all other symbols, all possible truth values are considered. Hence $w_0 \in extend\_succinct\_models(\{l\}, P)$.

Let us consider the conjunction $\phi_1 \wedge \phi_2$ and assume that the theorem holds for subformulas ϕ_1 and ϕ_2. We thus have two sets of succinct models $\ddot{W}_1 = succinct\_models(\{\phi_1\})$ and $\ddot{W}_2 = succinct\_models(\{\phi_2\})$. Two cases may occur:

$Case_1$, there is no inconsistencies between ϕ_1 and ϕ_2. Let us focus on the subcase where $w_0 \models_{\mathcal{LP}} \phi_1 \wedge \phi_2, \exists \ddot{w}_1 \in \ddot{W}_1$ **s.t.** $\ddot{w}_1 \subseteq w_0$ and $\exists \ddot{w}_2 \in \ddot{W}_2$ **s.t.** $\ddot{w}_2 \subseteq w_0$. Lines 16 and 17 combined with Lines 5–7 of Algorithm 2 guarantee that $\ddot{w}_1 \cup \ddot{w}_2$ belongs to $succinct\_models(\{\phi_1 \wedge \phi_2\})$. Hence, $w \supseteq \ddot{w}_1 \cup \ddot{w}_2$. Alternative subcase where w_0 is not a superset of $\ddot{w}_1$ or $\ddot{w}_2$, i.e. Algorithm 3 changes and duplicates some truth values to the B value, also holds.

$Case_2$, there are some inconsistent literals between ϕ_1 and ϕ_2. Assume succinct models $\ddot{w}_1$ and $\ddot{w}_2$ for ϕ_1 and ϕ_2. Lines 16 and 17 combined with Line 9 of Algorithm 2 guarantee that each symbol getting an inconsistent value between $\ddot{w}_1$ and $\ddot{w}_2$ is set to value B when performing the "union" of $\ddot{w}_1$ and $\ddot{w}_2$. If $w_0 \supseteq \ddot{w}_1 \ddot{\cup} \ddot{w}_2$ then $w_0 \in extend\_succinct\_models(\phi_1 \wedge \phi_2, P)$; other cases where function $extend\_succinct\_models$ creates additional models with other symbols set to B also holds.

Disjunction case can also be proved in a similar way.

So, when K contains only one formula, the theorem holds. Let us now consider the case where K contains more than one formula.

(Induction on formulas) Suppose that K contains i formulas and the theorem holds up to $i - 1$ formulas. Let $\ddot{W}_{i-1} = succinct\_models(K_{i-1})$, we thus have: $\forall w \models_{\mathcal{LP}} \cup_{j=1...i-1} K(j), \exists \ddot{w} \in \ddot{W}_{i-1}$ such that $w \supseteq \ddot{w}$. Let us focus on the i-th formula: Algorithm 1 calls function $add\_literals(\ddot{W}_{i-1}, K(i))$. Again, the different subcases defined by the structure of formula $K(i)$ should be considered. We do not detail the proof as it is similar to the base case.

5   Conclusion

We propose a tableau method for enumerating the paradoxical models of a knowledge base. We provide the corresponding algorithm and prove that it is sound and complete. This algorithm is decomposed in two main steps. First step uses Algorithm 2 which computes the models with minimal number of inconsistent symbols to satisfy the input formulas, which are the succinct models constituting $\ddot{W}$. Then Algorithm 3 calculates the full models constituting W, by extending the succinct models with all the symbols of the language. The second step is mandatory if we need the algorithm to be complete. However, this second step should not be done when computing the inconsistency measure I_{LP_m}.

Indeed, the succinct models allow us to make the difference between the symbols really inconsistent in the formulas and symbols taking B in order to get models. The succinct models of knowledge base K can be reused for computing the succinct models of K', where $K \subset K'$ by calling function $add\_literals(\ddot{W}, \phi)$ for every formula $\phi \in K'$ such that $\phi \notin K$. So, this allows us complexity reduction when we already know the succinct models of some frequently used formulas.

The complexity of our approach is exponential, however, we need to implement only 10% of the number of decomposition rules compared with [3] which needs to implement 20 rules.

References

1. Avron, A., Konikowska, B., Zamansky, A.: Cut-free sequent calculi for C-systems with generalized finite-valued semantics. J. Log. Comput. **23**(3), 517–540 (2013)
2. Ben-Ari, M.: Propositional logic: formulas, models, tableaux. In: Ben-Ari, M. (ed.) Mathematical Logic for Computer Science, pp. 7–47. Springer, London (2012). https://doi.org/10.1007/978-1-4471-4129-7_2
3. Bloesch, A.: A tableau style proof system for two paraconsistent logics. Notre Dame J. Form. Logic. **34**(2), 295–301 (1993)
4. Carnieli, W.A., Marcos, J.: Tableau systems for logics of formal inconsistency. In: Proceedings of the 2001 International Conference on Artificial Intelligence (IC-AI 2001), vol. 2, pp. 848–852. CSREA Press (2001)
5. Cholvy, L., Perrussel, L., Thevenin, J.M.: Using inconsistency measures for estimating reliability. Int. J. Approx. Reason. **89**, 41–57 (2017)
6. D'Agostino, M.: Tableau methods for classical propositional logic. In: D'Agostino, M., Gabbay, D.M., Hähnle, R., Posegga, J. (eds.) Handbook of Tableau Methods, pp. 45–123. Springer, Dordrecht (1999). https://doi.org/10.1007/978-94-017-1754-0_2
7. Grant, J., Hunter, A.: Measuring inconsistency in knowledgebases. J. Intell. Inf. Systems. **27**, 159–184 (2006)
8. Grant, J., Hunter, A.: Measuring consistency gain and information loss in stepwise inconsistency resolution. In: Liu, W. (ed.) ECSQARU 2011. LNCS (LNAI), vol. 6717, pp. 362–373. Springer, Heidelberg (2011). https://doi.org/10.1007/978-3-642-22152-1_31
9. Hunter, A., Konieczny, S.: Approaches to measuring inconsistent information. In: Bertossi, L., Hunter, A., Schaub, T. (eds.) Inconsistency Tolerance. LNCS, vol. 3300, pp. 191–236. Springer, Heidelberg (2005). https://doi.org/10.1007/978-3-540-30597-2_7
10. Hunter, A., Konieczny, S.: On the measure of conflicts: shapley inconsistency values. Artif. Intell. **174**(14), 1007–1026 (2010)
11. Kleene, S.C.: Introduction to Metamathematics. North-Holland Publishing Co., Amsterdam (1952)
12. Lin, Z., Li, W.: A note on tableaux of logic of paradox. In: Nebel, B., Dreschler-Fischer, L. (eds.) KI 1994. LNCS, vol. 861, pp. 296–307. Springer, Heidelberg (1994). https://doi.org/10.1007/3-540-58467-6_26
13. Oller, C.: Measuring coherence using LP-models. J. Appl. Logic. **2**(4), 451–455 (2004)

14. Priest, G.: Logic of paradox. J. Philos. Logic. **8**(1), 219–241 (1979)
15. Priest, G.: Minimally inconsistent LP. Stud. Logica. **50**(2), 321–331 (1991)
16. Smullyan, M.: First-Order Logic. Springer, Heidelberg (1968). https://doi.org/10.1007/978-3-642-86718-7
17. Thimm, M.: On the expressivity of inconsistency measures. Artif. Intell. **234**(C), 120–151 (2016)

On Instantiating Generalised Properties of Gradual Argumentation Frameworks

Antonio Rago[1]([⊠])(iD), Pietro Baroni[2](iD), and Francesca Toni[1](iD)

[1] Imperial College London, London, UK
{a.rago15,f.toni}@imperial.ac.uk
[2] Universitá degli Studi di Brescia, Brescia, Italy
pietro.baroni@unibs.it

Abstract. Several gradual semantics for abstract and bipolar argumentation have been proposed in the literature, ascribing to each argument a value taken from a scale, i.e. an ordered set. These values somewhat match the arguments' dialectical status and provide an indication of their dialectical strength, in the context of the given argumentation framework. These research efforts have been complemented by formulations of several properties that these gradual semantics may satisfy. More recently a synthesis of many literature properties into more general groupings based on parametric definitions has been proposed. In this paper we show how this generalised parametric formulation enables the identification of new properties not previously considered in the literature and discuss their usefulness to capture alternative requirements coming from different application contexts.

Keywords: Abstract argumentation · Bipolar argumentation
Gradual argumentation semantics · Parametric properties

1 Introduction

Abstract Argumentation Frameworks (AFs) and their semantics [9] are a well-known formalism to represent at an abstract level conflicts between *arguments*, expressed as a binary relation of *attack*, and to evaluate the acceptance status of the arguments on the basis of this relation. Several extensions or modifications of AFs and their semantics have been proposed. Notably, a *support* relation between arguments may be used alongside or instead of the attack relation, as in Bipolar Argumentation Frameworks (BAFs) [4]/Quantitative Argumentation Debates (QuAD frameworks) [7] and as in Support Argumentation Frameworks (SAFs) [2], respectively. Moreover, several *gradual* (rather than *extension-based* [9]) semantics for AFs, BAFs and SAFs have been proposed in the literature, e.g. as in [4,7,10–12], ascribing to arguments values from one ordered set or another. These values somewhat match the arguments' dialectical status and provide an indication of their dialectical strength, in the context of the given argumentation framework. For example, the *Game-Theoretical* strengths of [5,12] assign

© Springer Nature Switzerland AG 2018
D. Ciucci et al. (Eds.): SUM 2018, LNAI 11142, pp. 243–259, 2018.
https://doi.org/10.1007/978-3-030-00461-3_17

values in $[0,1]$ to arguments in an AF or a BAF, by assessing the arguments' influence on players' rewards when included in strategies of suitable classes of games of argumentation . The *Social Models* of [11] also assign strengths from $[0,1]$ to arguments in an AF, aggregating attacks so as to reduce an argument's strength in proportion with its attackers' strength, while taking into account users' votes in a social context. The approach of [4] is similar in that it also aggregates attackers and supporters in a proportional manner, but for BAFs where arguments have an initial *intrinsic strength* (or *base score*): attackers and supporters reduce and increase, respectively, an argument's strength (in this case a value in $[-1,1]$).

These research efforts have been complemented by formulations of several properties that these gradual semantics may satisfy. For example, *Maximality* [1] states that any argument in an AF without any attackers should have a strength equal to its base score . Analogously, for SAFs, *Minimality* [2] states that any argument without any supporters should have a strength equal to its base score. For BAFs, *Equation 4* in [7] combines both of these concepts to state that an argument with no attackers and no supporters has a strength equal to its base score. These properties are defined for different types of argumentation frameworks but clearly share the same intuition.

Recently, the study of specific properties for specific types of frameworks has been extended by critical syntheses within more general groupings based on parametric formulations [6]. In particular, [6] introduces eleven *group properties*, generalising twenty-nine properties from the literature (including the three mentioned earlier), and four *principles*, generalising the group (and thus the literature) properties. The principles are *(strict) balance* and *(strict) monotonicity*, where balance states that a difference between strength and base score of an argument must correspond to some imbalance between the strengths of its attackers and supporters, while monotonicity requires that the strength of an argument depends monotonically on its base score and on the strengths of its attackers and supporters. Both group properties and principles are defined for a generalised version of argumentation frameworks, called *Quantitative Bipolar Argumentation Frameworks* (QBAFs), in terms of a number of parameters, that can be instantiated in several different ways. These parametric formulations, besides recovering existing literature properties as special instances, reveal the existence of a larger spectrum of possible properties and thus enable the identification of novel concrete properties, for AFs, SAFs, BAFs and QuAD frameworks alike.

In this paper we demonstrate the potential of this line of investigation by focusing on three of the group properties identified in [6], there called GP2, GP3 and GP9, and instantiate them to obtain a number of novel properties, not already considered in the literature. These group properties amount to the following:

- GP2 - In the absence of supporters, if there is at least one attacker then the strength of an argument is lower than its base score. This property is proven to generalise *Weakening* [1,3] and to be implied by balance [6].

– GP3 - In the absence of attackers, if there is at least one supporter then the strength of an argument is higher than its base score. This property is proven to generalise *Strengthening* [2] and to be implied by balance [6].
– GP9 - A higher base score gives a higher strength. This property is proven to generalise *Proportionality* [3] and to be implied by strict monotonicity [6].

Our new instantiations can be seen as "relatives" of the previous literature properties and provide an illustration of how parametric group properties can be exploited to identify a comprehensive set of actual variants. We discuss the practical relevance of each variant with examples of application contexts where its features turn out to be appropriate.

The paper is organised as follows. In Sect. 2 we give the necessary background from [6]. Section 3 describes the concept of instantiating the group properties before Sects. 4, 5 and 6 give the instantiations of GP2, GP3 and GP9, respectively. In Sect. 7 we conclude.

2   Background

This section is adapted from [6].

Let $\mathbb{I}$ be a set equipped with a preorder $\leq$ where, as usual, $a < b$ denotes $a \leq b$ and $b \not\leq a$. $\mathbb{I}$ may (but is not required to) contain top ($\top$) and bottom ($\bot$) values, such that if $\top \in \mathbb{I}$, then $i < \top$ for all $i \in \mathbb{I}\backslash\{\top\}$ and, if $\bot \in \mathbb{I}$, then $\bot < i$ for all $i \in \mathbb{I}\backslash\{\bot\}$. For example, $\mathbb{I} = [0, 1]$ with $\top = 1$, $\bot = 0$, or $\mathbb{I} = \{low, medium, high\}$ with $low \leq medium \leq high$ and $\top = high$, $\bot = low$, or $\mathbb{I} = (0, \infty)$.

Given some $\mathbb{I}$, a QBAF assigns attackers, supporters and an initial evaluation (base score) in $\mathbb{I}$ to arguments.

Definition 1. *A Quantitative Bipolar Argumentation Framework (QBAF) is a quadruple $\langle \mathcal{X}, \mathcal{R}^-, \mathcal{R}^+, \tau \rangle$ consisting of a set $\mathcal{X}$ of arguments, a binary (attack) relation $\mathcal{R}^-$ on $\mathcal{X}$, a binary (support) relation $\mathcal{R}^+$ on $\mathcal{X}$ and a total function $\tau : \mathcal{X} \to \mathbb{I}$. For any $\alpha \in \mathcal{X}$:*
  $\mathcal{R}^-(\alpha) = \{\beta \in \mathcal{X} | (\beta, \alpha) \in \mathcal{R}^-\}$ *is the set of attackers of α,*
  $\mathcal{R}^+(\alpha) = \{\beta \in \mathcal{X} | (\beta, \alpha) \in \mathcal{R}^+\}$ *is the set of supporters of α, and*
  $\tau(\alpha)$ *is the base score of α.*

QBAFs can be visualised as graphs, e.g. Figure 1 visualises $\langle\{a, b, c, d\}, \{(c, a), (c, b)\}, \{(d, b)\}, \tau\rangle$ (for any τ).

In the remainder of the paper, unless specified otherwise, we will assume as given a generic QBAF $\mathcal{Q} = \langle \mathcal{X}, \mathcal{R}^-, \mathcal{R}^+, \tau \rangle$. We will also consider the following restricted instances of QBAFs:

aQBAFs	where	$\forall \alpha \in \mathcal{X}, \mathcal{R}^+(\alpha) = \emptyset$
sQBAFs	where	$\forall \alpha \in \mathcal{X}, \mathcal{R}^-(\alpha) = \emptyset$

aQBAFs correspond to AFs, as introduced in [9] and sQBAFs correspond to SAFs, as introduced in [2]. Whereas SAFs include a base score (called *basic*

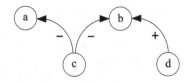

Fig. 1. Example QBAF visualised as a graph (ignoring τ).

strength in [2]) for all arguments, AFs do not. In order to capture existing (non-gradual) semantics for AFs, such as the grounded extension semantics [9], specific choices of base scores for arguments may be needed (e.g. $\forall \alpha \in \mathcal{X}$, $\tau(\alpha) = \top$, as in [7]). BAFs, as introduced in [4], do not include base scores either, but, analogously to the case of AFs, can be seen as special cases of QBAFs where the base score is the same for every argument. Finally, Social AFs [11] can be seen as QBAFs where base scores are determined by votes, whereas Weighted Argumentation Graphs [3] and Quantitative Argumentation Debate (QuAD) frameworks [7] are generic QBAFs.

Arguments in a QBAF have a final evaluation (strength):

Definition 2. *For any $\alpha \in \mathcal{X}$, the strength of α is given by $\sigma(\alpha)$ where $\sigma : \mathcal{X} \to \mathbb{I}$ is a total function.*

If $\bot \in \mathbb{I}$, it may or not play a role in evaluating arguments and hence arguments with bottom strength may or not be counted as effective attackers/supporters. This is captured by the following definition where $*$ is either $\sigma\bot$ (indicating that bottom strength arguments count) or $\sigma \not\bot$ (indicating that they can be ignored as attackers/supporters).

Definition 3. *For any $\alpha \in \mathcal{X}$, we define $\mathcal{R}_*^-(\alpha) = \mathcal{R}^-(\alpha)$ if $* = \sigma\bot$ and $\mathcal{R}_*^-(\alpha) = \mathcal{R}^-(\alpha)\backslash\{\beta \in \mathcal{R}^-(\alpha)|\sigma(\beta) = \bot\}$ if $* = \sigma \not\bot$. Similarly, $\mathcal{R}_*^+(\alpha) = \mathcal{R}^+(\alpha)$ if $* = \sigma\bot$ and $\mathcal{R}_*^+(\alpha) = \mathcal{R}^+(\alpha)\backslash\{\beta \in \mathcal{R}^+(\alpha)|\sigma(\beta) = \bot\}$ if $* = \sigma \not\bot$.*

Thus, we use $\mathcal{R}_*^-(\alpha)$ and $\mathcal{R}_*^+(\alpha)$ to denote, respectively, the set of attackers and supporters of an argument α including or excluding, depending on the choice of $*$, those with bottom strength.

The definition of group properties (and principles) in [6] is also parametric w.r.t. an operator over $\mathbb{I}$, denoted $\ll$, expressing a form of strict comparison between values in $\mathbb{I}$. The basic requirement for this operator is that $< \subseteq \ll \subseteq \leq$ and, naturally, $m \gg n$ iff $n \ll m$.

In this paper we consider the following instances of $\ll$:

- $\ll = <$,
- $\ll = \leq$,
- $\ll = <_\times$, where $m <_\times n$ iff $m < n$ or $m = n = \times$, where $\times$ is one of $\bot$ or $\top$.

In this paper we focus on the following group properties from [6]:

Definition 4. *Let $\alpha \in \mathcal{X}$.*

GP2: If $\mathcal{R}_*^-(\alpha) \neq \emptyset$ and $\mathcal{R}_*^+(\alpha) = \emptyset$ then $\sigma(\alpha) \ll \tau(\alpha)$.

GP3: If $\mathcal{R}_*^-(\alpha) = \emptyset$ and $\mathcal{R}_*^+(\alpha) \neq \emptyset$ then $\tau(\alpha) \ll \sigma(\alpha)$.

GP9: If $\mathcal{R}_*^-(\alpha) = \mathcal{R}_*^-(\beta)$, $\mathcal{R}_*^+(\alpha) = \mathcal{R}_*^+(\beta)$ and $\tau(\alpha) < \tau(\beta)$ then $\sigma(\alpha) \ll \sigma(\beta)$.

Note that GP2 is not applicable to sQBAFs, where $\mathcal{R}_*^-(\alpha) \neq \emptyset$ is impossible, and similarly GP3 is not applicable to aQBAFs.

3 Instantiating Group Properties

The following instances of GP2, GP3, GP9, already present in the literature as indicated, are identified in [6]:

- GP2 for aQBAFs with $* = \sigma \perp$ and $\ll = <_\perp$ and for $\mathbb{I} = [0, 1]$ implies the following:
  Weakening (Weak.) [1,3] - For any $\alpha \in \mathcal{X}$, if $\tau(\alpha) > \perp$ and $\exists \beta \in \mathcal{R}^-(\alpha)$ such that $\sigma(\beta) > \perp$ then $\sigma(\alpha) < \tau(\alpha)$.
- GP3 for sQBAFs with $* = \sigma \perp$ and $\ll = <_\top$ and for $\mathbb{I} = [0, 1]$ implies the following:
  Strengthening (Str.) [2] - For any $\alpha \in \mathcal{X}$, if $\tau(\alpha) < \top$ and $\exists \beta \in \mathcal{R}^+(\alpha)$ such that $\sigma(\beta) > \perp$ then $\tau(\alpha) < \sigma(\alpha)$.
- GP9 for aQBAFs with $* = \sigma \perp$ and $\ll = <_\perp$ and for $\mathbb{I} = [0, 1]$ implies the following:
  Proportionality (Prop.) [3] - For any $\alpha, \beta \in \mathcal{X}$, if $\mathcal{R}^-(\alpha) = \mathcal{R}^-(\beta)$, $\tau(\beta) > \tau(\alpha)$ and either $\sigma(\alpha) > \perp$ or $\sigma(\beta) > \perp$, then $\sigma(\beta) > \sigma(\alpha)$.

In this paper we consider the novel instances indicated as hyphens in Table 1, given in the remainder. Note that in the instantiations we do not make any commitment on τ nor on $\mathbb{I}$, thus ensuring that each novel property is still generic with respect to actual formalisms which may differ in the assignment of base scores and the scale adopted. We remark however that some instances (in particular

Table 1. The GPs that we consider and their implied instances in the literature.

		$\sigma\perp$				$\sigma\perp\!\!\!\!\perp$			
		$<$	$<_\perp$	$<_\top$	$\leq$	$<$	$<_\perp$	$<_\top$	$\leq$
GP2	aQBAF	-	-	-	-	-	Weak.	-	-
	QBAF	-	-	-	-	-	-	-	-
GP3	sQBAF	-	-	-	-	-	-	Str.	-
	QBAF	-	-	-	-	-	-	-	-
GP9	aQBAF	-	Prop.	-	-	-	-	-	-
	sQBAF	-	-	-	-	-	-	-	-
	QBAF	-	-	-	-	-	-	-	-

in the case $\ll=<$) imply some additional conditions, which will be discussed case by case, exemplifying some application contexts where such conditions are appropriate.

4   Novel Instances of GP2

GP2 expresses the general idea that, in absence of supporters and in presence of some attackers, the strength of an argument cannot be higher than its base score. As recalled in Sect. 3, in the literature this property has been explicitly considered, with the name *Weakening*, only for aQBAFs and for the choices $* = \sigma \perp$ and $\ll = <_\perp$

In this section we provide an explicit definition for the hyphens in Table 1 in the rows for GP2, thus identifying in a systematic way a set of novel properties, arising from the combinations of possible values of the parameters.

In this section and the two which follow, for each novel property we provide a short explanation and a comment on its possible relevance from the application point of view.

The first group of parameter choices which we assess are those with $* = \sigma \perp$, taking on the first notion for attackers or supporters with bottom strength stating that they may have an effect on their targeted argument. Generally, considering attackers or supporters with bottom strength may make sense in contexts where the presence itself of an argument, as weak as it may be, will always have some effect. This may make sense in trusted settings where malicious behaviour is not expected in the construction of the argumentation framework and so this choice of $*$ may be more appropriate when some inherent quality control is present. This is especially relevant in cases like GP2 where there are attackers but no supporters.

For GP2, GP3 and GP9, the selection of $\ll$ may allow a strength to *saturate* when the operator is relaxed, e.g. at $\perp$ when $\ll=<_\perp$. These properties are meaningful for semantics where base scores are uniform for all arguments.

GP2 for $\sigma \perp$ and $<$:

- For any $\alpha \in \mathcal{X}$ in an aQBAF, if $\mathcal{R}^-(\alpha) \neq \emptyset$ then $\sigma(\alpha) < \tau(\alpha)$.
- For any $\alpha \in \mathcal{X}$ in a QBAF, if $\mathcal{R}^-(\alpha) \neq \emptyset$ and $\mathcal{R}^+(\alpha) = \emptyset$ then $\sigma(\alpha) < \tau(\alpha)$.

In this first property (for aQBAFs and QBAFs) the selection of $\ll=<$ gives rise to the requirement of strict decrease and implies that $\tau(\alpha) > \perp$, i.e. that no argument is completely discarded from the beginning and the bottom value is attainable only in presence of attacks.

Comment. *This property and its implied condition may be appropriate when argumentatively modelling a trial in the court of law (somewhat subjectively), where a person is assumed to be innocent (their argument holds and its base score is $\top$) until there is sufficient evidence proving them to be guilty (arguments attacking their argument of innocence). Then any attacker of their argument which is introduced may affect the jury's presumption of their innocence* *negatively and therefore reduce the strength of their argument claiming their innocence.*

GP2 for $\sigma\bot$ and $<_\bot$:

- For any $\alpha \in \mathcal{X}$ in an aQBAF, if $\mathcal{R}^-(\alpha) \neq \emptyset$ then $\sigma(\alpha) < \tau(\alpha)$ or $\tau(\alpha) = \sigma(\alpha) = \bot$.
- For any $\alpha \in \mathcal{X}$ in a QBAF, if $\mathcal{R}^-(\alpha) \neq \emptyset$ and $\mathcal{R}^+(\alpha) = \emptyset$ then $\sigma(\alpha) < \tau(\alpha)$ or $\tau(\alpha) = \sigma(\alpha) = \bot$.

These instances of GP2 use $\ll = <_\bot$, which requires a strict decrease unless $\tau(\alpha) = \bot$, i.e. the argument is already as weak as it can possibly be. This may be useful in cases where an argument can have a base score equal to the minimum strength available, e.g. it is *rejected*, meaning that attackers may have no effect as they cannot reduce it further.

Comment. *An application of this may be in the modelling of an engineering discussion (e.g. see [14]) when a base score of $\bot$ may correspond to a critically dangerous solution to a problem, which would obtain the minimum base score and therefore attackers could not reduce it further.*

GP2 for $\sigma\bot$ and $<_\top$:

- For any $\alpha \in \mathcal{X}$ in an aQBAF, if $\mathcal{R}^-(\alpha) \neq \emptyset$ then $\sigma(\alpha) < \tau(\alpha)$ or $\tau(\alpha) = \sigma(\alpha) = \top$.
- For any $\alpha \in \mathcal{X}$ in a QBAF, if $\mathcal{R}^-(\alpha) \neq \emptyset$ and $\mathcal{R}^+(\alpha) = \emptyset$ then $\sigma(\alpha) < \tau(\alpha)$ or $\tau(\alpha) = \sigma(\alpha) = \top$.

This property may seem somewhat counter-intuitive in that it allows arguments with top base score to have top strength even when they are attacked. However, this behaviour may be useful when arguments with a top base score need to have a special "tautological" status.

Comment. *Considering once again the engineering discussion setting with the base scores corresponding to some predefined level of importance to the project, an argument may be given a fixed strength of $\top$ if it is required by the stakeholders, such that there is no need to consider the dialectical reasoning against the argument and so the attackers of the argument may be ignored.*

GP2 for $\sigma\bot$ and $\leq$:

- For any $\alpha \in \mathcal{X}$ in an aQBAF, if $\mathcal{R}^-(\alpha) \neq \emptyset$ then $\sigma(\alpha) \leq \tau(\alpha)$.
- For any $\alpha \in \mathcal{X}$ in a QBAF, if $\mathcal{R}^-(\alpha) \neq \emptyset$ and $\mathcal{R}^+(\alpha) = \emptyset$ then $\sigma(\alpha) \leq \tau(\alpha)$.

When $\ll = \leq$ is selected for GP2 we have a less strict property which does not require an argument's strength to be reduced in the presence of attackers for any given base score. If the two previous properties' exceptions are both required in the engineering discussion setting as stated, this property would allow such a situation.

Comment. *Ignoring rejected attackers or supporters may be beneficial in less trusted environments, e.g. those containing malicious users. Considering the dialectical context of social media, the effect of so-called trolls' arguments may be negated by sufficiently strong attacks from other arguments or simply flagging from users, both of which may restrict the arguments to the bottom strength, which lead to them being ignored when $* = \sigma \bot$. This is again especially relevant for GP2 where there are attackers but no supporters.*

The next group of four sets of instantiations take on the second notion for attackers or supporters with bottom strength, i.e. $* = \sigma \bot$. This means that attackers and supporters with bottom strength will not have an effect on their targeted argument.

GP2 for $\sigma \bot$ and $<$:

- For any $\alpha \in \mathcal{X}$ in an aQBAF, if $\exists \beta \in \mathcal{R}^-(\alpha)$ such that $\sigma(\beta) > \bot$ then $\sigma(\alpha) < \tau(\alpha)$.
- For any $\alpha \in \mathcal{X}$ in a QBAF, if $\exists \beta \in \mathcal{R}^-(\alpha)$ such that $\sigma(\beta) > \bot$ and $\not\exists \gamma \in \mathcal{R}^+(\alpha)$ such that $\sigma(\gamma) > \bot$ then $\sigma(\alpha) < \tau(\alpha)$.

The selection of $* = \sigma \bot$ and $\ll = <$ gives a flexibility in how strict an attacker's effect is, since it allows possibility of ignoring rejected attackers but implies that $\tau(\alpha) > \bot$, i.e. that no non-rejected argument is completely discarded from the beginning and the bottom value is attainable only in presence of attacks

Comment. *In argumentation-based aggregation of user reviews (e.g. see [8]) in a context such as* Trip Advisor, *where the base score may correspond to the number of users voting for an argument, this affords a guarantee to users that any attacker or supporter they add will have an effect on a targeted argument's evaluation unless their argument is rejected, for example if it is flagged for a particular reason. This behaviour seems to be ideal for a setting such as this in which malicious arguments may be present (and flagged), but it is nonetheless important for non-flagged arguments to have an effect to entice users to participate.*

GP2 for $\sigma \bot$ and $<_\bot$:

- For any $\alpha \in \mathcal{X}$ in an aQBAF, if $\exists \beta \in \mathcal{R}^-(\alpha)$ such that $\sigma(\beta) > \bot$ then $\sigma(\alpha) < \tau(\alpha)$ or $\tau(\alpha) = \sigma(\alpha) = \bot$.
- For any $\alpha \in \mathcal{X}$ in a QBAF, if $\exists \beta \in \mathcal{R}^-(\alpha)$ such that $\sigma(\beta) > \bot$ and $\not\exists \gamma \in \mathcal{R}^+(\alpha)$ such that $\sigma(\gamma) > \bot$ then $\sigma(\alpha) < \tau(\alpha)$ or $\tau(\alpha) = \sigma(\alpha) = \bot$.

With $\ll = <_\bot$, this property requires a strict decrease unless $\tau(\alpha) = \bot$. Once again we have a situation where an argument may be rejected based on its base score.

Comment. *We now imagine a real world example in which the government is using argumentation-based e-polling (e.g. see [13]) for its policy planning. Here, the base score of an argument may correspond to the proportion of the public who agree with it, while an argument's strength depends its base score and the strengths of its attackers and supporters. For this property we assume that the default is for the government to continue with an already-implemented policy unless a certain threshold for amount of people who disagree with this argument is reached, i.e. all voters disagree with the argument and its base score is $\bot$. At this point the policy may be abandoned regardless of its attackers since the base score takes precedence.*

Note also that the former of the two instantiations implies the literature property Weakening.

GP2 for $\sigma \not{L}$ and $<_\top$:

- For any $\alpha \in \mathcal{X}$ in an aQBAF, if $\exists \beta \in \mathcal{R}^-(\alpha)$ such that $\sigma(\beta) > \bot$ then $\sigma(\alpha) < \tau(\alpha)$ or $\tau(\alpha) = \sigma(\alpha) = \top$.
- For any $\alpha \in \mathcal{X}$ in a QBAF, if $\exists \beta \in \mathcal{R}^-(\alpha)$ such that $\sigma(\beta) > \bot$ and $\nexists \gamma \in \mathcal{R}^+(\alpha)$ such that $\sigma(\gamma) > \bot$ then $\sigma(\alpha) < \tau(\alpha)$ or $\tau(\alpha) = \sigma(\alpha) = \top$.

This property states that a first non-rejected attacker will cause a strict decrease in strength is required unless $\tau(\alpha) = \top$, which at first seem counter-intuitive but can be justified in the case of the analogous example to that of the previous property.

Comment. *In this case, the government may be using e-polling to determine whether to implement an as of yet untested policy, the default for which is to decline to do so. In this case the votes causing the base score to reach the $\top$ threshold, i.e. all voters agreeing with the argument, may mean the policy is implemented, regardless of its attackers since the base score takes precedence.*

GP2 for $\sigma \not{L}$ and $\leq$:

- For any $\alpha \in \mathcal{X}$ in an aQBAF, if $\exists \beta \in \mathcal{R}^-(\alpha)$ such that $\sigma(\beta) > \bot$ then $\sigma(\alpha) \leq \tau(\alpha)$.
- For any $\alpha \in \mathcal{X}$ in a QBAF, if $\exists \beta \in \mathcal{R}^-(\alpha)$ such that $\sigma(\beta) > \bot$ and $\nexists \gamma \in \mathcal{R}^+(\alpha)$ such that $\sigma(\gamma) > \bot$ then $\sigma(\alpha) \leq \tau(\alpha)$.

The final instantiations for GP2 cover the cases where $\ll = \leq$.

Comment. *The two exceptions in the two previous properties for $\bot$ and $\top$ may be required in an online debate setting where a group of people are to be convinced one way (in favour of an argument) or the other (in opposition an argument). Here, the base score may once again correspond to the proportion of the votes in favour of an argument and when the whole group agree (base score is $\top$) or disagree (base score is $\bot$) with an argument, this may obtain the maximum or minimum, respectively, strength, regardless of its attackers.*

5   Novel Instances of GP3

GP3 expresses the general idea that, in absence of attackers and in presence of some supporters, the strength of an argument cannot be lower than its base score. As recalled in Sect. 3, in the literature this property has been explicitly considered only for sQBAFs and for the choices $* = \sigma \perp$ and $\ll = <_\top$ with the name *Strengthening*.

In this section we provide an explicit definition for the hyphens in Table 1 in the rows for GP3 in the same manner as in the previous section.

For the choices of $*$, we use examples relating to the same two groupings as was done for GP2, i.e. $* = \sigma \perp$ corresponding to those in trusted settings and $* = \sigma \measuredangle$ to those in untrusted settings where malicious users may be present.

GP3 for $\sigma \perp$ and $<$:

– For any $\alpha \in \mathcal{X}$ in an sQBAF, if $\mathcal{R}^+(\alpha) \neq \emptyset$ then $\tau(\alpha) < \sigma(\alpha)$.
– For any $\alpha \in \mathcal{X}$ in a QBAF, if $\mathcal{R}^-(\alpha) = \emptyset$ and $\mathcal{R}^+(\alpha) \neq \emptyset$ then $\tau(\alpha) < \sigma(\alpha)$.

The selection of $\ll = <$ gives the requirement of strict increase and corresponds to the impossibility of ignoring the existence of any support, implying that $\tau(\alpha) < \top$, i.e. that no argument is completely accepted from the beginning and the top value is attainable only in presence of supports.

Comment. *This property and its implied condition may be appropriate when all non-rejected supports are taken into account, for example in game-theoretical contexts such as optimising debate strategy, as suggested in the games of argumentation strategy of [5], where the base score is some uniform midpoint.*

GP3 for $\sigma \perp$ and $<_\perp$:

– For any $\alpha \in \mathcal{X}$ in an sQBAF, if $\mathcal{R}^+(\alpha) \neq \emptyset$ then $\tau(\alpha) < \sigma(\alpha)$ or $\tau(\alpha) = \sigma(\alpha) = \perp$.
– For any $\alpha \in \mathcal{X}$ in a QBAF, if $\mathcal{R}^-(\alpha) = \emptyset$ and $\mathcal{R}^+(\alpha) \neq \emptyset$ then $\tau(\alpha) < \sigma(\alpha)$ or $\tau(\alpha) = \sigma(\alpha) = \perp$.

Comment. *Here we have another property which seems counter-intuitive at first glance but it can be justified in the engineering discussion setting once again: arguments that are assigned the $\perp$ base score may take on the special status of being rejected outright, i.e. their strength is also $\perp$, for example if they cause a critically dangerous situation. Here this special base score takes precedence over the presence of supporters.*

GP3 for $\sigma \perp$ and $<_\top$:

– For any $\alpha \in \mathcal{X}$ in an sQBAF, if $\mathcal{R}^+(\alpha) \neq \emptyset$ then $\tau(\alpha) < \sigma(\alpha)$ or $\tau(\alpha) = \sigma(\alpha) = \top$.
– For any $\alpha \in \mathcal{X}$ in a QBAF, if $\mathcal{R}^-(\alpha) = \emptyset$ and $\mathcal{R}^+(\alpha) \neq \emptyset$ then $\tau(\alpha) < \sigma(\alpha)$ or $\tau(\alpha) = \sigma(\alpha) = \top$.

This property requires that supporters increase an argument's strength to a value higher than the base score except in the case of the base score being maximal.

Comment. *We once again look to the engineering discussion setting in which an argument having stakeholder approval gives it a special, 'indispensable' status, i.e. a base score, and therefore strength, of $\top$.*

GP3 for $\sigma\bot$ and $\leq$:

- For any $\alpha \in \mathcal{X}$ in an sQBAF, if $\mathcal{R}^+(\alpha) \neq \emptyset$ then $\tau(\alpha) \leq \sigma(\alpha)$.
- For any $\alpha \in \mathcal{X}$ in a QBAF, if $\mathcal{R}^-(\alpha) = \emptyset$ and $\mathcal{R}^+(\alpha) \neq \emptyset$ then $\tau(\alpha) \leq \sigma(\alpha)$.

For the final property for GP3 with $* = \sigma\bot$, the choice $\ll=\leq$ applies so we have a less strict property which does not require supporters to increase an argument's strength above any given base score.

Comment. *If the two exceptions from the previous two properties are both required in the engineering discussion setting as stated, this property may be suitable.*

GP3 for $\sigma\measuredangle$ and $<$:

- For any $\alpha \in \mathcal{X}$ in an sQBAF, if $\exists \beta \in \mathcal{R}^+(\alpha)$ such that $\sigma(\beta) > \bot$ then $\tau(\alpha) < \sigma(\alpha)$.
- For any $\alpha \in \mathcal{X}$ in a QBAF, if $\nexists \beta \in \mathcal{R}^-(\alpha)$ such that $\sigma(\beta) > \bot$ and $\exists \gamma \in \mathcal{R}^+(\alpha)$ such that $\sigma(\gamma) > \bot$ then $\tau(\alpha) < \sigma(\alpha)$.

The selection of $* = \sigma\measuredangle$ and $\ll=<$ gives a flexibility in how strict a supporter's effect is, since it allows one to ignore rejected supporters, but implies that $\tau(\alpha) < \top$, i.e. that no argument is completely accepted from the beginning and the top value is attainable only in presence of supports.

Comment. *This behaviour once again seems to be ideal for the review aggregation/Trip Advisor setting in which supporting arguments (e.g. from an institution itself, attempting to increase its rating), may be malicious in the same way as attacking arguments. This property allows the malicious supports to be rejected and ignored, while still enticing users by guaranteeing that non-rejected supporters will have an effect on their targeted argument.*

GP3 for $\sigma\measuredangle$ and $<_\bot$:

- For any $\alpha \in \mathcal{X}$ in an sQBAF, if $\exists \beta \in \mathcal{R}^+(\alpha)$ such that $\sigma(\beta) > \bot$ then $\tau(\alpha) < \sigma(\alpha)$ or $\tau(\alpha) = \sigma(\alpha) = \bot$.
- For any $\alpha \in \mathcal{X}$ in a QBAF, if $\nexists \beta \in \mathcal{R}^-(\alpha)$ such that $\sigma(\beta) > \bot$ and $\exists \gamma \in \mathcal{R}^+(\alpha)$ such that $\sigma(\gamma) > \bot$ then $\tau(\alpha) < \sigma(\alpha)$ or $\tau(\alpha) = \sigma(\alpha) = \bot$.

Comment. *For this (seemingly unintuitive) property, we again turn to the setting of argumentation-based e-polling situation in which the default is for the government to continue with a policy unless the base score of the corresponding argument reaches the threshold $\perp$, in which case its supporters' effects are neglected.*

GP3 for $\sigma \underline{\perp}$ and $<_{\top}$:

- For any $\alpha \in \mathcal{X}$ in an sQBAF, if $\exists \beta \in \mathcal{R}^{-}(\alpha)$ such that $\sigma(\beta) > \perp$ then $\tau(\alpha) < \sigma(\alpha)$ or $\tau(\alpha) = \sigma(\alpha) = \top$.
- For any $\alpha \in \mathcal{X}$ in a QBAF, if $\nexists \beta \in \mathcal{R}^{-}(\alpha)$ such that $\sigma(\beta) > \perp$ and $\exists \gamma \in \mathcal{R}^{+}(\alpha)$ such that $\sigma(\gamma) > \perp$ then $\tau(\alpha) < \sigma(\alpha)$ or $\tau(\alpha) = \sigma(\alpha) = \top$.

Comment. *This analogous property of the previous one uses the same e-polling setting but with the default position from the government of not implementing a policy unless the base score of the corresponding argument reaches the threshold $\top$, in which case its supporters' effects are neglected.*

Note also that the former of the two instantiations implies the literature property Strengthening.

GP3 for $\sigma \underline{\perp}$ and $\leq$:

- For any $\alpha \in \mathcal{X}$ in an sQBAF, if $\exists \beta \in \mathcal{R}^{+}(\alpha)$ such that $\sigma(\beta) > \perp$ then $\tau(\alpha) \leq \sigma(\alpha)$.
- For any $\alpha \in \mathcal{X}$ in a QBAF, if $\nexists \beta \in \mathcal{R}^{-}(\alpha)$ such that $\sigma(\beta) > \perp$ and $\exists \gamma \in \mathcal{R}^{+}(\alpha)$ such that $\sigma(\gamma) > \perp$ then $\tau(\alpha) \leq \sigma(\alpha)$.

Comment. *The final instantiations for GP3 may be applied to the online debate setting of the final instantiations of GP2, such that when all of the group agree (base score is $\top$) or disagree (base score is $\perp$) with an argument, it may obtain the maximum or minimum, respectively, strength, regardless of its supporters.*

6   Novel Instances of GP9

GP9 expresses the general idea that, when two otherwise equivalent (in terms of attackers and supporters) arguments have a difference in base score, that with the higher base score will have a higher strength. As recalled in in Sect. 3, in the literature this property has been explicitly considered only for aQBAFs and for the choices $* = \sigma\perp$ and $\ll = <_{\perp}$ with the name *Proportionality* [3].

In this section we provide an explicit definition for the hyphens in Table 1 in the rows for GP9 in the same manner as in the previous section.

For GP9, the choice of $*$ determines whether or not the equality between attackers and supporters includes those with the bottom strength or not. We therefore once again use examples relating to the same two groupings as were used for GP2 and GP3, i.e. trusted (when $* = \sigma\perp$) and malicious (when $* = \sigma\underline{\perp}$) environments. It should be noted that GP9 is not a meaningful property for semantics with uniform base scores. Moreover, in the cases where $\ll = <$ or $\ll = <_\top$, to ensure that a strict inequality is always possible, it must be the case that for any argument α $\sigma(\alpha) > \bot$ if $\tau(\alpha) > \bot$, a property called Resilience in [3].

GP9 for $\sigma\bot$ and $<$:

- For any $\alpha, \beta \in \mathcal{X}$ in an aQBAF, if $\mathcal{R}^-(\alpha) = \mathcal{R}^-(\beta)$ and $\tau(\alpha) < \tau(\beta)$ then $\sigma(\alpha) < \sigma(\beta)$.
- For any $\alpha, \beta \in \mathcal{X}$ in an sQBAF, if $\mathcal{R}^+(\alpha) = \mathcal{R}^+(\beta)$ and $\tau(\alpha) < \tau(\beta)$ then $\sigma(\alpha) < \sigma(\beta)$.
- For any $\alpha, \beta \in \mathcal{X}$ in a QBAF, $\mathcal{R}^-(\alpha) = \mathcal{R}^-(\beta)$, $\mathcal{R}^+(\alpha) = \mathcal{R}^+(\beta)$ and $\tau(\alpha) < \tau(\beta)$ then $\sigma(\alpha) < \sigma(\beta)$.

In the first property for GP9 the selection of $\ll = <$ gives the requirement of a strict decrease in the strength when the base score is lowered, thereby giving the base score a high importance.

Comment. *We envisage this property's use when argumentation is used for conflict resolution in meeting support, for example the engineering discussion setting. If an argument's base score corresponds to the stakeholders' votes on the issues that arguments represent, in the case of equal reasoning (attackers and supporters) the votes would always be used to resolve the conflict, i.e. ranking one argument higher than another using their strengths, justifying the use of this property.*

GP9 for $\sigma\bot$ and $<_\bot$:

- For any $\alpha, \beta \in \mathcal{X}$ in an aQBAF, if $\mathcal{R}^-(\alpha) = \mathcal{R}^-(\beta)$ and $\tau(\alpha) < \tau(\beta)$ then $\sigma(\alpha) < \sigma(\beta)$ or $\sigma(\alpha) = \sigma(\beta) = \bot$.
- For any $\alpha, \beta \in \mathcal{X}$ in an sQBAF, if $\mathcal{R}^+(\alpha) = \mathcal{R}^+(\beta)$ and $\tau(\alpha) < \tau(\beta)$ then $\sigma(\alpha) < \sigma(\beta)$ or $\sigma(\alpha) = \sigma(\beta) = \bot$.
- For any $\alpha, \beta \in \mathcal{X}$ in a QBAF, if $\mathcal{R}^-(\alpha) = \mathcal{R}^-(\beta)$, $\mathcal{R}^+(\alpha) = \mathcal{R}^+(\beta)$ and $\tau(\alpha) < \tau(\beta)$ then $\sigma(\alpha) < \sigma(\beta)$ or $\sigma(\alpha) = \sigma(\beta) = \bot$.

With $\ll = <_\bot$, the strength of an argument is reduced when the base score is lowered unless it is already minimal. This may be required for situations where attacking arguments are able to reject their target arguments irrespective of their base score.

Comment. *This seems to be fairly important in the engineering discussion situation if an attacker states that its targeted argument is critically dangerous and therefore must take prevalence over their (unsupported) targeted argument's base score.*

Note also that the first of the three instantiations implies the literature property Proportionality.

GP9 for $\sigma\bot$ and $<_\top$:

- For any $\alpha, \beta \in \mathcal{X}$ in an aQBAF, if $\mathcal{R}^-(\alpha) = \mathcal{R}^-(\beta)$ and $\tau(\alpha) < \tau(\beta)$ then $\sigma(\alpha) < \sigma(\beta)$ or $\sigma(\alpha) = \sigma(\beta) = \top$.
- For any $\alpha, \beta \in \mathcal{X}$ in an sQBAF, if $\mathcal{R}^+(\alpha) = \mathcal{R}^+(\beta)$ and $\tau(\alpha) < \tau(\beta)$ then $\sigma(\alpha) < \sigma(\beta)$ or $\sigma(\alpha) = \sigma(\beta) = \top$.
- For any $\alpha, \beta \in \mathcal{X}$ in a QBAF, if $\mathcal{R}^-(\alpha) = \mathcal{R}^-(\beta)$, $\mathcal{R}^+(\alpha) = \mathcal{R}^+(\beta)$ and $\tau(\alpha) < \tau(\beta)$ then $\sigma(\alpha) < \sigma(\beta)$ or $\sigma(\alpha) = \sigma(\beta) = \top$.

Comment. *As with the previous property (though this one may seem more counter-intuitive), the exception here for a base score of $\top$ could be justified in the engineering discussion setting in which the stakeholder's opinions gives some supporters a special status, with base score and strength equal to $\top$, meaning their targeted (unattacked) arguments take on the $\top$ strength regardless of their base scores.*

GP9 for $\sigma\bot$ and $\leq$:

- For any $\alpha, \beta \in \mathcal{X}$ in an aQBAF, if $\mathcal{R}^-(\alpha) = \mathcal{R}^-(\beta)$ and $\tau(\alpha) < \tau(\beta)$ then $\sigma(\alpha) \leq \sigma(\beta)$.
- For any $\alpha, \beta \in \mathcal{X}$ in an sQBAF, if $\mathcal{R}^+(\alpha) = \mathcal{R}^+(\beta)$ and $\tau(\alpha) < \tau(\beta)$ then $\sigma(\alpha) \leq \sigma(\beta)$.
- For any $\alpha, \beta \in \mathcal{X}$ in a QBAF, $\mathcal{R}^-(\alpha) = \mathcal{R}^-(\beta)$, $\mathcal{R}^+(\alpha) = \mathcal{R}^+(\beta)$ and $\tau(\alpha) < \tau(\beta)$ then $\sigma(\alpha) \leq \sigma(\beta)$.

The instantiation of GP9 for $\ll=\leq$ relaxes the inequality and therefore reduces the base score's importance.

Comment. *This could be used in a meeting support setting similar to the engineering discussion but one in which the voting on arguments is less crucial for the exercise than the reasoning. One example of this is classroom debating, where children can be taught to think critically, i.e. add attacking and supporting arguments to the framework, and the voting is less important.*

GP9 for $\sigma\not\bot$ and $<$:

- For any $\alpha, \beta \in \mathcal{X}$ in an aQBAF, if $\mathcal{R}^-_{\sigma\not\bot}(\alpha) = \mathcal{R}^-_{\sigma\not\bot}(\beta)$ and $\tau(\alpha) < \tau(\beta)$ then $\sigma(\alpha) < \sigma(\beta)$.
- For any $\alpha, \beta \in \mathcal{X}$ in an sQBAF, if $\mathcal{R}^+_{\sigma\not\bot}(\alpha) = \mathcal{R}^+_{\sigma\not\bot}(\beta)$ and $\tau(\alpha) < \tau(\beta)$ then $\sigma(\alpha) < \sigma(\beta)$.
- For any $\alpha, \beta \in \mathcal{X}$ in a QBAF, $\mathcal{R}^-_{\sigma\not\bot}(\alpha) = \mathcal{R}^-_{\sigma\not\bot}(\beta)$, $\mathcal{R}^+_{\sigma\not\bot}(\alpha) = \mathcal{R}^+_{\sigma\not\bot}(\beta)$ and $\tau(\alpha) < \tau(\beta)$ then $\sigma(\alpha) < \sigma(\beta)$.

This first GP9 property for $* = \sigma\not\bot$ uses $\ll=<$ and therefore requires a strict reduction in strength if the base score is lowered but allows for rejected arguments to be ignored.

Comment. *This seems ideal for the Trip Advisor setting once again since it is desirable for users' votes, and therefore changes in arguments' base scores, to always have an effect on the arguments' strengths but malicious arguments can still be flagged, rejected and neglected.*

GP9 for $\sigma \perp\!\!\!\perp$ and $<_\perp$:

- For any $\alpha, \beta \in \mathcal{X}$ in an aQBAF, if $\mathcal{R}^-_{\sigma\perp\!\!\!\perp}(\alpha) = \mathcal{R}^-_{\sigma\perp\!\!\!\perp}(\beta)$ and $\tau(\alpha) < \tau(\beta)$ then $\sigma(\alpha) < \sigma(\beta)$ or $\sigma(\alpha) = \sigma(\beta) = \perp$.
- For any $\alpha, \beta \in \mathcal{X}$ in an sQBAF, if $\mathcal{R}^+_{\sigma\perp\!\!\!\perp}(\alpha) = \mathcal{R}^+_{\sigma\perp\!\!\!\perp}(\beta)$ and $\tau(\alpha) < \tau(\beta)$ then $\sigma(\alpha) < \sigma(\beta)$ or $\sigma(\alpha) = \sigma(\beta) = \perp$.
- For any $\alpha, \beta \in \mathcal{X}$ in a QBAF, if $\mathcal{R}^-_{\sigma\perp\!\!\!\perp}(\alpha) = \mathcal{R}^-_{\sigma\perp\!\!\!\perp}(\beta)$, $\mathcal{R}^+_{\sigma\perp\!\!\!\perp}(\alpha) = \mathcal{R}^+_{\sigma\perp\!\!\!\perp}(\beta)$ and $\tau(\alpha) < \tau(\beta)$ then $\sigma(\alpha) < \sigma(\beta)$ or $\sigma(\alpha) = \sigma(\beta) = \perp$.

When GP9 is instantiated for $* = \sigma \perp\!\!\!\perp$ and $\ll = <_\perp$ it requires a strict reduction in strength except when the strength is equal to $\perp$.

Comment. *In the setting of online debates on social media, this could correspond to a special status given to an attacker with which the whole group of users agrees, such that its base score and strength are $\top$ and therefore its (unsupported) target arguments are rejected irrespective of their base scores.*

GP9 for $\sigma \perp\!\!\!\perp$ and $<_\top$:

- For any $\alpha, \beta \in \mathcal{X}$ in an aQBAF, if $\mathcal{R}^-_{\sigma\perp\!\!\!\perp}(\alpha) = \mathcal{R}^-_{\sigma\perp\!\!\!\perp}(\beta)$ and $\tau(\alpha) < \tau(\beta)$ then $\sigma(\alpha) < \sigma(\beta)$ or $\sigma(\alpha) = \sigma(\beta) = \top$.
- For any $\alpha, \beta \in \mathcal{X}$ in an sQBAF, if $\mathcal{R}^+_{\sigma\perp\!\!\!\perp}(\alpha) = \mathcal{R}^+_{\sigma\perp\!\!\!\perp}(\beta)$ and $\tau(\alpha) < \tau(\beta)$ then $\sigma(\alpha) < \sigma(\beta)$ or $\sigma(\alpha) = \sigma(\beta) = \top$.
- For any $\alpha, \beta \in \mathcal{X}$ in a QBAF, if $\mathcal{R}^-_{\sigma\perp\!\!\!\perp}(\alpha) = \mathcal{R}^-_{\sigma\perp\!\!\!\perp}(\beta)$, $\mathcal{R}^+_{\sigma\perp\!\!\!\perp}(\alpha) = \mathcal{R}^+_{\sigma\perp\!\!\!\perp}(\beta)$ and $\tau(\alpha) < \tau(\beta)$ then $\sigma(\alpha) < \sigma(\beta)$ or $\sigma(\alpha) = \sigma(\beta) = \top$.

This (seemingly unintuitive) property requires a strict reduction in strength except when the strength is equal to $\top$.

Comment. *As with the previous property, this could be useful in the setting of online debates on social media. Here we may have the analogous case that a special status is given to a supporter with which the whole group of users agrees, such that its base score and strength are $\top$ and therefore its (unattacked) targeted arguments have $\top$ strength irrespective of their base scores.*

GP9 for $\sigma \perp\!\!\!\perp$ and $\leq$:

- For any $\alpha, \beta \in \mathcal{X}$ in an aQBAF, if $\mathcal{R}^-_{\sigma\perp\!\!\!\perp}(\alpha) = \mathcal{R}^-_{\sigma\perp\!\!\!\perp}(\beta)$ and $\tau(\alpha) < \tau(\beta)$ then $\sigma(\alpha) \leq \sigma(\beta)$.
- For any $\alpha, \beta \in \mathcal{X}$ in an sQBAF, if $\mathcal{R}^+_{\sigma\perp\!\!\!\perp}(\alpha) = \mathcal{R}^+_{\sigma\perp\!\!\!\perp}(\beta)$ and $\tau(\alpha) < \tau(\beta)$ then $\sigma(\alpha) \leq \sigma(\beta)$.
- For any $\alpha, \beta \in \mathcal{X}$ in a QBAF, if $\mathcal{R}^-_{\sigma\perp\!\!\!\perp}(\alpha) = \mathcal{R}^-_{\sigma\perp\!\!\!\perp}(\beta)$, $\mathcal{R}^+_{\sigma\perp\!\!\!\perp}(\alpha) = \mathcal{R}^+_{\sigma\perp\!\!\!\perp}(\beta)$ and $\tau(\alpha) < \tau(\beta)$ then $\sigma(\alpha) \leq \sigma(\beta)$.

Comment. *This property could be useful in the same setting as the previous two properties when both of the exceptions to the strict decrease in strength are required.*

7 Conclusions

This paper provides a contribution to the study of properties of (various types of) argumentation frameworks by instantiating, in several different ways, three *group properties* presented in [6]. As discussed in Sects. 4, 5 and 6, the parametric definition of the group properties turns out to support a systematic exploration of the spectrum of potential properties by allowing one to consider several actual variants of the same basic idea. In particular, for each of the three group properties considered in this paper only one instance had been previously considered in the literature while we have shown that a much larger set of alternatives (altogether 56 instances) can be considered. To illustrate the practical utility of this detailed investigation, examples of application contexts where each specific instance appears to be appropriate have been discussed. Altogether this line of investigation aims at building a comprehensive catalogue of fine-tuned properties where, in the phase of design of an argument-based system, one can identify those which are most appropriate for the considered application domain. In this line, a future work direction consists in extending this systematic analysis to the other group properties considered in [6].

References

1. Amgoud, L., Ben-Naim, J.: Axiomatic foundations of acceptability semantics. In: Principles of Knowledge Representation and Reasoning (KR): Proceedings of the Fifteenth International Conference, pp. 2–11 (2016). http://www.aaai.org/ocs/index.php/KR/KR16/paper/view/12855
2. Amgoud, L., Ben-Naim, J.: Evaluation of arguments from support relations: axioms and semantics. In: Proceedings of the Twenty-Fifth International Joint Conference on Artificial Intelligence (IJCAI), pp. 900–906 (2016)
3. Amgoud, L., Ben-Naim, J., Doder, D., Vesic, S.: Acceptability semantics for weighted argumentation frameworks. In: Proceedings of the Twenty-Sixth International Joint Conference on Artificial Intelligence (IJCAI), pp. 56–62 (2017). https://doi.org/10.24963/ijcai.2017/9
4. Amgoud, L., Cayrol, C., Lagasquie-Schiex, M., Livet, P.: On bipolarity in argumentation frameworks. Int. J. Intell. Syst. **23**(10), 1062–1093 (2008)
5. Baroni, P., Comini, G., Rago, A., Toni, F.: Abstract games of argumentation strategy and game-theoretical argument strength. In: An, B., Bazzan, A., Leite, J., Villata, S., van der Torre, L. (eds.) PRIMA 2017. LNCS (LNAI), vol. 10621, pp. 403–419. Springer, Cham (2017). https://doi.org/10.1007/978-3-319-69131-2_24
6. Baroni, P., Rago, A., Toni, F.: How many properties do we need for gradual argumentation? In: Proceedings of the Thirty-Second AAAI Conference on Artificial Intelligence, pp. 1736–1743 (2018)
7. Baroni, P., Romano, M., Toni, F., Aurisicchio, M., Bertanza, G.: Automatic evaluation of design alternatives with quantitative argumentation. Argum. Comput. **6**(1), 24–49 (2015). https://doi.org/10.1080/19462166.2014.1001791
8. Cocarascu, O., Toni, F.: Detecting deceptive reviews using argumentation. In: Proceedings of the 1st International Workshop on AI for Privacy and Security, PrAISe@ECAI 2016, The Hague, Netherlands, 29–30 August, pp. 9:1–9:8 (2016). http://doi.acm.org/10.1145/2970030.2970031

9. Dung, P.M.: On the acceptability of arguments and its fundamental role in non-monotonic reasoning, logic programming and n-person games. Artif. Intell. **77**(2), 321–358 (1995). https://doi.org/10.1016/0004-3702(94)00041-X

10. Gabbay, D.M.: Equational approach to argumentation networks. Argum. Comput. **3**(2–3), 87–142 (2012). https://doi.org/10.1080/19462166.2012.704398

11. Leite, J., Martins, J.: Social abstract argumentation. In: Proceedings of the 22nd International Joint Conference on Artificial Intelligence (IJCAI), pp. 2287–2292 (2011). http://ijcai.org/papers11/Papers/IJCAI11-381.pdf

12. Matt, P.-A., Toni, F.: A game-theoretic measure of argument strength for abstract argumentation. In: Hölldobler, S., Lutz, C., Wansing, H. (eds.) JELIA 2008. LNCS (LNAI), vol. 5293, pp. 285–297. Springer, Heidelberg (2008). https://doi.org/10.1007/978-3-540-87803-2_24

13. Rago, A., Toni, F.: Quantitative argumentation debates with votes for opinion polling. In: An, B., Bazzan, A., Leite, J., Villata, S., van der Torre, L. (eds.) PRIMA 2017. LNCS (LNAI), vol. 10621, pp. 369–385. Springer, Cham (2017). https://doi.org/10.1007/978-3-319-69131-2_22

14. Rago, A., Toni, F., Aurisicchio, M., Baroni, P.: Discontinuity-free decision support with quantitative argumentation debates. In: Principles of Knowledge Representation and Reasoning (KR): Proceedings of the Fifteenth International Conference, pp. 63–73 (2016). http://www.aaai.org/ocs/index.php/KR/KR16/paper/view/12874

Lower and Upper Probability Bounds for Some Conjunctions of Two Conditional Events

Giuseppe Sanfilippo[✉]

Department of Mathematics and Computer Science,
University of Palermo, Via Archirafi 34, 90123 Palermo, Italy
giuseppe.sanfilippo@unipa.it

Abstract. In this paper we consider, in the framework of coherence, four different definitions of conjunction among conditional events. In each of these definitions the conjunction is still a conditional event. We first recall the different definitions of conjunction; then, given a coherent probability assessment (x, y) on a family of two conditional events $\{A|H, B|K\}$, for each conjunction $(A|H) \wedge (B|K)$ we determine the (best) lower and upper bounds for the extension $z = P[(A|H) \wedge (B|K)]$. We show that, in general, these lower and upper bounds differ from the classical Fréchet-Hoeffding bounds. Moreover, we recall a notion of conjunction studied in recent papers, such that the result of conjunction of two conditional events $A|H$ and $B|K$ is (not a conditional event, but) a suitable conditional random quantity, with values in the interval $[0, 1]$. Then, we remark that for this conjunction, among other properties, the Fréchet-Hoeffding bounds are preserved.

Keywords: Coherence · Conditional event
Conditional random quantity
Kleene-Lukasiewicz-Heyting conjunction · Lukasiewicz conjunction
Bochvar internal conjunction · Sobocinski conjunction
Lower and upper bounds · Fréchet-Hoeffding bounds

1 Introduction

In probability theory and in probability logic a relevant problem, largely discussed by many authors (see, e.g., [2,8,9,26]), is that of suitably defining logical operations among conditional events. In this paper we consider four different notions of conjunction among conditional events such that in all cases the result of conjunction is a conditional event too: Kleene-Lukasiewicz-Heyting conjunction, Lukasiewicz conjunction, Bochvar internal conjunction, Sobocinski conjunction. For each conjunction $(A|H) \wedge (B|K)$, given the conditional probabilities

---

G. Sanfilippo was partially supported by the National Group for Mathematical Analysis, Probability and their Applications (GNAMPA – INdAM).

© Springer Nature Switzerland AG 2018
D. Ciucci et al. (Eds.): SUM 2018, LNAI 11142, pp. 260–275, 2018.
https://doi.org/10.1007/978-3-030-00461-3_18

$x = P(A|H)$ and $y = P(B|K)$, we determine the (best) lower and upper bounds for the conditional probability $z = P[(A|H) \wedge (B|K)]$, that is, given a coherent assessment (x, y) on the family $\{A|H, B|K\}$, we determine the lower and upper bounds z', z'' such that the extension (x, y, z) on $\{A|H, B|K, (A|H) \wedge (B|K)\}$, with $z = P[(A|H) \wedge (B|K)]$, is coherent if and only if $z \in [z', z'']$. Of course, $z', z'' \in [0, 1]$, but the extension $(x, y, 0)$ (resp., $(x, y, 1)$) is coherent if only if $z' = 0$ (resp., $z'' = 1$). We verify that in all cases such probability bounds do not coincide with the classical Fréchet-Hoeffding bounds: $z' = \max\{x + y - 1, 0\}$ and $z'' = \min\{x, y\}$. In particular, we obtain $z' = 0$ and $z'' = \min\{x, y\}$ for the Kleene-Lukasiewicz-Heyting conjunction and for the Lukasiewicz conjunction. We obtain $z' = 0$ and $z'' = 1$ for the Bochvar internal conjunction. Finally, for the Sobocinski conjunction we obtain $z' = \max\{x + y - 1, 0\}$ and $z'' = S_0^H(x, y)$, where $S_0^H(x, y)$ is the Hamacher t-conorm with parameter $\lambda = 0$. Then, we examine a notion of conjunction introduced in some recent papers, where the result of conjunction in general is not a conditional event, but a conditional random quantity. We remark that this notion of conjunction, differently from the previous of notions of conjunction, satisfy many properties. In particular the classical Fréchet-Hoeffding bounds are satisfied. We also recall a dual notion of disjunction by showing that within this approach the *prevision sum rule* is satisfied, that is $\mathbb{P}[(A|H) \vee (B|K)] = \mathbb{P}(A|H) + \mathbb{P}(B|K) - \mathbb{P}[(A|H) \wedge (B|K)]$.

2   Preliminary Notions and Results

An event A is a (non ambiguous) logical proposition which describes an uncertain fact; hence A is a two-valued logical entity which can be true (T), or false (F). The indicator of A, denoted by the same symbol, is a two-valued numerical quantity which is 1, or 0, according to whether the event A is true, or false, respectively. The sure event is denoted by Ω and the impossible event is denoted by $\emptyset$. Moreover, we denote by $A \wedge B$ the logical conjunction and by $A \vee B$ the logical disjunction. In many cases the conjunction between A and B is simple denoted as the product AB. The negation of an event A is denoted by $\overline{A}$. Given two events A and B, the inclusion relation between A and B, that is $A\overline{B} = \emptyset$, is denoted by $A \subseteq B$. We recall that n events are logically independent when the number of atoms, or constituents, generated by them is 2^n. In case of some logical dependencies among the events, the number of atoms is less than 2^n. Given two events A and B, with $A \neq \emptyset$, the conditional event $B|A$ is looked at as a three-valued logical entity which is true (T), or false (F), or void (V), according to whether AB is true, or $A\overline{B}$ is true, or $\overline{A}$ is true.

Coherent Conditional Probability Assessments. We recall that, using the betting scheme of de Finetti [12], if you assess $P(B|A) = p$, then you agree to pay an amount p, by receiving 1, or 0, or p, according to whether AB is true, or $A\overline{B}$ is true, or $\overline{A}$ is true (bet called off). Then, the random gain associated with the assessment $P(B|A) = p$ is $G = sH(E - p)$, where s is a non zero real number. More in general, let be given a real function $P : \mathcal{F} \to \mathbb{R}$, where $\mathcal{K}$ is an arbitrary family of conditional events. Given any subfamily $\mathcal{F} = \{E_1|H_1, \ldots, E_n|H_n\} \subseteq \mathcal{K}$, the restriction of P to $\mathcal{F}$ is the vector $\mathcal{P}_n = (p_1, \ldots, p_n)$, where $p_i = P(E_i|H_i)$, $i = 1, \ldots, n$. We denote by $\mathcal{H}_n$ the disjunction $H_1 \vee \cdots \vee H_n$. As $E_iH_i \vee \overline{E_i}H_i \vee \overline{H}_i = \Omega$, $i = 1, \ldots, n$, by expanding the expression $\bigwedge_{i=1}^{n}(E_iH_i \vee \overline{E_i}H_i \vee \overline{H}_i)$, we can represent Ω as the disjunction of 3^n logical conjunctions, some of which may be impossible. The remaining ones are the atoms, or constituents, generated by the family $\mathcal{F}$ and, of course, are a partition of Ω. We denote by $C_1, \ldots, C_m$ the constituents contained in $\mathcal{H}_n$ and (if $\mathcal{H}_n \neq \Omega$) by C_0 the remaining constituent $\overline{\mathcal{H}}_n = \overline{H}_1 \cdots \overline{H}_n$, so that

$$\mathcal{H}_n = C_1 \vee \cdots \vee C_m, \quad \Omega = \overline{\mathcal{H}}_n \vee \mathcal{H}_n = C_0 \vee C_1 \vee \cdots \vee C_m, \quad m+1 \leq 3^n.$$

In the betting metaphor, $G = \sum_{i=1}^{n} s_iH_i(E_i - p_i)$ is the random gain associated with $(\mathcal{F}, \mathcal{P})$, where $s_1, \ldots, s_n$ are n arbitrary real numbers, which is the difference between the amount that you receive, $\sum_{i=1}^{n} s_i(E_iH_i + p_i\overline{H}_i)$, and the amount that you pay, $\sum_{i=1}^{n} s_ip_i$. Let g_h be the value of G when C_h is true; of course $g_0 = 0$. Denoting by $\mathcal{G}_{\mathcal{H}_n}$ the set of possible values of G restricted to $\mathcal{H}_n$, it is $\mathcal{G}_{\mathcal{H}_n} = \{g_1, \ldots, g_m\}$. Then, we have

Definition 1. The function P defined on an arbitrary family of conditional events $\mathcal{K}$ is coherent if and only if, for every finite subfamily $\mathcal{F} = \{E_1|H_1, \ldots, E_n|H_n\}$ of $\mathcal{K}$, one has: $\min \mathcal{G}_{\mathcal{H}_n} \leq 0 \leq \max \mathcal{G}_{\mathcal{H}_n}$.

As shown by Definition 1, a probability assessment is coherent if and only if, in any finite combination of n bets, it may not happen that the values $g_1, \ldots, g_m$ are all positive, or all negative (*no Dutch Book*).

Given any integer n we set $J_n = \{1, 2, \ldots, n\}$; for each $h \in J_m$ with the constituent C_h we associate a point $Q_h = (q_{h1}, \ldots, q_{hn})$, where $q_{hj} = 1$, or 0, or p_j, according to whether $C_h \subseteq E_jH_j$, or $C_h \subseteq \overline{E}_jH_j$, or $C_h \subseteq \overline{H}_j$.

Denoting by $\mathcal{I}$ the convex hull of $Q_1, \ldots, Q_m$, by a suitable alternative theorem [13, Theorem 2.9], the condition $\mathcal{P} \in \mathcal{I}$ is equivalent to the condition $\min \mathcal{G}_{\mathcal{H}_n} \leq 0 \leq \max \mathcal{G}_{\mathcal{H}_n}$ given in Definition 1 (see, e.g., [17,21]). Moreover, the condition $\mathcal{P} \in \mathcal{I}$ amounts to the solvability of the following system (Σ) in the unknowns $\lambda_1, \ldots, \lambda_m$

$$(\Sigma): \qquad \sum_{h=1}^{m} q_{hj}\lambda_h = p_j, \quad j \in J_n; \quad \sum_{h=1}^{m} \lambda_h = 1; \quad \lambda_h \geq 0, \ h \in J_m.$$
$$(1)$$

We say that system (Σ) is associated with the pair $(\mathcal{F}, \mathcal{P})$. Hence, the following result provides a characterization of the notion of coherence given in Definition 1 ([14, Theorem 4.4], see also [15,20,21])

Theorem 1. *The function P defined on an arbitrary family of conditional events $\mathcal{K}$ is coherent if and only if, for every finite subfamily $\mathcal{F} = \{E_1|H_1, \ldots, E_n|H_n\}$ of $\mathcal{K}$, denoting by $\mathcal{P}$ the vector $(p_1, \ldots, p_n)$, where $p_j = P(E_j|H_j)$, $j = 1, 2, \ldots, n$, the system (Σ) associated with the pair $(\mathcal{F}, \mathcal{P})$ is solvable.*

Coherence Checking. We recall now some results on the coherence checking of a probability assessment on a finite family of conditional events. Given a probability assessment $\mathcal{P} = (p_1, \ldots, p_n)$ on $\mathcal{F} = \{E_1 | H_1, \ldots, E_n | H_n\}$, let S be the set of solutions $\Lambda = (\lambda_1, \ldots, \lambda_m)$ of the system (Σ). Then, assuming $S \neq \emptyset$, we define

$$\Phi_j(\Lambda) = \Phi_j(\lambda_1, \ldots, \lambda_m) = \sum_{r: C_r \subseteq H_j} \lambda_r , \quad j \in J_n ; \; \Lambda \in S;$$
$$M_j = \max_{\Lambda \in S} \Phi_j(\Lambda) , \quad j \in J_n ; \quad I_0 = \{j : M_j = 0\} .$$

If $S \neq \emptyset$, then S is a closed bounded set and the maximum M_j of the linear function $\Phi_j(\Lambda) = \sum_{r: C_r \subseteq H_j} \lambda_r$ there exists for every $j \in J_n$. Assuming $\mathcal{P}$ coherent, each solution $\Lambda = (\lambda_1, \ldots, \lambda_m)$ of system (Σ) is a coherent extension of the assessment $\mathcal{P}$ on $\mathcal{F}$ to the family $\{C_1 | \mathcal{H}_n, C_2 | \mathcal{H}_n, \ldots, C_m | \mathcal{H}_n\}$. Then, for each solution Λ of system (Σ) the quantity $\Phi_j(\Lambda)$ is the conditional probability $P(H_j | \mathcal{H}_n)$. Moreover, the quantity M_j is the upper probability $P''(H_j | \mathcal{H}_n)$ over all the solutions Λ of system (Σ). Of course, $j \in I_0$ if and only if $P''(H_j | \mathcal{H}_n) = 0$. Notice that I_0 is a strict subset of J_n. If I_0 is nonempty, we set $\mathcal{F}_0 = \{E_i | H_i \in \mathcal{F} : i \in I_0\}$ and $\mathcal{P}_0 = (P(E_i | H_i), i \in \mathcal{I}_0)$. We say that the pair $(\mathcal{F}_0, \mathcal{P}_0)$ is associated with I_0. Then, we have [16, Theorem 3.3].

Theorem 2. *The assessment $\mathcal{P}$ on $\mathcal{F}$ is coherent if and only if the following conditions are satisfied: (i) the system (Σ) associated with the pair $(\mathcal{F}, \mathcal{P})$ is solvable; (ii) if $I_0 \neq \emptyset$, then $\mathcal{P}_0$ is coherent.*

By Theorem 2, the following algorithm checks in a finite number of steps the coherence of the probability assessment $\mathcal{P}$ on a finite family of conditional events $\mathcal{F}$.

Algorithm 1. Let be given the pair $(\mathcal{F}, \mathcal{P})$.

1. Construct the system (Σ) associated with the pair $(\mathcal{F}, \mathcal{P})$ and check its solvability;
2. If the system (Σ) is not solvable then $\mathcal{P}$ is not coherent and the procedure stops, otherwise compute the set I_0;
3. If $I_0 = \emptyset$ then $\mathcal{P}$ is coherent and the procedure stops, otherwise set $(\mathcal{F}, \mathcal{P}) = (\mathcal{F}_0, \mathcal{P}_0)$ and go to step 1.

In the next definition we recall the notion of inclusion relation between two conditional events (see, e.g., [25]).

Definition 2. *Given two conditional events $A|H$ and $B|K$ we define that $A|H$ logically implies $B|K$, which we denote by $A|H \subseteq B|K$, if and only if AH is true implies BK is true and $\overline{BK}$ is true implies $\overline{AH}$ is true; i.e., $AH \subseteq BK$ and $\overline{BK} \subseteq \overline{AH}$.*

By coherence, it holds that (see, e.g., [19, Theorem 1]) $P(A|H) \leq P(B|K)$ when $A|H \subseteq B|K$. It can be also verified that [21, Remark 3]

$$A|H \subseteq B|K \iff AH\overline{B}K = \overline{H}\,\overline{B}K = AH\overline{K} = \emptyset . \tag{2}$$

We recall that the notion of implication given in [11] introduces a suitable conditional event associated with $A|H$ and $B|K$ (see also [27,28]). Between the notions of implication and inclusion relation there exists a relationship: as shown by (2), when the inclusion relation holds the implication between $A|H$ and $B|K$ is void or true. It is void if $\overline{H}\,\overline{K}$ is true; it is true in all the other cases.

3   Computation of Lower and Upper Bounds for Different Notions of the Conjunction

In this section we examine four different notions of conjunction between three valued events ([4], see also [5]), named in our approach conditional events: Kleene-Lukasiewicz-Heyting conjunction ($\wedge_K$), Lukasiewicz conjunction ($\wedge_L$), Bochvar internal conjunction, also known as Kleene weak conjunction ($\wedge_B$), and Sobocinski conjunction ($\wedge_S$). In all these definitions the conjunction of two conditional events is still a conditional event. We observe that, differently from other definitions of conjunctions, this four logical operations are all commutative and associative. The truth values of the four conjunctions are given in Table 1. We recall that a conditional event $A|H$, where A, H are two events with $H \neq \emptyset$, can be looked as a three valued random quantity $A|H = AH + x\overline{H} \in \{1, 0, x\}$, where $x = P(A|H)$. Let A, H, B, K be logical independent events, with $H \neq \emptyset, K \neq \emptyset$. Assuming that $P(A|H) = x$, $P(B|K) = y$ and $P[(A|H) \wedge (B|K)] = z$ the table of logical values for the different notions of conjunction are given in Table 2. We list below in an explicit way the four conjunctions as conditional events.

1. $(A|H) \wedge_K (B|K) = AHBK|(HK \vee \overline{A}H \vee \overline{B}K)$;
2. $(A|H) \wedge_L (B|K) = AHBK|(HK \vee \overline{A}\,\overline{B} \vee \overline{A}\,\overline{K} \vee \overline{B}\,\overline{H} \vee \overline{H}\,\overline{K})$;
3. $(A|H) \wedge_B (B|K) = AHBK|HK$;
4. $(A|H) \wedge_S (B|K) = (AH \vee \overline{H}) \wedge (BK \vee \overline{K})|(H \vee K)$.

Table 1. Truth values of the conjunctions. The values T, F, V denote *True*, *False*, and *Void*, respectively.

	C_h	$A|H$	$B|K$	$\wedge_K$	$\wedge_L$	$\wedge_B$	$\wedge_S$
C_1	$AHBK$	T	T	T	T	T	T
C_2	$AH\overline{B}K$	T	F	F	F	F	F
C_3	$AH\overline{K}$	T	V	V	V	V	T
C_4	$\overline{A}HBK$	F	T	F	F	F	F
C_5	$\overline{A}H\overline{B}K$	F	F	F	F	F	F
C_6	$\overline{A}H\overline{K}$	F	V	F	F	V	F
C_7	$\overline{H}BK$	V	T	V	V	V	T
C_8	$\overline{H}\,\overline{B}K$	V	F	F	F	V	F
C_0	$\overline{H}\,\overline{K}$	V	V	V	F	V	V

Table 2. Numerical values of the conjunctions. The values x, y, z denote $P(A|H)$, $P(B|K)$ and $P[(A|H) \wedge (B|K)]$, respectively.

	C_h	$A|H$	$B|K$	$\wedge_K$	$\wedge_L$	$\wedge_B$	$\wedge_S$
C_1	$AHBK$	1	1	1	1	1	1
C_2	$AH\overline{B}K$	1	0	0	0	0	0
C_3	$AH\overline{K}$	1	y	z	z	z	1
C_4	$\overline{A}HBK$	0	1	0	0	0	0
C_5	$\overline{A}H\overline{B}K$	0	0	0	0	0	0
C_6	$\overline{A}H\,\overline{K}$	0	y	0	0	z	0
C_7	$\overline{H}BK$	x	1	z	z	z	1
C_8	$\overline{H}\,\overline{B}K$	x	0	0	0	z	0
C_0	$\overline{H}\,\overline{K}$	x	y	z	0	z	z

3.1   The Kleene-Lukasiewicz-Heyting Conjunction

The Kleene-Lukasiewicz-Heyting conjunction is represented in Table 1 by the symbol $\wedge_K$. This notion coincides with the logical product between tri-events given in [11] (see also [27]). As shown in Table 2, based on the betting scheme the conjunction of two conditional events $A|H$ and $B|K$ in our approach coincides with the random quantity

$$(A|H) \wedge_K (B|K) = \begin{cases} 1, & \text{if } AHBK \text{ is true,} \\ 0, & \text{if } \overline{A}H \vee \overline{B}K \text{ is true,} \\ z, & \text{if } AH\overline{K} \vee \overline{H}BK \vee \overline{H}\,\overline{K} \text{ is true,} \end{cases} \quad (3)$$

where $z = P(A|H \wedge_K B|K)$. Then,

$$(A|H) \wedge_K (B|K) = 1 \cdot AHBK + z(\overline{H}BK \vee AH\overline{K} \vee \overline{H}\,\overline{K}). \quad (4)$$

Notice that the quantity $z = \mathbb{P}[(A|H) \wedge_K (B|K)]$ represents the value that you assess, with the proviso that, you will pay the amount sz by receiving the random quantity $s[(A|H) \wedge_K (B|K)]$. In particular, if $s = 1$, then you agree to pay z with the proviso that you will receive: 1, if both conditional events are true; 0, if at least one of the conditional events is false; z, otherwise. Based on (3) and (4), the Kleene-Lukasiewicz-Heyting conjunction is the following conditional event

$$(A|H) \wedge_K (B|K) = AHBK|(AHBK \vee \overline{A}H \vee \overline{B}K). \quad (5)$$

Remark 1. We observe that $(A|H) \wedge_K (B|K) \subseteq A|H$ and $(A|H) \wedge_K (B|K) \subseteq B|K$. Then, coherence requires that

$$P[(A|H) \wedge_K (B|K)] \leq \min\{P(A|H), P(B|K)\}.$$

Moreover, Table 1 shows that $(A|H) \wedge_K (B|K)$ and $A|H$ only differ when $AH\overline{B}K$ is true, or $\overline{H}\,\overline{B}K$ is true, or $AH\overline{K}$ is true. Then, from (2) and Table 1 we recover the definition of inclusion relation in the form given in [25], that is

$$A|H \subseteq B|K \iff (A|H) \wedge_K (B|K) = A|H.$$

Remark 2. We recall that given two conditional events $\{A|H, B|K\}$, with A, H, B, K logically independent, and with $H \neq \emptyset, K \neq \emptyset$, the set of all coherent assessments (x, y) on $\{A|H, B|K\}$ is the unit square $[0, 1]^2$.

In the next result we give the values of $z = P[(A|H) \wedge_K (B|K)]$ which are coherent extensions of a probability assessment (x, y) on $\{A|H, B|K\}$.

Theorem 3. Given any coherent assessment (x, y) on $\{A|H, B|K\}$, with A, H, B, K logically independent, and with $H \neq \emptyset, K \neq \emptyset$, the probability assessment $z = P[(A|H) \wedge_K (B|K)]$ is a coherent extension if and only if $z \in [z', z'']$, where $z' = 0$ and $z'' = \min\{x, y\}$.

Proof. We recall that, by Remark 2, any assessment $(x, y) \in [0, 1]^2$ is coherent. The constituents C_h's and the points Q_h's associated with $(\mathcal{F}, \mathcal{P})$, where $\mathcal{F} = \{A|H, B|K, (A|H) \wedge_K (B|K)\}$, $\mathcal{P} = (x, y, z)$, are given in Table 3. The constituents C_h's contained in $\mathcal{H}_3 = H \vee K$ are $C_1, \ldots, C_8$. We recall that coherence of the probability assessment $\mathcal{P} = (x, y, z)$ on $\mathcal{F}$ requires that the condition $\mathcal{P} \in \mathcal{I}$ be satisfied, where $\mathcal{I}$ is the convex hull of $Q_1, \ldots, Q_8$. This amounts to the solvability of the following system

$$(\Sigma) \qquad \mathcal{P} = \sum_{h=1}^{8} \lambda_h Q_h, \ \sum_{h=1}^{8} \lambda_h = 1, \ \lambda_h \geq 0, \ h = 1, \ldots, 8;$$

that is

$$\begin{cases} \lambda_1 + \lambda_2 + \lambda_3 + x\lambda_7 + x\lambda_8 = x, \ \lambda_1 + y\lambda_3 + \lambda_4 + y\lambda_6 + \lambda_7 = y \\ \lambda_1 + z\lambda_3 + z\lambda_7 = z, \ \sum_{h=1}^{8} \lambda_h = 1, \ \lambda_h \geq 0, \ h = 1, \ldots, 8. \end{cases}$$

We first prove that $z' = 0$, by verifying that the assessment $(x, y, 0)$ is coherent for any pair $(x, y) \in [0, 1]^2$. We distinguish the following cases: (i) $x < 1, y < 1$; (ii) $x = 1, y < 1$; (iii) $x < 1, y = 1$; (iv) $x = y = 1$.
Case (i). We observe that $(x, y, 0) \in \mathcal{I}$. Indeed, the system (Σ) is solvable, with a solution given by

$$\lambda_1 = 0, \lambda_2 = 0, \lambda_3 = \frac{x(1-y)}{1-xy}, \lambda_4 = 0, \lambda_5 = \frac{(1-x)(1-y)}{1-xy}, \lambda_6 = 0, \lambda_7 = \frac{y(1-x)}{1-xy}, \lambda_8 = 0,$$

that is

$$(x, y, 0) = \frac{x(1-y)}{1-xy}Q_3 + \frac{(1-x)(1-y)}{1-xy}Q_5 + \frac{y(1-x)}{1-xy}Q_7.$$

Moreover, this solution is such that $\lambda_5 = \frac{(1-x)(1-y)}{1-xy} > 0$, then

$$\sum_{h:C_h \subseteq H} \lambda_h = \lambda_1 + \lambda_2 + \lambda_3 + \lambda_4 + \lambda_5 + \lambda_6 = \frac{x(1-y)}{1-xy} + \frac{(1-x)(1-y)}{1-xy} = \frac{1-y}{1-xy} > 0,$$

$$\sum_{h:C_h \subseteq K} \lambda_h = \lambda_1 + \lambda_2 + \lambda_4 + \lambda_5 + \lambda_7 + \lambda_8 = \frac{(1-x)(1-y)}{1-xy} + \frac{y(1-x)}{1-xy} = \frac{1-x}{1-xy} > 0,$$

$$\sum_{h:C_h \subseteq AHBK\vee\overline{A}H\vee\overline{B}K} \lambda_h = \lambda_1 + \lambda_2 + \lambda_4 + \lambda_5 + \lambda_6 + \lambda_8 = \lambda_5 = \frac{(1-x)(1-y)}{1-xy} > 0.$$

Table 3. Constituents C_h's and points Q_h's associated with the prevision assessment $\mathcal{P} = (x, y, z)$ on $\mathcal{F} = \{A|H, B|K, (A|H) \wedge_K (B|K)\}$.

	C_h	Q_h	
C_1	$AHBK$	$(1,1,1)$	Q_1
C_2	$AH\overline{B}K$	$(1,0,0)$	Q_2
C_3	$AH\overline{K}$	$(1,y,z)$	Q_3
C_4	$\overline{A}HBK$	$(0,1,0)$	Q_4
C_5	$\overline{A}H\overline{B}K$	$(0,0,0)$	Q_5
C_6	$\overline{A}H\overline{K}$	$(0,y,0)$	Q_6
C_7	$\overline{H}BK$	$(x,1,z)$	Q_7
C_8	$\overline{H}\,\overline{B}K$	$(x,0,0)$	Q_8
C_0	$\overline{H}\,\overline{K}$	(x,y,z)	$Q_0 = \mathcal{P}$

Thus, $I_0 = \emptyset$ and from Theorem 2 it follows that $(x, y, 0)$ is coherent.

Case (ii). It holds that $\mathcal{P} = (1, y, 0) = \frac{1}{2}(1, y, 0) + \frac{y}{2}(1, 1, 0) + \frac{(1-y)}{2}(1, 0, 0) = \frac{1}{2}Q_3 + \frac{y}{2}Q_7 + \frac{1-y}{2}Q_8$, therefore the vector $(0, 0, \frac{1}{2}, 0, 0, 0, \frac{y}{2}, \frac{1-y}{2})$ is a solution of (Σ) such that $\lambda_8 = 1 - y > 0$, $\sum_{h:C_h \subseteq H} \lambda_h = \lambda_1 + \lambda_2 + \lambda_3 + \lambda_4 + \lambda_5 + \lambda_6 = \frac{1}{2} > 0$, $\sum_{h:C_h \subseteq K} \lambda_h = \lambda_1 + \lambda_2 + \lambda_4 + \lambda_5 + \lambda_7 + \lambda_8 = \frac{1}{2} > 0$, and $\sum_{h:C_h \subseteq AHBK \vee \overline{A}H \vee \overline{B}K} \lambda_h = \lambda_1 + \lambda_2 + \lambda_4 + \lambda_5 + \lambda_6 + \lambda_8 = \frac{1-y}{2} > 0$. Then $I_0 = \emptyset$ and from Theorem 2 it follows that $(x, y, 0)$ is coherent.

Case (iii). The analysis is similar to the case (ii).

Case (iv). The system (Σ) becomes

$$\lambda_1 = 0, \ \lambda_2 + \lambda_3 + \lambda_7 + \lambda_8 = 1, \ \lambda_3 + \lambda_4 + \lambda_6 + \lambda_7 = 1, \sum_{h=1}^{8} \lambda_h = 1, \ \lambda_h \geq 0, \ \forall h,$$

or equivalently: $\lambda_3 + \lambda_7 = 1$, $\lambda_3 \geq 0$, $\lambda_7 \geq 0, \lambda_1 = \lambda_2 = \lambda_4 = \lambda_5 = \lambda_6 = \lambda_8 = 0$. Then, the set S of solutions of (Σ) is $S = \{(0, 0, \lambda, 0, 0, 0, 1 - \lambda, 0) : 0 \leq \lambda \leq 1\}$. We observe that $\sum_{h:C_h \subseteq H} \lambda_h = \lambda_1 + \lambda_2 + \lambda_3 + \lambda_4 + \lambda_5 + \lambda_6 = \lambda$, $\sum_{h:C_h \subseteq K} \lambda_h = \lambda_1 + \lambda_2 + \lambda_4 + \lambda_5 + \lambda_7 + \lambda_8 = 1 - \lambda$, and $\sum_{h:C_h \subseteq AHBK \vee \overline{A}H \vee \overline{B}K} \lambda_h = \lambda_1 + \lambda_2 + \lambda_4 + \lambda_5 + \lambda_6 + \lambda_8 = 0$. For $0 < \lambda < 1$ it holds that $\sum_{h:C_h \subseteq H} \lambda_h > 0$ and $\sum_{h:C_h \subseteq K} \lambda_k > 0$. Then, $\mathcal{I}_0 = \{3\}$ and by Algorithm 1, from coherence of the assessment $P(AHBK|(AHBK \vee \overline{A}H \vee \overline{B}K)) = z = 0$, the assessment $(1, 1, 0)$ is coherent too.

Thus, $(x, y, 0)$ is coherent for every $(x, y) \in [0, 1]$ and hence $z' = 0$.

Concerning the upper bound, by Remark 1 coherence requires that $z \leq \min\{x, y\}$. We prove that $z'' = \min\{x, y\}$, by verifying that the assessment $(x, y, \min\{x, y\})$ is coherent for every $(x, y) \in [0, 1]^2$. We first show that the point $(x, y, \min\{x, y\})$ is always a linear convex combinations of a subset of the set $\{Q_1, Q_2, Q_4, Q_5\}$. By assuming $\min\{x, y\} = x$, we have

$$(x, y, x) = xQ_1 + (y - x)Q_4 + (1 - y)Q_5 = x(1, 1, 1) + (y - x)(0, 1, 0) + (1 - y)(0, 0, 0).$$

Then $(x, y, x) \in \mathcal{I}$ and the system (Σ) is solvable, with $\lambda_1 = x, \lambda_4 = y - x, \lambda_5 = 1 - y, \lambda_2 = \lambda_3 = \lambda_6 = \lambda_7 = \lambda_8 = 0$. Moreover, $\sum_{h:C_h \subseteq H} \lambda_h = \lambda_1 + \lambda_2 + \lambda_3 +$

$\lambda_4 + \lambda_5 + \lambda_6 = 1 > 0$, $\sum_{h:C_h \subseteq K} \lambda_h = \lambda_1 + \lambda_2 + \lambda_4 + \lambda_5 + \lambda_7 + \lambda_8 = 1 > 0$, and $\sum_{h:C_h \subseteq AHBK \vee \overline{AH} \vee \overline{BK}} \lambda_h = \lambda_1 + \lambda_2 + \lambda_4 + \lambda_5 + \lambda_6 + \lambda_8 = 1 > 0$. Therefore $I_0 = \emptyset$ and $(x, y, \min\{x, y\})$ is coherent for every $(x, y) \in [0, 1]^2$ such that $x \leq y$. By a similar reasoning $(x, y, \min\{x, y\})$ is coherent for every $(x, y) \in [0, 1]^2$ such that $y < x$.

Thus $(x, y, \min\{x, y\})$ is coherent for every $(x, y) \in [0, 1]^2$ and hence $z'' = \min\{x, y\}$. □

We recall that in [26, p. 161] lower and upper bounds for $(A|H) \wedge_K (B|K)$ have been obtained based on different premises.

3.2   The Lukasiewicz Conjunction

The Lukasiewicz conjunction is represented in Table 1 by the symbol $\wedge_L$. As shown in Table 2, based on the betting scheme the conjunction of two conditional events $A|H$ and $B|K$ in our approach coincides with the random quantity

$$(A|H) \wedge_L (B|K) = \begin{cases} 1, & \text{if } AHBK \text{ is true,} \\ 0, & \text{if } \overline{AH} \vee \overline{BK} \vee \overline{H}\,\overline{K} \text{ is true,} \\ z, & \text{if } AH\overline{K} \vee \overline{H}BK \text{ is true,} \end{cases} \qquad (6)$$

where $z = P(A|H \wedge_L B|K)$. Then,

$$(A|H) \wedge_L (B|K) = 1 \cdot AHBK + z(AH\overline{K} \vee \overline{H}BK). \qquad (7)$$

Based on (6) and (7), by observing that

$$\overline{AH\overline{K} \vee \overline{H}BK} =$$
$$= AHBK \vee AH\overline{B}K \vee \overline{H}\,\overline{K} \vee \overline{H}\,\overline{B}K \vee \overline{A}HBK \vee \overline{A}H\overline{K} \vee \overline{A}H\overline{B}K =$$
$$= HK \vee \overline{A}\,\overline{B} \vee \overline{A}\,\overline{K} \vee \overline{B}\,\overline{H} \vee \overline{H}\,\overline{K},$$

the Lukasiewicz conjunction is the following conditional event

$$(A|H) \wedge_L (B|K) = AHBK | (HK \vee \overline{A}\,\overline{B} \vee \overline{A}\,\overline{K} \vee \overline{B}\,\overline{H} \vee \overline{H}\,\overline{K}). \qquad (8)$$

As we can see from (3) and (6) (see also Table 1), the conjunctions $(A|H) \wedge_K (B|K)$ and $(A|H) \wedge_L (B|K)$ only differ when $\overline{H}\,\overline{K}$ is true; indeed in this case $(A|H) \wedge_K (B|K) = z$, where $z = P[(A|H) \wedge_K (B|K)]$, and $(A|H) \wedge_K (B|K) = 0$. Then, by observing that $z \geq 0$, it holds that

$$(A|H) \wedge_K (B|K) \geq (A|H) \wedge_L (B|K) \qquad (9)$$

and hence

$$P[(A|H) \wedge_K (B|K)] \geq P[(A|H) \wedge_L (B|K)]. \qquad (10)$$

On the other hand, as it can be verified, it holds that $(A|H) \wedge_L (B|K) \subseteq (A|H) \wedge_K (B|K)$, from which it follows (10). Concerning the lower and upper bounds on $(A|H) \wedge_L (B|K)$, we have

Theorem 4. Given any coherent assessment (x, y) on $\{A|H, B|K\}$, with A, H, B, K logically independent, and with $H \neq \emptyset, K \neq \emptyset$, the probability assessment $z = P[(A|H) \wedge_L (B|K)]$ is a coherent extension if and only if $z \in [z', z'']$, where $z' = 0$ and $z'' = \min\{x, y\}$.

Proof. By Theorem 3 the lower bound on $P[(A|H) \wedge_K (B|K)]$ is 0; then, from (10) the lower bound z' on $P[(A|H) \wedge_L (B|K)]$ is still 0. Moreover, from (10) it also follows that $z'' \leq \min\{x, y\}$. We will prove that $z'' = \min\{x, y\}$. The constituents C_h's and the points Q_h's associated with $(\mathcal{F}, \mathcal{P})$, where $\mathcal{F} = \{A|H, B|K, (A|H) \wedge_L (B|K)\}$, $\mathcal{P} = (x, y, z)$, are given in Table 4. We observe that $\mathcal{H}_3 = H \vee K \vee HK \vee \overline{A}\,\overline{B} \vee \overline{A}\,\overline{K} \vee \overline{B}\,\overline{H} \vee \overline{H}\,\overline{K} = \Omega$. Then, the constituents C_h's contained in $\mathcal{H}_3 = \Omega$ are $C_1, \ldots, C_9$, that is all constituents $C_1, \ldots, C_8$ associated with the conjunction $\wedge_K$ plus the constituent $C_9 = \overline{H}\,\overline{K}$. Then, as shown in Table 4, with respect to $\wedge_K$ the set of points Q_h's contains the further point $Q_9 = (x, y, 0)$. In order to prove that the assessment $(x, y, \min\{x, y\})$ is coherent, it is enough to repeat the same reasoning used in the proof of Theorem 3 by only considering the points $Q_1, \ldots, Q_8$ (this amounts to set $\lambda_9 = 0$ in the current system (Σ)). $\qquad\square$

Table 4. Constituents C_h's and points Q_h's associated with the prevision assessment $\mathcal{P} = (x, y, z)$ on $\mathcal{F} = \{A|H, B|K, (A|H) \wedge_L (B|K)\}$.

	C_h	Q_h	
C_1	$AHBK$	$(1, 1, 1)$	Q_1
C_2	$AH\overline{B}K$	$(1, 0, 0)$	Q_2
C_3	$AH\overline{K}$	$(1, y, z)$	Q_3
C_4	$\overline{A}HBK$	$(0, 1, 0)$	Q_4
C_5	$\overline{A}H\overline{B}K$	$(0, 0, 0)$	Q_5
C_6	$\overline{A}H\overline{K}$	$(0, y, 0)$	Q_6
C_7	$\overline{H}BK$	$(x, 1, z)$	Q_7
C_8	$\overline{H}\,\overline{B}K$	$(x, 0, 0)$	Q_8
C_9	$\overline{H}\,\overline{K}$	$(x, y, 0)$	Q_9

3.3   The Bochvar Internal Conjunction

The Bochvar internal conjunction [3] is represented in Table 1 by the symbol $\wedge_B$. As shown in Table 2, based on the betting scheme the conjunction of two conditional events $A|H$ and $B|K$ in our approach coincides with the random quantity

$$(A|H) \wedge_B (B|K) = \begin{cases} 1, & \text{if } AHBK \text{ is true,} \\ 0, & \text{if } \overline{A}HBK \vee AH\overline{B}K \vee \overline{A}H\overline{B}K \text{ is true,} \\ z, & \text{if } AH\overline{K} \vee \overline{H}BK \vee \overline{H}\,\overline{K} \vee \overline{A}H\overline{K} \vee \overline{H}\,\overline{B}K \text{ is true,} \end{cases}$$

(11)

where $z = P(A|H \wedge_B B|K)$. We observe that $AH\overline{K} \vee \overline{H}BK \vee \overline{H}\,\overline{K} \vee \overline{A}H\overline{K} \vee \overline{H}\,\overline{B}K = \overline{H} \vee \overline{K}$, then

$$(A|H) \wedge_B (B|K) = \begin{cases} 1, & \text{if } AHBK \text{ is true,} \\ 0, & \text{if } \overline{A}HBK \vee AH\overline{B}K \vee \overline{A}H\overline{B}K \text{ is true,} \\ z, & \text{if } \overline{H} \vee \overline{K} \text{ is true.} \end{cases}$$

Thus,

$$(A|H) \wedge_B (B|K) = 1 \cdot AHBK + z(\overline{H} \vee \overline{K}). \tag{12}$$

Based on (11) and (12), the Bochvar internal conjunction is the following conditional event

$$(A|H) \wedge_B (B|K) = AHBK|HK = AB|HK. \tag{13}$$

We have

Theorem 5. Given any coherent assessment (x, y) on $\{A|H, B|K\}$, with A, H, B, K logically independent, and with $H \neq \emptyset, K \neq \emptyset$, the probability assessment $z = P[(A|H) \wedge_B (B|K)]$ is a coherent extension if and only if $z \in [z', z'']$, where $z' = 0$ and $z'' = 1$.

Proof. We recall that every $(x, y) \in [0,1]^2$ is a coherent assessment on $\{A|H, B|K\}$. We will prove that every assessment $(x, y, z) \in [0,1]^3$ on $\{A|H, B|K, (A|H) \wedge_B (B|K)\}$ is coherent. The constituents C_h's and the points Q_h's associated with $(\mathcal{F}, \mathcal{P})$, where $\mathcal{F} = \{A|H, B|K, (A|H) \wedge_B (B|K)\}$, $\mathcal{P} = (x, y, z)$, are given in Table 5. We observe that $\mathcal{P}$ belongs to the segment with vertices Q_3, Q_6; indeed $(x, y, z) = x(1, y, z) + (1 - x)(0, y, z)$. $\mathcal{P}$ also belongs to the segment with vertices Q_7, Q_8; indeed $(x, y, z) = y(x, 1, z) + (1 - y)(x, 0, z)$. Then,

$$(x, y, z) = \frac{x}{2}(1, y, z) + \frac{1-x}{2}(0, y, z) + \frac{y}{2}(x, 1, z) + \frac{1-y}{2}(x, 0, z),$$

that is the vector $(\lambda_1, \ldots, \lambda_8) = (0, 0, \frac{x}{2}, 0, 0, \frac{1-x}{2}, \frac{y}{2}, \frac{1-y}{2})$ is a solution of system (Σ), with $\sum_{h:C_h \subseteq H} \lambda_h = \lambda_1 + \lambda_2 + \lambda_3 + \lambda_4 + \lambda_5 + \lambda_6 = \frac{1}{2} > 0$, $\sum_{h:C_h \subseteq K} \lambda_h = \lambda_1 + \lambda_2 + \lambda_4 + \lambda_5 + \lambda_7 + \lambda_8 = \frac{1}{2} > 0$, and $\sum_{h:C_h \subseteq HK} \lambda_h = \lambda_1 + \lambda_2 + \lambda_4 + \lambda_5 = 0$. Therefore $I_0 \subseteq \{3\}$. If $I_0 = \emptyset$, then from Theorem 2 it follows that $(x, y, z) \in [0,1]^3$ is coherent. If $I_0 = \{3\}$ from coherence of the assessment $P[(A|H) \wedge_B (B|K)] = z \in [0,1]$ and by Algorithm 1 it follows the coherence of the assessment $(x, y, z) \in [0,1]^3$. Thus, for every given $(x, y) \in [0,1]^2$ the lower and upper bounds on the extension $z = P[(A|H) \wedge_B (B|K)]$ are $z' = 0$ and $z'' = 1$, respectively. □

Remark 3. We observe that $(A|HK) \wedge (B|HK) = AB|HK$, so that $P(A|HK)$, $P(B|HK)$, and $P(AB|HK)$ satisfy the Fréchet-Hoeffding bounds, that is

$$\max\{P(A|HK)+P(B|HK)-1, 0\} \leq P(AB|HK) \leq \min\{P(A|HK), P(B|HK)\}.$$

Table 5. Constituents C_h's and points Q_h's associated with the prevision assessment $\mathcal{P} = (x, y, z)$ on $\mathcal{F} = \{A|H, B|K, (A|H) \wedge_B (B|K)\}$.

	C_h	Q_h	
C_1	$AHBK$	$(1,1,1)$	Q_1
C_2	$AH\overline{B}K$	$(1,0,0)$	Q_2
C_3	$AH\overline{K}$	$(1,y,z)$	Q_3
C_4	$\overline{A}HBK$	$(0,1,0)$	Q_4
C_5	$\overline{A}H\overline{B}K$	$(0,0,0)$	Q_5
C_6	$\overline{A}H\overline{K}$	$(0,y,z)$	Q_6
C_7	$\overline{H}BK$	$(x,1,z)$	Q_7
C_8	$\overline{H}\,\overline{B}K$	$(x,0,z)$	Q_8
C_0	$\overline{H}\,\overline{K}$	(x,y,z)	$Q_0 = \mathcal{P}$

Then, by assuming $P(A|H) = P(A|HK) = x$ and $P(B|K) = P(B|HK) = y$, that is by requiring suitable conditional independence hypotheses, it holds that $z = P(AB|HK)$ is coherent if and only if

$$\max\{P(A|H) + P(B|K) - 1, 0\} \leq P(AB|HK) \leq \min\{P(A|H), P(B|K)\}.$$

A discussion on this aspect related with conditional independence is given in [7], where a general definition of intersection and union between two fuzzy subsets is introduced in the framework of conditional probabilities (see also [6]).

3.4   The Sobocinski Conjunction or Quasi Conjunction

The Sobocinski conjunction or quasi conjunction [1] is represented in Table 1 by the symbol $\wedge_S$. We recall that the link between conditional events and Sobocinski conjunction was studied in [10]. As shown in Table 2, the conjunction of $A|H$ and $B|K$ is the following conditional event

$$(A|H) \wedge_S (B|K) = (AH \vee \overline{H}) \wedge (BK \vee \overline{K})|(H \vee K).$$

Based on the betting scheme the conjunction of two conditional events $A|H$ and $B|K$ in our approach coincides with the random quantity

$$(A|H) \wedge_S (B|K) = \begin{cases} 1, & \text{if } AHBK \vee AH\overline{K} \vee BK\overline{H} \text{ is true,} \\ 0, & \text{if } \overline{A}H \vee \overline{B}K \text{ is true,} \\ z, & \text{if } \overline{H}\,\overline{K} \text{ is true,} \end{cases} \tag{14}$$

From (14), the conjunction $(A|H) \wedge_S (B|K)$ is the following random quantity

$$(A|H) \wedge_S (B|K) = 1 \cdot AHBK + AH\overline{K} + \overline{H}BK + z\overline{H}\,\overline{K}. \tag{15}$$

We recall the following result (see, e.g., [18,21])

Theorem 6. *Given any coherent assessment (x, y) on $\{A|H, B|K\}$, with A, H, B, K logically independent, and with $H \neq \emptyset, K \neq \emptyset$, the probability assessment $z = P[(A|H) \wedge_S (B|K)]$ is a coherent extension if and only if $z \in [z', z'']$, where $z' = \max\{x + y - 1, 0\}$ and $z'' = \begin{cases} \frac{x+y-2xy}{1-xy}, & (x,y) \neq (1,1), \\ 1, & (x,y) = (1,1). \end{cases}$*

We recall that $z'' = S_0^H(x, y)$, where $S_0^H(x, y)$ is the Hamacher t-conorm with parameter $\lambda = 0$. It can be easily verified that

$$S_0^H(x, y) \geq \min\{x, y\}, \quad \forall (x, y) \in [0, 1]^2. \tag{16}$$

4   Conjunction as a Conditional Random Quantity

We recall that the extension $z = P(AB)$ of the assessment (x, y) on $\{A, B\}$, with A, B logically independent, is coherent if and only if z satisfies the Fréchet-Hoeffding bounds, that is $\max\{x + y - 1, 0\} \leq z \leq \min\{x, y\}$. As we have seen in the previous sections, no one of the given definitions of conjunction between two conditional events preserves both of these lower and upper bounds. A definition of conjunction which satisfies the Fréchet-Hoeffding bounds has been given in recent papers (see, e.g., [22,23]). Based on this definition the conjunction of two conditional events $A|H$ and $B|K$, with $P(A|H) = x$ and $P(B|K) = y$, is the following conditional random quantity

$$(A|H) \wedge (B|K) = \min\{A|H, B|K\}|(H \vee K) = \begin{cases} 1, & \text{if } AHBK \text{ is true,} \\ 0, & \text{if } \overline{A}H \vee \overline{B}K \text{ is true,} \\ x, & \text{if } \overline{H}BK \text{ is true,} \\ y, & \text{if } AH\overline{K} \text{ is true,} \\ z, & \text{if } \overline{H}\,\overline{K} \text{ is true,} \end{cases} \tag{17}$$

where z is the prevision $\mathbb{P}[(A|H) \wedge (B|K)]$ of $(A|H) \wedge (B|K)$. This notion of conjunction satisfies the Fréchet-Hoeffding bounds [22, Theorem 7], that is: given any coherent assessment (x, y) on $\{A|H, B|K\}$, with A, H, B, K logically independent, $H \neq \emptyset, K \neq \emptyset$, the extension $z = \mathbb{P}[(A|H) \wedge (B|K)]$ is coherent if and only if $\max\{x + y - 1, 0\} = z' \leq z \leq z'' = \min\{x, y\}$. In case of some logical dependencies among the events A, H, B, K the interval $[z', z'']$ of coherent extensions may be smaller and/or the conjunction may reduce to a conditional event.

For instance, the conjunction $(A|B) \wedge (B|A)$ reduces to the conditional event $AB|(A \vee B)$ [29, Theorem 7]. The disjunction of $A|H$ and $B|K$ is defined as the following conditional random quantity

$$(A|H) \vee (B|K) = \max\{A|H, B|K\}|(H \vee K) = \begin{cases} 1, & \text{if } AH \vee BK \text{ is true,} \\ 0, & \text{if } \overline{A}H\overline{B}K \text{ is true,} \\ x, & \text{if } \overline{H}\,\overline{B}K \text{ is true,} \\ y, & \text{if } \overline{A}H\overline{K} \text{ is true,} \\ w, & \text{if } \overline{H}\,\overline{K} \text{ is true,} \end{cases} \tag{18}$$

where $w = \mathbb{P}[(A|H) \vee (B|K)]$. Given any coherent assessment (x, y) on $\{A|H, B|K\}$, with A, H, B, K logically independent, $H \neq \emptyset, K \neq \emptyset$, the extension $w = \mathbb{P}[(A|H) \vee (B|K)]$ is coherent if and only if $\max\{x, y\} = w'' \leq w \leq w'' = \min\{x + y, 1\}$; that is the classical lower and upper bound for disjunction still hold. Other properties which are satisfied by these notions are the following ones:

- $(A|H) \wedge (B|K) = (B|K) \wedge (A|H)$ and $(A|H) \vee (B|K) = (B|K) \vee (A|H)$;

- $(A|H) \wedge (B|K) \leq A|H$ and $(A|H) \wedge (B|K) \leq B|K$;

- $(A|H) \wedge (A|H) = (A|H) \vee (A|H) = A|H$;

- $A|H \subseteq B|K \iff (A|H) \wedge (B|K) = A|H$ and $(A|H) \vee (B|K) = B|K$;

- $P(A|H) = P(B|K) = 1 \iff \mathbb{P}[(A|H) \wedge (B|K)] = 1$;

- $(A|H) \wedge (B|K) \leq (A|H) \wedge_S (B|K)$, with $(A|H) \wedge (B|K) = (A|H) \wedge_S (B|K)$ when $x = y = 1$.

Definitions (17) and (18) have been generalized to the case of n conditional events in [23] (see also [24]), where it has been shown the validity of the associative properties. Moreover, based on a suitable definition of the negation, it has been shown that De Morgan's Laws hold and that $(A|H) \vee (B|K) = (A|H) + (B|K) - (A|H) \wedge (B|K)$, from which it follows the prevision sum rule: $\mathbb{P}[(A|H) \vee (B|K)] = \mathbb{P}(A|H) + \mathbb{P}(B|K) - \mathbb{P}[(A|H) \wedge (B|K)]$.

Remark 4. Notice that, from Remark 1, Definitions 3 and 17, it follows that $(A|H) \wedge_K (B|K) \leq (A|H) \wedge (B|K)$. Then, it is not surprising that, given (x, y), the lower bound on $(A|H) \wedge_K (B|K)$ (equal to 0) is less than or equal to the lower bound on $(A|H) \wedge (B|K)$, which is $\max\{x + y - 1, 0\}$. A similar comment holds for the Lukasiewicz conjunction, because from (9) it follows that $(A|H) \wedge_L (B|K) \leq (A|H) \wedge (B|K)$. Moreover, as shown in Table 2, in general the inequality $(A|H) \wedge_B (B|K) \leq \min\{A|H, B|K\}$ is not satisfied. Then, it is not surprising that the upper bound on $P[(A|H) \wedge_B (B|K)]$ is greater than or equal to $\min\{x, y\}$; indeed it is equal to 1. Concerning the lower bound, the inequality $(A|H) \wedge_B (B|K) \geq (A|H) \wedge (B|K)$ in general is not satisfied. Then, it is not surprising that the lower bound on $P[(A|H) \wedge_B (B|K)]$ is less than or equal to $\max\{x + y - 1, 0\}$; indeed it is equal to 0. Finally, as $(A|H) \wedge (B|K) \leq (A|H) \wedge_S (B|K)$, it is not surprising that the upper bound on $P[(A|H) \wedge_S (B|K)]$ is greater than or equal to $\min\{x, y\}$; indeed it is $S_0^H(x, y)$ (see formula 16).

5 Conclusions

In this paper we examined four different notions of conjunction among conditional events given in literature such that the result of conjunction is still a conditional event. For each conjunction $(A|H) \wedge (B|K)$, given the conditional probabilities $x = P(A|H)$ and $y = P(B|K)$, we have determined the lower and upper bounds for the conditional probability $z = P[(A|H) \wedge (B|K)]$. We have verified that in all cases such probability bounds do not coincide with the classical Fréchet-Hoeffding bounds. Moreover, we examined a notion of conjunction introduced in some recent papers, where the result of conjunction is (not a conditional event, but) a conditional random quantity. With this notion of conjunction, among other properties, the Fréchet-Hoeffding bounds are satisfied. Further work should concerns the generalization of the results concerning lower and upper bounds for the different definitions of conjunction of n conditional events. A similar study should be made for the notions of disjunction.

Acknowledgments. We thank Angelo Gilio and the three anonymous reviewers for their useful comments and suggestions.

References

1. Adams, E.W.: The Logic of Conditionals. Reidel, Dordrecht (1975)
2. Benferhat, S., Dubois, D., Prade, H.: Nonmonotonic reasoning, conditional objects and possibility theory. Artif. Intell. **92**, 259–276 (1997)
3. Bochvar, D., Bergmann, M.: On a three-valued logical calculus and its application to the analysis of the paradoxes of the classical extended functional calculus. Hist. Philos. Log. **2**(1–2), 87–112 (1981)
4. Ciucci, D., Dubois, D.: Relationships between connectives in three-valued logics. In: Greco, S., Bouchon-Meunier, B., Coletti, G., Fedrizzi, M., Matarazzo, B., Yager, R.R. (eds.) IPMU 2012. CCIS, vol. 297, pp. 633–642. Springer, Heidelberg (2012). https://doi.org/10.1007/978-3-642-31709-5_64
5. Ciucci, D., Dubois, D.: A map of dependencies among three-valued logics. Inf. Sci. **250**, 162–177 (2013)
6. Coletti, G., Gervasi, O., Tasso, S., Vantaggi, B.: Generalized bayesian inference in a fuzzy context: from theory to a virtual reality application. Comput. Stat. Data Anal. **56**(4), 967–980 (2012)
7. Coletti, G., Scozzafava, R.: Conditional probability, fuzzy sets, and possibility: A unifying view. Fuzzy Sets Syst. **144**, 227–249 (2004)
8. Coletti, G., Scozzafava, R., Vantaggi, B.: Coherent conditional probability, fuzzy inclusion and default rules. In: Yager, R., Abbasov, A., Reformat, M., Shahbazova, S. (eds.) Soft Computing: State of the Art Theory and Novel Applications. Studies in Fuzziness and Soft Computing, vol. 291, pp. 193–208. Springer, Heidelberg (2013). https://doi.org/10.1007/978-3-642-34922-5_14
9. Coletti, G., Scozzafava, R., Vantaggi, B.: Possibilistic and probabilistic logic under coherence: Default reasoning and System P. Math. Slovaca **65**(4), 863–890 (2015)
10. Dubois, D., Prade, H.: Conditional objects as nonmonotonic consequence relationships. IEEE Trans. Syst. Man Cybern. **24**(12), 1724–1740 (1994)
11. de Finetti, B.: La Logique de la Probabilité. In: Actes du Congrès International de Philosophie Scientifique, Paris, 1935, pp. IV 1–IV 9. Hermann et C.ie, Paris (1936)
12. de Finetti, B.: Teoria delle probabilità. In: Einaudi (ed.) 2 vol., Torino (1970). English version: Theory of Probability 1 (2). Chichester, Wiley, 1974 (1975)
13. Gale, D.: The Theory of Linear Economic Models. McGraw-Hill, New York (1960)
14. Gilio, A.: Criterio di penalizzazione e condizioni di coerenza nella valutazione soggettiva della probabilità. Boll. Un. Mat. Ital. **4**–**B**(3, Serie 7), 645–660 (1990)

15. Gilio, A.: C_0-Coherence and extension of conditional probabilities. In: Bernardo, J.M., Berger, J.O., Dawid, A.P., Smith, A.F.M. (eds.) Bayesian Statistics 4, pp. 633–640. Oxford University Press (1992)
16. Gilio, A.: Probabilistic consistency of knowledge bases in inference systems. In: Clarke, M., Kruse, R., Moral, S. (eds.) ECSQARU 1993. LNCS, vol. 747, pp. 160–167. Springer, Heidelberg (1993). https://doi.org/10.1007/BFb0028196
17. Gilio, A.: Algorithms for conditional probability assessments. In: Berry, D.A., Chaloner, K.M., Geweke, J.K. (eds.) Bayesian Analysis in Statistics and Econometrics: Essays in Honor of Arnold Zellner, pp. 29–39. Wiley, New York (1996)
18. Gilio, A.: Generalizing inference rules in a coherence-based probabilistic default reasoning. Int. J. Approx. Reason. 53(3), 413–434 (2012)
19. Gilio, A., Sanfilippo, G.: Quasi conjunction and p-entailment in nonmonotonic reasoning. In: Borgelt, C. (ed.) Combining Soft Computing and Statistical Methods in Data Analysis. AISC, vol. 77, pp. 321–328. Springer, Berlin, Heidelberg (2010). https://doi.org/10.1007/978-3-642-14746-3_40
20. Gilio, A., Sanfilippo, G.: Coherent conditional probabilities and proper scoring rules. In: Proceedings of ISIPTA 2011, Innsbruck, pp. 189–198 (2011)
21. Gilio, A., Sanfilippo, G.: Quasi conjunction, quasi disjunction, t-norms and t-conorms: probabilistic aspects. Inf. Sci. 245, 146–167 (2013)
22. Gilio, A., Sanfilippo, G.: Conditional random quantities and compounds of conditionals. Stud. Log. 102(4), 709–729 (2014)
23. Gilio, A., Sanfilippo, G.: Conjunction and disjunction among conditional events. In: Benferhat, S., Tabia, K., Ali, M. (eds.) IEA/AIE 2017. LNCS (LNAI), vol. 10351, pp. 85–96. Springer, Cham (2017). https://doi.org/10.1007/978-3-319-60045-1_11
24. Gilio, A., Sanfilippo, G.: Generalized logical operations among conditional events. Applied Intelligence (in press). https://doi.org/10.1007/s10489-018-1229-8
25. Goodman, I.R., Nguyen, H.T.: Conditional objects and the modeling of uncertainties. In: Gupta, M.M., Yamakawa, T. (eds.) Fuzzy Computing, pp. 119–138. North-Holland, Amsterdam (1988)
26. Goodman, I.R., Nguyen, H.T., Walker, E.A.: Conditional Inference and Logic for Intelligent Systems: A Theory of Measure-Free Conditioning. North-Holland, Amsterdam (1991)
27. Milne, P.: Bruno de Finetti and the logic of conditional events. Br. J. Philos. Sci. 48(2), 195–232 (1997)
28. Pelessoni, R., Vicig, P.: The Goodman-Nguyen relation within imprecise probability theory. Int. J. Approx. Reason. 55(8), 1694–1707 (2014)
29. Sanfilippo, G., Pfeifer, N., Over, D.E., Gilio, A.: Probabilistic inferences from conjoined to iterated conditionals. Int. J. Approx. Reason. 93, 103–118 (2018)

Qualitative Probabilistic Relational Models

Linda C. van der Gaag[1(✉)] and Philippe Leray[2]

[1] Department of Information and Computing Sciences, Utrecht University,
Princetonplein 5, 3584 CC Utrecht, The Netherlands
`L.C.vanderGaag@uu.nl`
[2] LS2N UMR CNRS 6004, DUKe research group,
University of Nantes, Nantes, France
`philippe.leray@univ-nantes.fr`

Abstract. Probabilistic relational models (PRMs) were introduced to extend the modelling and reasoning capacities of Bayesian networks from propositional to relational domains. PRMs are typically learned from relational data, by extracting from these data both a dependency structure and its numerical parameters. For this purpose, a large and rich data set is required, which proves prohibitive for many real-world applications. Since a PRM's structure can often be readily elicited from domain experts, we propose manual construction by an approach that combines qualitative concepts adapted from qualitative probabilistic networks (QPNs) with stepwise quantification. To this end, we introduce qualitative probabilistic relational models (QPRMs) and tailor an existing algorithm for qualitative probabilistic inference to these new models.

Keywords: Probabilistic relational models
Qualitative notions of probability · Qualitative probabilistic inference

1 Introduction

The formalism of *Bayesian networks* (BNs) is generally considered an intuitively appealing and powerful formalism for capturing knowledge from a problem domain along with its uncertainties [10,17]. A BN is composed of a directed graphical structure, encoding the random variables of the domain and the probabilistic (in)dependencies between them, and an associated set of conditional probability distributions, capturing the strengths of the dependencies. For computing probabilities from a Bayesian network, tailored algorithms are available that derive their efficiency from exploiting the independencies read from the graphical structure through the well-known d-separation criterion [17].

Bayesian networks are propositional in nature, as they describe a problem domain from the perspective of a single object class. Modelling a domain involving multiple interacting classes by such a representation, in fact, may result in unacceptable loss of information [15]. The formalism of *probabilistic relational*

© Springer Nature Switzerland AG 2018
D. Ciucci et al. (Eds.): SUM 2018, LNAI 11142, pp. 276–289, 2018.
https://doi.org/10.1007/978-3-030-00461-3_19

models (PRMs) extends on the propositional modelling capacity of BNs by allowing the representation of relational information [6,9,14]. A PRM describes the object classes of a relational schema by a graphical dependency structure, with relational dependencies across classes, and supplements this structure again with conditional probability distributions. By instantiating object classes with sets of concrete objects, a *ground Bayesian network* is obtained in which inference is performed using any standard Bayesian-network algorithm.

A propositional Bayesian network for a real-world application is constructed automatically from data, using tailored machine-learning techniques [2,7,11,16], or manually, with the help of experts [13]. Automatically learning a Bayesian network typically requires a large amount of sufficiently rich data, which may prove prohibitive for various real-world applications. Since probabilistic relational models are more involved than Bayesian networks, this observation holds unabatedly for the construction of a PRM. In fact, although tailored algorithms for learning probabilistic relational models are available [5,6,15,25], many real-world relational contexts resist automated model construction. Now, experiences with building propositional Bayesian networks with domain experts show that, while assessing the typically large number of probabilities required is quite daunting for the experts involved [3], configuring the dependency structure is quite doable. These experiences have given rise to a stepwise approach to building Bayesian networks [19], that exploits qualitative notions of probability available from the framework of qualitative probabilistic networks. In this paper, we adapt and extend these qualitative notions for use with relational models, and thereby define the framework of *qualitative probabilistic relational models*.

A *qualitative probabilistic network* (QPN) is a qualitative abstraction of a propositional Bayesian network [22]. Having the same dependency structure as its numerical counterpart, a QPN summarises the probabilistic relationships between its variables by qualitative signs instead of by conditional probability tables; these signs in essence indicate the direction in which a distribution changes with observed values for the variables involved. For qualitative probabilistic inference with a QPN, an elegant message-passing algorithm is available that amounts to propagating signs throughout the network [4]. This algorithm again exploits the independencies read from the dependency structure through the d-separation criterion, and is known to have a polynomial runtime complexity.

The advocated stepwise approach to building a Bayesian network with domain experts now amounts to first building a qualitative probabilistic network and then stepwise replacing signs with numerical information. After configuring the dependency structure of a BN in the making, a domain expert assigns qualitative signs to the modelled dependencies to arrive at a QPN; specifying qualitative signs is known to require considerably less effort from experts than specifying numbers [4,8]. This qualitative network is then used to perform an initial study of the reasoning behaviour of the BN under construction. With quantification efforts yielding conditional probability distributions for (some of) the variables involved, signs are replaced with this numerical information. Before proceeding with the quantification, the reasoning behaviour of the intermediate network, with both signs and numbers, is studied [19]. Possible inadequacies in the dependency structure can thus be detected, and amended, early on in the quantification process. This process of quantifying parts of the network and studying reasoning behaviour is continued until the network is fully quantified.

Building upon the experience that the dependency structure of a PRM can often be readily elicited from domain experts, we envision a similar approach to stepwise quantification of probabilistic relational models as for propositional Bayesian networks. In view of this goal, we introduce in this paper QPRMs, that is, *qualitative probabilistic relational models*. We will argue that the notion of qualitative sign from the framework of QPNs readily fits a relational framework. Since not all probabilistic notions employed in PRMs have qualitative counterparts in the framework of QPNs, we define additional qualitative probabilistic notions, such as *qualitative aggregation functions*, for use with QPRMs. We then tailor the basic algorithm for qualitative probabilistic inference with QPNs, to application with ground models constructed from a QPRM.

The paper is organised as follows. In Sect. 2 we review PRMs and introduce our notational conventions. Throughout the paper we will use the example introduced in Sect. 3. Section 4 then introduces QPRMs, focusing on qualitative signs and qualitative aggregation functions. In Sect. 5 we will present our algorithm for inference with a QPRM and in Sect. 6 we will demonstrate its application to our example. The paper closes in Sect. 7 with our future plans.

2   Preliminaries

Relational data involve different types of object and multiple relationships between them. For modelling and reasoning about such data, we build in this paper on the framework of probabilistic relational models [9]. In this section, we briefly review the notions involved and thereby introduce our notational conventions; we assume basic knowledge of propositional Bayesian networks throughout.

2.1   A Relational Schema and Its Instances

A *relational schema* is a pair $S = (\mathcal{X}, \mathcal{R})$, where $\mathcal{X}$ is a finite non-empty set of object classes and $\mathcal{R}$ models the reference relationships between them. Each class $X \in \mathcal{X}$ has associated a finite, possibly empty, set $\mathcal{F}(X)$ of descriptive features. A *feature* F of a class X will be denoted as $X.F$ and its set of possible values will be indicated by $\Omega(X.F)$. We assume that $\Omega(X.F) = \{false, true\}$ for all $F \in \mathcal{F}(X)$, $X \in \mathcal{X}$. A known value *true* for the feature F of the class X will be indicated as $X.F = true$, or $X.f$ for short; $X.\bar{f}$ is used to indicate that F has the value *false* in X. A class $X \in \mathcal{X}$ can be *instantiated* with a finite, possibly empty, set of concrete *objects* $\mathcal{I}_S(X) = \{x_1, \ldots, x_{n(X)}\}$, $n(X) \geq 0$.

The set $\mathcal{R}(X)$ of *reference slots* of a class X describes the relationships of objects in X with objects in other classes of the relational schema. A slot

R of a class X, indicated by $X.R$, has associated a class pair, composed of a domain class and a range class. The *domain class* of the slot $X.R$ is denoted by $Dom(X.R)$ and equals X by definition. The *range class* of $X.R$ is indicated by $Range(X.R)$ and equals some class $X' \in \mathcal{X}$ with $X' \neq X$, that is, we do not allow self-referencing slots. Each reference slot $X.R$ further has associated an inverse slot $X'.R^{-1}$ with $Dom(X'.R^{-1}) = Range(X.R) = X'$ and $Range(X'.R^{-1}) = Dom(X.R) = X$. Upon instantiation of the slot $X.R$, the reference is a set of pairs (x, x') of related objects with $x \in X$ and $x' \in X'$. The slot is a *one-to-one* reference slot if for each object $x \in X$ there exists at most one object $x' \in Range(x.r)$ for which $(x, x') \in x.r$; the slot is of type *one-to-many* if for each object $x \in X$ there may be multiple objects $x'_i \in X'$ for which $(x, x'_i) \in Range(x.r)$. *Many-to-one* and *many-to-many* reference slots are defined analogously.

While a reference slot describes a direct relationship between two classes, references can also be indirect through other classes. A *slot chain* of length k is a sequence $X.\rho = X.R_1 \ldots R_k$, $k \geq 1$, of (possibly inversed) reference slots R_i with $Range(R_i) = Dom(R_{i+1})$. The domain class of the slot chain $X.\rho$ equals the domain of the first reference slot $X.R_1$, that is, $Dom(X.\rho) = X$; the range of $X.\rho$ equals the range class of the chain's final slot, that is, $Range(X.\rho) = Range(X'.R_k)$ with $X' = Range(X.R_1 \ldots R_{k-1})$. A slot chain $X.\rho$ describes a relation between the objects from its domain and range classes and, upon instantiation, also is a set of pairs of related objects. A slot chain can again be of type one-to-one, one-to-many, many-to-one and many-to-many.

An *instance* $\mathcal{I}_S$ of a relational schema $\mathcal{S}$ includes an assignment of sets of objects to the classes in $\mathcal{S}$, as described above. The instance may further include assignments to the features and reference slots of the classes. In the sequel, we assume that instances have *feature uncertainty* only, that is, while features may not have an assigned value, the values of all reference slots are known.

2.2 The PRM and Its Ground Bayesian Network

A relational schema $\mathcal{S}$ has associated a dependency structure $\mathcal{G}_S$, which essentially is a directed acyclic graph composed of nodes, modelling the classes' features, and their interrelationships; this structure is a *meta-model*, accommodating possible instances of the relational schema [6].

The *dependency structure* $\mathcal{G}_S$ associated with a schema $\mathcal{S}$ includes a node $X.F_i$ for every feature $F_i \in \mathcal{F}(X)$ of each class $X \in \mathcal{X}$. The arcs of $\mathcal{G}_S$ capture the dependencies between the modelled features. The set of all parents of a node $X.F_i$ in $\mathcal{G}_S$ is denoted as $\Pi_{\mathcal{G}_S}(X.F_i)$. A parent of $X.F_i$ can be either a feature $X.F_j$ of the same class or a feature $X.\rho.F_j$ of the class reachable through the slot chain $X.\rho$; the associated arc is called a *type-I arc* in the former case and a *type-II arc* in the latter case. When the slot chain underlying a type-II arc is of type many-to-one or many-to-many, the value of the feature $x.F_i$ of some concrete object x upon instantiation may depend on the values of the features $x'.F_j$ of multiple objects $x' \in \mathcal{I}_S(X')$ with $X' = Range(X.\rho)$. The exact dependency then is described by an *aggregation function* γ which takes a multiset of values from $\Omega(X'.F_j)$ and outputs a single such value; the function is indicated over the corresponding arc in the dependency structure. Common aggregation functions include the *minimum, maximum* and *mode* functions for numerical features [9].

A probabilistic relational model (PRM) for a relational schema $\mathcal{S}$ now is composed of the dependency structure $\mathcal{G}_{\mathcal{S}}$ of the schema, supplemented with a conditional probability table $\Pr(X.F_i \mid \Pi_{\mathcal{G}_{\mathcal{S}}}(X.F_i))$ for each node $X.F_i$. The conditional probability table for the node $X.F_i$ thereby includes conditional probability distributions over $X.F_i$ given each possible value combination for its parents, as in a propositional Bayesian network. For an instance $\mathcal{I}_{\mathcal{S}}$ of the relational schema, a *ground Bayesian network* is constructed by replicating the PRM for every concrete object in $\mathcal{I}_{\mathcal{S}}$. The dependency structure of this ground network includes, for each node $X.F_i$ from $\mathcal{G}_{\mathcal{S}}$ and for every object $x \in \mathcal{I}_{\mathcal{S}}(X)$, a replicate node $x.F_i$, along with copies of the original incident arcs from $\mathcal{G}_{\mathcal{S}}$. For the aggregation function associated with a many-to-one type-II arc $X'.F_j \to X.F_i$ in $\mathcal{G}_{\mathcal{S}}$, an auxiliary node is included in the dependency structure of the ground network, with incoming arcs from all replicate nodes $x'_k.F_j$ with $x'_k \in \mathcal{I}_{\mathcal{S}}(X')$ and a single emanating arc to $x.F_i$ with $x \in \mathcal{I}_{\mathcal{S}}(X)$; for the aggregation function associated with a many-to-many type-II arc, a similar construct is included. The nodes modelling features are further assigned copies of the conditional probability tables from the PRM, and the auxiliary nodes are assigned a conditional probability table conform the semantics of their associated aggregation function.

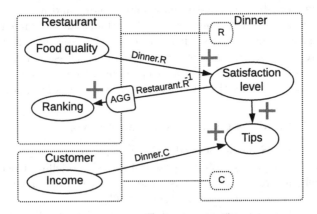

Fig. 1. The dependency structure of the example PRM, with five features (ellipses) for three classes (rounded boxes); the structure is supplemented with qualitative signs (the product synergies involved and their signs are not shown).

3   An Example

To illustrate a PRM and an associated ground Bayesian network, we introduce a small fictitious model for our example. The example pertains to customer satisfaction with restaurants. We assume that the quality of the food served at a restaurant is a determinant of a customer's level of satisfaction. This level can be read from the tip left by the customer after dining, although the size of the tip is dependent of her income as well. For the various restaurants involved, an internet ranking is maintained, reflecting customer satisfaction.

The example corresponds to a relational schema S which includes the three object classes Restaurant, Dinner and Customer, with five features in all, and the two reference slots R and C, defining the restaurant and customer, respectively, related to a specific dinner. The dependency structure of the PRM, shown in Fig. 1, includes a single type-I arc, between the Satisfaction level and Tips nodes of the Dinner class. It further includes three type-II arcs: the arc labelled Dinner.C between the Income and Tips nodes describes the dependency between the Dinner.Tips and Dinner.C.Income features; the many-to-one type-II arc between the Restaurant.Ranking and Restaurant.R^{-1}.Satisfaction level features expresses the information that the ranking of a restaurant is dependent of the satisfaction levels of all dinners taken there, with their joint effect captured by the aggregation function AGG; the third type-II arc describes the dependency between the satisfaction level of a specific dinner and the food quality at that specific restaurant.

Figure 2 shows an example ground Bayesian network constructed from the relational schema and dependency structure of Fig. 1, for an instance with two restaurants, four dinners and two customers. We note that instantiation of the many-to-one arc between the Restaurant.R^{-1}.Satisfaction level and Restaurant.Ranking features in the dependency structure has resulted, in the ground Bayesian network, in an auxiliary node with a single emanating arc to the ranking of the restaurant and incoming arcs from the satisfaction levels of all dinners taken there; we will return to the semantics of the auxiliary node in Sect. 4.2.

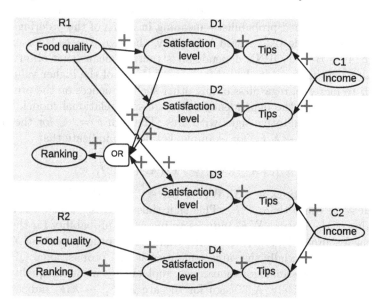

Fig. 2. The dependency structure of the ground Bayesian network constructed for an instance with two restaurants, two customers and four dinners; the structure is supplemented with qualitative signs (product synergies and their signs are not shown).

4 Qualitative Probabilistic Relational Models

Building upon the relational concepts reviewed above, we introduce in this section *qualitative probabilistic relational models* (QPRMs). Like a PRM, a qualitative relational model takes a relational schema and has an associated dependency structure. It differs from a PRM in its associated uncertainty specifications: while a PRM specifies a conditional probability table for each node in the dependency structure, a QPRM specifies a qualitative sign for each arc. These signs are taken from the propositional framework of *qualitative probabilistic networks* (QPNs) [22] and are now fitted into the relational framework.

4.1 Qualitative Signs for QPRMs

The qualitative signs of propositional QPNs assume an ordering on the value sets of all variables involved [22]. For using these signs in qualitative probabilistic relational models therefore, we have to define orderings on the value sets of the variables of such a model. We recall from Sect. 2.1 that we assumed all features of a relational model to be binary. Without loss of generality, we now assume the total ordering on the value set $\Omega(X.F)$ of a feature $X.F$ to be $\bar{f} \leq f$. The orderings per feature induce a partial ordering $\preceq$ over the set of value combinations for any (sub)set of features. We say that a value combination $\mathbf{f}$ for some feature set $\mathbf{F} \subseteq \bigcup_{X \in \mathcal{X}} \mathcal{F}(X)$ is *higher-ordered* than a value combination $\mathbf{f'}$ for $\mathbf{F}$, if $\mathbf{f'} \preceq \mathbf{f}$; if $\mathbf{f} \preceq \mathbf{f'}$, we say that $\mathbf{f}$ is *lower-ordered* than $\mathbf{f'}$.

A propositional QPN has associated with each arc in its dependency structure a *qualitative sign* $\delta \in \{+, -, 0, ?\}$. Although these signs are qualitative in nature, they have a well-defined probabilistic meaning in terms of the orderings on the value sets per variable [22]; stated informally, the sign $\delta = $ '+', for example, for an arc $A \to B$ in a QPN's dependency structure means that observing the higher value for the variable A causes the probability of the higher value of the variable B to increase, regardless of any other such influences on the probability distribution of B. In the dependency structure of a relational model, we now likewise associate qualitative signs with arcs. The sign $\delta = $ '+' for the (type-I or type-II) arc $X'.F_j \to X.F_i$, for example, is taken to indicate that

$$\Pr(X.f_i \mid X'.f_j, \mathbf{w}) \geq \Pr(X.f_i \mid X'.\bar{f}_j, \mathbf{w})$$

for all value combinations $\mathbf{w} \in \Omega(\mathbf{W})$ for the set $\mathbf{W} = \Pi_{\mathcal{G}_S}(X.F_i) \setminus \{X'.F_j\}$ of parents of $X.F_i$ other than $X'.F_j$. This sign thus expresses that observing the higher value of the feature $X'.F_j$ induces an increased probability for the higher value of the feature $X.F_i$, regardless of the values of the other parents of $X.F_i$; the sign thereby has essentially the same semantics as in a propositional QPN. The signs '−' and '0' have analogous meanings, replacing $\geq$ in the inequality above by $\leq$ and $=$, respectively. A '?'-sign for the arc $X'.F_j \to X.F_i$ indicates that there are value combinations $\mathbf{w}_1, \mathbf{w}_2 \in \Omega(\mathbf{W})$, $\mathbf{w}_1 \neq \mathbf{w}_2$, such that observing the higher value of $X'.F_j$ induces an increased probability of the higher value of $X.F_i$ in the context of $\mathbf{w}_1$ and causes a decrease of this probability in the context of $\mathbf{w}_2$. The sign thus indicates that the direction of change in the probability distribution of $X.F_i$ depends on the context and, hence, is not unambiguous.

In addition to the signs per arc, a propositional QPN associates two qualitative signs with each pair of parents per node. As in a numerical Bayesian network, an observed value for a multi-parent node A of a QPN induces a mutual dependency, called a *product synergy*, between each parent pair of A [4,23]. The sign of this dependency is dependent of the specific value observed for A. The two qualitative signs associated with a parent pair of the node A now are the signs of the dependencies induced by the two possible values of A. In the dependency structure of a relational qualitative model, we similarly associate two signs with each parent pair $X'.F_j$, $X''.F_k$ of a node $X.F_i$. The parent pair is said to exhibit a *product synergy* of sign $\delta = \text{`}-\text{'}$ for the value f_i of $X.F_i$ if

$$\Pr(X.f_i \mid X'.f_j, X''.f_k, \mathbf{w}) \cdot \Pr(X.f_i \mid X'.\bar{f}_j, X''.\bar{f}_k, \mathbf{w}) -$$
$$\Pr(X.f_i \mid X'.f_j, X''.\bar{f}_k, \mathbf{w}) \cdot \Pr(X.f_i \mid X'.\bar{f}_j, X''.f_k, \mathbf{w}) \leq 0$$

for all value combinations $\mathbf{w}$ of $\mathbf{W} = \Pi_{g_S}(X.F_i) \setminus \{X'.F_j, X''.F_k\}$. Product synergies of signs '+', '0' and '?' for f_i again are defined analogously; a negative product synergy for the value $\bar{f}_i$ of $X.F_i$ is defined by replacing f_i by $\bar{f}_i$ in the formula above. Since product synergies in essence are induced between pairs of parents of a node with an observed value, their signs are defined for parent pairs [4,23]; an extended concept encompassing all parents of a node has been proposed, however [1].

Qualitative signs exhibit various useful properties upon combination within a propositional QPN [22]. The *symmetry* property states that, if a node A exerts a qualitative influence on a node B over the arc $A \rightarrow B$, then B exerts an influence of the same sign over this arc on A. The property of *transitivity* states that, if a node A exerts a qualitative influence of sign δ_i on a node B which in turn has an influence of sign δ_j on node C, then the net influence of A on C over the trail of associated arcs is of sign $\delta_i \otimes \delta_j$, with the $\otimes$-operator as in Table 1. The *composition* property states that influences along multiple parallel trails convening at a node A combine into a net influence on A of a sign given by the $\oplus$-operator from Table 1. The three properties of symmetry, transitivity and composition underlie the basic algorithm for qualitative probabilistic inference with a propositional QPN; we will return to this observation in Sect. 5, where we adapt this algorithm to the relational context of QPRMs.

4.2 Qualitative Aggregation Functions

Thus far, we fitted the notion of qualitative sign from propositional QPNs into the framework of QPRMs. A relational dependency structure with such signs can now in essence be instantiated to a ground QPN. This ground network is constructed in a similar way as a ground Bayesian network is constructed from a numerical PRM, that is, by replicating the qualitative relational model for every concrete object of the instance at hand. Each replicate arc thereby inherits the qualitative sign of the original arc from the dependency structure and each

Table 1. The sign-product operator $\otimes$ and the sign-plus operator $\oplus$ for combining qualitative signs.

$\otimes$	+	−	0	?		$\oplus$	+	−	0	?
+	+	−	0	?		+	+	?	+	?
−	−	+	0	?		−	?	−	−	?
0	0	0	0	0		0	+	−	0	?
?	?	?	0	?		?	?	?	?	?

parent pair per node inherits the two signs of the induced product synergies. Upon constructing a ground Bayesian network from a numerical PRM however, instantiation of a many-to-one or many-to-many type-II arc is more involved than simple replication. We recall from Sect. 2.2 that such an arc has associated an aggregation function that takes a multiset of values from the value set of some feature and outputs a single such value. Upon instantiation, an auxiliary node is created with a conditional probability table encoding the semantics of this aggregation function. Since propositional QPNs include just type-I arcs, the notion of aggregation function does not have a qualitative counterpart as yet. We now define qualitative aggregation functions and detail the instantiation of a type-II arc upon construction of a ground qualitative model from a QPRM.

While a ground Bayesian network encodes the semantics of an aggregation function from a PRM in the conditional probability table of an auxiliary node, a ground QPN only offers qualitative signs and their combination for this purpose. Based upon this observation, we take a *qualitative aggregation function* to be a function that takes for its input a multiset of qualitative signs and outputs a single such sign. We say that the function is '+'-*preserving* if it returns a '+' whenever its input includes at least one '+'; it is called '−'-*preserving* if it returns a '−' whenever its input includes at least one '−'. If a qualitative aggregation function is neither '+'- nor '−'-preserving, it is called *non-preserving*. An example non-preserving function is the parity function which returns a '+' if the number of '+'s in its input is even and a '−' otherwise. In Sect. 5, we will see that non-preserving aggregation functions like this parity function are undesirable in QPRM applications as these will lead to ambiguous results upon inference. To accommodate possible dependencies induced between its input features upon inference, a qualitative aggregation function has further associated two qualitative signs to describe the product synergies involved.

We now address instantiation of a type-II arc of a QPRM. We consider to this end the many-to-one type-II arc $X'.F_j \rightarrow X.F_i$, with an associated qualitative aggregation function γ and sign δ. For the ground QPN given an instance $\mathcal{I}_\mathcal{S}$, again an auxiliary node A is created, with multiple incoming arcs $x'_k.F_j \rightarrow A$ from all objects $x'_k \in \mathcal{I}_\mathcal{S}(X')$, $k = 1, \ldots, n(X')$, and a single emanating arc $A \rightarrow x.F_i$ for $x \in \mathcal{I}_\mathcal{S}(X)$. The sign δ of the original arc from the QPRM is assigned to this latter arc. For embedding the semantics of the aggregation function γ, we build on the idea underlying the composition property of qualitative influences and introduce a new operator $\oplus_\gamma$ for combining signs of multiple parallel trails.

Where the $\oplus$-operator gives a combined sign for multiple trails convening at any feature node in general, the $\oplus_\gamma$-operator does so for trails convening at the auxiliary node A specifically. All arcs $x'_k.F_j \rightarrow A$ are further assigned the '+'-sign to mimic the partial order $\preceq$ on the value combinations of the set $\{x'_k.F_j \mid k = 1, \ldots, n(X')\}$. The thus specified signs now guarantee that, for example, with a '+'-preserving aggregation function, a higher-ordered value combination $\mathbf{f}$ for the feature $x'_k.F_j$ from all objects x'_k, results in an influence of sign '+' $\otimes \delta$ on $x.F_i$. To conclude, for each parent pair of the auxiliary node A, the signs of the two possible product synergies are copied from the aggregation function γ.

procedure PropagateSign($trail,from,to,messagesign$):

sign[to] $\leftarrow$ sign[to] $\oplus$ $messagesign$;
$trail \leftarrow trail \cup \{to\}$;
for each active neighbour $x'.F_j$ of to
do $linksign \leftarrow$ sign of (induced) influence between to and $x'.F_j$;
$\quad messagesign \leftarrow$ sign[to] $\otimes$ $linksign$;
$\quad$ **if** $x'.F_j \notin trail$ **and** sign[$x'.F_j$] $\neq$ sign[$x'.F_j$] $\oplus$ $messagesign$
$\quad$ **then** PropagateSign($trail,to,x'.F_j,messagesign$).

Fig. 3. A sketch of the sign-propagation algorithm for qualitative probabilistic inference with ground QPNs.

5   Qualitative Inference with a Ground QPN

Probabilistic inference with an instance of a numerical PRM essentially amounts to inference in the ground Bayesian network constructed from the instance. Since this ground network is a BN, essentially any standard inference algorithm can be used for this purpose, although various tailored algorithms are available as well [12,18,24]. In this section, we detail an algorithm for qualitative probabilistic inference with an instance of a QPRM, which builds more or less directly on the basic sign-propagation algorithm available for propositional QPNs.

The basic idea of the sign-propagation algorithm for inference with QPNs is to trace the effect of observing the value of a specific node on the other nodes in the dependency structure, by passing messages between neighbours. The algorithm thereby establishes qualitative signs for the nodes, which indicate the direction of change in the node's probability distribution, occasioned by the new observation [4]. A sketch of the basic algorithm is given in Fig. 3, stated in terms of the feature nodes of a ground QPN. Initially, all node signs are set to '0'. To enter a new observation, the appropriate sign is sent to the observed node, that is, either a '+' for the value $true$ or a '−' for $false$. In a ground QPN, this observed node is taken as the $perspective$ from which probabilistic effects will be traced along trails in the dependency structure. Each feature node receiving a message updates its own sign with this message, by application of the $\oplus$-operator; dependent of the origin of the message, an auxiliary node applies the

$\oplus_\gamma$-operator for its aggregation function γ to this end. If its sign changes, the node sends a sign-message to each active neighbour that is not yet on the trail being traversed from the perspective. An *active* neighbour of a node $x.F_i$ is a non-instantiated node $x'.F_j$ that satisfies one of the following conditions:

(i) $x'.F_j$ is directly connected to $x.F_i$ either by an emanating arc or an incoming arc; or,

(ii) $x'.F_j$ is connected to $x.F_i$ indirectly through a head-to-head trail $x'.F_j \rightarrow x''.F_k \leftarrow x.F_i$ such that either $x''.F_k$ or one of its descendants is observed.

In the first case, the sign of the message sent by $x.F_i$ to $x'.F_j$ is computed through application of the $\otimes$-operator to the (possibly new) sign of $x.F_i$ and the sign of the arc over which the message is sent. In the second case, the message will be sent over the dependency that is induced by observation of the value of (a descendant of) $x''.F_k$; we recall that the resulting product synergy has an associated sign that is available from the observed node or from the appropriate auxiliary node. The sign of the message sent over the induced dependency to the node $x'.F_j$ now is established through application of the $\otimes$-operator to the sign of $x.F_i$ and the sign of the product synergy involved. This process of message passing is repeated throughout the network, visiting nodes along trails starting at the perspective, until nodes no longer change sign. The algorithm is known to run in polynomial time as node signs can change at most twice, from the initial '0' to either '+' or '−', and then to '?' [4].

The original sign-propagation algorithm for propositional QPNs was designed for propagating the observed value of a single node, possibly in a context of previously entered evidence; multiple observations were dealt with essentially by entering them one at a time and combining results by means of the $\oplus$-operator. Later it was shown that multiple observations could be propagated simultaneously throughout a QPN by determining, for each observed node separately, to which nodes its effects should be traced; for further details, we refer to [21].

6   The Example Revisited

To illustrate qualitative probabilistic inference in a ground QPN, we consider again the dependency structure of our example QPRM, from Fig. 1. We further consider the ground network from Fig. 2, which resulted from the instance described in Sect. 3; we note that the OR function modelled in the ground network is a '+'-preserving qualitative aggregation function.

We suppose that we know that the quality of the food served at restaurant R1 is good, which is expressed by the value *true*. After entering this observation into the ground network, the sign-propagation algorithm traces its effects throughout the dependency structure. It thereby establishes the sign '+' for the nodes R1.Food quality, R1.Ranking, Di.Satisfaction level and Di.Tips, $i = 1, 2, 3$; all other nodes retain their initial node sign '0'. Although a high food quality will serve to increase the probability of large tips being left at the three dinners at this restaurant, it does not affect the probability distributions of the customers' incomes nor that of the food quality at the other restaurant.

We now suppose that customer C1 left large tips (expressed by the value *true*) at her first dinner at restaurant R1, and that customer C2 was less satisfied with his dinner at this restaurant and left just small tips (*false*). We further suppose that the product synergies given either value of the Tips node are equal to '0', which expresses that regardless of the size of the tips left, a customer's income and satisfaction with a specific dinner are independent. The two findings are entered into the ground network and are propagated throughout the dependency structure to the appropriate nodes by the sign-propagation algorithm; the resulting node signs are given in Table 2. The propagation results demonstrate, for example for the node R1.Food quality, that combining parallel qualitative influences by the $\oplus$-operator can yield ambiguous signs. Such an ambiguity, in fact, results whenever influences with opposite signs are combined; we say that the *trade-off* that is reflected by the conflicting influences cannot be resolved at the level of detail offered by the qualitative signs. In contrast, the $\otimes$-operator cannot introduce ambiguities by itself; it will cause ambiguous signs to spread throughout the network once they have arisen, however, as can be seen by the propagation result for, for example, the node D1.Satisfaction level.

Table 2. The node signs returned by the sign-propagation algorithm for our example ground QPN, given D1.Tips = *true* and D3.Tips = *false*.

Node name	Node sign	Node name	Node sign
R1.Food quality	?	C1.Income	+
R1.Ranking	?		
R2.Food quality	0	C2.Income	−
R2.Ranking	0		
D1.Satisfaction level	?	D3.Satisfaction level	?
D1.Tips	+	D3.Tips	−
D2.Satisfaction level	?	D4.Satisfaction level	0
D2.Tips	?	D4.Tips	−

7   Conclusions and Future Research

Real-world application of PRMs is often prohibited by the need of a large amount of sufficiently rich data for their automated construction. For practical construction with the help of domain experts, we envision an approach to stepwise quantification of probabilistic relational models similar to the approach proposed before for propositional Bayesian networks. As this approach builds upon qualitative probabilistic notions, we introduced in this paper the framework of qualitative probabilistic relational models. For this purpose, we adapted and extended available qualitative notions of probability to the relational framework and have adapted an existing algorithm for qualitative probabilistic inference to ground qualitative networks. Our qualitative probabilistic relational models are expected to allow ready construction by domain experts; in addition, they provide for efficiently studying reasoning behaviour, as the associated inference algorithm in essence is polynomial in the number of variables involved.

Straightforward use of qualitative probabilistic networks, be they propositional or relational, in real-world applications is associated with some disadvantages originating from their lack of representation detail. It is well known, for example, that qualitative probabilistic inference shows a tendency to lead to weak, and even uninformative, results. Researchers have attributed this tendency to the granularity of the qualitative signs employed and have proposed solutions such as including a notion of strength of qualitative signs [20]. It is expected that these and other extensions of the framework of propositional QPNs can be incorporated in our framework of qualitative probabilistic relational models. Another approach to forestalling uninformative results upon inference may be to further tailor the propagation algorithm to the prospective use of QPRMs. As the next step in our research, we intend to now first study the practicability of our new relational framework in a real-world application, in animal ecology.

References

1. Bolt, J.H.: Bayesian networks: the parental synergy. In: Jaeger, M., Nielsen, T.D. (eds.) Proceedings of the Fourth European Workshop on Probabilistic Graphical Models, Hirtshals, pp. 33–40 (2008)
2. Daly, R., Shen, Q., Aitken, S.: Learning Bayesian networks: approaches and issues. Knowl. Eng. Rev. **26**, 99–157 (2011)
3. Druzdel, M.J., van der Gaag, L.C., Henrion, M., Jensen, F.V.: Building probabilistic networks: where do the numbers come from? Guest editors' introduction. IEEE Trans. Knowl. Data Eng. **12**, 481–486 (2000)
4. Druzdzel, M.J., Henrion, M.: Efficient reasoning in qualitative probabilistic networks. In: Fikes, R., Lehnert, W. (eds.) Proceedings of the 11th National Conference on Artificial Intelligence, pp. 548–553. AAAI Press, Menlo Park (1993)
5. Ettouzi, N., Leray, Ph., Ben Messaoud, M.: An exact approach to learning probabilistic relational model. In: Antonucci, A., Corani, G., de Campos, C. (eds.) Proceedings of the 8th Conference on Probabilistic Graphical Models, vol. 52, pp. 171–182. PMLR (2016)
6. Friedman, N., Getoor, L., Koller, D., Pfeffer, A.: Learning probabilistic relational models. In: Dean, T. (ed.) Proceedings of the 16th International Joint Conference on Artificial Intelligence, pp. 1300–1307. Morgan Kaufmann, San Francisco (1999)
7. Friedman, N., Nachman, I., Peér, D.: Learning Bayesian network structure from massive datasets: the sparse candidate algorithm. In: Laskey, K., Prade, H. (eds.) Proceedings of the 15th Conference on Uncertainty in Artificial Intelligence, pp. 206–215. Morgan Kaufmann, San Francisco (1999)
8. van der Gaag, L.C., Renooij, S., Schijf, H.J.M., Elbers, A.R., Loeffen, W.L.: Experiences with eliciting probabilities from multiple experts. In: Greco, S., Bouchon-Meunier, B., Coletti, G., Fedrizzi, M., Matarazzo, B., Yager, R.R. (eds.) IPMU 2012. CCIS, vol. 299, pp. 151–160. Springer, Heidelberg (2012). https://doi.org/10.1007/978-3-642-31718-7_16
9. Getoor, L.: Learning Statistical Models from Relational Data. Ph.D. thesis, Stanford (2001)

10. Jensen, F.V., Nielsen, T.D.: Bayesian Networks and Decision Graphs. Springer, New York (2007). https://doi.org/10.1007/978-1-4757-3502-4
11. Jordan, M.I.: Learning in Graphical Models. MIT Press, Cambridge (1999)
12. Kaelin, F., Precup, D.: An approach to inference in probabilistic relational models using block sampling. J. Mach. Learn. Res. **13**, 315–330 (2010)
13. Kjaerulff, U.B., Madsen, A.L.: Bayesian Networks and Influence Diagrams: A Guide to Construction and Analysis. Springer, New York (2010). https://doi.org/10.1007/978-1-4614-5104-4
14. Koller, D., Pfeffer, A.: Probabilistic frame-based systems. In: Mostov, J., Rich, C. (eds.) Proceedings of the 15th National Conference on Artificial Intelligence, pp. 580–587. AAAI Press, Menlo Park (1998)
15. Maier, M.E., Marazopoulou, K., Jensen, D.D.: Reasoning about independence in probabilistic models of relational data (2013). arXiv: 1302.4381
16. Neapolitan, R.E.: Learning Bayesian Networks. Pearson Prentice Hall, Upper Saddle River (2004)
17. Pearl, J.: Probabilistic Reasoning in Intelligent Systems: Networks of Plausible Inference. Morgan Kaufmann, San Francisco (1988)
18. Pfeffer, A.J.: Probabilistic reasoning for complex systems. Ph.D. thesis, Stanford University, Stanford (2000)
19. Renooij, S., van der Gaag, L.C.: From qualitative to quantitative probabilistic networks. In: Darwiche, A., Friedman, N. (eds.) Proceedings of the 18th Conference on Uncertainty in Artificial Intelligence, pp. 422–429. Morgan Kaufmann, San Francisco (2002)
20. Renooij, S., van der Gaag, L.C.: Enhanced qualitative probabilistic networks for resolving trade-offs. Artif. Intell. **172**, 1470–1494 (2008)
21. Renooij, S., van der Gaag, L.C., Parsons, S.: Propagation of multiple observations in QPNs revisited. In: Proceedings of the 15th European Conference on Artificial Intelligence, pp. 665–669. IOS Press, Amsterdam (2002)
22. Wellman, M.P.: Fundamental concepts of qualitative probabilistic networks. Artif. Intell. **44**, 257–303 (1990)
23. Wellman, M.P., Henrion, M.: Qualitative intercausal relations, or explaining explaining away. In: Allen, J.F., Fikes, R., Sandewall, E. (eds.) KR91: Principles of Knowledge Representation and Reasoning, pp. 535–546. Morgan Kaufmann, San Francisco (1991)
24. Wuillemin, P.-H., Torti, L.: Structured probabilistic inference. Int. J. Approx. Reason. **53**, 946–968 (2012)
25. Xiao-Lin, L., Xiang-Dong, H.: A hybrid particle swarm optimization method for structure learning of probabilistic relational models. Inf. Sci. **283**, 258–266 (2014)

Rule-Based Conditioning of Probabilistic Data

Maurice van Keulen[1]([✉]), Benjamin L. Kaminski[2], Christoph Matheja[2], and Joost-Pieter Katoen[1,2]

[1] University of Twente, Enschede, The Netherlands
{m.vankeulen,j.p.katoen}@utwente.nl
[2] RWTH Aachen, Aachen, Germany
{benjamin.kaminski,matheja,katoen}@cs.rwth-aachen.de

Abstract. Data interoperability is a major issue in data management for data science and big data analytics. Probabilistic data integration (PDI) is a specific kind of data integration where extraction and integration problems such as inconsistency and uncertainty are handled by means of a probabilistic data representation. This allows a data integration process with two phases: (1) a quick partial integration where data quality problems are represented as uncertainty in the resulting integrated data, and (2) using the uncertain data and continuously improving its quality as more evidence is gathered. The main contribution of this paper is an iterative approach for incorporating evidence of users in the probabilistically integrated data. Evidence can be specified as hard or soft rules (i.e., rules that are uncertain themselves).

Keywords: Data cleaning · Data integration · Information extraction Probabilistic databases · Probabilistic programming

1 Introduction

Data interoperability is a major issue in data management for data science and big data analytics. It may be hard to extract information from certain kinds of sources (e.g., natural language, websites), it may be unclear which data items should be combined when integrating sources, or they may be inconsistent complicating a unified view, etc. *Probabilistic data integration* (PDI) [1] is a specific kind of data integration where extraction and integration problems such as inconsistency and uncertainty are handled by means of a probabilistic data representation. The approach is based on the view that data quality

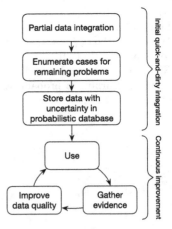

Fig. 1. Probabilistic data integration process [1,2]

© Springer Nature Switzerland AG 2018
D. Ciucci et al. (Eds.): SUM 2018, LNAI 11142, pp. 290–305, 2018.
https://doi.org/10.1007/978-3-030-00461-3_20

problems (as they occur in an integration process) can be modeled as uncertainty [3] and this uncertainty is considered an important result of the integration process [4].

The PDI process contains two phases (see Fig. 1):

- a quick partial integration where certain data quality problems are not solved immediately, but explicitly represented as uncertainty in the resulting integrated data stored in a probabilistic database;
- continuous improvement by using the data—a probabilistic database can be queried directly resulting in possible or approximate answers [5,6]—and gathering evidence (e.g., user feedback) for improving the data quality.

For details on the first phase, we refer to [2,3], as well as [7–9] for techniques on specific extraction and integration problems (merging semantic duplicates, merging grouping data, and information extraction from natural language text, respectively). This paper focuses on the second phase of this process, namely on the problem of how to incorporate evidence of users in the probabilistically integrated data with the purpose to continuously improve its quality as more evidence is gathered. We assume that evidence of users is obtained in the form of rules expressing what is necessary (in case of *hard rules*) or likely (in case of *soft rules*) to be true. Rules may focus on individual data items, individual query results, or may state general truths based on background knowledge of the user about the domain at hand. The paper proposes a method for incorporating the knowledge expressed by a rule in the integrated data by means of *conditioning* the probabilistic data on the observation that the rule is true.

In probabilistic programming and statistical relational learning, it is common to answer queries of the form $P(Q|E)$, where E denotes evidence [10,11], whereas probabilistic databases typically focus on scalable answering of top-k queries without considering evidence [12]. A notable exception is [13] which accounts for "improbable worlds" during query processing. Note that our approach to evidence is fundamentally different: instead of a query mechanism for computing $P(Q|E)$, we incorporate E in the database, such that computing a subsequent $P(Q)$ effectively determines $P(Q|E)$. This allows for an interative more scalable incorporation of accumulating evidence.

Contributions. This paper makes the following contributions:

- A technique to remap random variables (in this paper referred to as partitionings) to fresh ones in a probabilistic database.
- An extension to probabilistic query languages to specify evidence as hard and soft rules.
- An approach to incorporate such specified evidence in a probabilistic database by updating it.

Outlook. The paper is structured as follows. Section 1.1 presents a running example based on an information extraction scenario. Section 2 gives the background on probabilistic databases, the probabilistic datalog language (JudgeD), and how results from probabilistic data integration can be stored in a probabilistic database. Section 3 describes and explains all contributions, in particular how to rewrite (i.e., update) a probabilistic database with rule evidence into one in which the evidence is incorporated. Section 4 presents a sketch of the main proof: the semantics of a probabilistic database with evidence incorporated in it is equivalent to the semantics of a probabilistic database with its evidence still separate.

1.1   Running Example

Throughout the paper we use an information extraction scenario as running example: the *"Paris Hilton example"*. Although this scenario is from the Natural Language Processing (NLP) domain, note that it is equally applicable to other data integration scenarios such as semantic duplicates [7], entity resolution, uncertain groupings [8], etc.

Paris Hilton Example. This example and the problem of incorporating rule-based knowledge by means of conditioning was first described in [14]. We summarize it here.

Because natural language is highly ambiguous and computers are still incapable of 'real' semantic understanding, information extraction (IE) from natural language is an inherently imperfect process. We focus in this example on the sentence

"Paris Hilton stayed in the Paris Hilton."

A *named entity* (NE) is a phrase that is to be interpreted as a name refering to some entity in the real world. A specific task in IE is *Named Entity Recognition* (NER): detecting which phrases in a text are named entities, possibly also detecting the type of the NE. The resulting data of this task is typically in the form of *annotations*.

Here we have two NEs which happen to be the same phrase "Paris Hilton". It is ambiguous how to interpret it: it could be a person, a hotel, or even a fragrance. In fact, we as humans unconsciously understand that the first mention of "Paris Hilton" must refer to a person and the second to a hotel, because from the $3 \times 3 = 9$ combinations only 'person–stay in–hotel' seems logical (based on our background knowledge unknown to the IE algorithm).

Often ignored in NER, also the word "Paris" is a NE: it could be a first name or a city. Note that interpretations are correlated: if "Paris" is interpreted as a city, then "Paris Hilton" is more likely to be a hotel, and vice versa. The evidence a user may want to express is

- words contained in phrases interpreted as persons, should not be interpreted as cities, or
- 'stay-in' relationships between entities will not have buildings (such as hotels) on the lefthand side and no persons on the righthand side.

In this example, we assume that the initial information extraction produces a *probabilistic database with uncertain annotations* [9,15]: the type of the first "Paris Hilton" can be either a hotel, person, or fragrance with probabilities 0.5, 0.4, 0.1, respectively. The second "Paris Hilton" analogously. Both mentions of "Paris" are of type firstname or city. The contributions of this paper allow for expressing the evidence in a query language and update the database accordingly resulting in a database with less uncertainty and of higher quality (i.e., closer to the truth).

2   Background

2.1   Probabilistic Database

A common foundation for probabilistic databases is possible worlds theory [5]. We follow the formalization of [16] as it separates (a) the data model and the mechanism for handling uncertainty, and (b) the abstract notion of *worlds* and the data contained in them.

Probabilistic Database. We view a *database $DB \in \mathbb{P}A$* as a set of *assertions* $\{a_1, \ldots, a_n\}$. For the purpose of data model independence, we abstract from what an assertion is: it may be a tuple in a relational database, a node in an XML database, and so on. A *probabilistic database $PDB \in \mathbb{P}\mathbb{P}A$* is defined as a finite set of possible database states.

Partitionings and Descriptive Sentences. We postulate an infinite set of *worlds*. An assertion is contained only in a subset of all possible worlds. To describe this relationship, we introduce an identification mechanism, called *descriptive sentence*, to refer to a subset of the possible worlds. If two worlds contain the same assertions, they are said to be *indistinguishable* and we regard them as one possible world. As a consequence, this effectively defines a finite set of distinguishable possible worlds representing the possible database states. We use the symbols DB and w interchangeably.

Let Ω be the set of partitionings. A *partitioning $\omega^n \in \Omega$* introduces a set of n labels $l \in L(\omega^n)$ of the form $\omega = v$ (without loss of generality, we assume $v \in 1..n$). A partitioning splits the set of possible worlds into n disjunctive subsets $W(l)$. A *descriptive sentence φ* is a propositional formula over the labels. Let $\omega(\varphi)$ be the set of partitionings contained in formula φ. The symbols $\top$ and $\bot$ denote the *true* and *false* sentences. A sentence φ denotes a specific subset of worlds:

$$
W(\varphi) = \begin{cases}
PDB & \text{if } \varphi = \top \\
\emptyset & \text{if } \varphi = \bot \\
W(l) & \text{if } \varphi = l \\
W(\varphi_1) \cap W(\varphi_2) & \text{if } \varphi = \varphi_1 \wedge \varphi_2 \\
W(\varphi_1) \cup W(\varphi_2) & \text{if } \varphi = \varphi_1 \vee \varphi_2 \\
PDB \setminus W(\varphi_1) & \text{if } \varphi = \neg\varphi_1
\end{cases}
\tag{1}
$$

A *fully described sentence* $\bar{\varphi}$ over $\Omega = \{\omega_1^{n_1}, \ldots, \omega_k^{n_k}\}$ is a formula $\bigwedge_{i \in 1..k} l_i$ with $l_i \in L(\omega_i^{n_i})$. It denotes a set of exactly one world, hence can be used as the *name* or *identifier* for a world. Let $\Phi(\Omega)$ be the set of all fully described sentences over Ω. The following holds:

$$PDB = \bigcup_{\bar{\varphi} \in \Phi(\Omega)} W(\bar{\varphi}) \tag{2}$$

$$PDB = \bigcup_{l \in L(\omega^n)} W(l) \quad (\forall \omega^n \in \Omega) \tag{3}$$

Compact Probabilistic Database. A *compact probabilistic database* is a tuple $CPDB = \langle \widehat{DB}, \Omega, P \rangle$ where $\widehat{DB}$ is a set of *descriptive assertions* $\hat{a} = \langle a, \varphi \rangle$, Ω a set of partitionings, and P a probability assignment function for labels provided that $\sum_{v=1}^{n} P(\omega^n = v) = 1$. Figure 2 illustrates these notions. We consider $CPDB$ to be *well-formed* if all assertions a used in $CPDB$ occur only once. Well-formedness can always easily be obtained by 'merging duplicate assertions' using the transformation rule $\langle a, \varphi_1 \rangle, \langle a, \varphi_2 \rangle \mapsto \langle a, \varphi_1 \vee \varphi_2 \rangle$. We use the terms assertion and data item interchangeably. The possible worlds of $CPDB = \langle \widehat{DB}, \Omega, P \rangle$ are obtained as follows:

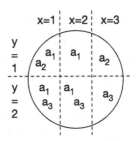

Fig. 2. Illustration of a compact probabilistic database $CPDB = \langle \widehat{DB}, \Omega, P \rangle$. $\widehat{DB} = \{\langle a_1, \neg x = 3 \rangle, \langle a_2, \neg x = 2 \wedge y = 1 \rangle, \langle a_3, y = 2 \rangle\}$. $\Omega = \{x^3, y^2\}$. $W(CPDB) = \{\{a_1\}, \{a_2\}, \{a_3\}, \{a_1, a_2\}, \{a_1, a_3\}\}$.

$$W(CPDB) = \{DB \mid \bar{\varphi} \in \Phi(\Omega) \wedge DB = \{a \mid \langle a, \varphi \rangle \in \widehat{DB} \wedge \bar{\varphi} \Rightarrow \varphi\}\} \tag{4}$$

This setup naturally supports to express several important dependency relationships:

- *Mutual dependence*: for $\langle a_1, \varphi \rangle$ and $\langle a_2, \varphi \rangle$, either a_1 and a_2 both exist in a world or none of them, but never only one of the two.
- *Mutual exclusivity*: for $\langle a_1, \varphi_1 \rangle$ and $\langle a_2, \varphi_2 \rangle$, it holds that a_1 and a_2 never noth occur in a world if $\varphi_1 \wedge \varphi_2 \equiv \bot$.
- *Independence*: Since each ω_i is a partitioning on its own, it can be considered as an independent random variable making an independent choice. For example, $\langle a_1, x = 1 \rangle$ and $\langle a_2, y = 1 \rangle$ use different partitionings, hence their existence in worlds is independent and a world can contain both a_1 and a_2, only one of the two, or none of them.

Probability Calculation. The probability of a sentence $P(\varphi)$ can be derived from label probabilities making use of properties like $P(\omega_1 = v_1 \wedge \omega_2 = v_2) = P(\omega_1 = v_1) \times P(\omega_2 = v_2)$ and $P(\omega_1 = v_1 \vee \omega_2 = v_2) = P(\omega_1 = v_1) + P(\omega_2 = v_2)$ if

$\omega_1 \neq \omega_2$. The probability of a world is defined as $P(w) = P(\bar{\varphi})$ with $W(\bar{\varphi}) = \{w\}$. The probability of a descriptive assertion is defined as $P(\langle a, \varphi \rangle) = P(\varphi)$. It holds that:

$$P(\langle a, \varphi \rangle) = \sum_{w \in PDB, a \in w} P(w) = \sum_{w \in W(\varphi)} P(w) = P(\varphi)$$

Probabilistic Querying. The concept of possible worlds means that querying a probabilistic database should produce the same answer as querying each possible world separately. Given traditional query results $Q(DB)$, let:

$$Q(PDB) = \{Q(DB) \mid DB \in PDB\}$$

As explained in [16], we abstract from specific operators analogously to the way we abstract from the form of the actual data items. Given a query language, for any query operator $\oplus$, we define an *extended operator* $\hat{\oplus}$ with an analogous meaning that operates on *CPDB*. It is defined by $\hat{\oplus} = (\oplus, \tau_\oplus)$ where $\tau_\oplus$ is a function that produces the descriptive sentence of a result based on the descriptive sentences of the operands in a manner that is appropriate for operation $\oplus$. Obviously, a thusly expressed query $\hat{Q}$ on a compact probabilistic database *CPDB* should adhere to the semantics above and Eq. 4:

$$\hat{Q}(CPDB) = \bigcup_{w \in W(CPDB)} Q(w) = \bigcup_{\bar{\varphi} \in \Phi(\Omega)} \{a \mid \langle a, \varphi \rangle \in \hat{Q}(\widehat{DB}) \wedge \bar{\varphi} \Rightarrow \varphi\} \quad (5)$$

2.2 Definition of JudgeD, a Probabilistic Datalog

As a representation formalism in which both probabilistic data as well as soft and hard rules can be expressed, we choose JudgeD, a probabilistic datalog [17]. Several probabilistic logics have been proposed in the last decades among others pD [18] and ProbLog [10]. In these logics probabilities can be attached to facts and rules. JudgeD is obtained by defining in the abovedescribed formalism that a data item is a fact or rule. Moreover, datalog entailment is extended with sentence manipulation [16]. The thus obtained probabilistic datalog is as expressive as ProbLog regarding dependency relationships.

Probabilistic Datalog. We base our definition of Datalog on [19, Ch. 6] (only positive Datalog for simplicity). We postulate disjoint sets *Const*, *Var*, *Pred* as the sets of *constants*, *variables*, and *predicate symbols*, respectively. Let $c \in Const$, $X \in Var$, and $p \in Pred$. A *term* $t \in Term$ is either a constant or variable where *Term* = *Const* $\cup$ *Var*. An *atom* $A = p(t_1, \ldots, t_n)$ consists of an n-ary predicate symbol p and a list of argument terms t_i. An atom is *ground* iff $\forall i \in 1..n : t_i \in Const$. A *clause* or *rule* $r = (A^h \leftarrow A_1, \ldots, A_m)$ is a Horn clause representing the knowledge that A^h is true iff all A_i are true. A *fact* is a rule without body $(A^h \leftarrow)$. A set KB of rules is called a *knowledge base* or *program*. The usual safety conditions of pure Datalog apply.

a_1 annot(id-ph,pos1-2,hotel) [x=1].   @p(x=1) = 0.5.
a_2 annot(id-ph,pos1-2,person) [x=2].   @p(x=2) = 0.4.
a_3 annot(id-ph,pos1-2,fragrance) [x=3].   @p(x=3) = 0.1.
a_4 annot(id-p,pos1,firstname) [y=1].   @p(y=1) = 0.3.
a_5 annot(id-p,pos1,city) [y=2].   @p(y=2) = 0.7.

a_6 contained(pos1,pos1-2).

a_7 hardrule :- annot(Ph1,P1,city), annot(Ph2,P2,person), contained(P1,P2).

Fig. 3. Paris Hilton example (simplified) in JudgeD (sentences in square brackets; '@p' syntax specifies probabilities).

Let $\theta = \{X_1/t_1, \ldots, X_n/t_n\}$ be a *substitution* where X_i/t_i is called a *binding*. $A\theta$ and $r\theta$ denote the atom or rule obtained by replacing (as defined by θ) each X_i occurring in A or r respectively by the corresponding term t_i.

Let $(A^{\mathrm{h}} \xleftarrow{\varphi} A_1, \ldots, A_m)$ denote the tuple $\langle A^{\mathrm{h}} \leftarrow A_1, \ldots, A_m, \varphi \rangle$. Note that this not only allows for the specification of uncertain facts, but also uncertain rules as well as dependencies between the existence of facts and rules using the sentences φ.

Probabilistic Entailment. Entailment is defined as follows:

$$\frac{\begin{array}{c} r \in \mathrm{KB} \quad r = (A^{\mathrm{h}} \xleftarrow{\varphi} A_1, \ldots, A_m) \\ \exists \theta : A^{\mathrm{h}}\theta \text{ is ground} \wedge \forall i \in 1..m : \mathrm{KB} \models \langle A_i\theta, \varphi_i \rangle \\ \varphi' = \varphi \wedge \bigwedge_{i \in 1..m} \varphi_i \quad \varphi' \not\equiv \bot \end{array}}{\mathrm{KB} \models \langle A^{\mathrm{h}}\theta, \varphi' \rangle}$$

In other words, given a rule r from the knowledge base and a substitution θ that makes the atoms A_i in the body true for sentences φ_i, then we can infer the substituted atom $A^{\mathrm{h}}\theta$ with a sentence that is a conjunction of all φ_i and the sentence φ of the rule r (unless this conjunction is inconsistent). This definition of probabilistic entailment is obtained from applying the querying framework of Sect. 2.1 to normal datalog entailment [16]. It can be proven to be consistent with Eq. (5).

2.3   Representing PDI Results in JudgeD

Probabilistic data integration (PDI) is a specific kind of data integration where extraction and integration problems are handled by means of a probabilistic data representation. In this section, we illustrate JudgeD by showing how to represent an information extraction result.

In the Paris Hilton example, the initial information extraction produces uncertain annotations: the type of the phrase "Paris Hilton" occuring as the first and second word of the sentence can be either a hotel, person, or fragrance with, for example, probabilities 0.5, 0.4, 0.1, respectively. Furthermore, the first word "Paris" can either be a firstname or a city with, for example, probabilities 0.3 and 0.7, respectively. We can represent this in JudgeD as in Fig. 3 (a_1–a_5).

Probabilities are obtained from classifiers or scoring or ranking functions used in information extraction and data integration.

A user may want to express evidence that words contained in phrases interpreted as persons, should not be interpreted as cities. If we absolutely trust this to be true, we express this as a *hard rule*. In contrast, a *soft rule* is a rule that is only partially trusted, i.e., the evidence is uncertain. In JudgeD we can express the evidence by rule hardrule in Fig. 3 (a_6-a_7). Executing this rule provides the information under which conditions the rule is true, in this case, $x = 2 \wedge y = 2$. In this case, it is a negative rule, i.e., we 'observe' the evidence that hardrule is false. As we will see in the next section, this evidence can be incorporated by conditioning and rewriting the database on $\neg(x = 2 \wedge y = 2)$.

3   Conditioning

As the example in Sect. 2.3 illustrates, our approach is to specify evidence by rules. Since a rule may only be true in a subset of worlds, the rule actually specifies which worlds are consistent with the evidence. By executing the rule, we obtain this information in terms of the *evidence sentence* φ_e. To incorporate such evidence means that the database[1] needs to be *conditioned*.

A common way of conditioning in probabilistic programming [10,11] is to extend inference with an **observe** capability. Here, we propose to *rewrite* the database into an equivalent one that no longer contains **observe** statements: the evidence is directly *incorporated* in the probabilistic data. By ensuring that evidence incorporation can be done iteratively, the "Improve data quality" step of Fig. 1 can be realized without an ever-growing set of **observe** statements.

The intuition of conditioning is to eliminate all worlds that are inconsistent with the evidence and redistribute the eliminated probability mass over the remaining worlds by means of normalization. This can be realized directly on the compact probabilistic database by constructing an adjusted set of partitionings Ω', rewriting the sentences of the data items, and removing any data items for which the sentence becomes inconsistent (i.e., $\bot$).

The approach is presented in several steps: Sect. 3.1 defines the semantics of a probabilistic database with evidence. Section 3.2 explains how to reduce conditioning with a complex set of evidences to one or more simple conditionings. Section 3.3 explains how to rewrite the original database into a conditioned one whereby we focus on hard rules first. Section 3.4 explains how to condition with soft rules. We conclude this section with a discussion on iterative conditioning.

3.1   Semantics of a Database with Evidence

We abstractly denote evidence as a set E of queries/rules that should be true (positive evidence). We extend the definition of $CPDB = \langle \widehat{DB}, \Omega, P \rangle$ to a *compact probabilistic database with evidence* $CPDBE = \langle \widehat{DB}, \Omega, P, E \rangle$ with semantics

---

[1] Note that we also refer to a JudgeD program as a database.

Worlds $\bar{\varphi}$	$W(\bar{\varphi})$	P	Remapped $\mapsto$ Renumbered	Consistent	P_e
$x=1 \wedge y=1$	$\{a_1, a_4, a_6, a_7\}$	0.15	$z=1 \mapsto z=1$	✓	0.2083
$x=2 \wedge y=1$	$\{a_2, a_4, a_6, a_7\}$	0.12	$z=2 \mapsto z=2$	✓	0.1667
$x=3 \wedge y=1$	$\{a_3, a_4, a_6, a_7\}$	0.03	$z=3 \mapsto z=3$	✓	0.0417
$x=1 \wedge y=2$	$\{a_1, a_5, a_6, a_7\}$	0.35	$z=4 \mapsto z=4$	✓	0.4861
$x=2 \wedge y=2$	$\{a_2, a_5, a_6, a_7\}$	0.28	$z=5$	✗	
$x=3 \wedge y=2$	$\{a_3, a_5, a_6, a_7\}$	0.07	$z=6 \mapsto z=5$	✓	0.0972

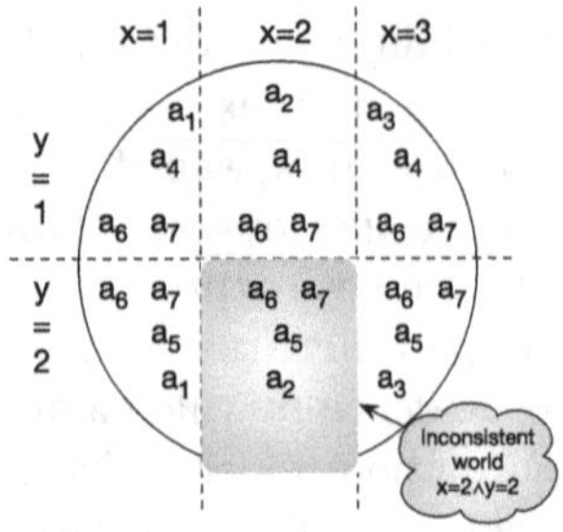

Fig. 4. Illustration of partitioning remapping

$$W(CPDBE) = \{w \mid w \in W(CPDB) \wedge \forall Q_e \in E : Q_e(w) \text{ is true}\}$$

Concrete probabilistic database formalisms may provide specific mechanisms for specifying evidence. For JudgeD, we extend the language with a specific kind of rule: $\mathsf{observe}(A_e)$. A program containing k observed atoms A_e^i ($i \in 1...k$) defines $E = \{A_e^1, \ldots, A_e^k\}$.

An evidence query $Q_e^i \in E$ has exactly two results: $Q_e^i(CPDB) = \{\langle true, \varphi_i \rangle,$ $\langle false, \neg\varphi_i \rangle\}$. Since evidence filters worlds that are inconsistent with it, we determine an *evidence sentence* $\varphi_e = \bigwedge_{i \in 1..k} \varphi_i$. We use E and φ_e interchangeably:

$$W(CPDBE) = \{w \mid w \in W(CPDB) \wedge \varphi_e\} \tag{6}$$

The probability mass associated with eliminated worlds is redistributed over the remaining worlds by means of normalization.

$$P_e(\varphi) = \frac{P(\varphi \wedge \varphi_e)}{P(\varphi_e)} \tag{7}$$

Querying is extended in a straightforward manner by adapting Eq. (5):

$$\hat{Q}(CPDBE) = \bigcup_{w \in W(CPDBE)} Q(w)$$

$$= \bigcup_{\bar{\varphi} \in \Phi(\Omega), \bar{\varphi} \Rightarrow \varphi_e} \{a \mid \langle a, \varphi \rangle \in \hat{Q}(\widehat{DB}) \wedge \bar{\varphi} \Rightarrow \varphi\} \tag{8}$$

3.2 Remapping Partitionings

Figure 4 illustrates that in the Paris Hilton example of Fig. 3, partitions x^3 and y^2 that were independent now become dependent because one of the six possible worlds is inconsistent with the evidence $\varphi_e = \neg(x=2 \wedge y=2)$. When this happens, we *remap* x and y, i.e., replace them with a fresh partitioning z^6 representing their combined possibilities. By simple logical equivalence, we can find formulas for the labels of the original partitionings, for example, $x=1 \Leftrightarrow (z=1 \vee z=4)$. These can be used to rewrite sentences based on x and y to sentences based on z.

Since worlds and their contents are determined by sentences and these sentences are replaced by equivalent ones, this remapping of two or more partitionings to a single fresh one is idempotent.

Remapping. For a sentence containing more than one partitioning, the partitionings may become dependent and remapping is necessary. Let $\Omega_e = \omega(\varphi_e) = \{\omega^{n_1}, \ldots, \omega^{n_k}\}$ be the set of partitionings to be remapped. We introduce a fresh partitioning $\bar{\omega}^n$ where $n = n_1 \times \ldots \times n_k$. Let the bijection $\lambda_{\Omega_e} : \Phi(\Omega_e) \leftrightarrow L(\bar{\omega}^n)$ be the *remapping function*. A valid remapping function can be constructed in a straightforward way by viewing the values in the labels of the partionings of a full sentence as a vector of numbers $v_1, \ldots, v_k$ and computing the value v in the label of $\bar{\omega}^n$ as $v = 1 + \sum_{i \in 1..k}(v_i - 1)\prod_{j \in i+1..k} n_j$. For example, $\lambda_{\Omega_e}(\mathsf{x}=3 \wedge \mathsf{y}=2) = (\mathsf{z}=6)$ because $1 + (3-1) \times 2 + (2-1) \times 1 = 6$.

A sentence φ can be rewritten into $\lambda_{\Omega_e}(\varphi)$ by replacing every label $l_{ij} = (\omega_i = v_i^j)$ with $\bigvee_{l \in L(\bar{\omega}^n), l_{ij} \in \lambda_{\Omega_e}^{-1}(l)} l$. For example, $\lambda_{\Omega_e}(\mathsf{x}=1 \wedge \mathsf{y}=2) = ((\mathsf{z}=1 \vee \mathsf{z}=4) \wedge (\mathsf{z}=4 \vee \mathsf{z}=5 \vee \mathsf{z}=6)) = (\mathsf{z}=4)$. Observe that, since all partitionings in a sentence are rewritten into a single one, the rewritten evidence sentence is of the form $\lambda_{\Omega_e}(\varphi_e) = (\bar{\omega}^n = v_1) \vee \ldots \vee (\bar{\omega}^n = v_m)$ for some m.

Finally, given φ_e, a compact probabilistic database $CPDB = \langle \widehat{DB}, \Omega, P \rangle$ can be rewritten into $\lambda_{\Omega_e}(CPDB) = \langle \widehat{DB'}, \Omega', P' \rangle$ where

$$\widehat{DB'} = \{\langle a, \lambda_{\Omega_e}(\varphi)\rangle \mid \langle a, \varphi \rangle \in \widehat{DB}\} \tag{9}$$

$$\Omega' = (\Omega \setminus \Omega_e) \cup \{\bar{\omega}^n\} \tag{10}$$

$$P'(l) = \begin{cases} P(\lambda_{\Omega_e}^{-1}(l)) & \text{if } l \in L(\bar{\omega}^n) \\ P(l) & \text{otherwise} \end{cases}$$

Splitting. If many partitionings are involved, remapping may introduce partitionings ω^n with large n. Note, however, that the procedure is only necessary if the partitionings become independent due to the evidence. For example, if the evidence would be $\varphi_e = \neg(\mathsf{x}=3) \wedge \mathsf{y}=2$, x and y remain independent. Therefore, we first *split* φ_e into independent components and treat them seperately.

First φ_e is brought into conjunctive normal form $\varphi_1 \wedge \ldots \wedge \varphi_n$ whose conjuncts are then 'clustered' into m independent *components* $\varphi_e^i = \varphi_{j_1} \wedge \ldots \wedge \varphi_{j_k}$ ($i \in 1...m$) such that for maximal m, every conjunct is in exactly one component, and for every pair of components φ_e^1 and φ_e^2, it holds $\omega(\varphi_e^1) \cap \omega(\varphi_e^2) = \emptyset$.

Note that, because of independence between partitionings, the components specify independent evidence that can be incorporated seperately. In the sequel, we denote with φ_e a single component of the evidence sentence. Furthermore, since remapping reduces an evidence sentence to one based on one partitioning, splitting and remapping togeher simplify conditioning to one or more conditionings on single partitionings.

3.3 Conditioning with Hard Rules by Means of Program Rewriting

Given $CPDBE = \langle \widehat{DB}, \Omega, P, \varphi_e \rangle$, let $CPDB = \langle \widehat{DB}'', \Omega'', P'' \rangle = \Lambda_{\varphi_e}(CPDBE)$ be a rewritten compact probabilistic database that incorporates φ_e in the probabilistic data itself. We define $\Lambda_{\varphi_e}(CPDBE)$ as follows. The partitionings $\Omega_e = \omega(\varphi_e)$ are remapped to fresh partitioning $\bar{\omega}$ using remapping function λ_{Ω_e}. Effectuating this remapping obtains $\langle \widehat{DB}', \Omega', P' \rangle = \lambda_{\Omega_e}(\langle \widehat{DB}, \Omega, P \rangle)$. The component φ_e itself can also be rewritten into $\bar{\varphi}_e = \lambda_{\Omega_e}(\varphi_e)$ which results in a sentence of the form $\bar{\varphi}_e = \bar{l}_1 \vee \ldots \vee \bar{l}_m$ where $\bar{l}_j = (\bar{\omega}{=}v_j)$ for some m.

The evidence sentence $\bar{\varphi}_e$ specifies which worlds $W(\langle \widehat{DB}', \Omega', P' \rangle)$ are valid, namely those identified by each $\bar{l}_j$. Let $L = \{\bar{l}_1, \ldots, \bar{l}_m\}$. The worlds identified by $\bar{L} = L(\bar{\omega}) \setminus L$ are inconsistent with $\bar{\varphi}_e$. This can be effectuated in $\widehat{DB}'$ by setting labels identifying inconsistent worlds to $\bot$ in all sentences occuring in $\widehat{DB}'$. A descriptive assertion for which the sentence becomes $\bot$ can be deleted from the database as it is no longer present in any remaining world.

Let $\lambda_{\bar{L}}(\varphi)$ be the sentence obtained by setting l to $\bot$ in φ for each $l \in \bar{L}$. We can now define $\widehat{DB}''$ as follows:

$$\widehat{DB}'' = \{\langle a, \lambda_{\bar{L}}(\varphi) \rangle \mid \langle a, \varphi \rangle \in \widehat{DB}' \wedge \lambda_{\bar{L}}(\varphi) \neq \bot\}$$

Finally, the probability mass of the inconsistent worlds needs to be redistributed over the remaining consistent ones. Furthermore, since some labels $\bar{l}_j$ representing these inconsistent worlds should obtain a probability $P''(\bar{l}_j) = 0$, these labels should be removed, and because we assume the values of a partitioning ω^n to range from 1 to n, we renumber them by replacing $\bar{\omega}^n$ with $\hat{\omega}^m$.

Let $\Omega'' = (\Omega' \setminus \{\bar{\omega}^n\}) \cup \{\hat{\omega}^m\}$. The bijection $f : L(\hat{\omega}^m) \leftrightarrow L$ uniquely associates each new 'renumbered' label with an original label of a consistent world. In $\widehat{DB}''$ replace every occurrence of a label $\bar{l}_j \in L$ with $f(\bar{l}_j)$. Note that labels from $\bar{L}$ will no longer occur in $\widehat{DB}''$. P'' is defined by setting the probabilities of the new labels as follows: $P''(\bar{l}_j) = \frac{1}{p}P'(f(\bar{l}_j))$ where $p = \sum_{\bar{l}_j \in L} P'(\bar{l}_j)$.

In the next section, we turn `hardrule` into a soft rule and show what the end result for the conditioned Paris Hilton example looks like (see Fig. 6).

3.4 Conditioning with Soft Rules

A soft rule is an uncertain hard rule, hence the same principle of probabilistic data can be used to represent a soft rule: with a partitioning ω_r^2 where labels $\omega_r^2 = 0$ and $\omega_r^2 = 1$ identify all worlds where the rule is false and true, respectively. For Fig. 3, we write a_7 softrule :- annot(Ph1,P1,city), annot(Ph2,P2,person), contained(P1,P2) [r=1]. which effectively means that $\langle a_7, \top \rangle$ is replaced with $\langle a_7, r{=}1 \rangle$ in the database. We now have 12 worlds in Fig. 4: the original 6 ones, and those 6 again but without a_7.

Executing `softrule` results in $\{\langle true, x{=}2 \wedge y{=}2 \wedge r{=}1 \rangle, \langle false, \neg(x{=}2 \wedge y{=}2 \wedge r{=}1) \rangle\}$. Since it is a negative rule, $\varphi_e = \neg(x{=}2 \wedge y{=}2 \wedge r{=}1)$. Instead of direct conditioning for this evidence, we strive for the possible worlds as illustrated in Fig. 5. Depicted here are the original worlds in case $r{=}0$ and the conditioned situation in case $r{=}1$. It can be obtained by conditioning the database as if it was a hard rule, but effectuate the result only for worlds for which $r{=}1$.

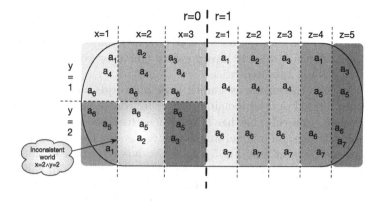

Fig. 5. Illustration of applying a soft rule.

Soft Rule Rewriting. Given $CPDBE = \langle \widehat{DB}, \Omega, P, \varphi_e \rangle$ and φ_e is a soft rule governed by partitioning ω_r. Let $\widehat{DB}'$ and φ_e' be the counterparts of $\widehat{DB}$ and φ_e where in all sentences $\omega_r = 1$ is set to $\top$ and $\omega_r = 0$ to $\bot$. Let $\Omega' = \Omega \setminus \{\omega_r\}$. Let P' be P restricted to the domain of Ω'. This effectively makes the rule a hard rule. Let $\langle \widehat{DB}'', \Omega'', P'' \rangle = \Lambda_{\varphi_e}(\langle \widehat{DB}', \Omega', P', \varphi_e' \rangle)$ be the database that incorporates the evidence as a hard rule.

From this result we construct a probabilistic database that contains both the data items from the original worlds when $\omega_r = 0$ and the data items from the rewritten worlds when $\omega_r = 1$. We define $\Lambda_{\varphi_e}(CPDBE) = \langle \widehat{DB}''', \Omega''', P''' \rangle$ where

$$\widehat{DB}''' = \{\langle a, (\varphi_1 \wedge \omega_r = 0) \vee (\varphi_2 \wedge \omega_r = 1) \rangle \mid \langle a, \varphi_1 \rangle \in \widehat{DB} \wedge (\omega_r = 0 \Rightarrow \varphi_1) \wedge \langle a, \varphi_2 \rangle \in \widehat{DB}'' \}$$
$$\cup \{\langle a, (\varphi_1 \wedge \omega_r = 0) \rangle \mid \langle a, \varphi_1 \rangle \in \widehat{DB} \wedge (\omega_r = 0 \Rightarrow \varphi_1) \wedge \langle a, \varphi_2 \rangle \notin \widehat{DB}'' \}$$
$$\cup \{\langle a, (\varphi_2 \wedge \omega_r = 1) \rangle \mid \langle a, \varphi_1 \rangle \in \widehat{DB} \wedge (\omega_r = 0 \not\Rightarrow \varphi_1) \wedge \langle a, \varphi_2 \rangle \in \widehat{DB}'' \}$$
$$\Omega''' = \Omega \cup \Omega''$$
$$P''' = P \cup P''$$

See Fig. 6 for the conditioned database of the Paris Hilton example.

a_1 `annot(id-ph,pos1-2,hotel)`
   `[(r=0 and x=1) or (r=1 and (z=1 or z=4))].`
a_2 `annot(id-ph,pos1-2,person)`
   `[(r=0 and x=2) or (r=1 and z=2)].`
a_3 `annot(id-ph,pos1-2,fragrance)`
   `[(r=0 and x=3) or (r=1 and (z=3 or z=5))].`
a_4 `annot(id-p,pos1,firstname)`
   `[(r=0 and y=1) or (r=1 and (z=1 or z=2 or z=3))].`
a_5 `annot(id-p,pos1,city)`
   `[(r=0 and y=2) or (r=1 and (z=4 or z=5))].`

a_6 `contained(pos1,pos1-2).`
a_7 `softrule :- annot(Ph1,P1,city), annot(Ph2,P2,person), contained(P1,P2) [r=1].`

`@p(x=1) = 0.5.`
`@p(x=2) = 0.4.`
`@p(x=3) = 0.1.`
`@p(y=1) = 0.3.`
`@p(y=2) = 0.7.`
`@p(z=1) = 0.2083.`
`@p(z=2) = 0.1667.`
`@p(z=3) = 0.0417.`
`@p(z=4) = 0.4861.`
`@p(z=5) = 0.0972.`
`@p(r=1) = 0.8.`
`@p(r=2) = 0.2.`

Fig. 6. Paris Hilton example with evidence of **softrule** incorporated as a soft rule.

3.5   Iterative Conditioning

The intention is to use this approach iteratively, i.e., whenever new evidence is specified, the evidence is directly incorporated. One may wonder what happens if the same rule is incorporated twice.

With hard rules the answer is simple: since all worlds inconsistent with the rule have been filtered out, all remaining rules are consistent with the rule, i.e., when the evidence is a rule that has already been incorporated $\varphi_e = \top$.

In case of soft rules, all original worlds, hence also the ones inconsistent with the rule, are still present (see Fig. 5). Observe, however, that all inconsistent worlds have r=0 in their full sentences. Applying the rule again, will leave all original worlds unaffected, because in those worlds the rule is not present. And where the rule is true, the worlds inconsistent with the rule have already been filtered out. Therefore, also for soft rules it holds that re-incorporating them leaves the database unaffected.

If, however, a soft rule $\langle r, r1 = 1 \rangle$ is incorporated again but governed by a different partitioning, i.e., $\langle r, r2 = 1 \rangle$, different probabilities for query answers are obtained. Note, however, that this pertains to a different situation: with both evidences based on $r = 1$, the evidence effectively comes from the same source twice, which provides no new evidence and the result is the same. With evidences based on different partitions, the evidence effectively comes from two different sources. Indeed, this provides extra independent evidence, hence probabilisties are conditioned twice.

Scalability. There are two main classes of probabilistic databases: relational PDBs and probabilistic logics. The first step of evaluating the evidence rule to obtain the evidence sentence φ_e has the same complexity as querying in such systems. Remapping and redistribution of probabilities depends on $|\Omega_e|$ which is exponential in the number of partionings involved in φ_e. We assume that uncertainty remains fairly local, i.e., after splitting only components with few partitionings remain. The same holds for simplification and normal form reduction of the sentences. Database rewriting affects all data items referring to remapped partitionings, which is worst case linear in the size of the database. The result is a probabilistic database again with at most the same size but possibly longer sentences, i.e., the complexity of querying the resulting database does not change. In short, assuming uncertainty remains local, algorithms implementing our approach are expected to be well-scalable.

4   Validation

The main proof obligation is that the database without evidence obtained by $\Lambda_{\varphi_e}(CPDBE)$ represents the same possible worlds as the original $CPDBE$.

Theorem 1. $W(\Lambda_{\varphi_e}(CPDBE)) = W(CPDBE)$

Proof Sketch. The proof sketch is based on showing that in each of the steps, the possible worlds remain the same. The first step splits the evidence sentence into independent components. Let $\varphi_e = \varphi_1 \wedge \varphi_2$. Since $W(CPDBE) = \{w \mid w \in W(CPDB) \wedge \varphi_e\}$ (see Eq. 6) and φ_1 and φ_2 share no partitionings, the filtering of worlds on $\varphi_1 \wedge \varphi_2$ is the same as filtering first on φ_1 and then on φ_2.

The second step is the remapping of the partitionings in the evidence sentence component. The remapping introduces a single fresh partitioning $\bar{\omega}^n$. Note that the remapping function λ_{Ω_e} is a bijection uniquely relating each full sentence $\bar{\varphi}$ constructed from $\Phi(\Omega_e)$ with one label $\bar{l} \in L(\bar{\omega}^n)$. In other words, $W(\bar{\varphi}) = W(\bar{l})$ hence the possible worlds remain the same (see Eqs. 2, 4 and 9)

$$W(CPDB) = \{DB \mid \bar{\varphi} \in \Phi(\Omega) \wedge DB = \{a \mid \langle a, \varphi \rangle \in \widehat{DB} \wedge \bar{\varphi} \Rightarrow \varphi\}\}$$
$$= \{DB \mid \bar{l} \in L(\bar{\omega}^n) \wedge DB = \{a \mid \langle a, \lambda_{\Omega_e}(\varphi) \rangle \in \widehat{DB} \wedge \bar{l} \Rightarrow \lambda_{\Omega_e}(\varphi)\}\}$$

Since $\lambda_{\Omega_e}(\varphi)$ replaces every label with an equivalent disjunction of fresh labels $\bar{\varphi} \Rightarrow \varphi$ is true whenever $\bar{l} \Rightarrow \lambda_{\Omega_e}(\varphi)$ is true. Therefore, remapping retains the same possible worlds. This can also be illustrated with Fig. 4. The six possible worlds in a 2-by-3 grid are remapped to a 1-by-6 grid containing the same distribution of assertions.

The above steps have transformed $W(CPDBE)$ into

$$W(CPDBE) = \{DB \mid \bar{l} \in L(\bar{\omega}^n)$$
$$\wedge DB = \{a \mid \langle a, \lambda_{\Omega_e}(\varphi) \rangle \in \widehat{DB} \wedge \bar{l} \Rightarrow \lambda_{\Omega_e}(\varphi)\}$$
$$\wedge \lambda_{\Omega_e}(\varphi_e)\}$$

It has already been noticed that, $\lambda_{\Omega_e}(\varphi_e)$ is of the form $\lambda_{\Omega_e}(\varphi_e) = (\bar{\omega}^n = v_1) \vee \ldots \vee (\bar{\omega}^n = v_m)$ for some m. The third step is setting labels identifying inconsistent worlds to $\bot$, i.e., labels $\bar{l} \notin \{(\bar{\omega}^n{=}v_1), \ldots, (\bar{\omega}^n{=}v_m)\}$. Figure 4 illustrates how the world identified by $z{=}5$ is eliminated, and the resulting database is

$$\{\langle a_1, z{=}1 \vee z{=}4 \rangle, \langle a_2, z{=}2 \rangle, \langle a_3, z{=}3 \vee z{=}6 \rangle, \langle a_4, z{=}1 \vee z{=}2 \vee z{=}3 \rangle, \langle a_5, z{=}4 \vee z{=}6 \rangle,$$
$$\langle a_6, z{=}1 \vee z{=}2 \vee z{=}3 \vee z{=}4 \vee z{=}6 \rangle, \langle a_7, z{=}1 \vee z{=}2 \vee z{=}3 \vee z{=}4 \vee z{=}6 \rangle\}$$

The label renumbering for $\bar{\omega}^n$ and redistribution of probability mass to labels $(\bar{\omega}^n{=}v_1), \ldots, (\bar{\omega}^n{=}v_m)$ in the remapped label space is equivalent with Eq. 7.

Figure 4 illustrates how the worlds remaining in $W(CPDBE) = \{w \mid w \in W(CPDB) \wedge \varphi_e\}$ (Eq. 6) after applying a soft rule are constructed by effectively taking the union of the $\omega_r{=}0$ *partition* of $W(CPDB)$ with the rewritten worlds of the $\omega_r{=}1$ partition of $W(CPDB)$.

5   Conclusions

The main contribution of this paper is an iterative approach for incorporating evidence of users in probabilistically integrated data, evidence which can be specified both as hard and soft rules. This capability makes the two-phase probabilistic data integration process possible where in the second phase, the use of integrated data could lead to evidence which can continuously improve the data quality. The benefit is that a data integration result can be more quickly obtained as it can be imperfect.

The first objective for future work is the engineering aspect of the approach: developing a software prototype with the purpose of investigating the scalability of the approach. Furthermore, more future work is needed to complete and improve aspects of the PDI process such as indeterministic approaches for other data integration problems, improving the scalability of probabilistic database technology, and application of PDI to real-world scenarios and data sizes.

References

1. van Keulen, M.: Probabilistic data integration. In: Sakr, S., Zomaya, A. (eds.) Encyclopedia of Big Data Technologies, pp. 1–9. Springer, Heidelberg (2018)
2. van Keulen, M., de Keijzer, A.: Qualitative effects of knowledge rules and user feedback in probabilistic data integration. VLDB J. **18**(5), 1191–1217 (2009)
3. van Keulen, M.: Managing uncertainty: The road towards better data interoperability. IT - Inf. Technol. **54**(3), 138–146 (2012)
4. Magnani, M., Montesi, D.: A survey on uncertainty management in data integration. JDIQ **2**(1), 5:1–5:33 (2010)
5. Dalvi, N., Ré, C., Suciu, D.: Probabilistic databases: diamonds in the dirt. Commun. ACM **52**(7), 86–94 (2009)
6. Suciu, D., Olteanu, D., Ré, C., Koch, C.: Probabilistic databases. Synth. Lect. Data Manage. **3**(2), 1–180 (2011)
7. Panse, F., van Keulen, M., Ritter, N.: Indeterministic handling of uncertain decisions in deduplication. JDIQ **4**(2), 9:1–9:25 (2013)
8. Wanders, B., van Keulen, M., van der Vet, P.: Uncertain groupings: probabilistic combination of grouping data. In: Chen, Q., Hameurlain, A., Toumani, F., Wagner, R., Decker, H. (eds.) DEXA 2015. LNCS, vol. 9261, pp. 236–250. Springer, Cham (2015). https://doi.org/10.1007/978-3-319-22849-5_17
9. Habib, M., Van Keulen, M.: TwitterNEED: a hybrid approach for named entity extraction and disambiguation for tweet. Nat. Lang. Eng. **22**, 423–456 (2016)
10. Raedt, L.D., Kimmig, A., Toivonen, H.: ProbLog: a probabilistic prolog and its application in link discovery. In: International Joint Conference on Artificial Intelligence (IJCAI), pp. 2468–2473. AAAI Press (2007)
11. Olmedo, F., Gretz, F., Jansen, N., Kaminski, B.L., Katoen, J.P., Mciver, A.: Conditioning in probabilistic programming. ACM Trans. Program. Lang. Syst. **40**(1), 4:1–4:50 (2018)
12. Theobald, M., De Raedt, L., Dylla, M., Kimmig, A., Miliaraki, I.: 10 years of probabilistic querying – what next? In: Catania, B., Guerrini, G., Pokorný, J. (eds.) ADBIS 2013. LNCS, vol. 8133, pp. 1–13. Springer, Heidelberg (2013). https://doi.org/10.1007/978-3-642-40683-6_1
13. Koch, C., Olteanu, D.: Conditioning probabilistic databases. Proc. VLDB Endow. **1**(1), 313–325 (2008)
14. van Keulen, M., Habib, M.: Handling uncertainty in information extraction. In: Proceedings of International Conference on Uncertainty Reasoning for the Semantic Web (URSW), vol. 778, pp. 109–112. CEUR-WS (2011)
15. Jayram, T.S., Krishnamurthy, R., Raghavan, S., Vaithyanathan, S., Zhu, H.: Avatar information extraction system. IEEE Data Eng. Bull. **29**(1), 40–48 (2006)
16. Wanders, B., van Keulen, M.: Revisiting the formal foundation of probabilistic databases. In: Conference of the International Fuzzy Systems Association and the European Society for Fuzzy Logic and Technology, IFSA-EUSFLAT, p. 47. Atlantis Press (2015)

17. Wanders, B., van Keulen, M., Flokstra, J.: JudgeD: a probabilistic datalog with dependencies. In: Proceedings of Workshop on Declarative Learning Based Programming, DeLBP, Number WS-16-07. AAAI (2016)
18. Fuhr, N.: Probabilistic datalog: a logic for powerful retrieval methods. In: International Conference on Research and Development in Information Retrieval (SIGIR), pp. 282–290. ACM (1995)
19. Ceri, S., Gottlob, G., Tanca, L.: Logic Programming and Databases. Springer, Heidelberg (1990). https://doi.org/10.1007/978-3-642-83952-8. ISBN 3-540-51728-6

Positional Scoring Rules with Uncertain Weights

Paolo Viappiani[(✉)]

Sorbonne Université, UMR7606 CNRS, LIP6, 4 pl. Jussieu, 75005 Paris, France
paolo.viappiani@lip6.fr

Abstract. Positional scoring rules are frequently used for aggregating rankings (for example in social choice and in sports). These rules are highly sensitive to the weights associated to positions: depending on the weights, a different winner may be selected. In this paper we explicitly consider the role of weight uncertainty in both the case of monotone decreasing weights and of convex decreasing weights. First we discuss the problem of finding possible winners (candidates that may win for a feasible instantiation of the weights) based on previous works that established a connection with the notion of stochastic dominance. Second, we adopt decision-theoretic methods (minimax regret, maximum advantage, expected value) to pick a winner based on the weight uncertainty and we provide a characterization of these methods. Finally, we show some applications of our methodology in real datasets.

Keywords: Scoring rules · Rank aggregation · Possible winners
Minimax regret · Stochastic dominance · Convex sequences

1 Introduction

In many contexts it is necessary to aggregate several rankings and either pick a winner or determine an output ranking. For example, rank aggregation emerges in recommender systems and social choice (preference aggregation), in information retrieval (aggregation of the output of several search engines), in sports (aggregation of the performance in several races into a single score). Positional scoring rules are frequently used due to their simplicity; in addition they satisfy a number of interesting properties (one of the most prominent results is that of Young [14]: a symmetric social choice function is continuous and consistent if and only if it is a scoring rule). Among positional scoring rules, Borda is a well-known method whose properties have been studied in depth [4,13].

A scoring rule assigns a scores to each candidate based on the rank obtained in each ranking. The output of a scoring rule crucially depends on the weights (attached to ranks); in general several different winners are possible with different weights. In this article we consider methods to generate a winner using positional

Work supported by the ANR project Cocorico-CoDec (ANR-14-CE24-0007-01).

© Springer Nature Switzerland AG 2018
D. Ciucci et al. (Eds.): SUM 2018, LNAI 11142, pp. 306–320, 2018.
https://doi.org/10.1007/978-3-030-00461-3_21

scoring rules under weight uncertainty. We assume non increasing weights; we also discuss the case of scoring rules where weights constitute a convex sequence.

We assume n agents express preferences in the form of rankings involving a set of m candidates $A = \{a, b, c, \ldots\}$; rankings are assumed to be linear orders (complete, transitive, asymmetric and irreflexive binary relation). Positional scoring rules discriminate between candidates by assigning a fixed score to each rank. Scoring rules assign a score to each alternative based on its rank distribution; let v_j^x be the number of times alternative x was ranked in the j-th position. Note that $\sum_{j=1}^{m} v_j^x = n$ for each $x \in A$ and $\sum_{x \in A} v_j^x = n$ for each $j = 1, \ldots, m$.

A scoring rule specifies the vector of weights $w_1, \ldots, w_m$ (also called *scoring vector*) to be assigned to each position. The score obtained by a candidate according to weight vector $w = (w_1, \ldots, w_m)$ is $s_w(x) = \sum_{j=1}^{m} w_j v_j^x$. These total scores can be used to pick a winner or to rank the alternatives. For example *plurality* is obtained by setting $w_1 = 1$ and $w_j = 0$ for all $j \in \{2, \ldots, m\}$.

By choosing a particular w, it is possible to specify some preferences on which kind of aggregation is desired, by giving more or less weight to the first positions compared to the positions that came afterwards in the ranking. First of all, we assume that not all weights are null, otherwise the alternatives are not discriminated (degenerated scoring rule). A natural hypothesis[1], that we adopt here, is to require that the sequence of weights is non-increasing: $w_i \geq w_{i+1}$ for all $i \in \{1, \ldots, m-1\}$; intuitively, in a ranking an alternative is (weakly) preferred to the alternatives that comes afterwards in the ranking.

A scoring rule is invariant to affine positive transformation of the scoring vector, that means that ranking the $w' = \alpha w + \beta$ with $\alpha > 0$ and arbitrary β gives the same output. Therefore, with no loss of generality, we let $w_1 = 1$ and $w_m = 0$ (therefore we have $m-2$ degrees of freedom). Given this assumption the *Borda* rule is given by setting $w_j = \frac{m-j}{m-1}$.

Moreover, it is often (but not always) assumed in practice that the positional weights constitute a convex sequence, meaning that the difference between the first and the second weight is not less than the difference between the second and of the third, and so on. In such a case the weights need to satisfy the following constraint, for each i between 1 and $m-2$:

$$w_i - w_{i+1} \geq w_{i+1} - w_{i+2} \iff w_i - 2w_{i+1} + w_{i+2} \geq 0. \tag{1}$$

Note that Borda and plurality are convex; furthermore convexity is often satisfied by the weights used when combining ranks in sports, races and other situations (e.g. formula one world championship, alpine skiing world cup).

We argue that setting a precise vector of weights for a scoring rule can be seen arbitrary; indeed the decision of which weights to use is critical since different weights often lead to different winners. Therefore there is interest to reason not just with a fixed vector w but about possible sets of parameters.

---

[1] This hypothesis is removed in Goldsmith et al. [5] where the authors allow preferences for intermediate positions.

Notation. We introduce some of the notation that we will use in the paper. We use $[\![m]\!]$ to denote the set $\{1, \ldots, m\}$. Given two vectors v^1 and v^2 we write $v^1 \succeq v^2$ iff $v_j^1 \geq v_j^2$ for all components j. If the inequalities are strict, $v_j^1 > v_j^2$ for all components j, then we write $v^1 \succ v^2$.

We will use W^D to denote the set of scoring vectors with non-increasing weights and W^C to denote the set of non-increasing (i.e. weakly decreasing) scoring vectors whose weights constitute a convex sequence[2] (with our boundary assumptions, $w_1 = 1$ and $w_m = 0$):

$$W^D = \left\{ (w_1, \ldots, w_m) \,\middle|\, 1 = w_1 \geq w_2 \geq \ldots \geq w_{m-1} \geq w_m = 0 \right\}, \tag{2}$$

$$W^C = \left\{ (w_1, \ldots, w_m) \,\middle|\, w \in W^D \wedge w_i - 2w_{i+1} + w_{i+2} \geq 0 \ \forall i \in [\![m-2]\!] \right\}. \tag{3}$$

2 Dominance and Possible Winners

In this section we discuss dominance relations and possible winners. The first step is to reformulate scoring rules in terms of cumulative ranks. This is useful in order to establish dominance relations between alternatives in the context of scoring rules. Dominance [11] holds between alternative x and alternative y iff the former has higher score than the latter alternative for any possible scoring rule; y is then said to be dominated. A rational decision maker will then never choose a dominated alternative; since, no matter how the weights are defined, there is another alternative that is at least as good (in case of weak dominance) or strictly better (strict dominance).

We will then refine undominated alternatives in order to identify which candidates are possible winners.

2.1 Reformulation Using Cumulative Standings

We consider *cumulative standings* that represent the fraction of times that a candidate was ranked over a certain point. Cumulative standings are defined as the cumulative sum of the rank vectors, starting from the first position. The vector $V^x = (V_1^x, \ldots, V_{m-1}^x)$ is such that $V_j^x = \sum_{l=1}^{j} v_l^x$ is the number of times that alternative x has been ranked in position j or better. Note that we consider in V^x only $m-1$ components: indeed V_m^x, the number of cumulative standings in the last place, would always equal to the number of voters n.

We now define the vector $\delta = (\delta_1, \ldots, \delta_{m-1})$, dubbed *differential weights*, as the vector of the differences between two successive positional weights of the original scoring vector:

$$\delta_j = w_j - w_{j+1} \quad \forall j \in [\![m-1]\!].$$

---

[2] There is some redundancy in the constraints: it is enough to assume convexity and $w_{m-1} \geq 0$ to ensure that the sequence is not increasing.

Remember that we assumed that $w_1 = 1$ and $w_m = 0$; this assumption means that we have $\delta_1 = 1 - w_2$ and $\delta_{m-1} = w_{m-1}$. The score obtained by an alternative x can now be expressed in function of δ and the cumulative standings V:

$$s_\delta(x) = \sum_{j=1}^{m-1} \delta_j^x V_j^x$$

where we use the subscript δ to underline the dependency on the differential weight. Decreasing weights in the original formulation correspond to positive differential weights. Note that the original rank vectors can be expressed in terms of cumulative standings: $v_j = V_j - V_{j-1}$; similarly the original weights can be recovered from the differential weights: $w_j = \sum_{l=j}^{m-1} \delta_l$. The requirement that the first weight should be equal to one can therefore be written as: $\sum_{j=1}^{m} \delta_j = 1$ (and this indirectly bounds all other weights to be lower than 1 due to monotonicity). The score of an alternative can therefore be seen as a convex combination of the cumulative ranks.

We now derive a reformulation that is useful for scoring vectors that constitute convex sequences. To do so we need to introduce the vector of *cumulative of the cumulative standings*:

$$\mathcal{V}_j^x = \sum_{l=1}^{j} V_l^x = \sum_{l=1}^{j} \sum_{o=1}^{l} v_o^x = \sum_{l=1}^{j} (j - l + 1) v_l^x \qquad j \in [\![m]\!]$$

We now define a new vector of parameters ϕ that represents the convexity of the weights w:

$$\begin{cases} \phi_j & = \delta_j - \delta_{j+1} = w_j - 2w_{j+1} + w_{j+2} \qquad j \in [\![m-2]\!] \\ \phi_{m-1} = \delta_{m-1} & = w_{m-1}. \end{cases} \tag{4}$$

Given that we assumed $w_1 = 1$, we have $\phi_1 = 1 - 2w_2 + w_3$ and, since $w_m = 0$, we have $\phi_{m-2} = \delta_{m-2} - \delta_{m-1} = w_{m-2} - 2w_{m-1}$. Stating that $\phi_j \geq 0$ for all components j is equivalent to require the weight vector w to be convex.

The score $s_w(x)$ of an alternative x under a scoring rule with weight vector w can now be expressed as $\sum_{j=1}^{m-1} \phi_j \mathcal{V}_j^x$ with ϕ obtained from w using Eq. 4. The constraint that all weights should be bounded and the highest weight, w_1 has value equal to one, becomes, in terms of convex weights, in assuming that $\sum_{l=1}^{m-1} l \phi_l = 1$.

We now show, in the following table, how some common scoring rules are expressed in terms of differential weights and in terms of convex weights.

Rule	Positional weights	Differential weights	Convex weights
Plurality	$w = (1, 0, \ldots, 0)$ $\underbrace{\qquad}_{m-1}$	$\delta = (1, 0, \ldots, 0)$ $\underbrace{\qquad}_{m-2}$	$\phi = (1, 0, \ldots, 0)$ $\underbrace{\qquad}_{m-2}$
k-approval	$w = (1, \ldots, 1, 0, \ldots, 0)$ $\underbrace{\quad}_{k}\ \underbrace{\quad}_{m-k}$	$\delta = (0, \ldots, 0, 1, 0, \ldots, 0)$ $\underbrace{\ }_{k-1}\ \underbrace{\ }_{m-k-1}$	$\phi = (0, \ldots, 0, -1, 1, 0, \ldots, 0)$ $\underbrace{\ }_{k-2}\ \underbrace{\ }_{m-k-2}$
Borda	$w = (1, \frac{m-2}{m-1}, \frac{m-3}{m-1}, \ldots, 0)$	$\delta = (\frac{1}{m-1}, \ldots, \frac{1}{m-1})$	$\phi = (0, \ldots, 0, \frac{1}{m-1})$ $\underbrace{\qquad}_{m-2}$
top-k Borda	$w = (1, \frac{k-2}{k-1}, \ldots, \frac{1}{k-1}, 0, \ldots, 0)$ $\underbrace{\qquad}_{m-k}$	$\delta = (\frac{1}{k}, \ldots, \frac{1}{k}, 0, \ldots, 0)$ $\underbrace{\ }_{k}\ \underbrace{\ }_{m-k-1}$	$\phi = (0, \ldots, 0, \frac{1}{k}, 0, \ldots, 0)$ $\underbrace{\ }_{k-1}\ \underbrace{\ }_{n-k-1}$

We highlight the following observations:

- k-approval is not convex when $k \geq 2$. The k-approval score of alternative x is exactly V_k^x.
- we call *top-K Borda* the scoring rule based on Borda restricted to the top k positions. The score of an alternative x with respect to top-k Borda is $\frac{V_k^x}{k}$.

2.2   Dominance Relations

The usefulness of the reformulations presented in Sect. 2.1 is that they can be used to discriminate candidates according to dominance relations, that allow to identify candidates that are less preferred than another one for any feasible scoring vector. Note that dominance only gives us with a partial order, so it is usually not enough to unambiguously define a winner.

Non-increasing Weights. When dealing with scoring vectors in W^D, i.e. with non increasing sequences of weights the set of possible scores obtained by a candidate x, with cumulative ranks V^x, are given by

$$\left\{ \sum_{j=1}^{m-1} \delta_j V_j^x \,\middle|\, \delta_1 \geq 0, \ldots, \delta_m \geq 0 \wedge \sum_{j=1}^{m} \delta_j = 1 \right\}.$$

Basically, all convex combinations of the components of V^x are possible. Since all elements of δ are non-negative, we can compare the cumulative ranks of two candidates componentwise to check if a dominance relation exists.

Proposition 1 [11]. *If $V^x \succeq V^y$ then the score x is necessarily at least as good than y for any scoring rule with non-increasing weights (and x is necessarily strictly better than x when $V^x \succ V^y$):*

- $V^x \succeq V^y \implies s_w(x) \geq s_w(y) \ \forall w \in W^D,$
- $V^x \succ V^y \implies s_w(x) > s_w(y) \ \forall w \in W^D.$

We say that x weakly dominates y in the first case, and that x strongly dominates y in the second case.

The previous statement can be seen as a form of *first-order stochastic dominance*. A candidate is said *dominated* if there exists another candidate that dominates the former.

Convex Weights. Assuming a non-increasing convex sequence (the scoring vector belongs to W^C), the space of possible scores associated to an alternative is given by

$$\Big\{ \sum_{j=1}^{m-1} \phi_j \mathcal{V}_j \Big| \phi_1 \geq 0, \dots, \phi_m \geq 0 \wedge \sum_{j=1}^{m-1} j\,\phi_j = 1 \Big\}.$$

All ϕ_j are non-negative since the sequence is convex. If each element of the vector $\mathcal{V}^x$ is at least as big as the corresponding element of $\mathcal{V}^y$, than x has at least the same score of y for any scoring rule with convex weight (and the analogous relation holds with strict inequalities).

Proposition 2 [11]. *If $\mathcal{V}^x \succeq \mathcal{V}^y$, then x is at least as good than y, for any scoring rule with a convex sequence of decreasing weights (and x is necessarily strictly better than x when $\mathcal{V}^x \succ \mathcal{V}^y$):*

- $\mathcal{V}^x \succeq \mathcal{V}^y \implies \forall w \in W^C \ s_w(x) \geq s_w(y),$
- $\mathcal{V}^x \succ \mathcal{V}^y \implies \forall w \in W^C \ s_w(x) > s_w(y).$

Again, the first case is referred as weak dominance *and the second as* strong dominance.

This is akin to *second order stochastic dominance*, but considering convex and not concave utility.

Example 1. Consider the following numeric example. The first table reports the distribution of the ranks, the second the cumulative ranks and the third the double cumulative ranks.

Candidate	v_1 v_2 v_3 v_4	Candidate	V_1 V_2 V_3	Candidate	$\mathcal{V}_1$ $\mathcal{V}_2$ $\mathcal{V}_3$
a	2 2 2 2	a	2 4 6	a	2 6 12
b	0 6 2 0	b	0 6 8	b	0 6 14
c	2 0 4 2	c	2 2 6	c	2 4 10
d	4 0 0 4	d	4 4 4	d	4 8 12

- When considering monotone weights, one can establish dominance by pairwise comparisons of rows in the table of cumulative ranks; for instance a weakly-dominates c since $V_1^a = V_1^c$, $V_2^a > V_2^c$, and $V_3^a = V_3^c$. The set of weakly-undominated candidates is then $\{a, b, d\}$. No strong domination holds.
- When considering convex weights, now d weakly dominates a; moreover d strongly dominates c since $V_j^d > V_j^c$ for all $j \in [\![m-1]\!]$. The set of weakly undominated candidates is $\{b, d\}$, while the set of strictly undominated is $\{a, b, d\}$.

2.3   Possible Winners

In the following we present the notions of *possible* and *necessary* winners under different assumptions about the scoring vector. The possible winners are all those candidates that may be winners under a realization of the weights; a necessary winner (if it exists) is a winner under any possible weight. Possible winners (and necessary winners) need to be undominated according to the relations described in the previous section. However it is important to note that there might be undominated alternatives that may not be a possible winner.

In what follows, let W be the set of feasible weights; we focus on W being either W^C or W^D.

Definition 1. *A candidate x is a possible co-winner iff there is a weight vector w such that the score of x under w is higher or equal than the score of all other candidates.*

$$\exists w \in W : s_w(x) \geq s_w(y) \ \forall y \in A$$

If the above formula holds with a strict inequality sign, x is a possible winner.

Definition 2. *A candidate x is a necessary co-winner iff for all weight vectors $w \in W$ the score of x under w is higher or equal than the score of all other candidates.*

$$\forall w \in W : s_w(x) \geq s_w(y) \ \forall y \in A$$

If the above holds with a strict inequality sign, x is a necessary winner.

Possible winners are a subset of maximal elements of the dominance relations seen in the previous sections; indeed a dominated candidate (in either sense, weak or strong) cannot be a possible winner for any scoring rule. A candidate that is weakly dominated by another candidate but it is not strongly dominated may be a possible co-winner. Moreover, note that if there is only a single candidate that is undominated, then it is a necessary winner.

We show that it may occur that an undominated alternative is not a possible winner with an example.

Example 2. Assume the rank distributions ($n = 12$, $m = 3$) associated to three candidates a, b, c presented in the following table (to be read as follows: a is ranked first 4 time, second 3 times and 5 times last).

	v_1	v_2	v_3
a	4	3	5
b	6	0	6
c	2	9	1

As usual we let $w_1 = 1$ and $w_3 = 0$, so the only free parameter is w_2. The score of a is $s(a) = 4 + 3w_2$; the score of b is $s(b) = 6$ and the score of c is $s(c) = 2 + 9w_2$. Can a be a winner for some values of w_2? The answer is no: for a to be better than b, we need $w_2 \geq \frac{2}{3}$ but for a to be better than c, w_2 should be less than $\frac{1}{3}$.

The cumulative ranks are $V^a = (4, 7)$, $V^b = (6, 6)$, $V^c = (2, 11)$ and no pairwise domination holds between a, b, or c. Therefore a is neither dominated by b or c but is not a possible winner (and not even a co-winner).

This kind of reasoning has been discussed in [1], dealing with combinatorial problems; we now present similar techniques for computing possible winners.

Computation of Possible Winners. We test whether candidate x is a possible winner by finding the maximum difference between its score and that of the best ranked alternative other than x. If this is positive, then candidate x is a possible winner, if it is zero is only a possible co-winner; otherwise it is not a possible winner. Formally, define the maximum advantage or margin MA(x) of x by:

$$\text{MA}(x) = \max_{w \in W} \min_{y \neq x} s_w(x) - s_w(y) = \max_{w \in W} \min_{y \neq x} \sum_{j=1}^{m} w_j v_j^x - \sum_{j=1}^{m} w_j v_j^y \quad (5)$$

$$= \max_{w \in W} \left\{ \sum_{j=1}^{m} w_j v_j^x - \max_{y \neq x} \sum_{j=1}^{m} w_j v_j^y \right\} \quad (6)$$

where $W \in \{W^D, W^C\}$ is the set of possible weight vectors (either the class of non-increasing or the class of convex weights). This can be achieved by the following optimization:

$$\max_{Z, w} Z \quad (7)$$

$$\text{s.t. } Z \leq \sum_{j=1}^{m} w_j v_j^x - \sum_{j=1}^{m} w_j v_j^y \quad \forall y \in A - \{x\} \quad (8)$$

$$w_j \geq w_{j+1} \quad (9)$$

$$w_j - 2w_{j+1} + w_{j+2} \geq 0 \quad \forall j \in [\![m-2]\!] \quad (10)$$

$$w_1 = 1; w_m = 0 \quad (11)$$

Equations. 7–11 represent a linear program that can be solved with standard optimization tools such as CPLEX or Gurobi. The alternative x is given in input. There are $m - 1$ decision variables, of which $m - 2$ represent the scoring vector (we have $m - 2$ degrees of freedom, since we assume $w_1 = 1$ and $w_m = 0$), and an additional decision variable Z (representing the margin) whose value is constrained (Eq. 8) to be less than the difference in score between the score of x and any other alternative $y \in A$. The resulting value of Z gives us the best margin with x being the winner when choosing w in W. If this value is positive then x is a possible winner. Constraint 10 refers to convex weights (region W^C) and should be removed when dealing with non-increasing weights (region W^D).

Example 3. Consider again the values of Example 1. The maximum advantage is given in the second table, once computed w.r.t. W^D and once w.r.t. W^C.

Candidate	v_1'	v_2'	v_3'	v_4'	Candidate	MA$_{W^D}$	MA$_{W^C}$
a	2	2	2	2	a	0	-0.29
b	0	6	2	0	b	2	0.66
c	2	0	4	2	c	0	-0.86
d	4	0	0	4	d	2	2

Note that when considering convex weights, we reduce the feasibility region and therefore the maximum margin will be less. Candidates b, d are possible winners in both cases, since they are associated with a positive maximum advantage. Note that even if a was undominated in W^D, it is not a possible winner; in W^D a and c are possible co-winners for scoring weight $w = (1, 0.5, 0.5, 0)$.

3   Aggregation of Scoring Rules with Uncertain Weights

In general there may be many possible winners, it is therefore often necessary to have a method to pick a single winner. We now discuss how to adapt classic criteria for decision-making under uncertainty for the case of scoring rules with uncertain weights.

Maximin and Maximax. Selecting the alternative according to *maximin* (criterion that picks the alternative whose worst score is highest) corresponds to setting each weight to zero, except w_1 that is equal to 1 by assumption; therefore this case corresponds to *plurality*. Instead maximizing the maximum possible score[3] (that means using *maximax* decision rule) corresponds to setting each weight to one, except w_m that is set to 0 by assumption; this case corresponds to $(m-1)$-*approval* for the case of non-increasing scoring rules. For convex rules, maximax is attained by Borda. Of course, these methods are trivial and do not actually support the idea that we should possibly consider a variety of scoring vectors (in W^m or in W^c depending on the case) for evaluating different candidates.

Minimax Regret. Minimax regret is a robust decision criterion classically used for optimization under uncertainty [6,10]; it has been more recently advocated to be used in decision-making where the uncertainty is over utility values [2,9]. In the context of voting, it has been proposed by Lu and Boutilier [8] as a way to deal with partially elicited rankings (but they assume that the social choice function is known precisely).

We consider now using minimax regret to identify a candidate (to be declared winner) in face of uncertainty over the values of the scoring vector $w_1, \ldots, w_m$. The idea is to associate each alternative with the maximum loss incurred in terms of score points (with respect to the "true" winner) assuming that an adversary can freely set the scoring vector; the minimax-regret alternative is the one that minimizes such loss. The max regret of alternative x is:

$$\mathrm{MR}(x) = \max_{w \in W} \max_{y \in A} s_w(y) - s_w(x) = \max_{w \in W} s_w^* - s_w(x) \qquad (12)$$

$$= \max_{w \in W} \left[\max_{y \in A} \sum_{j=1}^m w_j v_j^y \right] - \sum_{j=1}^m w_j v_j^x \qquad (13)$$

---

[3] Several authors, including [3], have proposed to take a similar *optimistic* approach, although the way the feasible set is defined makes the resulting rules quite different.

where W is either W^D (decreasing weights) or W^C (convex weights), depending on the context. We then pick the alternative with minimum max regret: $\arg\min_{x \in A} Z^{MR}(x)$; whatever the weight in W, the picked alternatives is at most $MR(x)$ score points from optimality.

We now show that we can characterize the minimax regret alternative in a way that it is not necessary to solve an optimization problem. As before, we slice out our analysis dealing with (1) non-decreasing weights and (2) convex sequences. Before presenting our results note that the winner according to k-approval is then just the alternative with highest value V_k (number of times x was ranked in position k or better); let $V_k^* = \max_{x \in A} V_k^x$ be such value. Similarly let $\mathcal{V}_k^* = \max_{x \in A} \mathcal{V}_k^x$ be the maximum value of the double cumulative ranks for a given position j.

Proposition 3. *In the case of non-increasing weights:*

$$MR(x) = \max_{j=1}^{m-1} V_j^* - V_j^x; \tag{14}$$

instead, assuming convex weights:

$$MR(x) = \max_{j=1}^{m-1} \frac{\mathcal{V}_j^* - \mathcal{V}_j^x}{j}. \tag{15}$$

Proposition 3 allows[4] us to compute max regret in an efficient way without solving an optimization problem. While in the original formulation of maximum regret (Eq. 13) the maximization is over a continuum of values (all feasible scoring vectors w), Eqs. 14 and 15 compute the maximum among a fixed number of alternatives (in the two cases, monotone and convex).

For non-increasing scoring vectors, note that that, since V_j^* is equivalent to the best score according to j-approval, the term $V_j^* - V_j^x$ is the loss occurred to x when considering j-approval; max regret can be seen as the maximum loss occurred by x with respect to the family of k-approval voting rules. *The minimax regret optimal alternative is then the candidate that is the least far away from the optimal score attained with any k-approval voting rule.*

Example 4. We provide an example of max regret computation using the first case of Proposition 3. One needs to consider the table of the cumulative standings; the max-regret of a given candidate is the the maximum shortfall between the values in the candidate's row compared to the best value (in bold below) in each column.

Candidate	v_1	v_2	v_3	v_4		Candidate	V_1	V_2	V_3		Candidate	MR_{W^D}
a	2	2	2	2		a	2	4	6		a	**2**
b	0	6	2	0		b	0	**6**	**8**		b	4
c	2	0	4	2		c	2	2	6		c	4
d	4	0	0	4		d	**4**	4	4		d	4

---

[4] Proofs are available from the author upon request.

Since we wish to minimize the maximum regret MR, the optimal alternative w.r.t. minimax regret is therefore a when considering non-increasing weights.

Now, note that, since $\frac{v_j^x}{j}$ represent the score associated to the top-j Borda rule, Eq. 15 basically states that the computation of max regret when w lies in W^C is equivalent to consider *the loss occurred by x with respect to the family of top-k Borda aggregators* (that includes plurality and Borda as a special case).

Example 5. We consider again the running example and compute the max regret values when considering that the scoring vector is a convex sequence, using Eq. 15. First, we compute the double cumulative distribution of the ranks; we then divide the second column by two, the third by three, etc. Then, for computing the maximum regret of a, we consider the maximum, among columns, between the value of a and the best (bold) values, i.e. $\max\{4 - 2, 4 - 3, 4.66 - 4\} = 2$. The minimax regret optimal candidate in this case is therefore d.

Candidate	$\frac{v_1^\cdot}{1}$	$\frac{v_2^\cdot}{2}$	$\frac{v_3^\cdot}{3}$	Candidate	MR_{W^C}
a	2	3	4	a	2
b	0	3	**4.66**	b	4
c	2	2	3.33	c	2
d	4	4	4	d	**0.66**

Expected Score. Assuming a distribution over the weights, it is possible to rank alternatives by the expected score. Let $P(w)$ be the such distribution. By the linearity of expectation, we can sort alternatives by using a scoring rules using the expected values of the uncertain weights: $\mathbb{E}_{w \sim P(w)}[s_w(x)] = \sum_{j=1}^m \mathbb{E}_{w \sim P(w)}[w_j]v_j^x$. Therefore it means that sorting the candidates by expected score – under distribution $P(w)$ – is equivalent to using a scoring rule whose weights are given by $\mathbb{E}[w_j]$, the expectations of the weights. The choice of the probability distribution over the weights is critical; we observe that if we assume an uniform distribution over non-increasing weights, sorting by expectation is equivalent to Borda.

Proposition 4. *Assume w uniformly distributed in W^D (non-increasing weights)*

$$\mathbb{E}_{w \sim U}[s_w(x)] = s_{Borda}(x)$$

For convex sequences, we don't have a closed formula for the expected weights in W^C. We rely on Monte Carlo methods based on Gibbs sampling to derive numerical values for the expected weights. In Fig. 1 we show the weights obtained by using expectation assuming an uniform distribution over all possible convex sequences. In the plot we compare the expected weights with that (normalized) used in official F1 car races (that adopt a fixed convex sequence). The F1 weights (dashed lines) can be considered rather "steep" since they only award points to a small number of top drivers while often there are about 15–20 drivers (cfr with expected weights with $m = 22$, in light blue); but if one wishes to award points only to the best 8 or 9 drivers, then the F1 weights are actually less "steep" than the ones obtained by expectation over convex sequence (in the plot consider the line representing $m = 9$, in violet).

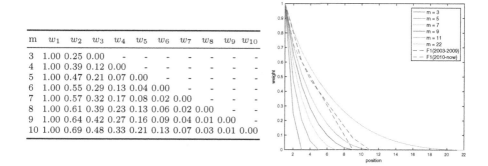

m	w_1	w_2	w_3	w_4	w_5	w_6	w_7	w_8	w_9	w_{10}
3	1.00	0.25	0.00	-	-	-	-	-	-	-
4	1.00	0.39	0.12	0.00	-	-	-	-	-	-
5	1.00	0.47	0.21	0.07	0.00	-	-	-	-	-
6	1.00	0.55	0.29	0.13	0.04	0.00	-	-	-	-
7	1.00	0.57	0.32	0.17	0.08	0.02	0.00	-	-	-
8	1.00	0.61	0.39	0.23	0.13	0.06	0.02	0.00	-	-
9	1.00	0.64	0.42	0.27	0.16	0.09	0.04	0.01	0.00	-
10	1.00	0.69	0.48	0.33	0.21	0.13	0.07	0.03	0.01	0.00

Fig. 1. Table (on the left): weights obtained by uniformly sampling a convex sequence of m positions; plot (right): sampled weights with the weights used by F1 races. (Color figure online)

4   Numerical Tests

We consider the rankings of the F1 race championship from 1961 to 2008[5]. In this context, a candidate is a driver and each ranking is the result of a race. In Table 1 below, we show the number of undominated alternatives and the number of possible winners in the case of monotone weights and that of convex weights. Several alternatives are Pareto dominated (e.g. a driver is Pareto dominated if there is another one that is better ranked in all races); the number of Pareto candidates is shown on the fourth column (for instance 16 drivers out of 54 are Pareto in 1961, while 17 out 22 in 2008).

We then analyze (using the propositions of Sect. 2) which drivers are undominated, and which are possible winners (Sect. 2.3). For monotone scoring vectors, there are often several possible winners but never more than 8. In just 5 years we have a necessary winner (1962, 1963, 1991, 1993, 2002), meaning that setting the scoring vector correctly is critical. When assuming convex scoring vectors, the number of possible winners is considerably reduced (and there is a necessary winner in several circumstances). Note that, although not frequently, there are sometimes undominated candidates that are not possible winners.

In Table 2 (top) we analyze how often the different rank aggregation methods (including ranking with the F1 point system and k-approval) disagree[6] on

---

[5] Obtained from the PREFLIB data repository (http://www.preflib.org/).

[6] To handle the case of ties (when a method returns multiple winners) we compute disagreement as the cardinality of the symmetric set difference normalized by the cardinality of the union.

Table 1. Number of undominated items and possible winners in actual rank data. (Legend: #Und. = number of undominated, #P.W.: = number of possible winners, m = number of drivers, n = number of races).

Dataset	m	n	Pareto	Monotone		Convex	
				#Und.	#P.W.	#Und.	#P.W.
1961	54	8	16	3	3	2	2
1962	51	9	12	1	1	1	1
1963	54	10	8	1	1	1	1
1964	42	10	17	6	5	4	4
1965	52	10	13	2	2	2	2
1966	43	9	20	4	4	2	2
1967	47	11	14	4	4	3	3
1968	44	12	22	4	4	2	2
1969	42	11	15	4	4	1	1
1970	43	13	21	8	8	5	4
1971	50	11	21	4	4	1	1
1972	44	12	25	7	7	1	1
1973	44	15	19	4	3	1	1
1974	62	15	23	3	3	2	2
1975	52	14	22	6	6	1	1
1976	57	16	23	6	6	3	3
1977	61	17	26	5	5	5	4
1978	49	16	26	6	6	1	1
1979	36	15	20	3	3	3	3
1980	41	14	23	2	2	1	1
1981	40	15	22	5	5	3	3
1982	40	16	26	6	6	4	4
1983	35	15	26	3	3	2	2
1984	35	16	24	3	3	2	2
1985	36	16	24	3	3	1	1
1986	32	16	20	4	4	2	2
1987	32	16	25	8	8	2	2
1988	36	16	16	2	2	2	2
1989	47	16	30	5	5	2	2
1990	40	16	20	5	5	3	3
1991	41	16	19	1	1	1	1
1992	39	16	23	5	5	1	1
1993	35	16	24	1	1	1	1
1994	46	16	24	4	4	2	2
1995	35	17	26	6	6	1	1
1996	24	16	20	2	2	1	1
1997	28	17	22	4	4	2	2
1998	23	16	17	3	3	2	2
1999	24	16	18	2	2	2	2
2000	23	17	21	4	4	2	2
2001	26	17	19	2	2	1	1
2002	23	17	5	1	1	1	1
2003	24	16	16	3	3	2	2
2004	25	18	16	2	2	1	1
2005	27	19	19	2	2	1	1
2006	27	18	19	2	2	1	1
2007	26	17	17	4	4	3	3
2008	22	18	17	4	4	2	2

Table 2. *Top:* average disagrement (with respect to winner) between the different rank aggregation methods. *Bottom:* average Spearman distance between the aggregated rankings.

Disagreement	F1	5-approval	Monotone		Convex	
			MR	$\mathbb{E}_U$	MR	$\mathbb{E}_U$
F1	-	0.30	0.17	0.29	0.17	0.05
5-approval	-	-	0.31	0.35	0.38	0.32
monotone-MR	-	-	-	0.28	0.18	0.15
m.-$\mathbb{E}_U$ (Borda)	-	-	-	-	0.39	0.27
convex-MR	-	-	-	-	-	0.11
convex-$\mathbb{E}_U$	-	-	-	-	-	-

Distance	F1	5-approval	Monotone		Convex	
			MR	$\mathbb{E}_U$	MR	$\mathbb{E}_U$
F1	-	0.15	0.17	0.22	0.16	0.15
5-approval	-	-	0.18	0.24	0.21	0.21
monotone-MR	-	-	-	0.20	0.14	0.15
m.-$\mathbb{E}_U$ (Borda)	-	-	-	-	0.16	0.14
convex-MR	-	-	-	-	-	0.11
convex-$\mathbb{E}_U$	-	-	-	-	-	-

picking the winner; while most of the time they pick the same winners, there is a considerable difference between. There is a considerable disagreement between Borda and the expected score assuming convexity (more than 1 out of 4 times the winner is different), and as well between Borda and convex MR. Since scoring rules can also be used to produce rankings, we also compare (Table 2 bottom) the rankings obtained by the different methods (e.g. sorting by ascending MR scores) using the Spearman distance.

5 Conclusions

In this article we have considered preference aggregation in the context of social choice when the scoring vectors are not defined apriori. The fact that using scoring rules with different weights may produce very different outputs have been noticed before; some of these methods are reviewed in [7]. In this paper we first considered the problem of discriminating which alternatives are possible winners, and the considered aggregation methods motivated by decision-theory (in particular, minimax regret and expected score).

In this paper we take the assumption that a committee specifies a generic settings for the scoring vector, that is using monotone or convex sequences; we believe that this is quite reasonable if one wish to use our proposed methods as social choice function. For minimax regret, the computation is quite easy using the cumulative (and double cumulative) ranks, therefore one could adopt such technique as an aggregation method. One could of course add additional desiderata about the weights or even consider elicitation protocols – in that case one would need to compute regret using optimization tools [12].

Note that the provided methods could be also used with an informative goal: for instance, for inspecting which candidates are possible winners (letting the decision maker discriminate between such alternatives in a second moment). The fact that the expectation of a uniformly distributed decreasing sequence gives the weights that are linearly spaced (as Borda), can be seen as an additional theoretical justification for using Borda.

References

1. Benabbou, N., Perny, P.: Incremental weight elicitation for multiobjective state space search. In: Proceedings of AAAI, pp. 1093–1099 (2015)
2. Boutilier, C., Patrascu, R., Poupart, P., Schuurmans, D.: Constraint-based optimization and utility elicitation using the minimax decision criterion. Artif. Intell. **170**(8–9), 686–713 (2006)
3. Cook, W.D., Kress, M.: A data envelopment model for aggregating preference rankings. Manag. Sci. **36**(11), 1302–1310 (1990)
4. Fishburn, P., Gehrlein, W.: Borda's rule, positional voting, and condorcet's simple majority principle. Public Choice **28**, 79–88 (1976)
5. Goldsmith, J., Lang, J., Mattei, N., Perny, P.: Voting with rank dependent scoring rules. In: Proceedings of AAAI, pp. 698–704 (2014)
6. Kouvelis, P., Yu, G.: Robust Discrete Optimization and Its Applications. Kluwer, Dordrecht (1997)
7. Llamazares, B., Peña, T.: Preference aggregation and DEA: an analysis of the methods proposed to discriminate efficient candidates. Eur. J. Oper. Res. **197**(2), 714–721 (2009)
8. Tyler, L., Boutilier, C.: Robust approximation and incremental elicitation in voting protocols. In: Proceedings of IJCAI, pp. 287–293 (2011)
9. Salo, A., Hämäläinen, R.P.: Preference ratios in multiattribute evaluation (PRIME)-elicitation and decision procedures under incomplete information. IEEE Trans. on Syst. Man Cybern. **31**(6), 533–545 (2001)
10. Savage, L.J.: The Foundations of Statistics. Wiley, New York (1954)
11. Stein, W.E., Mizzi, P.J., Pfaffenberger, R.C.: A stochastic dominance analysis of ranked voting systems with scoring. Eur. J. Oper. Res. **74**(1), 78–85 (1994)
12. Viappiani, P., Boutilier, C.: Regret-based optimal recommendation sets in conversational recommender systems. In: Proceedings of ACM RecSys, pp. 101–108 (2009)
13. Young, H.P.: An axiomatization of borda's rule. J. Econ. Theory **9**, 43–52 (1974)
14. Young, H.P.: Social choice scoring functions. SIAM J. Appl. Math. **28**(4), 824–838 (1975)

A New Measure of General Information on Pseudo Analysis

Doretta Vivona[1](✉)[iD] and Maria Divari[2]

[1] Faculty of Civil and Industrial Engineering, Department of Basic and Applied Sciences for Engineering, "Sapienza" University of Rome, Roma, Italy
doretta.vivona@sbai.uniroma1.it
[2] via Borgorose, Roma, Italy
maria.divari@alice.it

Abstract. The setting of this paper is the general information theory and the pseudo-analysis. We consider the general information measure J, defined without probability for crisp sets or without fuzzy measure for fuzzy sets and we propose a particular information measure for intersection of two sets. The pseudo-analysis is used to generalize the definition of independence and it leads to a functional equation. This equation belongs to a system of functional equations. We present some solutions of this system.

Keywords: Information theory · PseudoAnalysis Functional equations

1 Introduction

The setting of this research is the general information theory and the pseudo-analysis. On the one hand we introduce a general information measure J, on the other hand the pseudo-operations. The measure J is called *general* because it is defined without probability for crisp sets or without fuzzy measure for fuzzy sets. The pseudo-analysis involves two generalized operations $\oplus_g$ and $\ominus_g$: one of these, $\oplus_g$, is used to generalize the J-independence property.

Given two (crisp or fuzzy) sets A and B, the aim of this paper is to study the measure of $A \cap B$: the properties of the intersection lead to a system of functional equations. In this system it appears the following functional equation

$$\Phi(x, y) \oplus_g z = \Phi(x, y \oplus_g z) \qquad x, y \in (0, +\infty), \tag{1}$$

called by us *compatibility equation*.

The name *compatibility equation* has been used by other authors in different settings: for compositive information measures and entropies [2], for collectors of information measures [4], for conditional information [3].

Supported by GNFM of MIUR (ITALY).

© Springer Nature Switzerland AG 2018
D. Ciucci et al. (Eds.): SUM 2018, LNAI 11142, pp. 321–332, 2018.
https://doi.org/10.1007/978-3-030-00461-3_22

Also in this setting, we call (1) compatibility equation because it represents the link between the pseudo-analysis and the J−independence property in the general information theory.

First of all, we study the Eq. (1), then we apply the found results to the system and we are able to give some expression for the information measure of $JA \cap B$.

The paper is organized in the following way. In Sect. 2 we recall the preliminaries about pseudo-analysis and general information theory. The properties of this intersection are translated in a system of functional equations in Sect. 3 and the compatibility equation appears as a further equation in this system. The study of the compatibility equation is given in Sect. 4: the idempotent case and the general case are distinguished in order to look for some classes of solutions. In Sect. 5 the solutions of the whole system are shown and moreover some remarks are presented in Sect. 6. Section 7 is devoted to the conclusions.

2   Preliminaries

We begin with recalling some basic elements of the pseudo-analysis and the features of general information measures.

2.1   Pseudo-Analysis

The pseudo-analysis was introduced by Pap and his colleagues and it involves general operations. For the background we refer to [9–11]. We shall use only the pseudo-operations $\oplus$ and $\ominus$.

Definition 1. *The* pseudo-addition $\oplus$ *is a mapping*

$$\oplus : [0, M]^2 \longrightarrow [0, M], \ M \in (0, +\infty)$$

such that

(G_1) $a \leq a', b \leq b' \Longrightarrow a \oplus b \leq a' \oplus b';$
(G_2) $a \oplus b = b \oplus a;$
(G_3) $(a \oplus b) \oplus c = a \oplus (b \oplus c);$
(G_4) $a \oplus 0 = x;$
(G_5) $(a_n \longrightarrow b_n, a \longrightarrow b) \Longrightarrow a_n \oplus b_n \longrightarrow a \oplus b.$

We shall use only particular operations defined through a function g : $[0, M] \longrightarrow [0, +\infty]$ which is bijective, strictly increasing with $g(0) = 0$ and $g(M) = +\infty$. The function g is called *generator* function of the pseudo-addition, which is defined by

$$a \oplus_g b = g^{-1} \Big(g(a) + g(b) \Big). \tag{2}$$

Definition 2. *Let $\oplus_g$ be a pseudo-addition. The* pseudo-difference $\ominus_g$ *is a mapping*

$$\ominus : [0, M]^2 \longrightarrow [0, M], \; M \in (0, +\infty)$$

defined by

$$a \ominus_g b = \inf \left[x \in [0, M] : \; b \oplus_g x = a, \; a \geq b, \; b \oplus_g \left(a \ominus_g b \right) = a \right] \quad (3)$$

For all details see [5, 15].

2.2   General Information Measure

Now, we recall some preliminaries about general information measure.

Let Ω be an abstract space, $\mathcal{A}$ the σ–algebra of all subsets of Ω, such that $(\Omega, \mathcal{A})$ is measurable space and $\mathcal{K}$ a sub-family of $\mathcal{A} \times \mathcal{A}$.

Following Kampé de Feriét and Forte [6, 8], we recall that

Definition 3. *Measure of general information is a mapping*

$$J(\cdot) : \mathcal{A} \longrightarrow [0, +\infty]$$

such that $\forall \, A_1, A_2 \in \mathcal{A}$

(i) $A_1 \supset A_2 \Longrightarrow J(A_1) \leq J(A_2)$;
(ii) $J(\Omega) = 0, \;\; J(\emptyset) = +\infty$;
(iii) $J(K_1 \cap K_2) = J(K_1) + J(K_2), \; \forall \, (K_1, K_2) \in \mathcal{K}, \; K_1 \neq K_2, \; K_1 \cap K_2 \neq \emptyset$.

If the couple (K_1, K_2) satisfies (iii), we say that K_1 and K_2 are J–independent (independent each other with respect to J).

The J–independence property was generalized in [12].

2.3 J–Independence Property in Pseudo-Analysis

By using the pseudo-addition (2), in [13, 14], we have proposed the following:

Definition 4. *Two sets $K_1, K_2 \in \mathcal{K}, K_1 \neq K_2, K_1 \cap K_2 \neq \emptyset$ are J–independent in pseudo-analysis if*

$$J \left(K_1 \cap K_2 \right) = J(K_1) \oplus_g J(K_2). \quad (4)$$

3   Functional Equations in General Information Measure

In [13], given two sets $A, B \in \mathcal{A}$, we have considered the general information measure of their union: $J(A \cup B)$ as a function of $J(A)$ and $J(B)$, taking into account the J–independence property in pseudo-analysis (4).

3.1   Measure of Intersection

Now, we consider the subfamily of $\mathcal{A}$, defined by

$$\mathcal{A}_{O,X} = \{A \in \mathcal{A} : A \neq \emptyset, A \neq X\}. \tag{5}$$

Given two sets $A, B \in \mathcal{A}_{O,X}$, with $A \cap B \neq \emptyset, A \neq B$, we are going to consider the general information measure of their intersection: $J(A \cap B)$.

In particular, we suppose that this measure of the intersection depends only on $J(A)$ and $J(B)$: so we assume that there exists a function Φ such that

$$J(A \cap B) = \Phi\left(J(A), J(B)\right) \tag{6}$$

where $\Phi : V \to (0, +\infty)$ and $V = \Big\{(x, y) : \exists\, A, B \in \mathcal{A}_{O,X},$ with $x = J(A), y = J(B), \; x, y \in (0, +\infty)\Big\}$.

First of all, the intersection is commutative, associative, monotone with respect to ordinary inclusion and, as a consequence, $J(A \cap B) \geq J(A) \bigvee J(B)$.

In the Remark (**r2**) we shall examine the cases $x = 0$ and $x = +\infty$.

3.2   System of Functional Equations

In this paragraph, we are going to translate the properties of the intersection into a system of functional equations.

From now on, we suppose that:

(**ip**) for every $A, B \in \mathcal{A}_{O,X}$, there exists a set $K \in \mathcal{K}$ which is J-independent by $A, B, A \cap B$. Then by (4) we get:

$$J(A \cap K) = J(A) \oplus_g J(K), \;\; J(B \cap K) = J(B) \oplus_g J(K),$$

$$J((A \cap B) \cap K) = J(A \cap B) \oplus_g J(K).$$

Setting

$$x = J(A), \; y = J(B), \; z = J(C), \; x' = J(A'), \; t = J(K), \tag{7}$$

$$x, y, z, t, x' \in (0, +\infty), \;\; A, B, C, A' \in \mathcal{A}_{O,X}, \; K \in \mathcal{K},$$

the properties of (6) are translated into a system of functional equations:

$$
\begin{cases}
\textbf{(e1)} \; \Phi(x, y) = \Phi(y, x), \text{ (commutativity)} \\
\textbf{(e2)} \; \Phi(\Phi(x, y), z) = \Phi(x, \Phi(y, z),) \text{ (associativity)} \\
\textbf{(e3)} \; \Phi(x', y) \leq \Phi(x, y), \; x' \leq x, \text{ (monotonicity)} \\
\textbf{(e4)} \; \Phi(x, y) \geq x \bigvee y \\
\textbf{(e5)} \; \Phi(x, y) \oplus_g t = \Phi(x, y \oplus_g t), \\
\qquad \text{(\textbf{compatibility equation} between } \Phi \text{ and } \oplus_g).
\end{cases}
$$

We call (e5) compatibility equation because it represents the link between the pseudo-analysis and the J–independence property in the general information theory.

Let us spent few words to explain (e5). The associative property among A, B and K is

$$J((A \cap B) \cap K) = J(A \cap (B \cap K));$$

taking into account of (ip) and (6);

$$J((A \cap B) \cap K) = \Phi\left(J(A \cap B)\right) \oplus_g J(K) = \Phi(x, y) \oplus_g t, \qquad (8)$$

$$J((A \cap (B \cap K)) = \Phi\left(J(A), J(B) \oplus_g J(K)\right) = \Phi\left(x, y \oplus_g t\right). \qquad (9)$$

As $(8) = (9)$, we find the compatibility equation, which coincides with (e5).

4   Study of the Compability Equation

In the system (e1)–(e5) only the equation (e5) involves the pseudo-analysis. First, we are going to find some solutions of (e5), moreover we shall apply these results to the whole system.

Now, we are looking for function $\Phi : (0, +\infty)^2 \longrightarrow (0, +\infty)$ solution of (e5), which satisfies the following properties:

(p1) $x' \leq x \Longrightarrow \Phi(x', y) \leq \Phi(x, y) \quad \forall\, y;$
(p2) $\Phi(x, y) \geq x \bigvee y \quad \forall\, x, \, y \in (0, +\infty).$

From now on we shall indicate with $\mathcal{G} = \left\{\Phi : (0, +\infty)^2 \longrightarrow (0, +\infty),\right.$ continuous, satisfying (p1) and $\left.(p2)\right\}.$

In the application to general information measure, we shall see that the conditions (p1) and (p2) are satisfied.

We distinguish two main cases: *idempotent case* and *general case*.

4.1   Idempotent Case

From [7], we know that:

Definition 5. An element x^* is called *idempotent* element for any function Φ if

$$\Phi(x^*, x^*) = x^*.$$

We shall show the solution of (e5), step by step.

Lemma 1. *Let $\Phi \in \mathcal{G}$ be any solution of (e5). If there exists an idempotent element $x^* \in (0, +\infty)$ for Φ, then*

$$\Phi(x, x^*) = x \bigvee x^*, \ \forall\, x \in (0, +\infty).$$

Proof. Let $x < x^*$. For $(\mathbf{p_1})$: $\Phi(x, x^*) \leq \Phi(x^*, x^*) = x^*$, i.e.

$$\Phi(x, x^*) \leq x^*. \tag{10}$$

On the other hand, from $(\mathbf{p_2})$: $\Phi(x, x^*) \geq x \bigvee x^*$, i.e.

$$\Phi(x, x^*) \geq x^*. \tag{11}$$

From (10) and (11), we get

$$\Phi(x, x^*) = x^* = x \bigvee x^* \qquad \forall\, x \leq x^*. \tag{12}$$

Now, let $x \geq x^*$. Putting $y = x^*$ and $x = x^*$ in the equation $(\mathbf{e5})$, it becomes

$$\Phi(x^*, x^*) \oplus_g t = \Phi(x^*, x^* \oplus_g t), \text{ i.e. } x^* \oplus_g t = \Phi(x^*, x^* \oplus_g t),$$

as x^* is an independent element. Then,

$$\Phi(x^*, x^* \oplus_g t) = x^* \oplus_g t = x^* \bigvee (x^* \oplus_g t),$$

as $x^* \oplus_g t > x^*$. So, we have obtained

$$\Phi(x, x^*) = x \bigvee x^* \qquad \forall\, x \geq x^*. \tag{13}$$

The thesis

$$\Phi(x, x^*) = x \bigvee x^* \qquad \forall\, x \in (0, +\infty).$$

follows by (12) and (13). □

Lemma 2. Let $\Phi \in \mathcal{G}$ be any solution of $(\mathbf{e5})$. If there exists an idempotent element $x^* \in (0, +\infty)$ for Φ, then every $x < x^*$ is again an idempotent element for Φ.

Proof. Let $x < x^*$. Then, there exists $t > 0$ such that $x \oplus_g t = x^*$: by definition of pseudo difference (3) it is $x = x^* \ominus_g t$ and $(x^* \ominus_g t) \oplus_g t = x^*$.

Setting $y = x = x^* \ominus_g t$ in the equation $(\mathbf{e5})$, it becomes

$$\Phi\left(x^* \ominus_g t, x^* \ominus_g t \right) \oplus_g t = \Phi\left(x^* \ominus_g t, (x^* \ominus_g t) \oplus_g t \right).$$

But

$$\Phi\left(x^* \ominus_g t, (x^* \ominus_g t) \oplus_g t \right) = \Phi\left(x^* \ominus_g t, x^* \right) = (x^* \ominus_g t) \vee x^* = x^*$$

for Lemma 1. So,

$$\Phi\left(x^* \ominus_g t, x^* \ominus_g t \right) \oplus_g t = x^*$$

and then,

$$\left[\Phi\left(x^* \ominus_g t, x^* \ominus_g t \right) \oplus_g t \right] \ominus_g t = x^* \ominus_g t,$$

moreover, by (3)

$$\Phi\left(x^* \ominus_g t, x^* \ominus_g t\right) = x^* \ominus_g t,$$

that means that $x^* \ominus_g t$ is an idempotent element. As $x^* \ominus_g t < x^*, t > 0$ we have proved that every $x < x^*$ is an idempotent element. $\qquad \square$

Using the Lemmas 2 and 3, we are ready to give the *main proposition* in the idempotent case.

Proposition 1. *If there exists an idempotent element for the function* $\Phi \in \mathcal{G}$, *the solution of the compatibility equation is*

$$\Phi(x, y) = x \bigvee y \qquad \forall \, x, y \in (0, +\infty).$$

$\qquad \square$

4.2 General Case

In this section we present two classes of solutions of (**e5**), when an idempotent element doesn't exist.

We shall look for a class of solutions of the equation (**e5**) of the kind

$$\Phi_k(x, y) = x \oplus_k y \tag{14}$$

and we shall characterize the function k.

First of all, it is easy to see that

Proposition 2. *The class* (14) *is solution of* (**e5**) *if the function* k *coincides with the generator function* g *of the pseudo addition* (2).

Proof. The equation (**e5**) is reduced to associative property:

$$\Phi_k(x, y) \oplus_g z = \left(x \oplus_g y\right) \oplus_g z = x \oplus_g \left(y \oplus_g z\right) = \Phi_k(x, y \oplus_g z).$$

As the generator function g of the pseudo-addition $\oplus$ is associative (see (G_3)), then (14) is solution. $\qquad \square$

Proposition 3. *Let* $\oplus_g$ *be the pseudo-addition defined by a function* g *in* (2). *The function* (14) *is a solution of* (**e5**) *if the function* k *is:*

$$k(x) = \lambda \, g(x), \; \lambda > 0, \; k^{-1}(y) = g^{-1}\left(\lambda^{-1} y\right), \; \forall \, x, y \in (0, +\infty).$$

Proof. We have

$$x \oplus_k y = k^{-1}(k(x) + k(y)) = g^{-1}\left(\lambda^{-1}(\lambda g(x)) + \lambda^{-1}(\lambda g(y))\right) = x \oplus_g y.$$

So, we come back to Proposition 2. $\qquad \square$

Proposition 4. *Let g be any generator function of the pseudo addition $\oplus_g$, solution of the Cauchy equation* [1]

$$g(a) + g(b) = g(a \cdot b). \tag{15}$$

Then any function (14) *satisfies* (**e5**) *if k is an increasing solution of* (15).

Proof. First of all, taking into account of (15),

$$\Phi(x,y) = x \oplus_k y = k^{-1}(k(x) + k(y)) = k^{-1}(k(x \cdot y)) = x \cdot y. \tag{16}$$

Moreover,

$$\Phi(x,y) \oplus_g z = g^{-1}(g(x \cdot y) + g(z)) = g^{-1}(g(x \cdot y \cdot z)) = x \cdot y \cdot z; \tag{17}$$

on the other hand, by (16)

$$\begin{aligned}
\Phi(x, y \oplus_g z) &= x \cdot (y \oplus_g z) \\
&= x \cdot g^{-1}(g(y) + g(z)) = x \cdot g^{-1}(g(y \cdot z)) = x \cdot y \cdot z.
\end{aligned} \tag{18}$$

By (17) and (18), we get the assertion. □

Lemma 3. *Let g be any generator function of the pseudo addition $\oplus_g$, solution of the Cauchy Eq.* (15) *and h any increasing solution of functional equation*

$$h(a) \cdot h(b) = h(a \cdot b) \tag{19}$$

then the function $k = gh$ satisfies (19).

Proof. Applying (15) and (19), it is easy to see that

$$k(a) + k(b) = gh(a) + gh(b) = g(h(a)h(b)) = gh(a \cdot b) = k(a \cdot b).$$

□

Proposition 5. *Let g be any generator function of the pseudo addition $\oplus_g$, solution of the Cauchy Eq.* (15), *a class of solutions of* (**e5**) *is*

$$\Phi_h(x,y) = h^{-1}\left(h(x) \oplus_g h(y) \right), \tag{20}$$

where $h : (0, +\infty) \longrightarrow (0, +\infty)$, is any strictly increasing function, solution of (19).

Proof. The class of solutions (20) is of the kind (14) with $k = gh$ and $k^{-1} = h^{-1}g^{-1}$:

$$\Phi_h(x,y) = h^{-1}\left(h(x) \oplus_g h(y) \right) = h^{-1}g^{-1}\left(gh(x) + gh(y) \right) = x \oplus_k y.$$

The thesis follows by Lemma 3. □

Remark. When the generator function g of the pseudo-addition (2) is a linear function, the compatibility equation becomes

$$\Phi(x,y) + z = \Phi(x, y + z), \tag{21}$$

whose solution is $\Phi(x,y) = x + y$.

5   Solutions of the System (e1)–(e5)

In the previous paragraph we have studied the functional Eq. (1), finding some solutions when the unknown function Φ satisfies the properties $(\mathbf{p_1})$ and $(\mathbf{p_2})$.

However, the $(\mathbf{e3})$ and $(\mathbf{e4})$ above are just $(\mathbf{p_1})$ and $(\mathbf{p_2})$, so we get the following:

Proposition 6. *If there exists an idempotent element* (10) *for the function Φ, the solution of the system is*

$$\Phi(x, y) = x \bigvee y \qquad \forall\, x, y \in (0, +\infty). \tag{22}$$

Proof. It is easy to check that the function (14) satisfies $(\mathbf{e1})$–$(\mathbf{e4})$ and also $(\mathbf{e5})$ by Proposition 3. □

For the system $(\mathbf{e1})$–$(\mathbf{e5})$, we shall look for solutions of the kind (14).

Proposition 7. *If it doesn't exist an idempotent element* (10) *for the function Φ, the function*

$$\Phi(x, y) = x \oplus_g y,$$

where g is the generator function of $\oplus_g$, verifies the system $(\mathbf{e1})$–$(\mathbf{e5})$.

Proof. The thesis follows from Proposition 3 and the properties $(G_2), (G_3), (G_1)$ of the function g. □

As consequence of Propositions 4 and 8, there is

Proposition 8. *If there doesn't exist an idempotent element* (10) *for the function Φ, the class of the functions*

$$\Phi_k(x, y) = x \oplus_k y$$

verifies the system $(\mathbf{e1})$–$(\mathbf{e5})$ *if the function k is the following type:*

$$k(x) = \lambda\, g(x), \ \lambda > 0, \ k^{-1}(y) = g^{-1}\left(\lambda^{-1} y\right), \ \forall\, x, y \in (0, +\infty),$$

where g is the generator function of $\oplus_g$, and λ is any real positive number. □

For $\lambda = 1$ we come back to Proposition 8.

Now, we present other results.

Proposition 9. *Let g be any increasing generator function of the pseudo addition $\oplus_g$, solution of the Cauchy equation*

$$g(a) + g(b) = g(a \cdot b). \tag{23}$$

Then any class of the functions

$$\Phi_k(x, y) = x \oplus_k y = k^{-1}\left(k(x) + k(y)\right) \tag{24}$$

satisfies the system $(\mathbf{e1})$–$(\mathbf{e5})$ *if k is any strictly increasing solution of* (23).

Proof. First of all, the function Φ defined by (24) satisfies the (e5) by Proposition 5.

(e1) is valid by definition of (24).

As regards (e2), by using (G_3),

$$\Phi_k\left(\Phi_k(x,y),z\right) = k^{-1}\left\{k\Phi_k(x,y)+k(z)\right\} = k^{-1}\left\{kk^{-1}\left[k(x)+k(y)\right]+k(z)\right\} =$$
$$k^{-1}\left\{k(x)+\left[k(y)+k(z)\right]\right\} = k^{-1}\left\{k(x)+kk^{-1}\left[k(y)+k(z)\right]\right\} = \Phi_k\left(x,\Phi(y,z)\right).$$

(e3) is a consequence of monotonicity of the function k.

(e4) As the function k is positive, $k(x)+k(y) \geq k(x)$, then

$$k^{-1}\left(k(x)+k(y)\right) \geq k^{-1}k(x) = x.$$

So $\Phi_(x,y) \geq x$, $\Phi_k(x,y) \geq y$ and $\Phi_k(x,y) \geq x \bigvee y$. $\qquad\square$

Proposition 10. *Let g be any generator function of the pseudo addition $\oplus_g$, solution of the Cauchy Eq. (15), a class of solutions of the system (e1)–(e5) is*

$$\Phi_h(x,y) = h^{-1}\left(h(x)\oplus_g h(y)\right), \tag{25}$$

where $h : (0,+\infty) \longrightarrow (0,+\infty)$, is any strictly increasing function, solution of (19).

Proof. First of all, the (25) satisfies the (e5). The proof is the same of Proposition 9, by substituting $+$ with $\oplus_g$ and taking into account of the properties of the functions g and h. $\qquad\square$

6   Remarks

(r1) The associative property (8) of intersection among A, B and K is complete by writing

$$J((A\cap B)\cap K) = J(A\cap (B\cap K)) = J(B\cap (A\cap K));$$

from (7), the compatibility equation is

$$\Phi(x,y)\oplus_g z = \Phi(x,y\oplus_g z) = \Phi(y,x\oplus_g z). \tag{26}$$

It is easy to recognize that all propositions seen above are valid also for (26).

(r2) In the study of the information measure of $A\cap B$ we have considered A and B in the family (5).

Indeed, for $B = X$ it is $J(A\cap X) = J(A), \forall A$ and then, $\Phi : V_0 \to (0,+\infty)$, where $V_0 = \left\{(x,y) : \exists\, A, B \in \mathcal{A}, \text{ with } x = J(A), y = J(B), x, y \in [0,+\infty)\right\}$. So we must add another equation to the system (**e1**)–(**e6**). Setting $J(A) = x$, from (6), we get:

(**e7**) $\Phi(x, 0) = x, \ \forall \, x.$

Or, for $B = \emptyset$, it is $J(A \cap \emptyset) = J(\emptyset) = +\infty, \forall A$ and then, $\Phi : V_\infty \to (0, +\infty)$, where $V_{+\infty} = \Big\{ (x, y) : \exists \, A, B \in \mathcal{A}, \text{ with } x = J(A), y = J(B), x, y \in (0, +\infty] \Big\}$. So, setting $J(A) = x$, by (6), also in this case there is another equation in the system (**e1**)–(**e6**):

(**e8**) $\Phi(x, +\infty) = +\infty, \quad \forall \, x.$

The values 0 and $+\infty$ are separately idempotent elements for the function Φ which satisfies the systems (**e1**)–(**e6**)–(**e7**) or (**e1**)–(**e6**) and (**e8**), respectively.

Therefore the only solution of the systems (**e1**)–(**e6**)–(**e7**) or (**e1**)–(**e6**) and (**e8**), is

$$\Phi(x, y) = x \bigvee y \qquad \forall x, y \in [0, +\infty].$$

(**r3**) In general information theory, the (21) represents the independence axiom, for details see [3, 6, 8].

7   Conclusion

In this paper, by theoretical point of view, we have studied general information measures, i.e. without probability, of intersection set. We have supposed that the general information measure J of the intersection has the following form

$$J(A \cap B) = \Phi \Big(J(A), J(B) \Big).$$

Through the properties of the intersection we have obtained a system of functional equations in the sense of Aczel, system which involves the function Φ.

A meaning role is given by the so called "compatibility equation" (**e5**): it expresses the link between the J–independence property and the pseudo-analysis.

We have obtained the following results:

(1) If there is an idempotent element

$$J \Big(A \cap B \Big) = J(A) \bigvee J(B);$$

(2) If the idempotent element doesn't exist, it is possible to find some classes of solutions of the kind

$$J \Big(A \cap B \Big) = J(A) \oplus_k J(B),$$

where $\oplus_k$ is a pseudo-addition defined through a function k which satisfies some properties.

We think that it will be interesting to apply these results to particular general information measures.

References

1. Aczel, J.: Lectures on Functional Equations and Their Applications. Academic Press, New York (1966)
2. Benvenuti, P.: Sulle misure di informazione compositive con traccia compositiva unversale. Rendiconti di Matematica **2**(3–4), 481–505 (1969). Serie IV
3. Benvenuti, P.: Sur l'independence dans l'information, Colloquies Internationaux du C.N.R.S. N.276 - Theorie De l'Information, pp. 49–55 (1974)
4. Benvenuti, P., Divari, M., Pandolfi, M.: Su un sistema di equazioni funzionali provenienti dalla teoria soggettiva dell'informazione. Rendiconti di Matematica **5**(39), 529–540 (1972). Serie VI
5. Benvenuti, P., Mesiar, R., Vivona, D.: Handbook of Measure, chap. 33, Monotone set-functions-based integrals, pp. 1331–1379 (2002)
6. Forte, B.: Measures of information: the general axiomatic theory, R.A.I.R.O. Informatique Théorique et Appl. 63–90 (1969)
7. Halmos, P.R.: Measure Theory. Princeton, Van Nostrand Company (1966)
8. De Feriét, J.K., Forte, B.: Information et Probabilité. Comptes Rendue de L'Académie des Sci. Paris **265**, 110–114, 142–146, 350–353 (1967)
9. Mesiar, R., Rybarik, J.: Pseudo-aritmetical operations, Tatra. Mount. Math. Publ. **2**, 185–192 (1993)
10. Pap, E.: g-calculus Zr. Rad **23**(1), 145–156 (1993)
11. Pap, E.: Null-Additive Set Function. Kluver Academic Publishers, Dordrect (1995)
12. Vivona, D., Divari, M.: An independence property for genenral information. Nat. Sci. **8**, 66–69 (2016). https://doi.org/10.4236/ns.2016.82008
13. Vivona, D., Divari, M.: Pseudo-analysis: some measures of general information. In: IEEE 14th International Symposium on Intelligent Systems and Informatics, SISY 2016, 29–31 August 2016, Subotica, Serbia, pp. 141–144 (2016)
14. Vivona, D., Divari, M.: Pseudo-analysis: some measures of general conditional information. Adv. Sci. Technol. Eng. Syst. J. **2**(n2), 36–40 (2016)
15. Vivona, D., Divari, M.: On the functional equation: $F(x, y) \oplus z = F(x \oplus z, y \oplus z)$. Int. J. Pure Math. (2018, submitted)

A Formal Approach to Embedding First-Principles Planning in BDI Agent Systems

Mengwei Xu[✉], Kim Bauters, Kevin McAreavey, and Weiru Liu

University of Bristol, Bristol, UK
{mengwei.xu, kim.bauters, kevin.mcareavey, weiru.liu}@bristol.ac.uk

Abstract. The BDI architecture, where agents are modelled based on their beliefs, desires, and intentions, provides a practical approach to developing intelligent agent systems. However, these systems either do not include any capability for first-principles planning (FPP), or they integrate FPP in a rigid and ad-hoc manner that does not define the semantical behaviour. In this paper, we propose a novel operational semantics for incorporating FPP as an intrinsic planning capability to achieve goals in BDI agent systems. To achieve this, we introduce a declarative goal intention to keep track of declarative goals used by FPP and develop a detailed specification of the appropriate operational behaviour when FPP is pursued, succeeded or failed, suspended, or resumed in the BDI agent systems. Furthermore, we prove that BDI agent systems and FPP are theoretically compatible for principled integration in both offline and online planning manner. The practical feasibility of this integration is demonstrated, and we show that the resulting agent framework combines the strengths of both BDI agent systems and FPP, thus substantially improving the performance of BDI agent systems when facing unforeseen situations.

Keywords: BDI agent systems · First-principles planning
Decision making under uncertainty

1 Introduction

A well-studied and widely applied architecture for developing intelligent agents is the so-called Belief-Desire-Intention (BDI) paradigm. BDI builds a sound theoretical foundation to model an agent with an explicit representation of (B)eliefs, (D)esires, and (I)ntentions. This BDI paradigm has inspired a multitude of agent-oriented programming languages, such as AGENTSPEAK [1], CAN [2], and CAN-PLAN [3]. Notable BDI agent software platforms include, for example, Jack [4], Jason [5], and Jadex [6].

BDI agent systems are recognised for their efficiency and scalability in complex application domains, such as control systems [7] and power engineering [8]. However, they have often avoided the use of first-principles planning (FPP) in

© Springer Nature Switzerland AG 2018
D. Ciucci et al. (Eds.): SUM 2018, LNAI 11142, pp. 333–347, 2018.
https://doi.org/10.1007/978-3-030-00461-3_23

favour of a pre-defined plan library. While the use of a set of pre-defined plans simplifies the planning problem to an easier plan selection problem, obtaining a plan library that can cope with every possible eventuality requires adequate plan knowledge. This knowledge is not always available, particularly when dealing with uncertainty. Therefore, this limits the applicability and autonomy of BDI agent systems when there is no applicable plan for achieving a goal at hand. FPP can, on the other hand, synthesise a new plan to achieve a goal for which either no pre-defined plan worked or exists.

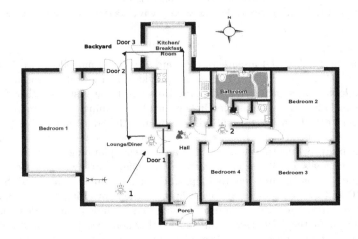

Fig. 1. Layout of a smart house with a domestic robot

To illustrate the problem, consider the following running example (see Fig. 1). In a smart home environment, there is an intelligent domestic robot whose job includes daily household chores (e.g. sweeping), security monitoring (e.g. burglary), and entertainment (e.g. playing music). The environment is dynamic and pervaded by uncertainty. When the robot does chores in the lounge, it may not be pre-encoded with plans to deal with an overturned clothes rack in the lounge, one of the doors to the hall being blocked unexpectedly, or urgent water overflow in a bathroom. Indeed, it is unreasonable to expect an agent designer to foresee all exogenous events and provide suitable pre-defined plans for all such eventualities. To address this weakness, a robot agent should be able to make use of FPP to generate novel plans to deal with such unforeseen events at design time in order to act intelligently.

Fortunately, to alleviate (some of) these issues, a large body of work on integrating various planning techniques with BDI have been proposed in recent years, as reviewed in [9]. For example, the work of [10] proposed an integration of AGENTSPEAK and a classical first-principles planner in which a new planning action in AGENTSPEAK is introduced to incorporate this planner. The BDI agent designer may include this new planning action at any point within a standard AGENTSPEAK plan to call a planner. In the work of [11], the authors provide a formal framework for FPP in BDI agent systems. This framework employs FPP to generate abstract plans, that is, plans that includes not only primitive actions, but also abstract actions summarised from the plan library. It allows for flexibility and robustness during the execution of these abstract plans. However, *none of these works have provided an operational semantics* that defines the behaviours of a BDI system with a built-in FPP. Moreover, the existing BDI systems (e.g. [10–12]) that integrate with FPP require the agent designer to define when the FPP is triggered. This limits the power and advantage of FPP to assist BDI agent systems to effectively accomplish their goals as the points of calling FPP can be unpredictable. Another important motivation of this paper is to respond to *the lack of work in strengthening the theoretical foundations of the BDI agent* pointed out by the comprehensive survey paper [9] as one of the future directions for planning in BDI agents. Therefore, the goal of this paper is to advance the state-of-art of planning in BDI agents by developing a rich and detailed specification of the appropriate operational behaviour when FPP is pursued, succeeded or failed, suspended, or resumed. In doing so, we introduce a novel operational semantics for FPP in BDI agent systems. This semantics specifies when and how FPP can be called, and articulates how a BDI agent system executes the new plan generated by FPP. To the best of our knowledge, we are not aware of any work on this problem so far.

The contributions of this paper are threefold. Firstly, we give a precise account of FPP within a typical BDI agent programming language, namely CAN. Secondly, the formal relationship between FPP and the BDI agent execution is established. Finally, a scenario case study is presented to highlight the usefulness and feasibility of the integration of FPP into BDI agent systems.

The remainder of the paper is organised as follows. In Sect. 2, we provide a brief overview of BDI and FPP. Sections 3.1 and 3.2 present the full operational semantics for integrating FPP into BDI. In Sect. 3.3, we establish the formal relationship between FPP and the BDI execution. In Sect. 4, the paper offers an intricate scenario discussion, which supports the feasibility of the resulting integrated framework and motivates the merits of the proposed framework to warrant future work on a fully implemented system. Section 5 discusses related work. Finally, in Sect. 6, we draw conclusions and outline future lines of research.

2 Preliminaries

CAN formalises the behaviours of a classical BDI agent, which is specified by a 5-tuple configuration $C = \langle \mathcal{B}, \Pi, \Lambda, \mathcal{A}, \Gamma \rangle$. The belief base $\mathcal{B}$ is a set of formulas encoding the current beliefs. The plan library Π is a collection of plan rules of the form $e : \varphi \leftarrow P$ with e the triggering event, φ the context condition, and P the plan-body program. The language used in the plan-body program P is defined by the following Backus-Naur grammar:

$$P ::= nil \mid act \mid ?\varphi \mid +b \mid -b \mid !e \mid P_1; P_2 \mid P_1 \triangleright P_2 \mid P_1 \parallel P_2 \mid$$
$$(|\psi_1 : P_1, \cdots, \psi_n : P_n|) \mid goal(\varphi_s, P, \varphi_f)$$

with *nil* an empty program, *act* a primitive action, $?\varphi$ a test for φ entailment in the belief base, $+b$ and $-b$ respectively belief addition and deletion, and $!e$ a subgoal. In addition, we use $P_1; P_2$ for sequence, $P_1 \triangleright P_2$ to execute P_2 only on failure of P_1, and $P_1 \parallel P_2$ for interleaved concurrency. A set of relevant plans is encoded by $(|\psi_1 : P_1, \cdots, \psi_n : P_n|)$. A goal program $goal(\varphi_s, P, \varphi_f)$ states that the declarative goal φ_s should be achieved through the procedural program P, failing when φ_f becomes true and retrying (alternatives) as long as neither φ_s nor φ_f is true (see [13]). The action library Λ is a collection of actions *act* in the form of $a : \psi \leftarrow \phi^-; \phi^+$. We have that ψ is the precondition, while ϕ^- and ϕ^+ denote respectively a delete and add set of belief atoms, i.e. propositional atoms. The sequence of actions executed so far by an agent is denoted as $\mathcal{A}$. The intention base Γ consists of a set of (partially) executed plans P. A basic configuration $\langle \mathcal{B}, \mathcal{A}, P \rangle$[1], with the plan-body program P being executed (i.e. the current intention), is also often used in notations to define what it means to execute a single intention.

The operational semantics for a BDI agent are defined in terms of configurations $\mathcal{C}$ and transitions $\mathcal{C} \to \mathcal{C}'$. A transition $\mathcal{C} \to \mathcal{C}'$ denotes that executing a single step in configuration $\mathcal{C}$ yields $\mathcal{C}'$. We write $\mathcal{C} \to$ (resp. $\mathcal{C} \nrightarrow$) to state that there is (resp. is not) a $\mathcal{C}'$ such that $\mathcal{C} \to \mathcal{C}'$, and $\xrightarrow{*}$ to denote the transitive closure of $\to$. A derivation rule specifies in which cases an agent can transition to a new configuration. Such a rule consists of a (possibly empty) set of premises p_i and a single transition conclusion c, denoted by

$$\frac{p_1 \quad p_2 \quad \cdots \quad p_n}{c} \quad l$$

where l is a label for reference. We refer the reader to [2,13] for a full overview of the semantics of CAN.

A FPP problem is defined as a 3-tuple $P = \langle \varphi_s, \mathcal{B}, \Lambda \rangle$, where φ_s is a set of successful goal states to be achieved, i.e. a set of formulas over some logical language, $\mathcal{B}$ stands for a set of initial belief states, and Λ represents the action library (defined as before). A first-principles planner takes as input the models of all known actions (i.e. action library Λ), a description of the state of the world (i.e. the initial state $\mathcal{B}$), and some objective (i.e. goals φ_s). It returns a sequence of actions σ which solves P, denoted $\sigma = sol(P)$.

3   First-Principles Planning in BDI Agent Systems

We now discuss how CAN agent systems and first-principles planning (FPP) can be integrated into a single framework. The resulting framework, called CAN(FPP), allows us to define agents that can perform FPP to provide new behaviours at runtime in an uncertain environment. We start by introducing the concept of declarative goal intention (used by FPP) and its semantical operation

---

[1] The plan and action libraries Π and Λ are omitted under the assumption that they are static entities, i.e. they remain unchanged as the agent moves between configurations.

in Sect. 3.1. The semantical behaviours of FPP within BDI execution presented in Sect. 3.2 is subsequently underpinned by the formal relationship between FPP and BDI execution in Sect. 3.3.

3.1 Declarative Goal Intention for First-Principles Planning

In a CAN agent, the intention set Γ is limited to just procedural goals. While valuable, procedural goals only describe *how* to achieve a given goal and do not answer the question as to which goals FPP should be trying to achieve in the BDI agent. To address this shortcoming, we modify the intention in this work to be a pair of sets, such that $\Gamma = \langle \Gamma_{pr}, \Gamma_{de} \rangle$ with Γ_{pr} and Γ_{de} a set of procedural and declarative goals, respectively. It allows us to keep track of both procedural goals (executed by the BDI engine) and declarative goals that tells us *what* we want to achieve (used by FPP). The set of declarative goals is furthermore partitioned into the subset of active goals Γ_{de}^{+}, and the suspended goals Γ_{de}^{-}. As a (slight) abuse of notation, we assume that adding an element to Γ_{de}^{+} ensures the element is removed from Γ_{de}^{-} and vice versa.

We start with the definition of a pure declarative goal and semantically enumerate three strategies, namely, *direct, belief-driven,* or *recovery-aid* strategy, to add such a pure declarative goal into the declarative goal intention to plan for by FPP.

A pure declarative goal $goal(\varphi_s, \varphi_f)$ is obtained from the ordinary declarative goal $goal(\varphi_s, P, \varphi_f)$ in CAN by dropping the procedural component P. It is read as "*achieve φ_s; failing if φ_f becomes true*" and defined to be the element of the declarative goal intention Γ_{de}. This new goal structure encodes the minimum information of what FPP needs to achieve (i.e. successful state φ_s) and when it is sensible to halt FPP (i.e. failure state φ_f).

The **first** *direct* strategy is to add a pure declarative goal into declarative goal intention when an ordinary declarative goal $goal(\varphi_s, nil, \varphi_f)$ is initially written as a part of the plan-body program. Here, $P = nil$ implies that there is no available procedural information on how to achieve the goal. Such a scenario occurs when either the procedural plan was not known during design time, or no efforts were made to create pre-defined plans (e.g. due to the priority of other part of plan library design tasks). Once the BDI agent selects goal $goal(\varphi_s, nil, \varphi_f)$ into procedural goal intention set Γ_{pre} (first premise), a pure declarative goal $P^{\uparrow} = goal(\varphi_s, \varphi_f)$ is automatically added to Γ_{de} by dropping nil if $goal(\varphi_s, \varphi_f)$ is not already in Γ_{de} (second premise):

$$\frac{P = goal(\varphi_s, nil, \varphi_f) \in \Gamma_{pre} \quad P^{\uparrow} \notin \Gamma_{de}}{\langle \mathcal{B}, \Pi, \Lambda, \mathcal{A}, \Gamma \rangle \xrightarrow{goal} \langle \mathcal{B}, \Pi, \Lambda, \mathcal{A}, \langle \Gamma_{pr} \setminus \{P\}, \Gamma_{de}^{+} \cup \{P^{\uparrow}\} \rangle \rangle} \; A_{goal}^{1}$$

The **second** *belief-driven* strategy is to allow adding a pure declarative goal to Γ_{de} in a proactive manner through the motivational library $\mathcal{M}$[2]. Inspired

---

[2] We only explicitly mention $\mathcal{M}$ in the agent configuration of A_{goal}^{2}; for all other rules, the library does not change and is omitted.

by conditionalised goals [14], a motivation planning library $\mathcal{M}$ is, a collection of rules of the form: $\psi \rightsquigarrow goal(\varphi_s, nil, \varphi_f)$, to add a declarative goal based on changes in beliefs. Semantically, we add a derivation rule for the motivational library $\mathcal{M}$ so that a pure declarative goal is added to Γ_{de} when the rule is triggered (second premise), the goal has an empty procedural component (third premise), and the goal $goal(\varphi_s, \varphi_f)$ is not already in Γ_{de} (fourth premise):

$$\frac{\psi \rightsquigarrow P \in \mathcal{M} \quad \mathcal{B} \models \psi\theta \quad P = goal(\varphi_s, nil, \varphi_f) \quad P^\uparrow\theta \notin \Gamma_{de}}{\langle \mathcal{B}, \Pi, \Lambda, \mathcal{A}, \Gamma \rangle \xrightarrow{goal} \langle \mathcal{B}, \Pi, \Lambda, \mathcal{A}, \langle \Gamma_{pr} \setminus \{P\}, \Gamma_{de}^+ \cup \{P^\uparrow\theta\} \rangle \rangle} A_{goal}^2$$

The **third** *recovery-aid* strategy is to overcome the limitations of the first and second strategy by recovering the unexpected failure of procedural plan program. It adopts $goal(\varphi_s, \varphi_f)$ into Γ_{de} when an ordinary declarative goal $goal(\varphi_s, P', \varphi_f)$ has a blocked procedural component, i.e. $P' \neq nil$ and $\langle \mathcal{B}, \mathcal{A}, goal(\varphi_s, P', \varphi_f) \rangle \nrightarrow$. The failure handling mechanism in [13] is already capable of (partially) dealing with such a situation. However, goals can still be blocked if failure handling mechanism failed. Since one of the properties of declarative goals held by a rational agent is that they should be persistent [2], it is rational to try and pursue these blocked declarative goals using FPP. We have:

$$\frac{P' = (P'' \triangleright P'') \vee nil \quad \langle \mathcal{B}, \mathcal{A}, goal(\varphi_s, P', \varphi_f) \rangle \nrightarrow \quad P^\uparrow \notin \Gamma_{de}}{\langle \mathcal{B}, \Pi, \Lambda, \mathcal{A}, \Gamma \rangle \xrightarrow{goal} \langle \mathcal{B}, \Pi, \Lambda, \mathcal{A}, \langle \Gamma_{pr} \setminus \{goal(\varphi_s, P', \varphi_f)\}, \Gamma_{de}^+ \cup \{P^\uparrow\} \rangle \rangle} A_{goal}^3$$

where $P^\uparrow = goal(\varphi_s, \varphi_f)$.

Finally, when either φ_s or φ_f is true, the pure declarative goal $goal(\varphi_s, \varphi_f)$ has been completed and it is dropped from Γ_{de}:

$$\frac{G \in \Gamma_{de} \quad G = goal(\varphi_s, \varphi_f) \quad \mathcal{B} \models \varphi_s \vee \varphi_f}{\langle \mathcal{B}, \Pi, \Lambda, \mathcal{A}, \Gamma \rangle \xrightarrow{drop} \langle \mathcal{B}, \Pi, \Lambda, \mathcal{A}, \Gamma_{de} \setminus \{G\} \rangle} G_{drop}$$

3.2   First-Principles Planning for Declarative Goals

We now consider how to invoke FPP and how it integrates with the BDI system and we use $plan(goal(\varphi_s, \varphi_f))$ to symbolise calling FPP.

The following two derivation rules G_s and G_f handle the cases where either the success condition φ_s or the failure condition φ_f holds.

$$\frac{\mathcal{B} \models \varphi_s}{\langle \mathcal{B}, \mathcal{A}, plan(goal(\varphi_s, \varphi_f)) \rangle \longrightarrow \langle \mathcal{B}, \mathcal{A}, nil \rangle} G_s \quad \frac{\mathcal{B} \models \varphi_f}{\langle \mathcal{B}, \mathcal{A}, plan(goal(\varphi_s, \varphi_f)) \rangle \longrightarrow \langle \mathcal{B}, \mathcal{A}, ?false \rangle} G_f$$

Intuitively, on success the plan is completed (replaced by nil). On failure, this is signalled to the BDI agent so that the basic CAN semantics can take over again. From now on, we will also distinguish between online planning [15] and offline planning [16] and give different derivation rules for accommodating each style of FPP due to their contrasting nature.

In *offline* planning, a complete sequence of actions σ to solve FPP problem $\langle \varphi_s, \mathcal{B}, \Lambda \rangle$ is generated first and executed afterwards.

The derivation rule for *offline* planning is defined as follows:

$$\frac{P = plan(goal(\varphi_s, \varphi_f)) \quad \langle \mathcal{B}, \mathcal{A}, P \rangle \xrightarrow{plan} \langle \mathcal{B}', \mathcal{A}', \sigma \rangle \quad \langle \mathcal{B}, \mathcal{A}, \sigma \rangle \xrightarrow{bdi_*} \langle \mathcal{B}'', \mathcal{A}'', nil \rangle}{\langle \mathcal{B}, \mathcal{A}, \Gamma \rangle \xrightarrow{bdi} \langle \mathcal{B}, \mathcal{A}, \Gamma' \rangle} \; P_{\mathcal{F}^{off}}$$

where $\Gamma = \langle \Gamma_{pr}, \Gamma_{de}^+ = \{goal(\varphi_s, \varphi_f)\} \rangle$ and $\Gamma' = \langle \Gamma_{pr} \cup \{\sigma\}, \Gamma_{de} \setminus \{goal(\varphi_s, \varphi_f)\} \rangle$.

It shows that the configuration $\langle \mathcal{B}, \mathcal{A}, \langle \Gamma_{pr}, \Gamma_{de}^+ = \{goal(\varphi_s, \varphi_f)\} \rangle \rangle$ will evolve to $\langle \mathcal{B}, \mathcal{A}, \langle \Gamma_{pr} \cup \{\sigma\}, \Gamma_{de} \setminus \{goal(\varphi_s, \varphi_f)\} \rangle \rangle$ if FPP can generate a sequence of actions σ (i.e. $\langle \mathcal{B}, \mathcal{A}, P \rangle \xrightarrow{plan} \langle \mathcal{B}, \mathcal{A}, \sigma \rangle$) that can achieve the successful state φ_s (i.e. $\langle \mathcal{B}, \mathcal{A}, \sigma \rangle \xrightarrow{bdi_*} \langle \mathcal{B}'', \mathcal{A}'', nil \rangle$). Once the successful state φ_s is met after the execution of the sequence of actions σ, the structure $plan(goal(\varphi_s, \varphi_f))$ will transition into nil and the goal $goal(\varphi_s, \varphi_f)$ will be dropped from Γ_{de} (i.e. $\Gamma_{de} \setminus \{goal(\varphi_s, \varphi_f)\}$) according to the above derivation rule G_s and G_{drop}.

In *online* planning, a single action is returned based on current belief states instead of generating the whole plan a priori, and executed immediately. The next action will be generated based on newly reached belief states. The loop of *"plan one action–execute one action"* will be iterated until the goal is reached.

The derivation rule for *online* planning is defined as follows:

$$\frac{P = plan(goal(\varphi_s, \varphi_f)) \quad \langle \mathcal{B}, \mathcal{A}, P \rangle \xrightarrow{plan} \langle \mathcal{B}', \mathcal{A}', act \rangle \quad act\theta = a}{\langle \mathcal{B}, \mathcal{A}, \Gamma \rangle \xrightarrow{bdi} \langle \mathcal{B}, \mathcal{A}, \Gamma' \rangle} \; P_{\mathcal{F}^{on}}$$

where $\Gamma = \langle \Gamma_{pr}, \Gamma_{de}^+ = \{goal(\varphi_s, \varphi_f)\} \rangle$ and $\Gamma' = \langle \Gamma_{pr} \cup \{I\}, \Gamma_{de}^- \cup \{goal(\varphi_s, \varphi_f)\} \rangle$ with $I = act; activate(goal(\varphi_s, \varphi_f))$. The intention $act; activate(goal(\varphi_s, \varphi_f))$ pursues the action act which was returned from FPP. When the action act is executed, it ensures that FPP is called again through reactivating the goal $goal(\varphi_s, \varphi_f)$. As such, FPP can take the new belief into consideration and plan for the next action. These two interleaved planning and execution will be repeated until the successful state is achieved if all possible.

The derivation rule to reactive the suspended goal is as follows:

$$\frac{P \in \Gamma_{pr} \quad P = activate(goal(\varphi_s, \varphi_f))}{\langle \mathcal{B}, \mathcal{A}, \Gamma \rangle \xrightarrow{bdi} \langle \mathcal{B}, \mathcal{A}, \langle \Gamma_{pr} \setminus \{P\}, \Gamma_{de}^+ \cup \{goal(\varphi_s, \varphi_f)\} \rangle \rangle} \; A_{re}$$

In addition, a trivial goal can be safely terminated:

$$\frac{}{\langle \mathcal{B}, \mathcal{A}, plan(nil) \rangle \xrightarrow{bdi} \langle \mathcal{B}, \mathcal{A}, nil \rangle} \; P_\top$$

Finally, when no solution is found to achieve goal state φ_s, the BDI agent will drop $plan(goal(\varphi_s, \varphi_f))$.

$$\frac{P = plan(goal(\varphi_s, \varphi_f)) \quad \langle \mathcal{B}, \mathcal{A}, P \rangle \xrightarrow{plan} \!\!\!\!\!/\;\; \langle \mathcal{B}', \mathcal{A}', nil \rangle}{\langle \mathcal{B}, \mathcal{A}, P \rangle \xrightarrow{bdi} \langle \mathcal{B}, \mathcal{A}, ?false \rangle} \; P_\perp$$

In Sect. 4, we will explore how the scenario from the introduction can be expressed using our CAN(FPP) framework.

3.3   Formal Relationship Between FPP and BDI Execution

In this section, the relationship between FPP and the BDI execution is formally established. The following theorem establishes the link between $plan(goal(\varphi_s, \varphi_f))$ and FPP in both offline and online setting so that $plan(goal(\varphi_s, \varphi_f))$ can – to some extent – indeed be seen as FPP. In order to distinguish between offline and online planning, we denote an **off**line solution for a FPP problem $\langle \varphi_s, \mathcal{B}, \Lambda \rangle$ as $sol^{off}(\varphi_s, \mathcal{B}, \Lambda)$ and an **on**line solution as $sol^{on}(\varphi_s, \mathcal{B}, \Lambda)$.

Theorem 1. *For any agent,*

(i) *For offline planning, we have* $\langle \mathcal{B}, \mathcal{A}, plan(goal(\varphi_s, \varphi_f)) \rangle \xrightarrow{bdi_*} \langle \mathcal{B}'', \mathcal{A}'', nil \rangle$ $\Longleftrightarrow sol^{off}(\varphi_s, \mathcal{B}, \Lambda) = \sigma \neq \emptyset$, $\langle \mathcal{B}, \mathcal{A}, \sigma \rangle \xrightarrow{bdi_*} \langle \mathcal{B}'', \mathcal{A}'', nil \rangle$ *such that* $\mathcal{B}'' \models \varphi_s$. *The BDI agent can evolve* $plan(goal(\varphi_s, \varphi_f))$ *to an empty program as long as offline FPP returns a non-empty solution which can be successfully executed to solve FPP problem* $(\varphi_s, \mathcal{B}, \Lambda)$.

(ii) *For online planning,* $\langle \mathcal{B}, \mathcal{A}, plan(goal(\varphi_s, \varphi_f)) \rangle \xrightarrow{bdi} \Longleftrightarrow sol^{on}(\varphi_s, \mathcal{B}, \Lambda) = act \neq \emptyset$ *and* $\langle \mathcal{B}, \mathcal{A}, act \rangle \xrightarrow{bdi}$. *The BDI agent can evolve* $plan(goal(\varphi_s, \varphi_f))$ *to a next step as long as online FPP returns an executable action.*

(iii) *For online planning,* $\langle \mathcal{B}_1, \mathcal{A}, plan(goal(\varphi_s, \varphi_f)) \rangle \xrightarrow{bdi_*} \langle \mathcal{B}_k, \mathcal{A} \cdot act_1 \cdot \ldots \cdot act_k, nil \rangle$ *with* $k \geq 1$ $\Longleftrightarrow$ *there exists a solution for each online stage planning, i.e.* $sol^{on}(\varphi_s, \mathcal{B}_1, \Lambda) = act_1$, $sol^{on}(\varphi_s, \mathcal{B}_2, \Lambda) = act_2$, $\cdots$, *and* $sol^{on}(\varphi_s, \mathcal{B}_k, \Lambda) = act_k$ *such that* $\langle \mathcal{B}_j, \mathcal{A} \cdot act_1 \cdot \ldots \cdot act_{j-1}, act_j \cdot \ldots \cdot act_k \rangle \xrightarrow{bdi}$ *for* $j \in \{1, \cdots, k\}$ *and* $\mathcal{B}_k \models \varphi_s$. *The BDI agent will successfully execute (i.e. will make the success condition* φ_s *true) if the goal can be achieved after the repetition of planning and execution.*

Proof. The proof of *(i)* relies on the derivation rule $P_{\mathcal{F}^{off}}$. In offline planning setting, the transition $\langle \mathcal{B}, \mathcal{A}, plan(goal(\varphi_s, \varphi_f)) \rangle \xrightarrow{bdi_*} \langle \mathcal{B}'', \mathcal{A}'', nil \rangle$ implies that there exists a complete sequence of actions which is generated from FPP (i.e. $sol^{off}(\varphi_s, \mathcal{B}, \Lambda) = \sigma \neq \emptyset$) such that it can then be successfully executed (i.e. $\langle \mathcal{B}, \mathcal{A}, \sigma \rangle \xrightarrow{bdi_*} \langle \mathcal{B}'', \mathcal{A}'', nil \rangle$) to achieve the goal state (i.e. $\mathcal{B}'' \models \varphi_s$). Hence, the right deduced from the left is proved. In order to prove from the right to the left, let us start from the derivation rule $P_{\mathcal{F}^{off}}$. Firstly, $sol^{off}(\varphi_s, \mathcal{B}, \Lambda) = \sigma \neq \emptyset$ means that $\langle \mathcal{B}, \mathcal{A}, plan(goal(\varphi_s, \varphi_f)) \rangle \xrightarrow{plan} \langle \mathcal{B}', \mathcal{A}', \sigma \rangle$ holds. Taking into account of that $\langle \mathcal{B}, \mathcal{A}, \sigma \rangle \xrightarrow{bdi_*} \langle \mathcal{B}'', \mathcal{A}'', nil \rangle$, the set of premises of the derivation rule $P_{\mathcal{F}^{off}}$ is satisfied. Therefore, a single transition conclusion derivable from these premises (i.e. $\langle \mathcal{B}, \mathcal{A}, \langle \Gamma_{pr}, \Gamma_{de}^+ = \{goal(\varphi_s, \varphi_f)\} \rangle \xrightarrow{bdi} \langle \mathcal{B}, \mathcal{A}, \langle \Gamma_{pr} \cup \{\sigma\}, \Gamma_{de} \setminus \{goal(\varphi_s, \varphi_f)\} \rangle \rangle$) holds according to the derivation rule $P_{\mathcal{F}^{off}}$. The final puzzle of the proof, i.e. $\langle \mathcal{B}, \mathcal{A}, plan(goal(\varphi_s, \varphi_f)) \rangle \xrightarrow{bdi_*} \langle \mathcal{B}'', \mathcal{A}'', nil \rangle$ is solved by $\langle \mathcal{B}, \mathcal{A}, \sigma \rangle \xrightarrow{bdi_*} \langle \mathcal{B}'', \mathcal{A}'', nil \rangle$ and $\mathcal{B}'' \models \varphi_s$ again. Hence, the right implying the left is proved. Therefore, *(i)* holds.

The proof of *(ii)* can be given similarly as *(i)* but depending on the rule $P_{\mathcal{F}^{on}}$ instead. In online setting, the semantics $\langle \mathcal{B}, \mathcal{A}, plan(goal(\varphi_s, \varphi_f))\rangle \xrightarrow{bdi}$ indicates that a single action is returned (i.e. $sol^{on}(\varphi_s, \mathcal{B}, \Lambda) = act \neq \emptyset$) and can be executed (i.e. $\langle \mathcal{B}, \mathcal{A}, act \rangle \xrightarrow{bdi}$). According to the rule $P_{\mathcal{F}^o}$, if an executable action is produced by FPP, the configuration $\langle \mathcal{B}, \mathcal{A}, plan(goal(\varphi_s, \varphi_f))\rangle$ will transition to $\langle \mathcal{B}, \mathcal{A}, act \rangle$. Hence *(ii)* holds.

The proof of *(iii)* can be presented by induction on the planning step k. So if $k = 1$, then $act_1 \cdot \ldots \cdot act_n = \emptyset$. It means that $\langle \mathcal{B}, \Lambda, plan(goal(\varphi_s, \varphi_f))\rangle \xrightarrow{bdi} \hspace{-1.2em}/\hspace{0.6em}$ is true if $sol^o(\langle \varphi_s, S_0, AS \rangle) = \emptyset$, which holds trivially. Therefore, *(iii)* holds. Next, suppose the claim holds for all numbers less than some $k \geq 1$. We show that *(iii)* holds for k. Since we have, by the hypothesis, that there exists a solution $act_2 \cdot act_3 \cdots act_k$ such that $\langle \mathcal{B}_j, \mathcal{A} \cdot act_1 \cdot \ldots \cdot act_{j-1}, act_j \cdot \ldots \cdot act_k \rangle \xrightarrow{bdi}$ for $j \in \{2, \cdots, k\}$ and $\mathcal{B}_k \models \varphi_s$ iff $\langle \mathcal{B}_2, \mathcal{A}, plan(goal(\varphi_s, \varphi_f))\rangle \xrightarrow{bdi_k} \langle \mathcal{B}_k, \mathcal{A} \cdot act_2 \cdot act_3 \cdots act_k, nil \rangle$. Clearly, we now only need to discuss the transition from $\langle \mathcal{B}_1, \mathcal{A}, plan(goal(\varphi_s, \varphi_f))\rangle$ to $\langle \mathcal{B}_2, \mathcal{A} \cdot act_1, plan(goal(\varphi_s, \varphi_f))\rangle$. If act_1 is a solution of FPP problem $\langle \varphi_s, \mathcal{B}_1, \Lambda \rangle$ for achieving the goal, then the problem is apparently solved already. Hence *(iii)* holds. If not, taking into consideration the hypothesis induction applying from 2 to k, *(iii)* holds still. Therefore, by induction, we have proved *(iii)*. □

This theorem underpins the theoretical foundation that a successful execution resulting from our operational rules for $plan(goal(\varphi_s, \varphi_f))$ corresponds directly to a sequence of actions from FPP. Concretely, *(i)* shows that if an offline planning step is able to start executing, then there is one solution for this FPP problem, provided there is no intervention from other concurrent intention of the BDI agent and the external environment. Both *(ii)* and *(iii)* unveil the dynamic aspect related to the online management of the plans.

4   Feasibility Study

In this section, we demonstrate the practical feasibility of integrating a BDI agent system with FPP. We show how the cleaning task scenario from the introduction can be expressed using our CAN(FPP) framework. Without the loss of generality and for the simplicity of discussions, we consider the offline FPP and assume that the environment is dynamic (i.e. exogenous events can occur) and deterministic (i.e. the effects of actions can be precisely predicted). We stress though that the purpose of this discussion is not to present an actual fully developed CAN(FPP) system, but rather to motivate the merits of the proposed framework to warrant future work on a fully implemented system. Therefore, we briefly discuss a prototype system which we designed to verify the feasibility of our approach as a basis for this future work.

We recall that in a cleaning task scenario in Fig. 1, a robot finished cleaning in the lounge, and needs to proceed to the hall to vacuum. There is a door labelled as door1 connecting the lounge and the hall. The straight-forward

```
1    // Initial beliefs
2
3    dirty(hall)
4    location(lounge)
5    open(door1)
6    open(door2)
7    open(door3)
8    connect(door1, lounge, hall)
9    connect(door2, lounge, backyard)
10   connect(door3, backyard, hall)
11
12   // Initial goals
13
14   !clean(hall)
15
16   // Plan library
17
18   +!clean(X) : dirty(X) & location(X) <- vacuum(X); ? not dirty(X)
19
20   +!clean(X) : dirty(X) & location(Y) & connect(D, Y, X) & open(D)<-
goal(at(X), move(D, Y, X), nil); ? location(X); vacuum(X); ? not dirty(X)
```

Fig. 2. BDI agent in domestic cleaning scenario

route to the hall is to go through door1 when it is open. There are also two doors, namely door2 and door3 which connect the backyard with the lounge and the hall, respectively. The design of this robot has been shown by its belief base, initial goal and plan library in Fig. 2. The initial beliefs of the robot are described on lines 3–10 and the initial goal to clean the hall is displayed on line 14 of Fig. 2. In this case, the achievement goal !clean(hall) is added to the event set of the robot as an external event. At this point, two plans in the plan library on lines 18–20 are stored as plans P_1 and P_2, and BDI agent reasoning cycle begins. Both of plans P_1 and P_2 are relevant plans for the event !clean(hall). After validating and unifying the pre-condition given the current belief base, plan P_2 (see line 20) is identified as an applicable plan and becomes an intention in the procedural intention Γ_{pr} adopted for the execution. The execution of the body of P_2 starts from the execution of an ordinary declarative goal goal(at(hall), move(door1,lounge,hall), nil) which purses action move(door1,lounge,hall) to achieve the successful state at(hall) with empty failure state nil. However, it is realistic to expect in a real life setting that some situation will block the execution of the robot (i.e. exogenous events can occur). For example, in a scenario where the door1 was slammed shut unexpectedly (i.e. -open(door1)) amidst the execution of the action move(door1,lounge,hall). As a consequence, the action of move(door1,lounge,hall) would be undesirably halted, thus eventually causing the failure of the whole cleaning task.

To address this problem, the derivation rule A^3_{goal} in Sect. 3.1 will elevate the pure declarative goal goal(at(hall), nil) into the declarative intention

Γ_{de} with *nil* being no failure condition specified. Semantically, a FPP problem $P = \langle \text{at(hall)}, \mathcal{B}, \Lambda \rangle$ for $plan(\text{goal(at(hall), nil)})$ is to indicate that a first-principles planner will be triggered to generate a sequence of actions from action library Λ to achieve the successful state at(hall) in the initial belief state $\mathcal{B}$. When a sequence of actions σ is successfully generated in the offline fashion, the configuration $\langle \mathcal{B}, \mathcal{A}, plan(\text{goal(at(hall), nil)}) \rangle$ transitions to the configuration $\langle \mathcal{B}, \mathcal{A}, \sigma \rangle$. It follows that the BDI agent starts to execute actions in σ in turn in order to reach a goal at(hall). The goal is achieved if and only if $\langle \mathcal{B}, \mathcal{A}, \sigma \rangle \xrightarrow{bdi_*} \langle \mathcal{B}'', \mathcal{A} \cdot \sigma, nil \rangle$ such that $\mathcal{B}'' \models$ at(hall). In practice, the BDI agent will need to pass along the successful state at(hall) it wants to achieve, the current belief $\mathcal{B}$, and a set of action Λ to the first-principles planner when calling the planner. We choose an offline first-principles planner called Fast-Forward planner[3] and employ the Planning Domain Definition Language (PDDL) [17] for specifying planning problems for the first-principles planner in this concrete example. Due to the syntactic knowledge difference, the transformation of knowledge (e.g. predicate, belief, and action)[4] between BDI and PDDL is required to be conducted to generate PDDL planning problem specification using our PDDL generator[5]. Afterward, the first-principles planner deliberates and generates a plan solution if all possible. Finally, a sequence of actions is returned from the planner to reach the successful state at(hall), denoted as $\sigma =$ move(door2, lounge, backyard); move(door3, backyard, hall). It states the robot can move to the **backyard** through the **door2** first and proceed to the **hall** through the **door3**. The route is depicted pictorially in Fig. 1.

This case study on the blocked plan-body program highlights a number of key benefits offered by the CAN(FPP) systems. Compared to classical BDI agent, we are able to improve the scalability of the BDI agent systems to tackle the problems beyond their current reach (e.g. due to incomplete plans and dynamic environment). Compared to a pure FPP, our formal framework ensures maximums reactiveness for most of the subgoals (tracked in the procedural goal intention Γ_{pr}) and only plans on-demand for the pure declarative goals in the declarative goal intention Γ_{de}.

5   Related Work

There have been various planning mechanisms studied in the context of BDI agent systems.

Some researchers focus on the declarative notion of goals as a means to associate FPP. This approach is appealing because a declarative goal gives a description of the desired state for FPP to achieve. One of the first works to look at integrating FPP in a BDI agent system is the Propice-plan framework [18]. It is

---

[3] https://fai.cs.uni-saarland.de/hoffmann/ff.html.

[4] Due to the lack of space, interested readers are referred to [17] for the full content. We also omit the detailed discussion of the knowledge transformation between BDI and PDDL as it is implementation-dependent.

[5] https://github.com/kevinmccareavey/ppddl.

the combination of the IPP planner [19] and an extended version of the PRS BDI agent system [20]. In Propice-plan, planning occurs only (and always) when no options are available for solving an achievement goal. Another early work is [21] which combines FPP with the IndiGolog agent system [22]. The contribution of [21] is to extend IndiGolog with a classical FPP via its achieve(G) component, where G is a goal formula to achieve. Interestingly, they approached the integration from the direction of translating the planning language, namely ADL, into Golog problem, contrary to our work. However, our approach is based on the typical BDI agent systems while they explore the integration in non-BDI agent architecture (i.e. situation calculus). Notable works on FPP in a BDI setting [10,11] had been ad-hoc approaches without a formal operational semantics for the integration of FPP in BDI. Another important work [23] studies the integration of an online risk-aware planner with a BDI agent. However, it is more concerned with how to calculate risk alongside utility in online planning algorithm instead of integrating online FPP with BDI agent systems in a semantics fashion as we do in this work.

Meanwhile, some researchers tackle the problem from hierarchical task network (HTN) planning perspective. It is revealed in [24] that there are many similarities between HTN planning and BDI agent systems, hence, making them suitable candidates for a principled integration. This principled integration is the semantics of CANPLAN [13]. It is an extension of CAN with a built-in HTN planning structure which performs a local offline plan search in the pre-defined plan library. In some sense, our work is close to the spirit of CANPLAN which provides strong theoretical underpinnings. However, this integration of the HTN planning in CANPLAN functions as an advanced plan selection tool which cannot generate new plans.

Some of the works approach the integration of automated planning in BDI agent paradigms by examining the relationship between BDI agent systems and probabilistic planning techniques. For example, [25,26] explore the relationships between certain components of the BDI agent architectures and those in the Markov Decision Processes (MDPs) and the Partially Observable Markov Decision Processes (POMDPs), respectively. More pragmatic approaches to the application of probabilistic planning techniques in BDI agent systems can be found in the works of [12,27]. Although the hybrid BDI and (PO)MDP frameworks provide good insights into the potential integration of probabilistic planning into BDI agent architectures, there is still significant work to be done in modelling and reasoning uncertainty in BDI paradigms beforehand.

6   Conclusions

In this work we proposed a framework with a strong theoretical underpinning for integrating first-principles planning (FPP) within BDI agent systems based on the intrinsic relationship between the two. We introduced a formal operational semantics that incorporates FPP and that lends power to BDI agents when the situation calls for it. We do this by extending the CAN language, and extending it with operational semantics to handle a tight integration with FPP. As such, a BDI agent can accomplish the goals beyond its own pre-defined capabilities. We have also established a theorem that the principled integration between FPP and the BDI execution is the one intuitively expected both in offline and online FPP style. We believe the work presented here lays a firm foundation for augmenting the range of behaviours of the agents by expanding the set of BDI plans available to the agent from FPP. More importantly, this paper is a significant step towards incorporating different types of advanced planning techniques into BDI agent systems in a principled manner. For future work, we plan to advance the state-of-art of the hybrid planning BDI agents by proposing a novel BDI plan library evolution architecture to improve the robustness of the BDI agents which operates in a fast-changing environment. To achieve this, we want to introduce the plan library expansion and contraction scheme. The plan library expansion is to adopt new plans generated from the first-principles planner for future reuse. The contraction scheme is accomplished by defining the plan library contraction operator regarding the rationality postulates to remove undesirable plans (e.g. obsolete or incorrect plans).

Acknowledgements. This work has received funding from the EU Horizon 2020 Programme through the DEVELOP project (under grant agreement No. 688127).

References

1. Rao, A.S.: AgentSpeak(L): BDI agents speak out in a logical computable language. In: Van de Velde, W., Perram, J.W. (eds.) MAAMAW 1996. LNCS, vol. 1038, pp. 42–55. Springer, Heidelberg (1996). https://doi.org/10.1007/BFb0031845
2. Winikoff, M., Padgham, L., Harland, J., Thangarajah, J.: Declarative and procedural goals in intelligent agent systems. In: The 8th International Conference on Principles of Knowledge Representation and Reasoning. Morgan Kaufman (2002)
3. Sardina, S., Padgham, L.: A BDI agent programming language with failure handling, declarative goals, and planning. Auton. Agents Multi-Agent Syst. **23**, 18–70 (2011)
4. Winikoff, M.: Jack^{TM}: intelligent agents: an industrial strength platform. In: Bordini, R.H., Dastani, M., Dix, J., El Fallah Seghrouchni, A. (eds.) Multi-Agent Programming. Multiagent Systems, Artificial Societies, and Simulated Organizations (International Book Series), vol. 15, pp. 175–193. Springer, Boston (2005). https://doi.org/10.1007/0-387-26350-0_7
5. Bordini, R.H., HüJomi, J.F., Wooldridge, M.: Programming Multi-Agent Systems in Agentspeak Using Jason, vol. 8. John Wiley & Sons, Chichester (2007)
6. Pokahr, A., Braubach, L., Jander, K.: The Jadex project: programming model. In: Ganzha, M., Jain, L. (eds.) Multiagent Systems and Applications. Intelligent Systems Reference Library, vol. 45, pp. 21–53. Springer, Heidelberg (2013). https://doi.org/10.1007/978-3-642-33323-1_2
7. Jennings, N.R., Bussmann, S.: Agent-based control systems. IEEE Control Syst. **23**, 61–73 (2003)
8. McArthur, S.D., et al.: Multi-agent systems for power engineering applications – Part I: concepts, approaches, and technical challenges. IEEE Trans. Power Syst. **22**, 1743–1752 (2007)

9. Meneguzzi, F., De Silva, L.: Planning in BDI agents: a survey of the integration of planning algorithms and agent reasoning. Knowl. Eng. Rev. **30**, 1–44 (2015)
10. Meneguzzi, F., Luck, M.: Composing high-level plans for declarative agent programming. In: Baldoni, M., Son, T.C., van Riemsdijk, M.B., Winikoff, M. (eds.) DALT 2007. LNCS (LNAI), vol. 4897, pp. 69–85. Springer, Heidelberg (2008). https://doi.org/10.1007/978-3-540-77564-5_5
11. De Silva, L., Sardina, S., Padgham, L.: First principles planning in BDI systems. In: The 8th International Conference on Autonomous Agents and Multiagent Systems. International Foundation for Autonomous Agents and Multiagent Systems, vol. 2, pp. 1105–1112 (2009)
12. Bauters, K., et al.: Probabilistic Planning in agentspeak using the POMDP framework. In: Hatzilygeroudis, I., Palade, V., Prentzas, J. (eds.) Combinations of Intelligent Methods and Applications. SIST, vol. 46, pp. 19–37. Springer, Cham (2016). https://doi.org/10.1007/978-3-319-26860-6_2
13. Sardina, S., Padgham, L.: Goals in the context of BDI plan failure and planning. In: The 6th International Joint Conference on Autonomous Agents and Multiagent Systems, pp. 16–23 (2007)
14. van Riemsdijk, M.B., Dastani, M., Dignum, F., Meyer, J.-J.C.: Dynamics of declarative goals in agent programming. In: Leite, J., Omicini, A., Torroni, P., Yolum, I. (eds.) DALT 2004. LNCS (LNAI), vol. 3476, pp. 1–18. Springer, Heidelberg (2005). https://doi.org/10.1007/11493402_1
15. Keller, T., Eyerich, P.: PROST: probabilistic planning based on UCT. In: The 22nd International Conference on Automated Planning and Scheduling (2012)
16. Hoffmann, J., Nebel, B.: The FF planning system: fast plan generation through heuristic search. J. Artif. Intell. Res. **14**, 253–302 (2001)
17. McDermott, D.: The AIPS-98 planning competition committee. PDDL – The Planning Domain Definition Language. (1998)
18. Despouys, O., Ingrand, F.F.: Propice-Plan: toward a unified framework for planning and execution. In: Biundo, S., Fox, M. (eds.) ECP 1999. LNCS (LNAI), vol. 1809, pp. 278–293. Springer, Heidelberg (2000). https://doi.org/10.1007/10720246_22
19. Koehler, J., Nebel, B., Hoffmann, J., Dimopoulos, Y.: Extending planning graphs to an ADL subset. In: Steel, S., Alami, R. (eds.) ECP 1997. LNCS, vol. 1348, pp. 273–285. Springer, Heidelberg (1997). https://doi.org/10.1007/3-540-63912-8_92
20. Ingrand, F.F., Georgeff, M.P., Rao, A.S.: An architecture for real-time reasoning and system control. IEEE Expert **7**, 34–44 (1992)
21. Claßen, J., Eyerich, P., Lakemeyer, G., Nebel, B.: Towards an integration of Golog and planning. In: The 20th International Joint Conferences on Artificial Intelligence, pp. 1846–1851 (2007)
22. Sardina, S., Giacomo, G.D., Lespérance, Y., Levesque, H.J.: On the semantics of deliberation in indiGolog-from theory to implementation. Ann. Math. Artif. Intell. **41**, 259–299 (2004)
23. Killough, R., Bauters, K., McAreavey, K., Liu, W., Hong, J.: Risk-aware planning in BDI agents. In: The 8th International Conference on Agents and Artificial Intelligence, pp. 322–329 (2016)
24. de Silva, L., Padgham, L.: A comparison of BDI based real-time reasoning and HTN based planning. In: Webb, G.I., Yu, X. (eds.) AI 2004. LNCS (LNAI), vol. 3339, pp. 1167–1173. Springer, Heidelberg (2004). https://doi.org/10.1007/978-3-540-30549-1_118
25. Simari, G.I., Parsons, S.: On the relationship between MDPs and the BDI architecture. In: the 5th International Joint Conference on Autonomous Agents and Multiagent Systems, pp. 1041–1048. ACM (2006)

26. Schut, M., Wooldridge, M., Parsons, S.: On partially observable MDPs and BDI models. In: d'Inverno, M., Luck, M., Fisher, M., Preist, C. (eds.) Foundations and Applications of Multi-Agent Systems. LNCS (LNAI), vol. 2403, pp. 243–259. Springer, Heidelberg (2002). https://doi.org/10.1007/3-540-45634-1_15

27. Chen, Y., Hong, J., Liu, W., Godoís, L., Sierra, C., Loughlin, M.: Incorporating PGMs into a BDI architecture. In: Boella, G., Elkind, E., Savarimuthu, B.T.R., Dignum, F., Purvis, M.K. (eds.) PRIMA 2013. LNCS (LNAI), vol. 8291, pp. 54–69. Springer, Heidelberg (2013). https://doi.org/10.1007/978-3-642-44927-7_5

Short Papers

Imprecise Sampling Models for Modelling Unobserved Heterogeneity? Basic Ideas of a Credal Likelihood Concept

Thomas Augustin[(✉)]

Foundations of Statistics and Their Applications Group, Department of Statistics,
Ludwig-Maximilians Universität Müchen (LMU Munich), 80539 Munich, Germany
augustin@stat.uni-muenchen.de
https://www.foundstat.statistik.uni-muenchen.de

Abstract. In this research note, we sketch the idea to use (aspects of) imprecise probability models to handle unobserved heterogeneity in statistical (regression) models. Unobserved heterogeneity (frailty) is a frequent issue in many applications, arising whenever the underlying probability distributions depend on unobservable individual characteristics (like personal attitudes or hidden genetic dispositions). We consider imprecise sampling models where the likelihood contributions depend on individual parameters, varying in an interval (cuboid). Based on this, and a hyperparameter controlling the amount of ambiguity, we directly fit a credal set to the data. We introduce the basic concepts of this credal maximum likelihood approach, sketch first aspects of practical calculation of the resulting estimators by constrained optimization, derive some first general properties and finally discuss some ideas of a data-dependent choice of the hyperparameter.

Keywords: Imprecise probabilities · Credal sets · Likelihood
Unobserved heterogeneity · Frailty · Imprecise sampling models

1 Introduction and Background

While there had been considerable progress in working with imprecise priors (for instance modelling prior ignorance (e.g. [2]) or prior-data conflict (e.g. [15]), imprecise sampling models have rarely been investigated by members of the imprecise probability community – although explicitly considered in [14]. Notable exceptions include research induced by the pioneering works of robust statistics like [7] (see also [1, Chap. 7.5.2] for a review of the work based on it) and some scattered, rarely explored contributions, like the work by [11] on what he called likelihood robustness.

The present paper utilizes parametrically constructed imprecise sampling models (cf. [1, Chap. 7.3.2]). The notion extends traditional parametric sampling models, which specify the joint distribution of the n stochastically independent

© Springer Nature Switzerland AG 2018
D. Ciucci et al. (Eds.): SUM 2018, LNAI 11142, pp. 351–358, 2018.
https://doi.org/10.1007/978-3-030-00461-3_24

random variables $Y_1, \ldots, Y_n$ describing a sample of size n up to a finite dimensional parameter ϑ that assumes values in a parameter space $\Theta \subseteq \mathbb{R}^q$ and has to be estimated from the data. This formulation includes the case of independently identically distributed (i.i.d.) data, where for each $i = 1, \ldots, n$ the variable Y_i follows the same distribution, as well as the case of regression models, where the distribution of Y_i, depends not only on ϑ but also on *observable* individual characteristics X_i (covariates). These characteristics are commonly treated as fixed, and the conditional distribution of Y_i given the value x_i of X_i is used in the modelling. Moreover, it is generally assumed that the distribution $p_\vartheta(\cdot)$ and $p_{\vartheta, X_i}(\cdot) = p_\vartheta(\cdot|x_i)$, respectively, of each Y_i can be described by $f_\vartheta(\cdot)$ and $f_\vartheta(\cdot|x_i)$, the (conditional) density function (with respect to the Lebesgue measure) or probability mass function on the domain of Y_i.

Imprecise sampling models specify a credal set instead of a single precise probability distribution. In the case of parametric models this means to consider a set[1] $\mathcal{P}_{\boldsymbol{\vartheta}}$ of traditional parametric models p_ϑ, described by a parameter ϑ lying in a (multidimensional) interval (cuboid) $\boldsymbol{\vartheta} \subseteq \Theta$. A natural interpretation of such models is the understanding that for each observation its data generating mechanism is an element p_ϑ of the set $\mathcal{P}_{\boldsymbol{\vartheta}}$ of distributions, where ϑ is selected by an unknown and indescribable mechanism.

The core idea of this paper is that such models

i. provide a very natural description of the frequent problem of unobserved heterogeneity and
ii. motivate a corresponding extension of the maximum likelihood principle, called *credal maximum likelihood* in the sequel.

Unobserved heterogeneity denotes the problem that the underlying distributions depend on – in the context of the study at hand – unobservable characteristics, like individual attitudes or genetic dispositions. The problem is quite frequent in all major areas of application, from econometrics and the social sciences to biometrics and epidemiology. Approaches to tackle this issue either explicitly assume latent variables/classes and corresponding mixture distributions (see, e.g., [8,12,13,17]) or alternatively work with random parameters (see, e.g., [3, Chap. 18] for an example from econometric literature or, e.g., [5] from biometrics, where the related term 'frailty' is particularly common). In the later class of models, a parametric distribution is assumed for the random parameter, in order to integrate it out. We argue that the specification of such a precise distribution for a unobservable quantity is a severe case of overprecision, where conclusions suffer from Manski's Law of Decreasing Credibility [9, p. 1].

In this work we suggest an alternative approach based on imprecise sampling models. It extends the traditional maximum likelihood approach (e.g. [10]), which yields a general estimation methodology for statistical models that are vastly applied in statistical modelling, like generalized linear models or survival

---

[1] We consider here directly the set of parametric models; often in the theory of imprecise the whole convex hull of the generating models is used.

models (e.g. [6]). Under mild regularity conditions, attractive asymptotic properties are guaranteed (e.g. [16, Chap. 9]). Then the maximum likelihood estimator $\hat{\vartheta}_{ML}$ of a parameter ϑ is consistent, asymptotically normally distributed and asymptotically efficient in the sense of minimal variance/covariance.

Given a sample $y_1, \ldots y_n$, the likelihood concept reinterprets the – in case of covariates conditional — joint density $f_\vartheta(y_1, \ldots, y_n)$ as a function $\text{Lik}(\vartheta \| y_1, \ldots y_n)$ in ϑ, describing how likely it is that the product measure of p_ϑ is the true probability underlying the data generating mechanism having produced $y_1, \ldots y_n$. Maximizing $\text{Lik}(\vartheta \| y_1, \ldots y_n)$ with respect to ϑ gives the most likely parameter value, the maximum likelihood estimate. In the case of i.i.d. normal data or a linear regression model with normally distributed error terms, maximum likelihood estimation of the main (i.e. location) parameters coincides with applying least squares estimation.

The remainder of the paper is organized as follows. Section 2 introduces the fundamental definition of the level δ–credal maximum likelihood estimator and formulates some basic properties of it. The concept is briefly illustrated in Sect. 3 by two major examples: the mean estimation in a normal model and the estimation of the coefficients of a linear regression. Section 4 concludes this research note by discussing in brief the choice of the hyperparameter δ and sketching some other aspects of further research.

2   Credal Maximum Likelihood: General Definition and First Properties

In this section we describe the basic general aspects of our approach to modelling and estimating under unobserved heterogeneity, expressed by individually varying parameters, denoted by ϑ_i, $i = 1, \ldots, n$.[2] The ideal aim of inference in our context would be to estimate each of the individual parameter values. Since, however, of course, it is not possible in any meaningful way (i.e. without not just reproducing the sample) to estimate n parameter values ϑ_i from n observations, some further structure has to be introduced into the model. Traditional modelling of unobserved heterogeneity assumes a certain distribution to describe the unknown individual variation, yielding eventually an estimator of its mean.[3] We instead refrain from any assumption on the form of the latent distribution and impose some structure by bounding the variation of the $\vartheta_i's$ by a hyperparameter δ; i.e. we fix[4] the interval length of the parameter of a para-

---

[2] In the case of multi-dimensional parameters typically only few components of the parameter vector are taken to vary individually. We do not explicitly distinguish the varying and the constant components of the parameter vector notationally.

[3] Typically the models are equivalently formulated in terms of a central parameter ν_{overall} and individual variations ν_i around it with zero mean such that $\vartheta_i = \nu_{\text{overall}} + \nu_i$.

[4] Some ideas how to choose δ in a data-dependent way are sketched in Sect. 4.

metrically constructed imprecise sampling model. These considerations lead to the following constrained optimization of the joint likelihood.[5]

Definition 1 (Level δ – Credal Maximum Likelihood Estimation). *Consider a parametric class of distributions p_θ with $\vartheta \in \Theta \subseteq \mathbb{R}^q$ and data $y_1, y_2, \ldots, y_n$. Let $\delta \in \mathbb{R}_0^{+q}$ be fixed and $\hat{\vartheta}_1, \ldots, \hat{\vartheta}_n, \widehat{L\vartheta}, \widehat{U\vartheta}$, be an optimal solution of the optimization problem*

$$\prod_{i=1}^{n} f_{\vartheta_i}(y_i) \rightarrow \max_{\vartheta_1, \ldots, \vartheta_n, L\vartheta, U\vartheta} \tag{1}$$

subject to[6]

$$L\vartheta \le \vartheta_i \le U\vartheta, \quad i = 1, \ldots, n, \quad and \quad U\vartheta - L\vartheta \le \delta, \tag{2}$$

then $\left[\widehat{L\vartheta}, \widehat{U\vartheta}\right]$ is called level-δ credal maximum likelihood estimator.

From this definition directly some basic properties of level-δ credal maximum likelihood estimators can be deduced.

Remark 1 i. Obviously, the credal maximum approach generalizes the traditional likelihood concept. Setting $\delta = 0$, yields $\widehat{L\vartheta} = \widehat{U\vartheta} = \hat{\vartheta}_{ML}$; the level-0 maximum likelihood estimator is the traditional maximum likelihood estimator.

ii. Often it is sufficient to focus on one of the two main quantities $\widehat{L\vartheta}$ and $\widehat{U\vartheta}$: Studying the constraints in (2) shows that there is always an optimal solution with $\widehat{U\vartheta} = \widehat{L\vartheta} + \delta$.

iii. Of course, it is – as in the traditional case – typically much more convenient to replace the objective function by the equivalent objective function

$$\sum_{i=1}^{n} \ln f_{\vartheta_i}(y_i). \tag{3}$$

iv. Especially in the multivariate case, credal maximum estimation does not necessarily lead to a unique maximum. For many applications it is then often recommendable to consider the credal set induced by the union of all optimal solutions.

3   Examples: Two Least Squares Problems

Now we apply our approach to two of the most fundamental settings, mean estimation of normally distributed data and standard linear regression with a single covariate.

---

[5] For the sake of conciseness, the formulation is given in terms of $f_\vartheta(\cdot)$ only; it is immediately adopted to the regression case relying on $f_\vartheta(\cdot|x_i)$.

[6] Throughout this paper vectorial inequalities are understood as inequalities for all components.

3.1   The Normal Model with Unknown Location Parameters

If we assume the random variables Y_i describing our sample to follow a normal distribution with mean μ_i and known variance σ^2, $i = 1, \ldots, n$, we have to consider, referring to (3), the quadratic optimization problem

$$\sum_{i=1}^{n} \frac{1}{\sigma^2} (y_i - \mu_i)^2 \rightarrow \min_{\mu_1, \ldots, \mu_n, L\mu, U\mu} \qquad (4)$$

subject to

$$L\mu \leq \mu_i \leq U\mu \quad i = 1, \ldots, n, \qquad \text{and} \qquad U\mu - L\mu \leq \delta, \qquad (5)$$

which can be solved by standard software.

In this situation, and by referring to ii. of Remark 1, a credal maximum likelihood estimator can also be obtained by an easy to handle one-dimensional grid search for that value $\widehat{L\vartheta}$ that minimizes the generalized residuals

$$\mathcal{E}(y_i, L\vartheta) = (y_i - L\vartheta)^2 \cdot I\{y_i \leq L\vartheta\} + \left(y_i - (L\vartheta + \delta)^2\right) \cdot I\{y_i \geq L\vartheta + \delta\}. \qquad (6)$$

To obtain a very first impression, a toy example may be helpful, which is easily implemented by using (6). For the four data points $y_1 = 1$, $y_2 = 2$, $y_3 = 3$, $y_4 = a$, the (credal) maximum likelihood estimator is given in Table 1 for different values of a. We see that the estimates are symmetric around the mean only in the symmetric situation $a = 4$. With $a > 4$ the interval length δ is asymmetrically partitioned around the mean value, with a higher proportion towards above.

Table 1. Credal maximum likelihood estimation of normal means in the toy example $y_1 = 1$; $y_2 = 2$; $y_3 = 3$; $y_4 = a$ for different values of a and δ.

δ	$\left[\widehat{L\mu}, \widehat{U\mu}\right]$	
	$a := 4$	$a := 14$
0	2.5	5
0.1	[2.45 , 2.55]	[4.975 , 5.075]
0.5	[2.25 , 2.75]	[4.875 , 5.375]
1	[2, 3]	[4.75 , 5.75]

3.2   Linear Regression

Now we turn to the basic model of regression analysis, where a linear functional relationship for the conditional mean of a generic variable Y given a generic covariate X is assumed. For the observed sample (x_i, y_i), $i = 1, \ldots, n$, in the traditional approach one relies on the relationship $y_i = \alpha + \beta * x_i + \epsilon_i$, where ϵ_i are realizations of normally distributed variables with mean zero and variance σ^2, such that ϵ_i and X_j are independent of each other for every pair i, j. In the following we take σ^2 to be fixed and known.

Then, the conditional distribution of the underlying random variables Y_i given $X_i = x_i$ is normal with mean $\alpha + \beta * x_i$, describing the regression line (, and variance σ^2). Generalizing this to the credal situation, one obtains sets of regressions lines, with varying intercept α_i and/or varying slope β_i. Applying the credal maximum likelihood approach, and denoting the components of δ by δ_α and δ_β, leads to considering

$$\sum_{i=1}^{n} (y_i - \alpha_i - \beta_i x_i) \rightarrow \min_{\alpha_1,\dots,\alpha_n,\beta_1,\dots,\beta_n,L\alpha,U\alpha,L\beta,U\beta} \tag{7}$$

subject to

$$\alpha_i \in [\widehat{L\alpha}, \widehat{U\alpha}], \ i = 1, \dots, n \quad \text{and} \quad \widehat{U_\alpha} - \widehat{L_\alpha} \leq \delta_\alpha \tag{8}$$

$$\beta_i \in [\widehat{L\beta}, \widehat{U\beta}], \ i = 1, \dots, n \quad \text{and} \quad \widehat{U\beta} - \widehat{L\beta} \leq \delta_\beta. \tag{9}$$

A first graphical illustration of the approach is given in Fig. 1, where a regression through the origin is considered ($\alpha_i \equiv 0$) for the centred variables (X^*, Y^*). We simulated $n = 200$ data points (x_i, y_i), $i = 1, \dots, 200$. The points are produced by the relationship $Y_i = \beta_i * X_i + \epsilon_i$, where ϵ_i and X_i are independently and identically standard normal such that the independence assumptions stated above are met. To introduce heterogeneity, we relied, as a very first setting, on a dichotomy: β_i was set equal to 1 for 3/4 of the units and to 10 else. The figure shows the centred data, and based on them the traditional least square estimate (dashed line), the lower and upper interval limits for $\delta = 5$ (dashed and dotted line) and for $\delta = 10$ (solid line).

4   First Concluding Remarks

Clearly, the development of the credal maximum likelihood approach sketched in this research note is still in its infancy. Nevertheless, the first results obtained so far are encouraging enough to motivate further work on it.

One issue of major practical importance is the choice of appropriate values of the level δ from observed data only.[7] Currently, three methods to choose δ are investigated.

---

[7] This corresponds to the determination of the variance of the distribution in the traditional random effects approaches to unobserved heterogeneity, while, from this perspective, in our approach the range of the random effects has to be specified. This analogy should however not hide the major difference: the traditional approach needs a fixed distributional class for the – unobservable (!) – random effect (typically normal distribution), while our approach refrains from any modelling of the hidden process.

Therefore, if one looks at an interpretation of our approach in the context of traditional theory, our approach could be termed 'nonparametric'.

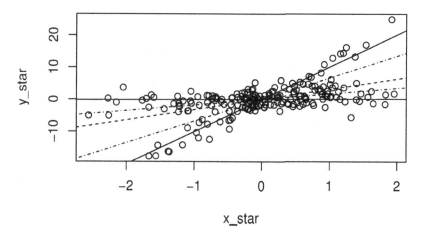

Fig. 1. Credal maximum likelihood and traditional least squares estimation for a regression through the origin (simulated data)

i. Choices based on goodness-of-fit: one calculates generalized residuals in dependence on δ and looks for a "natural change point" where a further increase of δ has decreasing marginal utility.

ii. Penalization: one adds a term to the objective function penalizing for the magnitude of δ.

iii. Holistic approach: one considers the whole family of estimators indexed by δ as the appropriate description, like a membership function.

Although intuitively appealing, the approach presented here, of course, also needs urgently further theoretical underpinning. Imprecise sampling models have not yet been sufficiently investigated, and estimation of imprecise models is an even more subtle issue (cf., e.g., [4]). Also, more concretely, the properties of the credal maximum likelihood estimators have to investigated much more deeply, the more so as standard likelihood theory may fail when the number of parameters to be estimated depends on the sample size.

Acknowledgement. I am very grateful to the two anonymous referees for very helpful comments and quite stimulating remarks.

References

1. Augustin, T., Walter, G., Coolen, F.: Statistical inference. In: Augustin, T., Coolen, F., de Cooman, G., Troffaes, M. (eds.) Introduction to Imprecise Probabilities, pp. 135–189. Wiley, Chichester (2014)
2. Benavoli, A., Zaffalon, M.: Prior near ignorance for inferences in the k-parameter exponential family. Statistics **49**, 1104–1140 (2014)
3. Cameron, A., Trivedi, P.: Microeconometrics: Methods and Applications. Cambridge University Press, Cambridge (1991)

4. Cattaneo, M.: Empirical interpretation of imprecise probabilities. In: Antonucci, A., Corani, G., Couso, I., Destercke, S. (eds.) Proceedings of the Tenth International Symposium on Imprecise Probability: Theories and Applications, ISIPTA 2017, Lugano. Proceedings of Machine Learning Research, vol. 62, pp. 61–72. PMLR (2017). http://proceedings.mlr.press/v62/cattaneo17a.html

5. Duchateau, L., Janssen, P.: The Frailty Model. Springer, New York (2007). https://doi.org/10.1007/978-0-387-72835-3

6. Fahrmeir, L., Kneib, T., Lang, S., Marx, B.: Regression: Models, Methods and Applications. Springer, Berlin (2013). https://doi.org/10.1007/978-3-642-34333-9

7. Huber, P., Strassen, V.: Minimax tests and the Neyman-Pearson lemma for capacities. Ann. Stat. **1**, 251–263 (1973)

8. Manrique-Vallier, D.: Bayesian population size estimation using Dirichlet process mixtures. Biometrics **72**, 1246–1254 (2016)

9. Manski, C.: Partial Identification of Probability Distributions. Springer, New York (2003). https://doi.org/10.1007/b97478

10. Pawitan, Y.: In All Likelihood: Statistical Modelling and Inference Using Likelihood. Oxford University Press, Oxford (2001)

11. Shyamalkumar, N.: Likelihood robustness. In: Insua, D.R., Ruggeri, F. (eds.) Robust Bayesian Analysis, pp. 127–143. Springer, New York (2000). https://doi.org/10.1007/978-1-4612-1306-2_7

12. Stanghellini, E., Vantaggi, B.: Identification of discrete concentration graph models with one hidden binary variable. Bernoulli **19**, 1920–1937 (2013)

13. Vicard, P.: On the identification of a single-factor model with correlated residuals. Biometrika **87**, 199–205 (2000)

14. Walley, P.: Statistical Reasoning with Imprecise Probabilities. Chapman & Hall, London (1991)

15. Walter, G., Augustin, T.: Imprecision and prior-data conflict in generalized Bayesian inference. J. Stat. Theor. Pract. **3**, 255–271 (2009)

16. Wasserman, L.: All of Statistics: A Concise Course in Statistical Inference. Springer, New York (2004). https://doi.org/10.1007/978-0-387-21736-9

17. Wermuth, N., Marchetti, G.: Star graphs induce tetrad correlations: for Gaussian as well as for binary variables. Electr. J. Stat. **8**, 253–273 (2014)

Representation of Multiple Agent Preferences
A Short Survey

Nahla Ben Amor[1], Didier Dubois[2], Henri Prade[2(✉)], and Syrine Saidi[1]

[1] LARODEC Laboratory, ISG de Tunis, 41 rue de la Liberté, 2000 Le Bardo, Tunisia
nahla.benamor@gmx.fr, syrine.saiidi@gmail.com
[2] IRIT - CNRS, 118, route de Narbonne, 31062 Toulouse Cedex 09, France
{dubois,prade}@irit.fr

Abstract. Different types of graphical representation for local preferences have been proposed in the literature. Graphs may be directed or not. Modeling may be quantitative or qualitative. Principles for extending local preferences to complete configurations may be based on different independence assumptions. Some extensions of such graphical representation settings to multiple agent preferences have been proposed, with different ways of handling agents: they may be viewed just as a set of individual agents, or described in terms of attribute values inducing a partition of the set of agents in terms of subcategories, or they may be reduced to some anonymous statistical counting. The fact that preferences pertain to multiple agents raises the question of either working with a collective graphical representation or aggregating individual preferences, the preferences of each agent being then represented as a graph. Moreover the multiple agent nature of the representation enriches the types of preference queries that can be addressed. The purpose of this short note is to start with a brief survey of the main graphical preference models found in the literature, such as CP-nets, π-pref nets, GAI networks, and to discuss their multiple agent extensions in an organized way, with a view to understand how the different representation options could be combined when possible.

Keywords: Multiple agent model · Preference model
Graphical representation · Possibilistic network

1 Introduction

The idea of representing individual local preferences in a graphical manner, together with an independence assumption for being able to compare complete solutions of decision problems, is attractive since it combines the benefits of a compact representation with the easiness of elicitation. Several models, which significantly differ, have been proposed and developed along this line: in the following we shall consider the main ones, CP-nets [11,12], GAI networks [15] and π-pref nets [2,4], and some of their variants. General surveys can be found

© Springer Nature Switzerland AG 2018
D. Ciucci et al. (Eds.): SUM 2018, LNAI 11142, pp. 359–367, 2018.
https://doi.org/10.1007/978-3-030-00461-3_25

in [3,18]. The graphical representation of the collective preferences of a group of agents is clearly of interest, and several multiple agent extensions of the previous settings have been proposed: mCP-nets [19], PCP-nets [9], mGAI networks [14] and ma π-pref nets [5]. This short paper intends to provide a survey and a discussion of these models going from qualitative to quantitative ones.

The paper is organized as follows. In the next section, we first review the different options underlying the graphical representation of individual preferences before providing a synthetic survey of the existing proposals. Section 3 first discusses how sets of agents can be handled, either on an individual basis, or in a collective anonymous manner, or in terms of subcategories described in terms of attribute values. Existing multiple agent representation proposals are then surveyed, and new options are also proposed.

2   Single Agent Preference Representation

Consider an agent expressing his/her preferences on the solutions of a decision problem, each solution being described by a set of features. A solution is thus represented by an instantiation of all the features, what is called a *configuration* in the following. Due to their combinatorial nature, complete configurations (also called outcomes) are difficult to compare. Then, models for preference representation rank configurations on the basis of (i) (conditional) local preferences pertaining to each feature domain and (ii) independence assumptions. This section provides a short description of the main graphical models for representing single agent preferences, comparing them on the basis of their underlying principles.

2.1   Building Principles of a Graphical Representation of Preferences

The basic idea is to represent local preferences of an agent by means of an acyclic graph. Some models represent preferences with *directed* structures, other ones with *undirected* ones. Within these structures, we can distinguish two main categories: quantitative models, where preferences are associated with numerical values, and purely ordinal models, where conditional preference statements are expressed by comparisons between feature instantiations and only the preference order between values matters. As a result, the preference relation between outcomes can be either complete, when all configurations can be compared, or partial, where some comparisons cannot be made. Each setting relies on a specific independence property between variables that allows us to construct the preference relation between configurations from the preference graph.

2.2   A Brief Review of the Main Graphical Preference Models for a Single Agent

CP-Nets. A Conditional Preference network (CP-net for short) [11,12] represents strict partial preference orderings of an agent under the form of local comparisons between values of each variable conditioned by other (parent) variables. It uses a directed acyclic graph $G= (V, E)$, where nodes in the set $V = \{X_1, ..., X_n\}$ represent variables (features) and E is a set of arcs. An arc from X_j to X_i expresses that the preference between values of variable X_i depends on the values of parent variable X_j. Each variable $X_i \in V$ is then associated to a conditional preference table $CPT(X_i)$ in the context of X_i's parents. CP-nets are based on the *ceteris paribus* principle, which enables the preference between values x_i and x_i' of variable X_i in the context of an instantiation of its parents to be extended to complete configurations assuming the other variables take the same values for those configurations. Let $par(X_i) = u$ be an instantiation of the set of X_i's parents, $d(X_i) = d$ an instantiation of the set of X_i's children, and $n(X_i) = n$ an instantiation of other X_j's. Actually, a direct preference comparison (denoted by $\succ$) between two complete configurations can only be done for those that differ by a single flip of one variable, in the context of an instantiation of its parents. In other words, due to ceteris paribus independence, we can conclude that $\forall d, \forall n, ux'dn \succ uxdn$ if and only if x' is preferred to x in the context u (denoted $u : x' \succ x$).

π-pref Nets. A possibilistic preference network [2,4,6] is a directed graph sharing the same structure as a CP-net. However, preferences are no longer represented by an ordering (of the form $u : x' \succ x$), but with a *symbolic* conditional possibility distribution over the domain of each variable in V. It means a possibility distribution where possibility degrees, taking values between $]0, 1]$, are not instantiated. Let us consider a variable X_i taking x and x' as possible values and having U_i as parents. If we are in a case of a strict preference $x \prec x'$ for an instantiation u_i of variables U_i, then $\pi(x|u_i) = \alpha < \pi(x'|u_i) = \beta$. In contrast, if $x \sim x'$ then $\pi(x|u_i) = \pi(x'|u_i) = \alpha \leq 1$ (in case of a binary-valued variable, π being normalized, $x \sim x'$ entails that the two configurations have a possibility 1, and $x \prec x'$ entails $\beta = 1$).

In the spirit of possibilistic belief networks [7,8], the degree of satisfaction of each configuration is computed as the product of symbolic weights using the chain rule associated to the product-based conditioning in possibility theory [2], namely, $\pi(X_i, ..., X_n) = \prod_{i=1,...,n} \pi(X_i|u(X_i))$. As an illustration, consider preferences described in terms of three binary variables X_1, X_2, X_3, based on a π-pref net of the form $X_1 \to X_2 \to X_3$. Let $\pi(X_1 X_2 X_3) = \pi(X_3|X_2)\pi(X_2|X_1)\pi(X_1)$. Suppose $\pi(x_1) = \pi(x_1') = 1$, $\pi(x_2|x_1) > \pi(x_2'|x_1) = \gamma$, $\pi(x_2'|x_1') > \pi(x_2|x_1') = \alpha$, $\pi(x_3|x_2) > \pi(x_3'|x_2) = \beta$. Then $\pi(x_2|x_1) = 1$ and when comparing configurations $x_1 x_2 x_3'$ and $x_1' x_2 x_3'$, we note that using the chain rule, $\pi(x_1 x_2 x_3') = \beta$ and $\pi(x_1' x_2 x_3') = \alpha\beta$, and we can see that $\beta > \alpha\beta$ whatever the values of α and β, which means that $x_1 x_2 x_3' \succ x_1' x_2 x_3'$. It would be equivalent to compare the vectors of the form $(\pi(X_3|X_2), \pi(X_2|X_1), \pi(X_1))$, here $(1, 1, \beta)$ with $(1, \alpha, \beta)$ by means of the Pareto ordering (in fact its symmetric version [2]).

The two partial orderings of configurations respectively obtained from a π-pref net and from a CP-net built on the basis of the same preference statements of an agent, do not contradict each other [4]. Constraints between symbolic weights of the π-pref net can be added so as to recover comparisons produced by the ceteris paribus property. As a consequence, π-pref nets look more flexible that CP-nets since they leave complete freedom for adding relative priorities between possibilistic weights. However, if no constraints are added, configurations may remain incomparable while they can be compared in the sense of the CP-net. Moreover, since π-pref nets offer the opportunity to switch from symbolic weights to instantiated numerical ones, this representation can be viewed as being halfway between qualitative models such as CP-nets and quantitative ones. Although CP-nets share the same graphical structure and level of simplicity as π-pref nets, they do not have the same expressive power. In fact, CP-nets are based on ceteris paribus independence, while π-pref nets rely on a Markov independence property namely, each variable X_i is independent from the remaining non-children nodes (N) in the context of its parents (U). With ceteris paribus independence, we can say that $uxdn \succ ux'dn$, i.e., the parent-dependent preference between local configurations of one variable are completed with the same instantiation of the other variables, while with Markov-based nets, when flipping a variable X from x to x' in the context u, first we choose the best instantiations for all variables that depend on the value of X, and next, we instantiate the other variables in the same manner in all possible ways; see [2] for an example.

GAI-Nets. Generalized Additive Independence (GAI) networks [15] are graphical quantitative models for representing preferences expressed by means of utilities. A GAI-net is composed of two components. The first is a graphical structure defined by an *undirected* graph $G = (C, E)$ where C denotes a set of cliques and E denotes the corresponding set of edges. Each clique $C_j \in C$, is a set of variables s.t. $C_j \subseteq V$ and $\cup_{i=1}^{k} C_i = V$. Each edge refers to overlapping cliques, and is labeled by a separator $S_{ij} = C_i \cap C_j \neq \emptyset$. The second component is a set of local numerical utility functions associated to cliques.

GAI-nets are based on a generalized additive independence decomposition [15]. This property allows us to associate to each clique of attributes a utility function and then to sum them in order to compare the different configurations. Then a *total* ordering between possible instantiations can be obtained. Thus the utility of a configuration σ can be expressed as the sum of partial utilities associated with clique configurations, namely $u(\sigma) = \sum_{j=1}^{k} u_j(\omega_{C_j})$. This is a form of decomposition that allows interaction between some variables, while preserving independence between other ones. In fact, the variables in $C_i \setminus S_{ij}$ are considered independent from the variables in $C_j \setminus S_{ij}$ when variables in S_{ij} are instantiated. A total preorder between outcomes is obtained.

UCP-Nets. Utility functions can be added to CP-nets in order to represent preferences in quantitative terms. In this context, a Utility CP-net (UCP-net for short) [10] is a DAG with quantified utility preferences associated to nodes. This representation combines some aspects of both CP-nets and GAI-nets. UCP-nets share the same graphical structure as CP-nets, and constraints between utility values are added so as to obey the ceteris paribus principle. However, unlike CP-nets, this model obtains a total preorder between configurations. Moreover, by adding utilities to CP-nets, the expressiveness and computational power are enhanced. Comparing different outcomes comes down to comparing their respective utility functions, additively computed as $u(x_1, ..., x_n) = \sum_{i=1}^{n} u_i(x_i | par(X_i))$. This function is very similar to the chain rule in Bayesian networks [17], namely $p(x_1, ... x_n) = \prod_{i=1}^{n} p(x_i | par(X_i))$, up to a logarithmic transformation.

3   Multiple Agent Preference Networks

In the existing studies on multiple agent representations, groups of agents are represented differently. In the simplest approaches, agents are represented as a set of individuals regardless of their characteristics; this is the case of several multiple agent models. Other representations view groups of agents as a whole and their preferences are summarized in statistical terms. By contrast, one may also describe subgroups of agents in terms of attribute values, e.g., gender, age, etc. Yet, only a single model has been proposed along this line, which are the ma π-pref nets, presented at the end of this section.

3.1   A Survey on Multiple Agent Preference Models

MCP-Nets. A multiple agent CP-net [19] is a collection of m distinct CP-nets with graphs $(V_i, E_i), i = 1, \ldots, m$, reflecting the preferences of m agents: each CP-net thus represents individual preferences. Graphically, an mCP-net is just the juxtaposition of m single agent CP-nets. Different ways of deriving collective preferences between configurations have been studied, using a collection of dominance (voting) semantics (Pareto, majority, Condorcet winner, etc.) [16].

Individual CP-nets in an mCP-net share a common set of nodes $V_C \subseteq V$, such that every variable in V is informed by at least one agent. A particular agent is allowed not to use all variables in V when stating his/her preferences. When an agent does not define his preferences over the domain of a variable in V_i, this is interpreted as an incomparability situation. It could be also interpreted as indifference between values of this variable. Interestingly, both cases are treated alike in m-CP nets. Moreover, even if preferences are explicitly provided about X, an agent might neglect dependencies between variables. This case can be represented by a CP-net associated to some empty conditional preference tables. However, such cases have not been fully explored yet.

Probabilistic CP-Nets. PCP-nets [9] represent collective preferences of a set of agents. Formally, a PCP-net is as usual a DAG $G = (V, E)$, where conditional preferences over variable domains are replaced by conditional probabilities assigned to each local preference. These probabilities reflect the proportions of agents that share these local preferences in their personal CP-nets. The preference relation over all possible configurations is replaced by a probability distribution on them.

For reasoning with this model, voting semantics can be used in order to determine the most preferred configuration. However, these methods are sequential and proceed locally. PCP-nets offer the alternative to aggregate the set of CP-nets specific to each agent into a single structure in order to perform globally. To proceed, another structure is used, called *induced* CP-net. A CP-net induced from a PCP-net P is a network that has the same variables, with the same corresponding domains, but with a subset of P's edges. Each induced CP-net has its associated probability, computed by taking the product of the orderings probabilities chosen in P. For example, let us consider a PCP-net over 2 variables, where $p(x_1 \succ x_1') = 0.6$ means that 60% of agents prefer x_1 to x_1', hence $p(x_1' \succ x_1) = 0.4$, $p(x_2' \succ x_2 | x_1 \succ x_1') = 0.7$, $p(x_2 \succ x_2' | x_1 \succ x_1') = 0.3$, $p(x_2' \succ x_2 | x_1' \succ x_1) = 0.2$, $p(x_2 \succ x_2' | x_1' \succ x_1) = 0.8$. An induced CP-net can be derived by assuming $x_1' \succ x_1$, $x_2' \succ x_2$ in context x_1, $x_2 \succ x_2'$ in context x_1' with the following probability: $0.4 \times (0.7 \times 0.8) = 0.224$. To determine the most probable optimal configuration, we have to compute the sum of probabilities of induced CP-nets that have each configuration σ optimal. On top of comparing two or different configurations, this model allows to deal with a new query: finding the optimal configuration of the most probable induced CP-net [13].

Multiple Agent GAI Nets. Multiple agent GAI nets are an extension of GAI-nets for representing collective preferences and reasoning with them. In this case, preferences are expressed by means of utility vectors, one component per agent. In order to compare different utility vectors, voting semantics can be used. The most widely used is Pareto optimality. However, this semantics reflects a partial weak order that leaves many configurations incomparable. This is easily overcome by means of aggregation operations. In this context, one such aggregation function is the Choquet integral. It is an aggregation function that associates weights to subset of agents (modelling their possible interactions) and then proceeds to a piecewise linear aggregation [14].

Multiple Agent π-pref Nets. The extension of π-pref nets to multiple agent preferences has been proposed recently [5], by combining π-pref nets with the multi-agent counterpart of possibilistic logic [1]. An *ma-π-pref net* shares the same graphical structure as π-pref nets, consequently, as CP-nets. Each node is associated to a multiple agent possibility distribution, where $(x_i \succ x_i', \alpha | A)$ (A is a subset of agents) is interpreted as: at least all agents in A prefer x_i to x_i' with minimal priority degree α. However, this model may display some contradictions inside subsets of agents, which leads to a non-normalized possibility distribution. Moreover, the independence property in this representation is yet to be studied. This model is of interest, since it allows for expressing new types of query, such as finding the configuration that best satisfies a given category of agents. This model uses logical descriptions of classes of agents, that can be simplified via a propositional logic machinery.

4   Perspectives

The above considerations suggest a number of research lines:

- Regarding single agent preference network representations, there is a need for a precise comparison of independence notions at work in CP-nets, GAI networks, UCP-nets and π-pref nets as well as the types of preference graphs between solutions obtained from these independence assumptions. This is related to the study of translations of directed structures into non-directed ones, well-known with quantitative representations, especially Bayesian networks.
- On the issue of multiple agent representations, we have seen that a set of agents can be represented in extension (as a list of agents) or in intension (describing subclasses of agents by means of suitable attributes). In the first case, some approaches compute statistics, evaluating the proportion of agents preferring one option to another. In other approaches, voting methods are used to compare options at the collective level. Finally, one may be interested in computing preferences of subgroups of agents from the known preferences of other subgroups described by attribute values. The three approaches are not mutually exclusive, since one may wish to go from a statistical description of the preferences of subsets of agents to a description of agents having those preferences, and conversely.
- Finally, it is interesting to consider open issues in the state of the art of multiple agent graphical preference representations. Is it possible to envisage applying the various representations of groups of agents to all graphical structures. For instance, we may consider a probabilistic GAI-net, where one would count the proportion of agents that prefer one option to another based on their utility values for each agent.

Table 1. Conditional tables of a Probabilistic ma π-pref net

$N(b_1 \succ b_2)$	$N(b_2 \succ b_1)$
$0.8 \setminus F \cap Y, 0.7$	$0.6 \setminus M, 0.9$

$N(.\vert.)$	b_1	b_2
$c_1 \succ c_2$	$0.3 \setminus F \cap O, 0.7$	$0.1 \setminus O, 0.8$
$c_2 \succ c_1$	$(0.5 \setminus F, 1) \wedge (0.85 \setminus M \cap O, 0.9)$	$0.8 \setminus M \cap Y, 0.7$

Especially, one could hybridize PCP-nets and ma-π-pref nets, the former summarizing the latter if we have knowledge about the number of agents having properties used in the ma-π-pref net. One could also make inferences involving both frequentist and logical computations, using probabilistic inference patterns, for instance, knowing the proportion of agents in group A that prefer p to its negation and of group B that prefer $\neg p \vee q$ (to its negation), describe the set of agents that prefer q to its negation and the proportion thereof.

Example of a Probabilistic ma π-pref Net. A group described by their gender: male (M) or female (F) and their age: old (O) or young (Y) along with proportions of agents for each category. Agents express their preferences over car colors (c_1 or c_2) in the context of their make (b_1 or b_2). Consider the multiple agent conditional preference tables in Table 1. A statement of the form $(\alpha \setminus C, \gamma)$ associated to a preference $x_i \succ x_i'$, means that at least $\alpha\%$ of agents in category C are satisfied with $x_i \succ x_i'$ with priority (necessity) at least γ. For instance, $(0.8 \setminus M \cap Y, 0.7)$ reads at least 80% of young men prefer x_i to x_i' with priority 0.7.

5   Conclusion

First, this paper has summarized the most important properties of different preference representation models, with some comparison between them. Several lines of research are proposed for further extensions. A new model for representing multiple agent possibilistic preferences under uncertainty has been suggested. Several queries for this model are yet to be analyzed and processed. We might also think of bridging the gap between this new model with Probabilistic CP-nets.

References

1. Belhadi, A., Dubois, D., Khellaf-Haned, F., Prade, H.: Multiple agent possibilistic logic. J. Appl. Non-Class.l Log. **23**(4), 299–320 (2013)
2. Ben Amor, N., Dubois, D., Gouider, H., Prade, H.: Preference modeling with possibilistic networks and symbolic weights: a theoretical study. In: Kaminka, G.A., et al. (eds.) Proceedings of the 22nd European Conference on Artificial Intelligence (ECAI 2016), The Hague, August 29-September 2, pp. 1203–1211. IOS Press (2016)
3. Amor, N.B., Dubois, D., Gouider, H., Prade, H.: Graphical models for preference representation: an overview. In: Schockaert, S., Senellart, P. (eds.) SUM 2016. LNCS (LNAI), vol. 9858, pp. 96–111. Springer, Cham (2016). https://doi.org/10.1007/978-3-319-45856-4_7
4. Ben Amor, N., Dubois, D., Gouider, H., Prade, H.: Expressivity of possibilistic preference networks with constraints. In: Moral, S., Pivert, O., Sánchez, D., Marín, N. (eds.) SUM 2017. LNCS (LNAI), vol. 10564, pp. 163–177. Springer, Cham (2017). https://doi.org/10.1007/978-3-319-67582-4_12
5. Ben Amor, N., Dubois, D., Gouider, H., Prade, H.: Graphical representations of multiple agent preferences. In: Benferhat, S., Tabia, K., Ali, M. (eds.) IEA/AIE 2017. LNCS (LNAI), vol. 10351, pp. 142–153. Springer, Cham (2017). https://doi.org/10.1007/978-3-319-60045-1_17
6. Ben Amor, N., Dubois, D., Gouider, H., Prade, H.: Possibilistic preference networks. Information Sciences, August 2017, in press
7. Benferhat, S., Dubois, D., Garcia, L., Prade, H.: On the transformation between possibilistic logic bases and possibilistic causal networks. Int. J. Approx. Reason. **29**(2), 135–173 (2002)
8. Benferhat, S., Ben Amor, N., Mellouli, K.: Anytime propagation algorithm for min-based possibilistic graphs. Soft Comput. **8**(2), 150–161 (2003)
9. Bigot, D., Zanuttini, B., Fargier, H., Mengin, J.: Probabilistic conditional preference networks. In: Nicholson, A., Smyth, P. (eds.) Proceedings of the 29th Conference on Uncertainty in Artificial Intelligence (UAI 2013), Bellevue, Washington, August 11-15, pp. 72–81 (2013)

10. Boutilier, C., Bacchus, F., Brafman, R.I.: UCP-networks: a directed graphical representation of conditional utilities. In: Proceedings of the 17th Conference on Uncertainty in Artificial Intelligence (UAI 2001), Seattle, Aug. 2–5, pp. 56–64. Morgan Kaufmann Publ. (2001)

11. Boutilier, C., Brafman, R.I., Hoos, H.H., Poole, D.: Reasoning with conditional ceteris paribus preference statements. In: Laskey, K.B., Prade, H. (eds.) Proceedings of the 15th Conference on Uncertainty in Artificial Intelligence (UAI 1999), Stockholm, July 30–August 1, pp. 71–80. Morgan Kaufmann (1999)

12. Boutilier, C., Brafman, R.I., Domshlak, C., Hoos, H.H., Poole, D.: CP-nets: a tool for representing and reasoning with conditional ceteris paribus preference statements. J. Artif. Intell. Res. **21**, 135–191 (2004)

13. Cornelio, C., Grandi, U., Goldsmith, J., Mattei, N., Rossi, F., Brent Venable, K.: Reasoning with PCP-nets in a multi-agent context. In: Weiss, G., Yolum, P., Bordini, R.H., Elkind, E. (eds.) Proceeding 14th International Conference on Autonomous Agents and Multiagent Systems (AAMAS 2015), Istanbul, May 4–8, 2, pp. 969–977. ACM (2015)

14. Dubus, J.-P., Gonzales, C., Perny, P.: Choquet optimization using gai networks for multiagent/multicriteria decision-making. In: Rossi, F., Tsoukias, A. (eds.) ADT 2009. LNCS (LNAI), vol. 5783, pp. 377–389. Springer, Heidelberg (2009). https://doi.org/10.1007/978-3-642-04428-1_33

15. Gonzales, C., Perny, P.: GAI networks for utility elicitation. In: Dubois, D., Welty, C.A., Williams, M.-A. (eds.) Proceedings of the 9th International Conference on Principles of Knowledge Representation and Reasoning (KR 2004), Whistler, June 2–5, pp. 224–234 (2004)

16. Lukasiewicz, T., Malizia, E.: On the complexity of mCP-nets. In: Schuurmans, D., Wellman, M.P. (eds.) Proceedings of the 30th AAAI Conference on Artificial Intelligence, February 12–17, Phoenix, pp. 558–564. AAAI Press (2016)

17. Pearl, J.: Probabilistic Reasoning in Intelligent Systems: Networks of Plausible Inference. Morgan Kaufmann, San Mateo (1998)

18. Rossi, F., Brent Venable, K., Walsh, T.: A short introduction to preferences: between artificial intelligence and social choice. In: Synthesis Lectures on Artificial Intelligence and Machine Learning, Morgan & Claypool Publ. (2011)

19. Rossi, F., Venable, K., Walsh, T.: mCP-nets: representing and reasoning with preferences of multiple agents. In: McGuinness, D.L., Ferguson, G. (eds.) Proceedings of the 19th National Conference on Artificial Intelligence (AAAI 2004), San Jose, July 25–29, pp. 729–734. AAAI Press / The MIT Press (2004)

Measuring and Computing Database Inconsistency via Repairs

Leopoldo Bertossi[1,2]([⊠])

[1] Carleton University, Ottawa, Canada
bertossi@scs.carleton.ca
[2] RelationalAI, Inc., Berkeley, USA

Abstract. We propose a generic numerical measure of inconsistency of a database with respect to a set of integrity constraints. It is based on an abstract repair semantics. A particular inconsistency measure associated to cardinality-repairs is investigated; and we show that it can be computed via answer-set programs.

Keywords: Integrity constraints in databases
Inconsistent databases · Database repairs · Inconsistency measures

Intuitively, a relational database may be more or less consistent than others databases for the same schema with the same integrity constraints (ICs). This comparison can be accomplished by assigning a *measure of inconsistency* to a database. The inconsistency degree of a database D with respect to (wrt.) a set of ICs Σ should depend on how complex it is to restore consistency; or more technically, on the class of *repairs* of D wrt. Σ. For this we can apply concepts and results on database repairs (cf. [2] for a survey and references). *Our stand on degrees of inconsistency is that they depend on how consistency is restored, i.e. involving the admissible repair actions and how close we want stay to the instance at hand.* This short communication shows preliminary research on possible ways to make these ideas concrete, by defining and analyzing a measure of inconsistency of a relational database instance, and providing mechanisms for computing this measure using *answer-set programming* (ASP) [5].

Database Repairs. When a database instance D does not satisfy its intended ICs, it is repaired, by deleting or inserting tuples from/into the database. An instance obtained in this way is a *repair* of D if it satisfies the ICs and departs in a minimal way from D [2]. In this work, just to fix ideas, we consider mostly ICs that can only be solved by tuple deletions, e.g. most prominently, *denial constraints* (DCs) and *functional dependencies* (FDs). DCs are logical formulas of the form $\neg \exists \bar{x}(P_1(\bar{x}_1) \wedge \cdots \wedge P_m(\bar{x}_m))$, where $\bar{x} = \bigcup \bar{x}_i$; and FDs are of the form $\neg \exists \bar{x}(P(\bar{v}, \bar{y}_1, z_1) \wedge P(\bar{v}, \bar{y}_2, z_2) \wedge z_1 \neq z_2)$, with $\bar{x} = \bar{y}_1 \cup \bar{y}_2 \cup \bar{v} \cup \{z_1, z_2\}$.

L. Bertossi—Member of the "Millenium Institute for Foundational Research on Data" (IMFD, Chile). Research supported by NSERC Discovery Grant #06148.

© Springer Nature Switzerland AG 2018
D. Ciucci et al. (Eds.): SUM 2018, LNAI 11142, pp. 368–372, 2018.
https://doi.org/10.1007/978-3-030-00461-3_26

We treat FDs as DCs. A database is *inconsistent* wrt. a set of ICs Σ when D does not satisfy Σ, denoted $D \not\models \Sigma$.

Example 1. The DB $D = \{P(a), P(e), Q(a, b), R(a, c)\}$ is inconsistent with respect to the (set of) *denial constraints* (DCs) $\kappa_1 : \neg \exists x \exists y (P(x) \wedge Q(x, y))$, and $\kappa_2 : \neg \exists x \exists y (P(x) \wedge R(x, y))$. Here, $D \not\models \{\kappa_1, \kappa_2\}$.

A *subset-repair*, in short an *S-repair*, of D wrt. the set of DCs is a $\subseteq$-maximal subset of D that is consistent, i.e. no proper superset is consistent. The following are S-repairs: $D_1 = \{P(e), Q(a, b), R(a, c)\}$ and $D_2 = \{P(e), P(a)\}$. Under this repair semantics, both repairs are equally acceptable. A *cardinality-repair*, in short a *C-repair*, is a maximum-cardinality S-repair. D_1 is the only C-repair. □

For an instance D and a set Σ of DCs, the sets of S-repairs and C-repairs are denoted with $Srep(D, \Sigma)$ and $Crep(D, \Sigma)$, resp. It holds: $Crep(D, \Sigma) \subseteq Srep(D, \Sigma)$. More generally, for a set Σ of ICs, not necessarily DCs, they can be defined by (cf. [2]): $Srep(D, \Sigma) = \{D' : D' \models \Sigma,$ and $D \bigtriangleup D'$ is minimal under set inclusion$\}$, and $Crep(D, \Sigma) = \{D' : D' \models \Sigma,$ and $D \bigtriangleup D'$ is minimal in cardinality$\}$. Here, $D \bigtriangleup D'$ is the symmetric set difference $(D \smallsetminus D') \cup (D' \smallsetminus D)$.

Repair Semantics and Inconsistency Degrees. In general terms, a *repair semantics* S for a schema $\mathcal{R}$ that includes a set Σ of ICs assigns to each instance D for $\mathcal{R}$ (which may not satisfy Σ), a class $Rep^S(D, \Sigma)$ of S-*repairs* of D wrt. Σ, which are instances of $\mathcal{R}$ that satisfy Σ and depart from D according to some minimization criterion. Several repair semantics have been considered in the literature, among them and beside those above, *prioritized repairs* [15], and *attribute-based repairs* that change attribute values by other data values, or by a null value, NULL, as in SQL databases (cf. [1,2]).

According to our take on how a database inconsistency degree depends on database repairs, we define the *inconsistency degree* of an instance D wrt. a set of ICs Σ in relation to a given repair semantics S, as the distance from D to the class $Rep^S(D, \Sigma)$:

$$inc\text{-}deg^S(D, \Sigma) := dist(D, Rep^S(D, \Sigma)). \tag{1}$$

This is an abstract measure that depends on S and a chosen distance function *dist*, from a world to a set of possible worlds. Under the assumption that any repair semantics should return D when D is consistent wrt. Σ and $dist(D, \{D\}) = 0$, a consistent instance D should have 0 as inconsistency degree.[1]

Notice that the class $Rep^S(D, \Sigma)$ might contain instances that are not sub-instances of D, for example, for different forms of *inclusion dependencies* (INDs) we may want to insert tuples;[2] or even under DCs, we may want to appeal to attribute-based repairs. *In the following we consider only repairs that are sub-instances of the given instance.* Still this leaves much room open for different

---

[1] Abstract distances between two point-sets are investigated in [8], with their computational properties. Our setting is a particular case.

[2] For INDs repairs based only on tuple deletions can be considered [6].

kinds of repairs. For example, we may prefer to delete some tuples over others [15]. Or, as in database causality [3,13], the database can be partitioned into *endogenous* and *exogenous* tuples, assuming we have more control on the former, or we trust more the latter; and we prefer *endogenous repairs* that delete preferably (only or preferably) endogenous tuples [1].

An Inconsistency Measure. Here we consider a concrete instantiation of (1), and to fix ideas, only DCs. For them, the repair semantics $Srep(D,\Sigma)$ and $Crep(D,\Sigma)$ are particular cases of repair semantics S where each $D' \in Rep^{\mathsf{S}}(D,\Sigma)$ is maximally contained in D. On this basis, we can define:

$$inc\text{-}deg^{\mathsf{S},g3}(D,\Sigma) := dist^{g3}(D, Rep^{\mathsf{S}}(D,\Sigma)) := \frac{|D| - max\{|D'| : D' \in Rep^{\mathsf{S}}(D,\Sigma)\}}{|D|}$$

$$= \frac{min\{|D \smallsetminus D'| : D' \in Rep^{\mathsf{S}}(D,\Sigma)\}}{|D|}, \tag{2}$$

inspired by distance g_3 in [11] to measure the degree of violation of an FD by a database, whose satisfaction is restored through tuple deletions.[3] This measure can be applied more generally as a "quality measure", not only in relation to inconsistency, but also whenever possibly several intended "quality versions" of a dirty database exist, e.g. as determined by additional contextual information [4].

Example 2.   (Example 1 cont.) Here, $Srep(D,\Sigma) = \{D_1, D_2\}$, and $Crep(D,\Sigma) = \{D_1\}$. They provide the inconsistency degrees:

$$inc - deg^{s,g3}(D,\Sigma) := \frac{4 - max\{|D'| : D' \in Srep(D,\Sigma)\}}{4} = \frac{4 - |D_1|}{4} = \frac{1}{4}, \tag{3}$$

$$inc - deg^{c,g3}(D,\Sigma) := \frac{4 - max\{|D'| : D' \in Crep(D,\Sigma)\}}{4} = \frac{4 - |D_1|}{4} = \frac{1}{4}, \tag{4}$$

respectively.   □

It holds $Crep(D,\Sigma) \subseteq Srep(D,\Sigma)$, but $max\{|D'| : D' \in Crep(D,\Sigma)\} = max\{|D'| : D' \in Srep(D,\Sigma)\}$, so it holds $inc\text{-}deg^{s,g3}(D,\Sigma) = inc\text{-}deg^{c,g3}(D,\Sigma)$. These measures always takes a value between 0 and 1. The former when D is consistent (so it itself is its only repair). The measure takes the value 1 only when $Rep^{\mathsf{S}}(D,\Sigma) = \emptyset$ (assuming that $max\{|D'| : D' \in \emptyset\} = 0$), i.e. the database is *irreparable*, which is never the case for DCs and S-repairs: there is always an S-repair. However, it could be irreparable with different, but related repair semantics. For example, when we accept only endogenous repairs and none of them exists [3].

Example 3. (Example 2 cont.) Assume D is partitioned into endogenous and exogenous tuples, say resp. $D = D^n \,\dot\cup\, D^x$, with $D^n = \{Q(a,b), R(a,c)\}$ and $D^x = \{P(a), P(e)\}$. In this case, the endogenous-repair semantics that allows only a minimum number of deletions of endogenous tuples, defines

---

[3] Other possible measures for single FDs and relationships between them can be found in [11].

the class of repairs: $Srep^{c,n}(D, \Sigma) = \{D_2\}$, with D_2 as above. In this case,[4] $inc\text{-}deg^{c,n,g_3}(D, \Sigma) = \frac{4-2}{4} = \frac{1}{2}$. Similarly, if now $D^n = \{P(a), Q(a,b)\}$ and $D^x = \{P(e), R(a,c)\}$, there are no endogenous repairs, and $inc\text{-}deg^{c,n,g_3}(D, \Sigma) = 1$. □

ASP-Based Computation of the Inconsistency Measure. We concentrate on measure $inc\text{-}deg^{c,g_3}(D, \Sigma)$ (cf. (4)). More generally, we can start from $inc\text{-}deg^{s,g_3}(D, \Sigma)$, which can be computed through the maximum cardinality of an S-repair for D wrt. Σ, or, equivalently, using the cardinality of a (actually, every) repair in $Crep(D, \Sigma)$. In its turn, this can be done[5] through compact specifications of repairs by means of ASPs. We just show an example.

Example 4. (Example 1 cont.) For technical convenience, we insert global tuple-ids in D, i.e. $D = \{P(1, e), Q(2, a, b), R(3, a, c), P(4, a)\}$. It is possible to write an answer-set program, a *repair program*, Π whose stable models $\mathcal{M}_1, \mathcal{M}_2$ are correspondence with the repairs D_1, D_2, resp., namely $\mathcal{M}_1 = \{P'(1, e, \mathsf{s}), Q'(2, a, b, \mathsf{s}), R'(3, a, c, \mathsf{s}), \; P'(4, a, \mathsf{d})\} \cup D$ and $\mathcal{M}_2 = \{P'(1, e, \mathsf{s}), P'(4, a, \mathsf{s}), Q'(2, a, b, \mathsf{d}), R'(3, a, c, \mathsf{d})\} \cup D$, where the primed predicates are nicknames for the original ones, and the annotations constants s, d indicate that the tuple stays or is deleted in/from the database, resp. [1,7]

Now, to compute $inc\text{-}deg^{c,g_3}(D, \Sigma)$, for the C-repair semantics, we can add rules to Π to collect the *tids* of tuples deleted from the database: $Del(t) \leftarrow R'(t, x, y, \mathsf{d})$, similarly for Q' and P'. And next, a rule to count the deleted tuples, say: $NumDel(n) \leftarrow \#count\{t : Del(t)\} = n$. For example, program Π with the new rules added will see the original stable model $\mathcal{M}_1$ extended with the atoms $Del(4), NumDel(1)$. Similarly for $\mathcal{M}_2$.

Since the stable models of the program capture the S-repairs, i.e. $\subseteq$-maximal and consistent sub-instances of D, we can add to Π *weak program constraints* [12], such as "$:\sim P(t, x), P'(t, x, \mathsf{d})$" (similarly for R and Q). They have the effect of eliminating the models of the original program that do not violate them in a minimum way. More precisely, they make us keep only the stable models of the original program that minimize the number of satisfactions of the constraint bodies. In our case, only the models (repairs) that minimize the number of tuple deletions are kept, i.e. models that correspond to C-repairs of D. In this example, only (the extended) $\mathcal{M}_1$ remains. The value for $NumDel$ in any of them can be used to compute $inc\text{-}deg^{c,g_3}(D, \Sigma)$. There is no need to explicitly compute all stable models, their sizes, and compare them. This value can be obtained by means of the query, "$NumDel(x)$?", answered by the program under the *brave semantics* (returning an answer from *some* stable model). □

Discussion. There are many open issues, among them exploring other inconsistency measures, e.g. based on the *Jaccard distance* [14]. Several measures have

---

[4] For certain forms of *prioritized repairs*, such as endogenous repairs, the normalization coefficient $|D|$ might be unnecessarily large. In this particular case, it might be better to use $|D^n|$.

[5] This approach was followed in [1] to compute maximum *responsibility degrees* of database tuples as causes for violations of DCs, appealing to a causality-repair connection [3].

been considered in knowledge representation [9,10,16], mostly for the propositional case. It would be interesting to analyze the general properties of those measures that are closer to database applications, along the lines of [8]; and their relationships. For each measure it becomes relevant to investigate the complexity of its computation, in particular, in data complexity (databases may have exponentially many repairs, in data [2]).[6] Actually, it is possible to prove that computing $inc\text{-}deg^{c,g_3}(D, \Sigma)$ is complete for the functional class $FP^{NP(log(n))}$ in data, and this both for sets Σ of DCs and of FDs.

References

1. Bertossi, L.: Characterizing and computing causes for query answers in databases from database repairs and repair programs. Corr Arxiv cs.DB/1712.01001 (2018). Proc. FoIKs. LNCS, vol. 10833, pp. 55–76. Springer (2018)
2. Bertossi, L.: Database Repairing and Consistent Query Answering. Synthesis Lectures on Data Management. Morgan & Claypool, San Rafael (2011)
3. Bertossi, L., Salimi, B.: From causes for database queries to repairs and model-based diagnosis and back. Theory Comput. Syst. **61**(1), 191–232 (2017)
4. Bertossi, L., Rizzolo, F., Jiang, L.: Data quality is context dependent. In: Castellanos, M., Dayal, U., Markl, V. (eds.) BIRTE 2010. LNBIP, vol. 84, pp. 52–67. Springer, Heidelberg (2011). https://doi.org/10.1007/978-3-642-22970-1_5
5. Brewka, G., Eiter, T., Truszczynski, M.: Answer set programming at a glance. Commun. ACM **54**(12), 93–103 (2011)
6. Chomicki, J., Marcinkowski, J.: Minimal-change integrity maintenance using tuple deletions. Inf. Comput. **197**(1–2), 90–121 (2005)
7. Caniupan-Marileo, M., Bertossi, L.: The consistency extractor system: answer set programs for consistent query answering in databases. Data Knowl. Eng. **69**(6), 545–572 (2010)
8. Eiter, T., Mannila, H.: Distance measures for point sets and their computation. Acta Inform. **34**, 109–133 (1997)
9. Grant, J., Martinez, M.V. (eds.): Measuring Inconsistency in Information. College Publications, Los Angeles (2018)
10. Grant, J., Hunter, A.: Analysing inconsistent information using distance-based measures. Int. J. Approx. Reason. **89**, 3–26 (2017)
11. Kivinen, J., Mannila, H.: Approximate inference of functional dependencies from relations. Theor. Comput. Sci. **149**, 129–149 (1995)
12. Leone, N., et al.: The DLV system for knowledge representation and reasoning. ACM Trans. Comput. Logic. **7**(3), 499–562 (2006)
13. Meliou, A., Gatterbauer, W., Moore, K.F., Suciu, D.: The complexity of causality and responsibility for query answers and non-answers. Proc. VLDB **4**, 34–41 (2010)
14. Rajamaran, A., Ullman, J.: Mining of Masssive Datasets. Cambridge University Press, Cambridge (2012)
15. Staworko, S., Chomicki, J., Marcinkowski, J.: Prioritized repairing and consistent query answering in relational databases. Ann. Math. Artif. Intell. **64**(2–3), 209–246 (2012)
16. Thimm, M.: On the compliance of rationality postulates for inconsistency measures: a more or less complete picture. Künstliche Intell. **31**(1), 31–39 (2017)

---

[6] Certain (or skeptical) reasoning with repair programs for DCs with weak constraints is $\Delta_2^P(log(n))$-complete in data complexity, i.e. in the size of the database [7,12].

Scalable Bounding of Predictive Uncertainty in Regression Problems with SLAC

Arno Blaas[1,2]([✉]), Adam D. Cobb[1]([✉]), Jan-Peter Calliess[2], and Stephen J. Roberts[1,2]

[1] Machine Learning Research Group, University of Oxford, Oxford, UK
{arno,acobb}@robots.ox.ac.uk
[2] Oxford-Man Institute, University of Oxford, Oxford, UK

Abstract. We propose SLAC, a sparse approximation to a Lipschitz constant estimator that can be utilised to obtain uncertainty bounds around predictions of a regression method. As we demonstrate in a series of experiments on real-world and synthetic data, this approach can yield fast and robust predictive uncertainty bounds that are as reliable as those of Gaussian Processes or Bayesian Neural Networks, while reducing computational effort markedly.

Keywords: Predictive uncertainty bounds · Regression Lipschitz Interpolation

1 Introduction and Background

Machine learning methods are typically utilised in regression tasks where little is known a priori and one prefers 'black-box' approaches that are endowed with the capacity to flexibly learn rich function classes. However, when regression methods are employed in decision making, quantifying the uncertainty around predictions can often be key. While many such methods exist, uncertainty bounds often rest on assumptions that are hard to establish *a priori*, necessitating either manual or optimisation based tuning approaches to yield sufficient black-box learning capabilities. Unfortunately, this renders the bounds less interpretable and typically, the computational effort intractable for many applications.

We consider two widely used methods that offer uncertainty quantifications, Gaussian Processes (GPs) [1] and Bayesian Neural Networks (BNNs) [2]. In regression, both approaches compute Gaussian posterior input-output relationships that allow the usage of subjective confidence intervals around predictions for a given input as uncertainty bounds. As with all Bayesian methods, the problem is that these subjective bounds, as well as the pertaining predictions, are contingent on a priori choices such as the nature of the probability space and the prior distribution. In GPs, these are encoded in the kernel and mean functions (as well as their hyper-parameters). In BNNs, the prior is encoded by

© Springer Nature Switzerland AG 2018
D. Ciucci et al. (Eds.): SUM 2018, LNAI 11142, pp. 373–379, 2018.
https://doi.org/10.1007/978-3-030-00461-3_27

the architecture of the network, as well as by the hyperparameters, particularly the dropout rate.

To facilitate black-box learning and to avoid hand-tuning, approaches for model selection involve optimising these a priori choices to maximise some data-dependent criterion function such as the marginal log-likelihood. Unfortunately, this pragmatic approach has undesirable side-effects. Firstly, it can greatly inflate the computational effort required for training and make these approaches too slow to use to derive uncertainty bounds in online, or quickly changing, settings where frequent model re-training is necessary. Secondly, since also the optimiser (and its initialisations) becomes part of the training algorithm, the uncertainty bounds become dependent on the optimiser's solution (with its performance often dependent on initialisations). This makes the interpretation of the uncertainty bounds questionable and their accuracy has to be assessed empirically on a case by case basis.

While we might argue that the tension between the desire to have reliable uncertainty bounds and fast, reliable black-box learning can never be fully reconciled, in this paper we present our early work on SLAC (**S**parse **L**azily **A**dapted **L**ipschitz-**C**onstant) uncertainty bounds as an approach to ameliorate this tension. Conceived from black-box learners whose training is magnitudes faster and simpler than for existing methods, SLAC bounds demonstrably serve their purpose in practice: on a range of benchmark comparisons against GPs and BNNs, they compare favourably against their competitors both in terms of computational time and the reliability of their uncertainty bounds. Code for the experiments described in this paper can be found online[1].

2   Task and Approach

In supervised Machine Learning applications, we typically desire to learn from a set of training points $\mathcal{D}_n := \{(s_i, f(s_i)) | i = 1, \ldots, n\} \subset (\mathcal{X}, \mathcal{Y})$, generated by some unknown target function f in order to predict its values for new test points $\mathcal{T}_q := \{(t_i | i = 1, \ldots, q\} \subset \mathcal{X}$. In many real-world applications it is necessary that these predictions can reliably quantify uncertainty. For standard regression problems, where $\mathcal{X} \subset \mathbb{R}^d$ and $\mathcal{Y} \subset \mathbb{R}$, this typically translates to finding two functions $l(\cdot) \leq u(\cdot)$ such that $\forall x \in \mathcal{X} : \quad Pr\left(l(x) \leq f(x) \leq u(x) | \mathcal{D}_n\right) = 1 - \delta$, with $\delta \in [0, 1)$. For safety critical applications $\delta \approx 0$ is the most important case as in these applications we ultimately wish to bound the uncertainty with high probability and thereby (almost) replace it with certainty. We thus use the phrase *uncertainty bounds* (or simply *bounds*) for $l(\cdot)$ and $u(\cdot)$ to indicate $\delta \approx 0$.

2.1   Definition of Our Model

Our model is originally inspired by the deterministic bounds (i.e. $\delta = 0$) that can be inferred if f is known to be Lipschitz-continuous with best Lipschitz-constant L^* with respect to some metrics $\partial : \mathcal{X}^2 \to \mathbb{R}_{\geq 0}, \partial y : \mathcal{Y}^2 \to \mathbb{R}_{\geq 0}$,

---

[1] https://github.com/arblox/SLAC.

which is defined as L^* being the smallest value for which it holds that $\forall x, x' \in \mathcal{X} : \partial y(f(x), f(x')) \leq L^* \partial(x, x')$. In this case, Sukharev [3] has shown that with bounds defined as

$$u_{L^*}(x; \mathcal{D}_n) := \min_{s_i \in \mathcal{X}_n} (f_i + L^* \partial(s_i, x)) \tag{1}$$

$$l_{L^*}(x; \mathcal{D}_n) := \max_{s_i \in \mathcal{X}_n} (f_i - L^* \partial(s_i, x)),$$

it holds for all $x \in \mathcal{X}$ that

$$l_{L^*}(x; \mathcal{D}_n) \leq f(x) \leq u_{L^*}(x; \mathcal{D}_n). \tag{2}$$

Unfortunately, in most regression problems L^* is unknown. Still, in light of this strong guarantee when L^* is known, we stick to the prior assumption that our target function f is Lipschitz-continuous to define our model. It is worth noting that this Lipschitz-continuity assumption is actually implicitly shared by many other models, including many popular neural network based models [4]. However, as opposed to such models, we are only interested in the quality of uncertainty bounds, and thus do not assume one potential function value to be more likely than another as long as they both lie inside the uncertainty bounds of our model. That is, we assume a uniform distribution between the prediction bounds of our model. With bounds of the form in Eqns. (1), this yields in the following model for inference at $x \in \mathcal{X}$:

$$f(x; L_n, \mathcal{D}_n) \sim U(l_{L_n}(x; \mathcal{D}_n), u_{L_n}(x; \mathcal{D}_n)), \tag{3}$$

where L_n acts as a hyperparameter to be learned from the data.

2.2   Uncertainty Bounds with Lazily Adapted Constants

The canonical approach to learn L_n for our model is to lazily adapt it, i.e. to set it to the smallest value that is compatible with the hypothesis that f is L_n−Lipschitz-continuous (hence LAC, Lazily Adapted Constant). This approach, which comes from Lipschitz Interpolation literature [5,6], yields

$$L_n^{\text{LAC}} := \max_{s_i, s_j \in \mathcal{X}_n} \frac{\partial y(f_i, f_j)}{\partial(s_i, s_j)}, \tag{4}$$

as resulting estimator of L^*, i.e. the maximum slope observable in the available training data. It can be easily shown that $l_{L_n^{\text{LAC}}}(x; \mathcal{D}_n) \leq u_{L_n^{\text{LAC}}}(x; \mathcal{D}_n)$ for all $x \in \mathcal{X}$, hence the resulting model is well-defined.

2.3   Computational Complexity and Sparse Approximation

Naturally, the computational complexity for computing L_n^{LAC} from scratch is $\mathcal{O}(n^2)$ [5]. While this natural quadratic scalability is better than the natural cubic computational cost of GPs [1], it is still prohibitive for truly large data sets. This is why we introduce a simple sparse approximation which is inspired by Strongin's estimator for the one-dimensional case [6]: after ordering all training points such that $s_1 < s_2 < \ldots < s_n$ it can be seen from elementary analysis that $L_n^{\mathrm{LAC}} = \max_{2 \leq i \leq n} \frac{\partial_y(f_i, f_{i-1})}{\partial(s_i, s_{i-1})}$, which only requires $\mathcal{O}(n)$ operations. For higher dimensions, we similarly apply the idea of only inspecting the slopes between ordered points. Randomly drawing J permutations $\pi_j : \{1, \ldots, n\} \to \{1, \ldots, n\}$, we evaluate $L_n^{\mathrm{LAC}}(j) := \max_{2 \leq i \leq n} \frac{\partial_y(f_{\pi_j(i)}, f_{\pi_j(i-1)})}{\partial(s_{\pi_j(i)}, s_{\pi_j(i-1)})}$ and set the sparse lazily adapted constant (SLAC) to be

$$L_n^{\mathrm{SLAC}} := \max_j L_n^{\mathrm{LAC}}(j) = \max_j \max_{2 \leq i \leq n} \frac{\partial_y(f_{\pi_j(i)}, f_{\pi_j(i-1)})}{\partial(s_{\pi_j(i)}, s_{\pi_j(i-1)})}. \tag{5}$$

Obtaining L_n^{SLAC} is $\mathcal{O}(nJ)$, where in our experiments (Sect. 3) we found that the bounds for $J = 100$ are comparably conservative to those of L_n^{LAC}, which massively speeds up training for larger data sets and made L_n^{SLAC} by far the fastest method overall.

By definition, it is clear that $L_n^{\mathrm{SLAC}} \leq L_n^{\mathrm{LAC}}$. A problem that naturally arises with this sparse approximation for values of $L_n^{\mathrm{SLAC}} < L_n^{\mathrm{LAC}}$ is that in these cases the model might not be not well-defined everywhere. In such cases it can happen that $\mathfrak{l}_{L_n^{\mathrm{SLAC}}}(x; \mathcal{D}_n) > \mathfrak{u}_{L_n^{\mathrm{SLAC}}}(x; \mathcal{D}_n)$ for some regions in $\mathcal{X}$. Therefore we define $\tilde{\mathcal{D}}_n \subset \mathcal{D}_n$ to be the subset of training points for which $\mathfrak{l}_{L_n^{\mathrm{SLAC}}}(x; \tilde{\mathcal{D}}_n) \leq \mathfrak{u}_{L_n^{\mathrm{SLAC}}}(x; \tilde{\mathcal{D}}_n)$ for all $x \in \mathcal{X}$. This makes the resulting sparse model:

$$f(x; L_n^{\mathrm{SLAC}}, \mathcal{D}_n) \sim U(\mathfrak{l}_{L_n^{\mathrm{SLAC}}}(x; \tilde{\mathcal{D}}_n), \mathfrak{u}_{L_n^{\mathrm{SLAC}}}(x; \tilde{\mathcal{D}}_n)). \tag{6}$$

3   Experiments

3.1   Data Sets

Real-World Data. We tested the performance of the SLAC uncertainty bounds on 8 multidimensional real-world regression data sets of between 300 and 45000 data points, that are publicly available in the UCI machine learning repository. Each data set was randomly split into training and testing set accounting for 80% and 20% of the data respectively and centred and normalised using the training set mean and standard deviation.

Synthetic Data. Additionally, we demonstrated the robustness of the performance of the uncertainty bounds of our model, by pressure-testing it on a set of 100 synthetic functions on the interval $[-1, 1]$ with Lipschitz-constants ranging between 14.7 and 129.8. These were generated by randomising the weights of a 2-hidden-layer neural network with 2886 units in the first hidden layer and 577 units in the second hidden layer. Each function was converted into a data set of 1000 data points on an equally spaced grid over $[-1, 1]$, from which we cut out test sets of 200 data points corresponding to the intervals $[-0.94, -0.86]$, $[-0.60, -0.52]$, $[-0.42, -0.34]$, $[-0.12, -0.04]$ and $[0.20, 0.28]$.

Table 1. Characteristics of data and comparison of training times across UCI data sets. We denote with * values for which sparse GPs were used. SLAC was consistently by far the fastest method on all data sets and its advantage over LAC is most evident on the biggest data sets. Earlier stopping of optimisation during BNN training was found to result in less consistent bounds across data sets.

Data set	Size of data		Average training time in seconds			
	Data points	Dimensions	GP	BNN	LAC	SLAC
Boston housing	504	13	23 (4)	724 (195)	1	**0.4 (0)**
Concrete strength	1,030	8	30 (7)	1.488 (430)	4	**1 (0)**
Energy efficiency	768	8	16 (3)	2,733 (1,100)	2	**1 (0)**
Kin8nm	8,192	8	177* (41)	23,879 (5,499)	242	**11 (0)**
Naval propulsion	11,934	16	159* (117)	28,049 (6,978)	536	**22 (1)**
Power plant	9,568	4	248* (27)	23,395 (7,390)	332	**14 (0)**
Protein structure	45,730	9	1,295* (134)	59,689 (11,180)	14,158	**43 (3)**
Yacht dynamics	308	6	47 (3)	982 (388)	0.4	**0.3 (0)**

3.2 Experimental Setup

Our Model. We used Euclidean distance as metric for $\mathcal{X}$ and $\mathcal{Y}$. For SLAC, J was set to 100 and training was done 10 times repeatedly on each training set to analyse its sensitivity to the random seed used for drawing the permutations π_j.

Baselines. For the GPs, we assumed Gaussian noise. We used the automatic relevance determination (ARD) versions of the squared exponential (SE), the Matern32 and the Matern52 covariance functions. Training was done by marginal likelihood maximisation as implemented in the GPflow package. For data sets with more than 2000 training points, we resorted to sparse approximations through variational inference as proposed in [7] using 250 inducing points. For the real-world data sets, we performed 10 random initialisations for each training process to inspect the stability of the prediction bounds with respect to optimisation initialisation. For the BNNs, we used three different architectures with ReLu activation functions, labelled *low*, *medium* and *high*, with respectively 2, 3 and 4 hidden layers of $1,024$ units each. We used MC dropout [2] to approximate the posterior. As the model precision τ is vital for calibrating the uncertainty bounds with MC dropout, we selected it together with the lengthscale l by maximising the log-likelihood employing the sample efficient technique of Bayesian optimisation implemented in GPyOpt. Furthermore, we set dropout to 0.1 after experimenting empirically over the training data. For both baselines, we analysed both the three standard deviation bounds ($\delta = 0.003$) and four standard deviation bounds ($\delta = 0.00006$).

Evaluation Metrics. In order to assess the quality of the uncertainty bounds, we apply two criteria. Firstly, we assess whether the bounds are conservative enough. To evaluate this we computed the ratio of test points inside the bounds,

Table 2. Comparison of uncertainty bounds across UCI data sets. For GPs and BNNs the uncertainty bounds are for $\delta = 0.003$. With more than 99% of test points in bounds for most data sets, SLAC bounds are about as conservative as the three standard deviation bounds of GPs and BNNs. Average distance between bounds for SLAC is larger on many data sets, but overall still comparable. On the Concrete Strength, Power Plant, and Protein Structure data set, we suspect noise in the data causes LAC bounds to be unacceptably wide. However, SLAC seems to be more robust to noise by not evaluating every possible slope.

Data set	% Test points in bounds				Avg. distance of bounds			
	GP	BNN	LAC	SLAC	GP	BNN	LAC	SLAC
Boston housing	97.1 (0.0)	98.0 (0.5)	100	100 (0.0)	1.5 (0.0)	2.3 (1.1)	5.7	5.2 (0.4)
Concrete strength	96.6 (33.6)	99.5 (0.2)	97.6	98.2 (0.8)	1.6 (0.6)	2.5 (1.5)	20.4	15.3 (3.4)
Energy efficiency	98.7 (0.5)	100 (0.0)	100	99.5 (1.2)	0.2 (69.1)	4.5 (1.9)	2.5	2.5 (0.1)
Kin8nm	99.5 (0.2)	100 (0.1)	99.9	99.6 (0.2)	2.2 (1.1)	2.5 (3.5)	4.7	3.6 (0.5)
Naval propulsion	100 (0.0)	100 (0.0)	100	99.4 (0.4)	5.9 (69.1)	5.1 (1.1)	1.7	1.3 (0.3)
Power plant	99.4 (0.1)	100 (0.1)	100	99.4 (0.6)	1.7 (1.1)	8.2 (2.4)	14.6	4.5 (3.5)
Protein structure	100 (0.1)	95.8 (0.8)	99.8	99.5 (0.3)	4.8 (0.1)	2.6 (0.2)	60.2	8.4 (2.0)
Yacht dynamics	98.4 (33.0)	100 (0.0)	98.4	97.7 (0.8)	0.1 (5583.7)	2.2 (0.1)	2.3	2.3 (0.1)

$\frac{|\{t_j \in T_q | l(t_j) \le f(t_j) \le u(t_j)\}|}{|T_q|}$ and compare this to $1 - \delta$. For the synthetic data, we additionally analysed the number of functions that are entirely inside the uncertainty bounds of a model over the test set intervals. Secondly, we assess how tight the bounds are. Whilst from a risk analysis perspective, conservativeness is most important, it is obvious that useful uncertainty bounds should be as close together as possible while achieving the desired conservativeness. We thus computed the average distance of the bounds on the test set, $\frac{\sum_{t_j \in T_q} |u(t_j) - l(t_j)|}{|T_q|}$, and prefer smaller average distance conservative bounds.

Table 3. Comparison of uncertainty bounds across 100 random functions. For GPs and BNNs the bounds are for $\delta = 0.003$. In the one-dimensional case, LAC = SLAC as the training inputs can be ordered by value. Again, training for SLAC is by far the fastest and yields bounds which are comparable to the three standard deviation bounds of GPs and BNNs in terms of conservativeness and tightness.

Method	Training time	Bound performance		
	Avg. time in s	% Test points in bounds	Avg. distance of bounds	Functions in bounds
GP^{SE}	29.5	98.6 (2.9)	1.96 (0.27)	73
$GP^{Matern32}$	28.6	90.3 (7.3)	0.30 (0.12)	13
$GP^{Matern52}$	29.2	97.5 (4.0)	0.50 (0.20)	56
$L^{LAC/SLAC}$	**0.03**	99.8 (0.8)	1.84 (0.92)	92
BNN^{low}	4.6	98.5 (5.2)	8.98 (8.17)	88
BNN^{medium}	11.6	99.7 (1.5)	5.80 (6.59)	93
BNN^{high}	17.8	100.0 (0.0)	18.98 (0.00)	93

4   Results and Discussion

The results of our experiments are shown in Tables 1, 2 and 3. For brevity, we only show the three standard deviation bounds ($\delta = 0.003$) for GPs and BNNs. The four standard deviation bounds ($\delta = 0.00006$), omitted here, had identical training times and were 33% wider, which resulted in $99\% - 100\%$ of test points being inside these bounds for all data sets. In summary, we found across both types of data that training our sparse model to calculate the SLAC bounds was magnitudes faster than training sophisticated models like GPs or BNNs (Table 1). The resulting SLAC bounds were comparable in performance to the three and four standard deviation bounds of GPs and BNNs, especially in terms of conservativeness, which seemed to be a reliable feature of SLAC bounds (Tables 2 and 3). Their average distance was roughly the same as for BNNs and usually about twice as large as for GPs. However, noisy data caused this distance to be unnecessarily large. Nevertheless, SLAC was able to cope with such data better than LAC due to its sparsity (Table 2), as not every slope in the training data was taken into account.

Overall this is still early work and there are a number of topics which we plan to investigate in the future. In particular, we desire to investigate the effect of J more rigorously. It seems beneficial to find theoretical guarantees for how big it needs to be for a given desired conservativeness and also to analyse its choice to make the SLAC bounds more robust to noise in the data. However, our results already provide empirical evidence that SLAC bounds with $J = 100$ can be a fast, stable and reliable alternative to three to four standard deviation bounds of GPs and BNNs.

References

1. Rasmussen, C.E.: Gaussian Processes for Machine Learning (2006)
2. Gal, Y., Ghahramani, Z.: Dropout as a Bayesian approximation: representing model uncertainty in deep learning. In: International Conference on Machine Learning, pp. 1050–1059 (2016)
3. Sukharev, A.G.: Optimal search of a root of a function that satisfies a Lipschitz condition. In: Zhurnal Vychislitel'noi Matematiki i Matematicheskoi Fiziki (1976)
4. Szegedy, C., et al.: Intriguing properties of neural networks. In: International Conference on Learning Representations (2014)
5. Calliess, J.P.: Lazily adapted constant kinky inference for nonparametric regression and model-reference adaptive control. ArXiv preprint arXiv:1701.00178 (2016)
6. Strongin, R.G.: On the convergence of an algorithm for finding a global extremum. Eng. Cybern. **11**, 549–555 (1973)
7. Titsias, M.: Variational learning of inducing variables in sparse Gaussian processes. In: Artificial Intelligence and Statistics, pp. 567–574 (2009)

Predicting the Possibilistic Score of OWL Axioms Through Support Vector Regression

Dario Malchiodi[1(✉)], Célia da Costa Pereira[2], and Andrea G. B. Tettamanzi[2]

[1] Dipartimento di Informatica, Università degli Studi di Milano, Milan, Italy
dario.malchiodi@unimi.it
[2] Université Côte d'Azur, CNRS, I3S, Sophia-Antipolis, France
{celia.pereira,andrea.tettamanzi}@unice.fr

Abstract. Within the context of ontology learning, we consider the problem of selecting candidate axioms through a suitable score. Focusing on subsumption axioms, this score is learned coupling support vector regression with a special similarity measure inspired by the Jaccard index and justified by semantic considerations. We show preliminary results obtained when the proposed methodology is applied to pairs of candidate OWL axioms, and compare them with an analogous inference procedure based on fuzzy membership induction.

Keywords: Support vector regression
Possibilistic OWL axiom scoring

1 Introduction and Related Works

Schema enrichment is one important ingredient of ontology learning [9]. In particular, in the context of the semantic Web, the increasing amount of Linked Data causes schema enrichment to be an emerging field of research. Its goal is that of automatizing the work of knowledge engineers by leveraging existing ontologies (typically expressed in OWL) and instance data (typically represented in RDF) [8]. The final aim is detecting meaningful patterns and learn schema axioms from existing instance data (facts) and their metadata, if available, using induction-based methods like the ones developed in inductive logic programming.

Many researchers are convinced of the benefit of using enriched schemas to improve the quality of this reasoning process. Fleischhacker and colleagues use statistical schema induction to enrich the schema of any RDF dataset with property axioms [6]. Töpper and colleagues propose an approach focused on the enrichment of the DBpedia ontology by using statistical methods [16]. Huitzil et al. explored the possibility of learning dataypes within ontologies [7]. Bühmann and colleagues proposed a light-weight method to enrich knowledge bases accessible via SPARQL endpoints with almost all types of OWL 2 axioms. The aim of their approach was to allow to create a schema in a semi-automatic way [4].

© Springer Nature Switzerland AG 2018
D. Ciucci et al. (Eds.): SUM 2018, LNAI 11142, pp. 380–386, 2018.
https://doi.org/10.1007/978-3-030-00461-3_28

All these approaches to schema enrichment critically rely on (candidate) axiom scoring. In practice, testing an axiom boils down to computing an acceptability score, measuring the extent to which the axiom is compatible with the recorded facts.

Methods to approximate the semantics of given types of axioms have been throughly investigated in the last decade (e.g., approximate subsumption [14]) and some related heuristics have been proposed to score concept definitions in concept learning algorithms [13]. The most popular candidate axiom scoring heuristics proposed in the literature are based on statistical inference (see, e.g., [4]). An alternative axiom scoring heuristics based on a formalization in possibility theory of the notions of logical content of a theory and of falsification and complying with an open-world semantics has recently been proposed [15]. While empirical evidence has been found that such a possibilistic scoring heuristics may lead to more accurate ontologies, the heavy computational cost of the heuristics makes it hard to apply in practice, unless some implementation tricks are devised (e.g., time capping).

A promising alternative to the direct computation of the possibilistic score consists in training a surrogate model on a sample of candidate axioms for which the score had already been computed or is otherwise available, in order to be capable of *predicting* the score of a novel, unseen candidate axiom. This idea was recently proposed in [10], using an adaptation of support vector clustering for learning the membership functions for fuzzy sets.

In this work, we follow the same general scheme, but we apply support vector regression to obtain a simpler surrogate model and we compare our preliminary results to those obtained with the modified support vector clustering on the same dataset of SubClassOf (i.e., subsumption) axioms, whose possibilistic score has been previously determined by direct application of the heuristics on the DBpedia RDF dataset [15].

The paper is structured as follows: Sect. 2 gives some background on the possibilistic axiom scoring heuristics and on how the similarity between axioms is computed. In Sect. 3 we illustrate the learning procedure having as input the above mentioned scores and similarities, and producing predictors as output. The performed numerical experiments are described and discussed in Sect. 4. Some concluding remarks end the paper.

2   Background on Axiom Scoring and Similarity

The possibilistic axiom score we wish to be able to predict was proposed in [15], to which the reader is referred to for the details. Given a candidate OWL 2 axiom ϕ, expressing a *hypothesis* about the relations holding among some entities of a domain, a degree of *possibility* $\Pi(\phi)$ and of *necessity* $N(\phi)$ for ϕ are computed based the *evidence* available contained in an RDF dataset $\mathcal{K}$.

The possibility and necessity of an axiom can then be combined into a single handy acceptance/rejection index

$$\mathrm{ARI}(\phi) = N(\phi) + \Pi(\phi) - 1 = N(\phi) - N(\neg\phi)$$
$$= \Pi(\phi) - \Pi(\neg\phi) \in [-1,1], \qquad (1)$$

because $N(\phi) = 1 - \Pi(\neg\phi)$ and $\Pi(\phi) = 1 - N(\neg\phi)$ (duality of possibility and necessity). A negative $\mathrm{ARI}(\phi)$ suggests rejection of ϕ ($\Pi(\phi) < 1$), whilst a positive $\mathrm{ARI}(\phi)$ suggests its acceptance ($N(\phi) > 0$), with a strength proportional to its absolute value. A value close to zero reflects ignorance about the status of ϕ.

One nice property of this acceptance/rejection index, which stems from the duality of possibility and necessity, is that, for all ϕ,

$$\mathrm{ARI}(\neg\phi) = -\mathrm{ARI}(\phi).$$

The idea proposed in [10] is that, if we can train a model to predict $\Pi(\phi)$ and $\Pi(\neg\phi)$, we have enough information to estimate $\mathrm{ARI}(\phi)$ without having to directly compute it.

Support vector regression, which we use to train a predictor of the possibility of OWL axioms, requires a kernel function which, for our purposes, may be viewed as a similarity measure between candidate axioms. To allow the comparison of the results, we adopt the semantic similarity measure, somehow reminiscent of the Jaccard index, proposed and justified in [10], to which the interested reader is referred. Given two subsumption axioms $A \sqsubseteq B$ and $C \sqsubseteq D$ and their negations $A \not\sqsubseteq B$ and $C \not\sqsubseteq D$, where A, B, C, and D are OWL class expressions, such similarity can be written as shown in Table 1, where $[E] = \{a : E(a)\}$ denotes the extension of class expression E in the RDF dataset at hand.

Table 1. A summary of the formulas to be used to compute the similarity $\mathrm{sim}(\phi, \psi)$ between positive or negated subsumption axioms ϕ and ψ.

	$\psi = C \sqsubseteq D$	$\psi = C \not\sqsubseteq D$
$\phi = A \sqsubseteq B$	$\mathrm{sim}(\phi,\psi) = \dfrac{\|[A] \cap [B] \cup [C] \cap [D]\|}{\|[A] \cup [C]\|}$	$\mathrm{sim}(\phi,\psi) = \dfrac{\|[A] \cap [B] \cup [C] \cap \overline{[D]}\|}{\|[A] \cup [C]\|}$
$\phi = A \not\sqsubseteq B$	$\mathrm{sim}(\phi,\psi) = \dfrac{\|[A] \cap \overline{[B]} \cup [C] \cap [D]\|}{\|[A] \cup [C]\|}$	$\mathrm{sim}(\phi,\psi) = \dfrac{\|[A] \cap \overline{[B]} \cup [C] \cap \overline{[D]}\|}{\|[A] \cup [C]\|}$

3   Support Vector Regression

In its simplest formulation, the extension of support-vector framework [5] to regression problems consists in considering a multiobjective optimization in which lines are scored according to their flatness and to their distance from a set of points. The tradeoff between these components is ruled via a parameter $C > 0$. Different flavors of this technique are defined in function of how the above mentioned distance is defined and penalized [3,12]. In particular:

- *ε-insensitive* regression fixes a parameter $\epsilon > 0$ representing the width of a tube centered around the line: the loss is null for all points falling within this tube and equal to the distance w.r.t. its frontier otherwise;
- *ridge regression* considers a quadratic loss in terms of the distance between line and points.

This basic scheme is extended to nonlinear regression using kernel methods, that is mapping the original points via a nonlinear transformation onto a higher-dimensional space, and trying to find a linear regression therein. In particular, as the original points occur in the problem formalization only in form of dot product computations, the only change amounts to replacing all such occurrences with invocations of a suitable kernel function amounting for computing the dot product of the images of its arguments. In the next section we detail how the heuristic of Sect. 2 can be used in order to build a kernel function whose arguments are axioms.

4   Experiments

We adopted the approaches to support vector regression resumed in Sect. 3 to the problem of building a predictor for the ARI value for candidate OWL axioms on the basis of its measurements on a limited set of formulas, using the methodology described in Sect. 2. In particular, we used as a reference the same settings as in [10], briefly summarized hereafter:[1]

- we considered $m = 722$ SubClassOf axioms involving atomic classes which were exactly scored against DBpedia,[2] as well as their negations, in a set A (thus a total of $n = 2m = 1444$ formulas), computing sim for each pair;
- we took the possibility of each formula $\phi_i \in A$ previously computed using the heuristic described in [15], henceforth identified as a value $\mu_i = \Pi(\phi_i)$ of the "acceptability" of ϕ_i as an axiom.

We used the sim values as kernel computations and the possibility values as target to be predicted using the two variants of support vector regression described in previous section. Given the small size of available data, we resorted to iterating ten times the following holdout scheme in order to assess the generalization ability of the inferred predictors. Training, validation, and test sets containing the 80%, 10%, and 10% of original data, respectively, have been obtained after shuffling the dataset.[3] Validation involved in both cases the tradeoff parameter C, and the tube width when considering ϵ-insensitive regression.[4] Model selection was guided by RMSE accuracy. Table 2 summarizes the results on test sets

---

[1] Code and data to replicate all experiments is available at https://github.com/dariomalchiodi/SUM2018.
[2] The computation of the exact score took 290 CPU days on a 12 6-core CPU machine.
[3] Each pair $(\phi, \neg\phi)$ was assigned to a same set in order to be able to compute its ARI.
[4] After some experimentation these parameters were selected within a grid considering all magnitudes between 10^{-3} and 10^4 and between 10^{-2} and 10^5, respectively.

in terms of RMSE, median, and standard deviation of the corresponding errors, both considering the ability of the inferred model to predict: (i) the acceptability $\mu_i = \Pi(\phi_i)$ of a formula, and (ii) its ARI according to (1). The ridge regression variant highlights better results w.r.t. both performance metrics with essentially comparable variability, although only ARI was deemed in [10] as a key indicator.

Table 2. Results of acceptability and ARI learning using ϵ-insensitive regression, ridge regression, and fuzzy inference in 10 repeated holdout experiments measuring root mean square (RMSE), median (Median), and standard deviation (STDEV) of errors.

Method	Acceptability			ARI			Time (mins.)
	RMSE	Median	STDEV	RMSE	Median	STDEV	
ϵ−insensitive	4.83e−01	2.50e−01	3.94e−02	8.46e−01	9.55e−01	4.11e−01	4
Ridge	3.89e−01	8.58e−02	2.67e−01	6.28e−01	1.91e−01	6.37e−01	2
Fuzzy	3.08e−01	0.00e+00	1.54e−01	4.86e−01	7.56e−04	3.34e−01	500

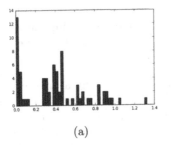 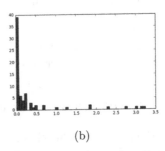

(a)   (b)

Fig. 1. Histograms of median errors for two iterations of the holdout scheme when using ridge regression, respectively highlighting the presence and the absence of a mix of two distributions.

Table 2 also compares the results with those of an analogous experiment based on a tailored procedure interpreting μ_is as membership values to a fuzzy set to be learned [10].[5] It is clear that the original results outperform the presented ones in terms of the proposed metrics. Things are radically different if we take into account also the time dimension: the eight hours needed to train and tune the fuzzy-based system—already incomparable to the 290 days required to produce the initial ARI labels—are further reduced to two minutes when using ridge regression. This amounts to a speed factor of roughly 0.004 with an equal factor in performance degradation. The ridge regression results also share with the original approach a better performance of the median w.r.t. RMSE, although

---

[5] This procedure is parametrized on the choice of different shapes for the fuzzy set membership function: the table reports for each column the best obtained result.

with a smaller intensity. This led to conjecturing the presence of *easy* and *hard* to learn formulas, but this conjecture is less evident now: indeed, only in around half of the holdout iterations the histogram of median errors highlights a mixture of two error distributions, and anyhow they appear not strongly separated (see Fig. 1 for an example of both cases). Puzzlingly, a very strong separation between two median error distribution is always obtained when considering ϵ-insensitive regression, despite its lower performances. Thus we repeated the experiment proposed in [10] with the aim of testing this conjecture, computing the average median error in all iterations and finding its best clusterization in terms of sil-houtette index [11]. More precisely, we considered: (i) both regression methods, (ii) only ϵ-insensitive regression, and (iii) only ridge regression. The best clus-terization consisted of two groups only in (ii) and (iii). This led us to further inspect the two classes of candidate axioms only in terms of ridge regression. Specifically, we found 30 hard axioms with an overlap of around 40% with those found in the original paper. Such axioms are listed in Table 3; those that were also in [10] are marked with an asterisk.

Table 3. Positive members of the detected "hard" axiom pairs.

```
SubClassOf(schema:Product dbo:MeanOfTransportation) *
SubClassOf(dbo:Chancellor dbo:Person)
SubClassOf(schema:School gml:_Feature) *
SubClassOf(dbo:BeautyQueen dbo:Person)
SubClassOf(dbo:InformationAppliance schema:Person)
SubClassOf(dbo:Racecourse gml:_Feature) *
SubClassOf(dbo:WomensTennisAssociationTournament skos:Concept)
SubClassOf(dbo:VolleyballCoach dbo:Person)
SubClassOf(dbo:VolleyballCoach owl:Thing)
SubClassOf(dbo:VolleyballCoach foaf:Person)
SubClassOf(dbo:Presenter dbo:RadioHost)
SubClassOf(dbo:Venue gml:_Feature) *
SubClassOf(dbo:YearInSpaceflight skos:Concept) *
SubClassOf(dbo:ComedyGroup schema:Organization)
SubClassOf(dbo:ComedyGroup foaf:Person) *
```

5   Conclusions

We have applied support vector regression to the task of predicting the pos-sibilistic score of candidate OWL subsumption axioms. Ridge regression gives better results than ϵ-insensitive regression. A comparison with a previous pro-posal using a modified support vector clustering for learning fuzzy sets shows that the regression approach allows for a faster training time fairly scaling with performance degradation. Our results also confirm the existence of a small sub-set of axioms that are much harder to score than the rest; however, this subset appears to depend, at least to some extent, on the method used to predict the score. Future work includes trying other prediction methods [1,2], but also reformulating the scoring problem as a binary classification problem, which would suit the needs of schema enrichment equally well.

References

1. Apolloni, B., Bassis, S., Malchiodi, D., Pedrycz, W.: Interpolating support information granules. Neurocomputing **71**, 2433–2445 (2008)
2. Apolloni, B., Iannizzi, D., Malchiodi, D., Pedrycz, W.: Granular regression. In: Apolloni, B., Marinaro, M., Nicosia, G., Tagliaferri, R. (eds.) NAIS/WIRN -2005. LNCS, vol. 3931, pp. 147–156. Springer, Heidelberg (2006). https://doi.org/10.1007/11731177_22
3. Apolloni, B., Malchiodi, D., Valerio, L.: Relevance regression learning with support vector machines. Nonlinear Anal. **73**, 2855–2867 (2010)
4. Bühmann, L., Lehmann, J.: Universal OWL axiom enrichment for large knowledge bases. In: ten Teije, A., et al. (eds.) EKAW 2012. LNCS (LNAI), vol. 7603, pp. 57–71. Springer, Heidelberg (2012). https://doi.org/10.1007/978-3-642-33876-2_8
5. Cortes, C., Vapnik, V.: Support-vector networks. Mach. Learn. **20**(3), 273–297 (1995)
6. Fleischhacker, D., Völker, J., Stuckenschmidt, H.: Mining RDF data for property axioms. In: Meersman, R., et al. (eds.) OTM 2012. LNCS, vol. 7566, pp. 718–735. Springer, Heidelberg (2012). https://doi.org/10.1007/978-3-642-33615-7_18
7. Huitzil, I., Straccia, U., Díaz-Rodríguez, N., Bobillo, F.: Datil: learning fuzzy ontology datatypes. In: Medina, J., et al. (eds.) IPMU 2018. CCIS, vol. 854, pp. 100–112. Springer, Cham (2018). https://doi.org/10.1007/978-3-319-91476-3_9
8. Lehmann, J., Völker, J. (eds.): Perspectives on Ontology Learning, Studies on the Semantic Web, vol. 18. IOS Press, Amsterdam (2014)
9. Maedche, A., Staab, S.: Ontology learning for the semantic web. IEEE Intell. Syst. **16**(2), 72–79 (2001)
10. Malchiodi, D., Tettamanzi, A.G.B.: Predicting the possibilistic score of OWL axioms through modified support vector clustering. In: SAC 2018: Symposium on Applied Computing, Pau, France, 9–13 April 2018. ACM, New York (2018)
11. Rousseeuw, P.J.: Silhouettes: A graphical aid to the interpretation and validation of cluster analysis. J. Comput. Appl. Math. **20**, 53–65 (1987). http://www.sciencedirect.com/science/article/pii/0377042787901257
12. Smola, A.J., Schölkopf, B.: A tutorial on support vector regression. Stat. Comput. **14**(3), 199–222 (2004)
13. Straccia, U., Mucci, M.: pFOIL-DL: learning (fuzzy) EL concept descriptions from crisp OWL data using a probabilistic ensemble estimation. In: Proceedings of the 30th Annual ACM Symposium on Applied Computing, pp. 345–352. ACM, New York (2015)
14. Stuckenschmidt, H.: Partial matchmaking using approximate subsumption. In: Proceedings of the Twenty-Second AAAI Conference on Artificial Intelligence, 22–26 July 2007, Vancouver, British Columbia, Canada, pp. 1459–1464. AAAI Press (2007)
15. Tettamanzi, A.G.B., Faron-Zucker, C., Gandon, F.: Possibilistic testing of OWL axioms against RDF data. Int. J. Approx. Reason. **91**, 114–130 (2017)
16. Töpper, G., Knuth, M., Sack, H.: Dbpedia ontology enrichment for inconsistency detection. In: I-SEMANTICS, pp. 33–40. ACM (2012)

Inferring Quantitative Preferences: Beyond Logical Deduction

Maria Vanina Martinez[1], Lluis Godo[2], and Gerardo I. Simari[1(✉)]

[1] Department of Computer Science and Engineering, Universidad Nacional del Sur (UNS), Institute for Computer Science and Engineering (UNS–CONICET), San Andres 800, 8000 Bahia Blanca, Argentina
{mvm,gis}@cs.uns.edu.ar

[2] Artificial Intelligence Research Institute (IIIA–CSIC), Campus de la University Autonoma de Barcelona, 08193 Bellaterra, Spain
godo@iiia.csic.es

Abstract. In this paper we consider a hybrid possibilistic-probabilistic alternative approach to Probabilistic Preference Logic Networks (PPLNs). Namely, we first adopt a possibilistic model to represent the beliefs about uncertain strict preference statements, and then, by means of a pignistic probability transformation, we switch to a probabilistic-based credulous inference of new preferences for which no explicit (or transitive) information is provided. Finally, we provide a tractable approximate method to compute these probabilities.

Keywords: Preferences · Possibilistic logic · Necessity degrees
Probabilistic transformation · Tractable approximation

1 Introduction

Understanding and being able to model people's preferences is of outmost importance for an intelligent decision support system, as adequately eliciting and using that kind of information can provide an experience tailored to specific needs and expectations. For this reason, modeling and reasoning about an agent's preferences has a long history in different areas of study such as philosophy, logic, computer science, and economics. In the databases literature, preferences are expressed over tuples or ground atoms rather than over mutually exclusive "outcomes" (such as truth assignments to formulas) as is traditional in philosophy.

Independently from where the preferences are obtained, in real world applications, preferences are bound to be incomplete (underspecified), inconsistent (overspecified), or inherently uncertain. Information is usually sparse and it is quite impossible to express preferences over all possible pairs of alternatives; users are much more likely to express preferences that are subject to exceptions than ones that always hold. Following this motivation, [7] proposed Probabilistic Preference Logic Networks (PPLNs), which combine preferences expressed as strict partial orders over elements within a knowledge base, with probabilistic

© Springer Nature Switzerland AG 2018
D. Ciucci et al. (Eds.): SUM 2018, LNAI 11142, pp. 387–395, 2018.
https://doi.org/10.1007/978-3-030-00461-3_29

uncertainty and a semantics based on Markov random fields. The work aims to leverage the incomplete and uncertain preference statements as much as possible in order to be able to reason about cases for which no explicit (or transitive) information is provided. Despite the flexibility that PPLNs provide, the basic reasoning problem of computing the probability of a world or a given query is #P-hard. Furthermore, there is also the question of how intuitive and/or practical it is to learn or elicit weights for preference statements so that meaningful probabilistic distributions are computed. To address this issue, in this work we adopt some elements of PPLNs but provide a semantics based on *possibility* theory, interpreting the uncertainty of preferences as subjective beliefs (a qualitative characteristic of the preference statement of being true).

Consider the following scenario, where we wish to model a user's preferences about movies given the analysis of her posts on her social media sites. We have learned that: (i) she usually prefers sci-fi over action movies; (ii) with greater likelihood, she prefers dramas over comedies; and (iii) she is less likely to prefer drama over action. We are particularly interested in reasoning about cases that are not *explicitly mentioned*, such as how likely she is to prefer comedy over action films, or what her preferences are about sci-fi and comedy movies. Note that for these cases no transitivity relation can be exploited.

2   Preliminaries: A Basic Preference Logic and Its Possibilistic Extension

We first introduce a very basic logic of preferences on which we base our knowledge representation formalism for expressing preferences under possibilistic uncertainty over a set of alternatives.

Language. Let $A = \{a_1, a_2, \ldots, a_n\}$ be a finite set of propositional constants, which correspond to the set of alternatives we are interested in establishing preferences over. The language of strict preference statements $\mathcal{L}_P$ based on A is defined as follows:

- for each $a, b \in A$, $a \succ b \in \mathcal{L}_P$ is an atomic preference statement,
- if $\varphi, \psi \in \mathcal{L}_P$, then $\varphi \wedge \psi, \neg\varphi \in \mathcal{L}_P$ (other connectives like $\vee, \rightarrow, \leftrightarrow$ are definable as usual).

Example 1. Going back to the example from the introduction, we can have $A = \{s, a, c, d\}$ representing the different film genra *sci-fi*, *action*, *comedy*, and *drama*, respectively. Based on what we learned about the user's preference we would have the following set of preference statements: $S = \{(d \succ c), (s \succ a), (d \succ a)\}$.   ■

Semantics. Models, or worlds, are ordered lists $\omega = [b_1, b_2, \ldots, b_n]$ of elements of A. Let us denote by Ω the set of all ordered lists from A. Note that Ω contains $n!$ elements. For each $a \in A$ and $\omega = [b_1, b_2, \ldots, b_n] \in \Omega$, define $\omega(a)$ as the index $1 \leq i \leq n$ such that $a = b_i$, *i.e.*, the position of alternative a in ω.

Intuitively, in the context of reasoning about preferences, the set of all models consists of all ways in which the alternatives can be linearly ordered. This is different from other preference languages in the literature [2], where models, or possible worlds, correspond to truth assignments to preference statements. The *satisfaction relation* $\models \subseteq \Omega \times \mathcal{L}_P$ between models and formulas is defined:

- $w \models a \succ b$ if $w(a) > w(b)$
- $w \models \varphi \wedge \psi$ if $w \models \varphi$ and $w \models \psi$
- $w \models \neg\varphi$ if $w \not\models \varphi$

Next, the notion of logical consequence is defined as usual: if $T \cup \{\varphi\}$ is a set of formulas, we define $T \models_P \varphi$ if for every model $w \in \Omega$, if $w \models \psi$ for every $\psi \in T$, then $w \models \varphi$. The set of models for an arbitrary preference statement φ is denoted by $[\varphi] = \{w \in \Omega \mid w \models \varphi\}$.

Axioms. Consider the following axiomatic system LP, the corresponding notion of proof will be denoted $\vdash_P$:

- **CPL** axioms of classical propositional logic governing $\wedge, \vee, \neg$
- **Ax1** $(a \succ b) \wedge (b \succ c) \rightarrow (a \succ c)$
- **Ax2** $\neg((a \succ b) \wedge (b \succ a))$
- **Ax3** $(a \succ b) \vee (b \succ a)$
- **MP (modus ponens)** from φ and $\varphi \rightarrow \psi$ derive ψ

Note that one can show that $\neg(a \succ b)$ is equivalent to $(b \succ a)$, that is, $\vdash_P (b \succ a) \leftrightarrow \neg(a \succ b)$. Moreover, by (Ax2), we also have $\vdash_P \neg(a \succ a)$.

Theorem 1 (Soundness and Completeness). *For any $T \cup \{\varphi\} \subseteq \mathcal{L}_P$, we have $T \models_P \varphi$ iff $T \vdash_P \varphi$.*

Proof. Soundness is easy. As for completeness, assume $T \not\vdash_P \varphi$. But $T \vdash_P \varphi$ iff $T \cup AX \vdash_{CPL} \varphi$ iff $T \cup AX \models_{CPL} \varphi$, where AX is the set of all possible instantiations of axioms $Ax1, Ax2$, and $Ax3$. Thus, if $T \not\vdash_P \varphi$, there is a classical logic evaluation e such that $e(T) = e(AX) = 1$ and $e(\varphi) = 0$. Since $e(AX) = 1$, e determines a complete ordering of all the variables $w : b_1 > b_2 > ... > b_n$ such that $w(a) > w(b)$ iff $e(a > b) = 1$. Then $w \not\models \varphi$.

Now, we extend LP by allowing to attach weights to preference statements of LP expressing degrees of belief over them, in a completely analogous way Possibilistic Logic does with formulas of classical propositional logic [4].

Language. Formulas of ΠLP are weighted preference statements of the form (φ, α) where $\varphi \in \mathcal{L}_P$ and $\alpha \in (0, 1]$. The intended semantics of weights are lower bounds for necessity values, so that a formula like $(\varphi, 1)$ stands for φ is certain, while (φ, α) with $\alpha < 1$ means φ is somewhat certain, at least to the degree α.

Semantics. A possibilistic model is a mapping $\pi : \Omega \rightarrow [0, 1]$. Such a mapping (possibility distribution) π induces a necessity measure on formulas of $\mathcal{L}_P$:

$$N_\pi(\varphi) = \min\{1 - \pi(w) \mid w \not\models \varphi\} = 1 - \max\{\pi(w) \mid w \not\models \varphi\}$$

We will denote by $\Pi(\Omega)$ the set of all possibility distributions on Ω.

The satisfaction relation between possibilistic models and formulas is defined as $\pi \models (\varphi, \alpha)$ if $N_\pi(\varphi) \geq \alpha$. The corresponding notion of possibilistic entailment is as follows: a set of ΠLP-formulas K entails a formula Φ, denoted $K \models_\Pi \Phi$, in case that, for every $\pi \in \Pi(\Omega)$, if $\pi \models \Psi$ for every $\Psi \in K$, then $\pi \models \Phi$.

Example 2. Returning to our running example, this model allows us to associate a weight to each preference statement. One possible set of ΠLP formulas can be the following, according to the interpretation of the elicitation of preferences described in the introduction: $K = \{(d \succ a, 0.9), (s \succ a, 0.6), (d \succ a, 0.2)\}$. ∎

Analogously to possibilistic logic, possibilistic entailment from set K of ΠLP formulas can be characterized in terms of *minimum specific* possibility distribution π_K underlying K. Since this will be used later, we recall the main facts (see [4] for more details):

(i) Given a set $K = \{(\varphi_i, \alpha_i) \mid 1 \leq i \leq m\}$ of ΠLP formulas, there always exists the minimum specific (i.e. the greatest) distribution π_K such that $\pi_K \models K$, that is, if $\pi \models K$ then $\pi \leq \pi_K$ (in the sense of $\pi(w) \leq \pi_K(w)$ for all $w \in W$).

(ii) π_K can be computed as the meet of the minimum specific distributions associated to each formula in K. Namely, for each weighted formula (φ, α), consider the minimum specifics distribution satisfying it, which is defined as:

$$\pi_{(\varphi,\alpha)} = \begin{cases} 1, & \text{if } w \models \varphi \\ 1 - \alpha, & \text{otherwise} \end{cases}$$

Then, $\pi_K = \bigwedge_{(\varphi,\alpha)\in K} \pi_{(\varphi,\alpha)}$, and can be computed as follows:

$$\pi_K(w) = \begin{cases} 1, & \text{if } w \models \varphi_1 \wedge \ldots \wedge \varphi_m \\ 1 - \max\{\alpha_i : w \not\models \varphi_i\}, & \text{otherwise} \end{cases} \quad (1)$$

(iii) Finally, it is shown that to check whether a weighted preference formula (ψ, β) can be entailed from K it is enough to check (ψ, β) is satisfied by π_K, i.e. it holds that $K \models_\Pi (\psi, \beta)$ iff $\pi_K \models (\psi, \beta)$.

Example 3. Let $K = \{(\varphi_1, \alpha_1), (\varphi_2, \alpha_2), (\varphi_3, \alpha_3)\}$ with $\alpha_1 > \alpha_2 > \alpha_3$. Then one can check that π_K is such that:

$$\pi_K(w) = 1 \text{ if } w \models \varphi_1 \wedge \varphi_2 \wedge \varphi_3, \pi_K(w) = 1 - \alpha_3 \text{ if } w \models \varphi_1 \wedge \varphi_2 \wedge \neg\varphi_3,$$
$$\pi_K(w) = 1 - \alpha_2 \text{ if } w \models \varphi_1 \wedge \neg\varphi_2, \text{ and } \pi_K(w) = 1 - \alpha_1 \text{ if } w \models \neg\varphi_1.$$

Put it in another way, in terms of the α-cuts:

$$\pi_K(w) = 1 \text{ if } w \models \varphi_1 \wedge \varphi_2 \wedge \varphi_3, \pi_K(w) \geq 1 - \alpha_3 \text{ if } w \models \varphi_1 \wedge \varphi_2,$$
$$\pi_K(w) \geq 1 - \alpha_2 \text{ if } w \models \varphi_1, \text{ and } \pi_K(w) \geq 1 - \alpha_1 \text{ for all } w.$$

∎

Axioms. An axiomatic system for ΠLP can be easily defined by combining the above axioms of LP with those of possibilistic logic governing the weights:

(LP) $(\varphi, 1)$, where φ is an axiom of LP
(W) Weight weakening rule: from (φ, α) derive (φ, β), whenever $\beta \leq \alpha$
(GMP) Weighted modus ponens: from (φ, α) and $(\varphi \rightarrow \psi, \beta)$, derive $(\psi, \min(\alpha, \beta))$

Denote by $\vdash_{\Pi}$ the corresponding syntactic notion of proof for ΠLP. Then, one can easily show that this axiomatic system is indeed *sound and complete* by directly combining the corresponding results for LP and Possibilistic logic. Using results by Hollunder [6], deductions in ΠLP from a set K of formulas can also be captured by (non-possibilistic) deductions in LP from suitable subsets of non-weighted formulas. These results are summarized in the next theorem.

Theorem 2 (Soundness and Completeness). *For any set of ΠLP formulas $K \cup \{\Phi\}$, the following are equivalent:*

(i) $K \vdash_{\Pi} \Phi$ (ii) $K \models_{\Pi} \Phi$
and, in the case of $\Phi = (\varphi, \alpha)$,
(iii) $K_\alpha \vdash_P \varphi$, where $K_\alpha = \{(\psi, \beta) \in K \mid \beta \geq \alpha\}$.

3   Beyond Logical Inference

The logical setting to reason about strict preferences described above is appealing at first sight; it is however, somewhat limited if the goal is to make some adventurous inferences on what plausible preferences can be beyond the ones logically implied (*i.e.*, those obtained by applying the transitive rule).

One possibility is to transform the possibility distribution π_K underlying a set K of weighted preference statements into a probability distribution. In the literature there are several proposals for such a possibility-to-probability transformation—one of the best behaved transformations is the one independently proposed by Yager [9] and Dubois-Prade [5], that amounts to: (1) first consider the possibility measure induced by π_K as a special case of plausibility function in the frame of the Dempster-Shafer evidential model, (2) consider the corresponding mass assignment, and (3) consider Smets's pignistic probability transformation [8] of the mass assignment into a probability distribution p_K. Next, we describe the necessary steps to compute the (pignistic) probability of any preference statement out of a finite set of ΠLP-formulas K:

1. Let $K = \{(\varphi_i, \alpha_i) \mid 1 \leq i \leq m\}$. We assume K is consistent.
2. Compute the minimum specific *possibility* distribution $\pi_K : \Omega \rightarrow [0, 1]$.
3. Transform π_K into a *probability* distribution $p_K : \Omega \rightarrow [0, 1]$ (Yager's transformation) according to the following procedure:
   (a) Let $\mathsf{Im}(\pi_K) = \{\pi_K(w) : w \in \Omega\} = \{\delta_0, \delta_1, \ldots, \delta_k\}$ such that $\delta_0 > \ldots > \delta_k$. (Under the consistency hypothesis, $\delta_0 = 1$.)

(b) Consider the α-cuts of π_K, i.e. the following nested sets: for each $i = 0, \ldots, k$, let $H_i = \{w \in \Omega : \pi_K(w) \geq \delta_i\}$. Note that $H_k = \Omega$.

(c) For each $i = 0, \ldots, k - 1$, let $\lambda_i = \delta_i - \delta_{i+1}$ and $\lambda_k = \delta_k$. Note that $\sum_i \lambda_i = \delta_0$ (= 1 under consistency).

(d) Finally, for each $w \in \Omega$, define $p_K(w) = \sum_{i:w \in H_i} \frac{\lambda_i}{|H_i|}$.

4. For any query ψ, compute its probability as $P_K(\psi) = \sum_{w \in \Omega : w \models \psi} p_K(w)$.

This allows to define a kind of entailment between sets of ΠLP-formulas K and probabilistic quantified LP-formulas $[\varphi, \alpha]$, with $\alpha \in (0, 1]$ as follows:

$$K \models_{pr} [\varphi, \alpha] \text{ whenever } P_K(\varphi) \geq \alpha.$$

Example 4. Consider the following encoding of the preference statements in Example 1: $K = \{(d > c, 0.9), (s > a, 0.6), (d > a, 0.2)\}$, where d stands for *drama*, s for *sci-fi*, a for *action*, and c for *comedy*. Then, the computations of the minimum specific possibility distribution π_K and its induced probability distribution p_K are shown in the following table, where we assume the set of alternatives reduced is to $A = \{a, c, d, s\}$ and worlds $[x, y, z, t]$ are simply denoted $xyzt$:

World	π_k	p_K	World	π_k	p_K	World	π_k	p_K	World	π_k	p_K
acds	0.1	0.004	cads	0.1	0.004	dacs	0.4	0.029	sacd	0.1	0.004
acsd	0.1	0.004	casd	0.1	0.004	dasc	0.4	0.029	sadc	0.8	0.095
adcs	0.4	0.029	cdas	0.1	0.004	dcas	0.4	0.029	scad	0.1	0.004
adsc	0.4	0.029	cdsa	0.1	0.004	dcsa	1	0.135	scda	0.1	0.004
ascd	0.1	0.004	csad	0.1	0.004	dsac	1	0.135	sdac	1	0.135
asdc	0.4	0.029	csda	0.1	0.004	dsca	1	0.135	sdca	1	0.135

Now, one can compute the probability of any preference statement. For instance, the probability that the agent prefers a sci-fi film over a comedy is: $P_K(s \succ c) = \sum\{p_k(w) \mid w \models s > c\} = 0.718$. Notice that $(s \succ c, \beta)$, with any $\beta > 0$, cannot be derived from K, however it has a relatively high probability. So, in this case we have $K \not\models_\Pi (s \succ c, \beta)$ with $\beta > 0$, but $K \models_{pr} [s \succ c, 0.718]$. ∎

The problem with this exact procedure is that the number of worlds is $n!$, where n is the number of alternatives, and thus it is intractable. In the next section we explore a tractable approximation algorithm.

4   Tractable Approximation Algorithm

We now investigate how to mitigate the intractability of computing P_K. In particular, we use tools that are results for the problem of counting linear extensions (topological sortings) of a strict partial order (SPO) as these correspond exactly to the models of a set of ΠLP-formulas. Inspired by [7], we first divide the set K of ΠLP-formulas into strata that will allow us to efficiently compute an approximation to the probabilities of a given query.

Given a consistent finite set $K = \{(\psi_i, \alpha_i) : i \in I\}$ of ΠLP-formulas, let $Val(K) = \{1 - \alpha_i \mid (\psi_i, \alpha_i) \in K\}$. For the sake of simplicity, in the following we assume K is such that all ψ_i's in it are atomic preference statements. In such a case, we denote by K' the closure of K under the transitivity rule:

$$\frac{(a \succ b, \alpha), (b \succ c, \beta)}{(a \succ c, \min(\alpha, \beta))}.$$

Still, for each $a \succ b$ appearing in K', there might be multiple instances in K' of the same preference statement with different weights. Then let $K'' \subseteq K'$ be such that $(a \succ b, \alpha) \in K''$ iff $\alpha = \max\{\beta : (a \succ b, \beta) \in K'\}$.

Proposition 1. $\mathsf{Im}(\pi_K) = Val(K'') \cup \{1\}$.

This proposition says that, under the current hypothesis, we can easily compute $\mathsf{Im}(\pi_K)$ directly from K'' without using Eq. 1.

Moroever, we can further assume that in K'' we have a single formula for every level α_i, since if there are more we can take their conjunction. Then let us assume $K'' = \{(\varphi_1, \alpha_1), \ldots, (\varphi_k, \alpha_k)\}$, with $\alpha_1 > \ldots > \alpha_k > 0$. In the following, assuming the consistency of K, we will consider $\mathsf{Im}(\pi_K) = Val(K'') \cup \{1\} = \{\delta_0 > \delta_1 > \ldots > \delta_k\}$, where $\delta_0 = 1, \delta_1 = 1 - \alpha_k, \ldots, \delta_k = 1 - \alpha_1$.

For each stratum $l = 0, 1, \ldots, k$, we define the set

$$D_l = \bigwedge_{i=1}^{k-l} \varphi_i,$$

where we take $D_k = \top$. Note that, for each l, the set of models of D_l is exactly the δ_l-cut of π_K, i.e. $[D_l] = H_l = \{w : \pi_K(w) \geq \delta_l\}$. Therefore, we have:

$$p_K(w) = \sum_{i:w \in H_i} \frac{\lambda_i}{|H_i|} = \sum_{l:w \models D_l} \frac{\delta_l - \delta_{l+1}}{|[D_l]|},$$

where we take $\delta_{k+1} = 0$. Now, for any query ψ, we can express the probability $P_K(\psi)$ as:

$$P_K(\psi) = \sum_{w \models \psi} p_K(w) = \sum_{w \models \psi} \sum_{l:w \models D_l} \frac{\delta_l - \delta_{l+1}}{|[D_l]|};$$

but this is equivalent to

$$P_K(\psi) = \sum_{l=0,k} \left(|[\psi \wedge D_l]| \cdot \sum_{m=l,k} \frac{\delta_m - \delta_{m+1}}{|[D_m]|} \right). \tag{2}$$

The first product of the outward summation is the proportion of models of formula D_l that are also a model for ψ; the second one, *i.e.*, the inner summation, adds up the probability mass that corresponds to stratum l. Therefore, if every model of D_l is a model of the query ψ, then the whole probability mass of stratum l contributes to the probability of ψ. On the other hand, if no model of D_l is a model for ψ, then stratum l does not contribute in the probability of ψ.

It is well known that the problem of counting linear extensions of an SPO is #P-hard [3]. Several algorithms have been developed in the literature to approximate this number; as done in [7], we will use TPA algorithm [1], a fully polynomial randomized approximation scheme (FPRAS) algorithm that, given an SPO O, approximates the number of linear extensions of O, denoted $\mathcal{L}(O)$ to within a factor of $1 + \epsilon$ with probability at least $1 - \delta$, in time $O((\ln \mathcal{L}(O))^2 \cdot n^3 \cdot (\ln n) \cdot \epsilon^{-2} \cdot \ln(1/\delta))$. Both the cardinalities of the sets $[\psi \wedge D_l]$ and $[D_l] = H_l$ can be approximated by TPA; we thus have the following result:

Proposition 2. *Let K be a set of atomic ΠLP-formulas expressing preferences over n elements, ψ be a preference statement, and ϵ, δ be positive real numbers. Then, $P_K(\psi)$ can be approximated to within a factor of $1 + \epsilon$ with probability at least $1 - \delta$ in time $O((|K| + |K''|^2) \cdot (\ln \mathcal{L}(O))^2 \cdot n^3 \cdot (\ln n) \cdot \epsilon^{-2} \cdot \ln(1/\delta))$.*

The running time expression essentially follows from first deriving $|K''|$ from $|K|$ and then directly applying TPA in Eq. 2; the query answering algorithm is therefore a FPRAS. Note that although a similar result holds for PPLNs [7], their algorithm is a FPRAS only under the assumption that the set of preference statements is fixed, since its cost has a term that is exponential in its size.

Acknowledgments. Martinez and Simari have been partially supported by EU H2020 research and innovation programme under the Marie Sklodowska-Curie grant agreement No. 690974 for the project MIREL: MIning and REasoning with Legal texts; and funds provided by Universidad Nacional del Sur (UNS), Agencia Nacional de Promocion Cientifica y Tecnologica, and CONICET, Argentina. Godo acknowledges the EU H2020 project SYSMICS (MSCA-RISE-2015 Project 689176) and the Spanish FEDER/MINECO project TIN2015-71799-C2-1-P.

References

1. Banks, J., Garrabrant, S., Huber, M., Perizzolo, A.: Using TPA to count linear extensions. ArXiv e-prints (2010)
2. Bienvenu, M., Lang, J., Wilson, N.: From preference logics to preference languages, and back. In: Proceedings of KR (2010)
3. Brightwell, G., Winkler, P.: Counting linear extensions is #P-complete. In: Proceedings of STOC, pp. 175–181 (1991)
4. Dubois, D., Lang, J., Prade., H.: Possibilistic logic. In: Handbook of Logic in Artificial Intelligence and Logic Programming, vol. 3, pp. 439–513 (1994)
5. Dubois, D., Prade, H.: Unfair coins and necessity measures: towards a possibilistic interpretation of histograms. Fuzzy Sets Syst. **10**, 15–20 (1983)
6. Hollunder, B.: An alternative proof method for possibilistic logic and its application to terminological logics. Int. J. Approximate Reasoning **12**, 85–109 (1995)

7. Lukasiewicz, T., Martinez, M.V., Simari, G.I.: Probabilistic preference logic networks. In: Proceedings of ECAI, pp. 561–566 (2014)
8. Smets, P.: Constructing the pignistic probability function in a context of uncertainty. In: Proceedings of UAI, vol. 89, pp. 29–40 (1989)
9. Yager, R.: Level sets for membership evaluation of fuzzy subset. In: Fuzzy Sets and Possibility Theory - Recent Developments, pp. 90–97 (1982)

Handling Uncertainty in Relational Databases with Possibility Theory - A Survey of Different Modelings

Olivier Pivert[1]([⊠]) and Henri Prade[2]

[1] ENSSAT-Lannion, IRISA, Rennes, France
pivert@enssat.fr
[2] IRIT – CNRS, 118, route de Narbonne, 31062 Toulouse Cedex 09, France
prade@irit.fr

Abstract. Mainstream approaches to uncertainty modeling in relational databases are probabilistic. Still some researchers persist in proposing representations based on possibility theory. They are motivated by the ability of this latter setting for modeling epistemic uncertainty and by its qualitative nature. Interestingly enough, several possibilistic models have been proposed over time, and have been motivated by different application needs ranging from database querying, to database design and to data cleaning. Thus, one may distinguish between four different frameworks ordered here according to an increasing representation power: databases with (i) layered tuples; (ii) certainty-qualified attribute values; (iii) attribute values restricted by general possibility distributions; (iv) possibilistic c-tables. In each case, we discuss the role of the possibility-necessity duality, the limitations and the benefit of the representation settings, and their suitability with respect to different tasks.

Keywords: Possibility theory · Relational databases · Uncertainty
Inconsistency · Data cleaning

1 Introduction

Many authors have made proposals to model and handle relational databases involving uncertain data. In particular, the last two decades have witnessed a blossoming of researches on this topic (cf. [29] for a survey of probabilistic approaches). Even though most of the literature about uncertain databases uses probability theory as the underlying uncertainty model, some approaches rather rest on possibility theory [30]. The initial idea of applying possibility theory to this issue goes back to the early 1980's [26,27]. This was short after the introduction of the idea of a "fuzzy database", for which various proposals were made, ranging from fuzzy relations (thus having weighted tuples) to ordinary relations with tuples of fuzzy values (represented by fuzzy sets), or more simply with tuples of weighted values. These different views developed by several authors

© Springer Nature Switzerland AG 2018
D. Ciucci et al. (Eds.): SUM 2018, LNAI 11142, pp. 396–404, 2018.
https://doi.org/10.1007/978-3-030-00461-3_30

were not necessarily referring to possibility theory; see [4] for references. Since this time, several possibilistic representations have been introduced, and it is useful to clarify their respective roles.

As we will discuss in Sect. 3, the possibilistic framework constitutes an interesting alternative to the probabilistic one, notably because of its qualitative nature. In this paper, we provide a survey of different modelings of uncertain data with possibility theory. The remainder is structured as follows. In Sect. 2, we recall some notions about uncertain databases and their interpretation in terms of possible worlds. Section 3 is devoted to a presentation of four possibilistic database models, with different levels of expressiveness. Section 4 discusses a specific topic where uncertain data management can play a role, namely data cleaning. Section 4.2 points out a sample of issues deserving further investigations. Finally, Sect. 5 concludes the paper and outlines some short-term research perspectives.

2   About Uncertain Databases and Possible Worlds

In the context of uncertain databases, two kinds of uncertainty are considered: tuple-level uncertainty (where the existence of some tuples in a relation is uncertain, i.e., is more or less probable/possible) and attribute-level uncertainty (where some attribute values in some tuples may be ill-known or uncertainly known). The latter case can be seen as more general than the former, since a tuple involving uncertain attribute values may be translated into a set of mutually exclusive uncertain tuples (involving only ordinary attribute values). An attribute value represented as a disjunctive weighted set can be interpreted as a probability distribution or a possibility distribution depending on the underlying uncertainty model considered. From a semantic point of view, an uncertain database D can be interpreted as a set of usual databases, called possible worlds $W_1, ..., W_p$, and the set of all interpretations of D is denoted by $rep(D) = \{W_1, ..., W_p\}$. Any world W_i is obtained by choosing a value in each disjunctive set appearing in D. One of these (regular) databases is supposed to correspond to the actual state of the universe modeled. The assumption of independence between the sets of candidates is usually made and then any world W_i corresponds to a conjunction of independent choices, thus the probability, or possibility, degree associated with a world is computed using a conjunction operator, namely, the product, or "min", respectively.

When processing a query, a naive way of doing would be to make explicit all the interpretations of D in order to query each of them. Such an approach is intractable in practice and it is of prime importance to find a more realistic alternative. To this end, the notion of a representation system was introduced by Imielinski and Lipski [14]. The basic idea is to represent both initial tables and those resulting from queries in such a way that the representation of the result of a query q against any database D denoted by $q(D)$, is equivalent (in terms of worlds) to the set of results obtained by applying q to every interpretation of D, i.e.: $rep(q(D)) = q(rep(D))$ where $q(rep(D)) = \{q(W) \mid W \in rep(D)\}$. If this property holds for a representation system ρ and a subset σ of the relational algebra, ρ is called a *strong representation system* for σ. From a querying point of view, this property enables a direct (or compact) calculus of a query q, which then applies to D itself without making the worlds explicit.

3 Possibilistic Uncertainty

We first recall some distinctive features of possibility theory before reviewing the different possibilistic representations.

3.1 Possibility Theory

Possibility theory departs from probability theory in several respects. Possibility theory involves two dual set functions: the possibility Π and the necessity N such that $N(A) = 1 - \Pi(\bar{A})$, while probability is self-dual, namely $P(A) = 1 - P(\bar{A})$. This provides room for modeling epistemic uncertainty, including total ignorance. Indeed, $\Pi(A) = 1$ does not prevent to have also $\Pi(\bar{A}) = 1$ in case of complete ignorance about A (while $\Pi(A) = P(\bar{A})(= 1/2)$ does not distinguish situations of genuine equiprobability from situations where, due to ignorance, one applies the Insufficient Reason Principle). Π (and N) are associated with a possibility distribution π, defined from a universe U to a scale such as scale $[0, 1]$, where $\forall A \subseteq U, \Pi(A) = \max_{u \in A} \pi(u)$. Due to the use of max and min operations, possibility and necessity functions are more "qualitative" than the probabilistic models involving sum and product.

Still, possibility theory may be quantitative or qualitative [8]. In the first case, the whole scale $[0, 1]$ is used, and possibility and necessity may be thought as upper and lower bounds of an unknown probability (then conditioning is based on product rather than "min"). However, possibility theory does not require the use of the scale $[0, 1]$, but can be defined with any linearly ordered chain (e.g., a finite subset $[0, 1]$ including 0 and 1), or more generally any lattice, and is then qualitative. Moreover, possibility theory has a logical counterpart, namely possibilistic logic [6] (which involves only lower bounds of necessity degrees, which can be viewed as certainty levels), and generalized possibilistic logic [11] (which involves both set functions). Besides, two other set functions are of interest in possibility theory, namely the guaranteed possibility, $\Delta(A) = \min_{u \in A} \delta(u)$, and the dual set function, where δ is a possibility distribution. In bipolar representations [9], one uses a pair of possibility distributions (δ, π) for distinguishing between values u such as $\pi(u) = 0$ that are excluded, from values u' such as $\delta(u') > 0$ that are guaranteed to be possible to some extent (since, e.g., they were observed), assuming the consistency condition $\delta \leq \pi$ (expressing that what is guaranteed to be possible cannot be excluded).

3.2 Possibilitistic Representations

There is not a unique possibilistic data model. The existing models serve different purposes. From the least to the most expressive, we can distinguish four possibilistic models for uncertain data which have been actually proposed:

- databases with layered tuples;
- tuples involving certainty-qualified attribute values;
- tuples involving attribute values restricted by possibility distributions;
- possibilistic c-tables.

Layered Tuples. The idea, here, is just to provide a complete ordering of the tuples in the database according to the more or less strong confidence we have in their truth. This can be easily encoded by associating a possibility level with each tuple. This results in a layered database: all the tuples having the same degree are in the same layer (and only them). Those tuples having a possibility level equal to 1 may also be associated with a certainty level equal to 1, while the others with a possibility level strictly less than 1 are not certain at all; this means that any possible world database contains all the tuples at level 1, while the other tuples may or may not be present in a particular possible world; see [17] for details. This modeling is not very expressive since it provides no indication on what attribute values in the tuple are particularly uncertain. In that respect, it may be considered as a modeling that is too poor from a querying perspective. Still, it has been shown useful for design purposes by providing a setting for attaching certainty levels to functional dependencies (FDs) (through a duality relation with the possibility levels of the tuples that are violating the FDs). Then, this enables the generalization of Armstrong's axioms by attaching certainty levels, and the extension of Boyce-Codd/3rd Normal Forms approaches to database design in the presence of uncertain tuples, by taking advantage of the levels [18]. Such a possibilistic model is also useful for handling keys [15] and cardinality constraints [12,28] in presence of uncertain data.

Certainty-Qualified Attribute Values. In this model [23], attribute values (or disjunctions thereof) are associated with a certainty level (which is the lower bound of the value of a necessity function). This amounts to associating each attribute value with a simplified type of possibility distribution restricting it[1]. Different attributes in a tuple may have different certainty levels associated with their respective values. Then a tuple may be associated with a certainty level, which is the minimum of the certainty levels associated with the attribute values of the tuple, in agreement with the minitivity of necessity functions. Still this global certainty level should not be confused with the possibility level of the

---

[1] Then the attribute value, or more generally the disjunction of possible values is/are considered as fully possible, while any other value in the attribute domain is all the less possible as the certainty level is higher. In case of full certainty these other values are all impossible. This is a particular case of the certainty qualification of a fuzzy set, here reduced to a singleton, or in any case to a classical subset. There are other basic qualifications of a fuzzy set in possibility theory, for instance in terms of guaranteed possibility (rather than in terms of necessity as in certainty qualification), or which lead to enlarge the core, or to reduce the support of the fuzzy set, see [7] for the four canonic transformations; see also [20] for hybrid transformations combining enlargement with uncertainty.

previous approach. In terms of possible worlds, a tuple associated with such a certainty level correspond to *several* tuples with a possibility level. Indeed consider the simple example of a tuple made of two attribute values a, and b, associated respectively with certainty α and β: this yields as possible worlds $\langle a, b \rangle$ with possibility 1, $\langle a', b \rangle$ with possibility $1 - \alpha$, $\langle a, b' \rangle$ with possibility $1 - \beta$, $\langle a', b' \rangle$ with possibility $\min(1 - \alpha, 1 - \beta)$, where a' (resp. b') is any value distinct from a (resp. b) in the attribute domain to which a (resp. b) belongs.

This model has some advantages with respect to querying: (i) it constitutes a strong representation system for the whole relational algebra (up to some minor restrictions); (ii) it does not require the use of any lineage mechanism and the query complexity is close to the classical case; (iii) the approach seems more robust with respect to small changes in the value of degrees than a probabilistic handling of uncertainty (see the last section of [23]). Moreover, there exists a simplified version of this model, see [24], that uses a scale with only three certainty levels ("completely certain","somewhat certain", "not at all certain"). This makes the assessment of certainty particularly easy. Besides, another approach with the same formal type of modeling, but where certainty is evaluated in terms of subsets of sources (together with their reliability level) makes it possible to rank-order the answers to a query also on such a basis [22].

Attribute Values Restricted by General Possibility Distributions. In this "full possibilistic model" [3], any attribute value can be represented by any possibility distribution. Moreover, representing the result of some relational operations (in particular the join) in this model requires the expression of dependencies between candidate values of different attributes in the same tuple, which leads to the use of nested relations. In [3], it is shown that this model is a strong representation system for selection, projection and foreign-key join only. The handling of the other relational operations requires the use of a lineage mechanism as in the probabilistic approaches. This model makes it possible to compute not only the more or less certain answers to a query (as in the previous model), but also the answers which are only possible to some extent.

Possibilistic *c*-tables. This model is outlined in [25]. The possibilistic extension of *c*-tables preserves all the advantages of classical *c*-tables (for expressing constraints linking attribute values) while the attribute values are restricted by any kind of possibility distribution. This model generalizes the two previous ones. In fact, possibilistic *c*-tables, as probabilistic *c*-tables, can be encompassed in the general setting of the semiring framework proposed by Val Tannen *et al.*

4   Data Cleaning

This section first provides a brief overview of two approaches that respectively (i) allow you to query inconsistent databases, and (ii) take advantage of a possibilistic modeling for cleaning the data, before suggesting new lines of research.

4.1   Some Existing Approaches

In the presence of inconsistent data, two points of view may be taken. The first one consists in cleaning the database so as to make it consistent, either by means of an automated process [13], or by an interactive approach. The second one, such as Consistent Query Answering (CQA) approaches [2], takes into account the inconsistencies at query processing time.

An approach corresponding to this second line of thought is described in [21]. It aims at warning the user about the presence of suspect answers in a selection query result, in the context of a classical database (that may include data inconsistent with some functional dependencies). Roughly speaking, the idea is that such elements can be identified inasmuch as they can also be found in the result of negative associated queries. The notion of a suspect answer can be refined by introducing some gradedness in terms of cardinality (number of functional dependency violations in which the tuple is involved) or similarity (by relaxing the equality constraint of a functional dependency into an approximate equality). However, this approach, for the moment, does not involve any uncertainty degree associated with attribute values or tuples. In other words, it handles only inconsistency but not uncertainty.

A possibilistic approach to data cleaning has been recently proposed in [16]. This approach belongs to the research trend aimed at restoring a form of consistency in the database. Still, the approach identifies tuples that are suspect or even fraudulent. This is done independently from any particular query. This relies on a model closely related to the layered-tuple-based model reviewed above. However, it is used in the reverse way, since it starts with certainty-valued constraints (called business rules) from which one computes the confidence levels associated with the tuples (on a qualitative scale: "normal"/"suspect"/"fraud"), by solving a minimal possibilistic vertex cover (taking into account the number of violations in which the tuples are involved). Here these are the possibility levels of the tuples that are revised in order to restore the (graded) consistency.

4.2   Some Issues Deserving Further Investigations

A first extension we may think of is to introduce certainty degrees in the first of the two approaches reviewed in the preceding section (reference [21]). This means extending the querying method keeping the data as they are and indicating which answers are suspect, to the setting of the certainty-based model described in Sect. 3. In the original model, an answer is suspect as soon as there exists a repair (w.r.t. a functional dependency) of the query result to which it does not belong. In the extended context, the notion of repair becomes naturally graded[2], as well as the concept of suspiciousness (now appreciated both in terms of the certainty degrees attached to the values of the concerned tuples and in terms of the number of functional dependencies violated by the tuples).

---

[2] For example, assume Peter has two ages, each with a certainty level, the levels being denoted by α and β respectively. Then the FD *name* $\rightarrow$ *age* is violated with a certainty degree that is equal to $\min(\alpha, \beta)$.

Another interesting issue is to unify the above view with the possibilistic approach to data cleaning reviewed in the previous section (reference [16]). We can observe that, although the outputs of the two approaches are quite similar (tuples assigned with a certainty degree expressing different levels of suspiciousness), the inputs are completely different: in one case, constraints with certainty levels, in the other case, attribute values with certainty levels. However, it seems clear that the approach [21] can also be extended by introducing functional dependencies with certainty levels and keeping all of the attribute values completely certain (rather than the opposite as suggested in the paragraph above). Then, this will make the two approaches easier to compare.

5   Conclusion

In this brief survey, we have tried to make clear that there exist different possibilistic models, with different levels of expressiveness, but also dedicated to different database tasks (design, data cleaning, querying). Other worth mentioning issues are the modeling of null values [1] and the extrapolation of missing data [5]. Two kinds of tasks, in our opinion, are particularly worth investigating: (i) a practical comparison of the certainty-based model (which offers a rather good simplicity/expressivity compromise) with probabilistic approaches; (ii) the comparison and the cooperation between different possibilistic data cleaning tools and probabilistic ones. Another line of thought which, we think, might be of interest, is to consider causality issues for evaluating the responsibility in inconsistencies, for which AI probabilistic models have been considered in a database perspective [19], while there also exist possibilistic counterparts to these AI models [10].

References

1. Arrazola, I., Plainfossé, A., Prade, H., Testemale, C.: Extrapolation of fuzzy values from incomplete data bases. Inf. Syst. **14**(6), 487–492 (1989)
2. Bertossi, L.E.: Database Repairing and Consistent Query Answering. Synthesis Lectures on Data Management. Morgan & Claypool Publishers, San Rafael (2011)
3. Bosc, P., Pivert, O.: About projection-selection-join queries addressed to possibilistic relational databases. IEEE Trans. Fuzzy Syst. **13**(1), 124–139 (2005)
4. Bosc, P., Prade, H.: An introduction to the fuzzy set and possibility theory-based treatment of flexible queries and uncertain or imprecise databases. In: Motro, A., Smets, P. (eds.) Uncertainty Management in Information Systems. From Needs to Solutions, pp. 285–324. Kluwer Academic Publishers, Dordrecht (1997)
5. De Tré, G., De Caluwe, R.M.M., Prade, H.: Null values in fuzzy databases. J. Intell. Inf. Syst. **30**(2), 93–114 (2008)
6. Dubois, D., Lang, J., Prade, H.: Automated reasoning using possibilistic logic: semantics, belief revision, and variable certainty weights. IEEE Trans. Knowl. Data Eng. **6**, 64–71 (1994)
7. Dubois, D., Prade, H.: What are fuzzy rules and how to use them. Fuzzy Sets Syst. **84**(2), 169–185 (1996)

8. Dubois, D., Prade, H.: Possibility theory: qualitative and quantitative aspects. In: Gabbay, D.M., Smets, P. (eds.) Quantified Representation of Uncertainty and Imprecision. Handbook of Defeasible Reasoning and Uncertainty Management Systems, vol. 1, pp. 169–226. Kluwer, Dordrecht (1998)

9. Dubois, D., Prade, H.: An overview of the asymmetric bipolar representation of positive and negative information in possibility theory. Fuzzy Sets Syst. **160**(10), 1355–1366 (2009)

10. Dubois, D., Prade, H.: A glance at causality theories for artificial intelligence. In: A Guided Tour of Artifial Intelligence, vol. 1: Knowledge Representation, Reasoning and Learning. Springer (2018)

11. Dubois, D., Prade, H., Schockaert, S.: Generalized possibilistic logic: foundations and applications to qualitative reasoning about uncertainty. Artif. Intell. **252**, 139–174 (2017)

12. Hall, N., Köhler, H., Link, S., Prade, H., Zhou, X.: Cardinality constraints on qualitatively uncertain data. Data Knowl. Eng. **99**, 126–150 (2015)

13. Ilyas, I.F., Chu, X.: Trends in cleaning relational data: consistency and deduplication. Found. Trends Databases **5**(4), 281–393 (2015)

14. Imielinski, T., Lipski, W.: Incomplete information in relational databases. J. ACM **31**(4), 761–791 (1984)

15. Koehler, H., Leck, U., Link, S., Prade, H.: Logical foundations of possibilistic keys. In: Fermé, E., Leite, J. (eds.) JELIA 2014. LNCS (LNAI), vol. 8761, pp. 181–195. Springer, Cham (2014). https://doi.org/10.1007/978-3-319-11558-0_13

16. Köhler, H., Link, S.: Qualitative cleaning of uncertain data. In: Mukhopadhyay, S., et al. (eds.) Proceedings of the 25th ACM International Conference on Information and Knowledge Management, CIKM 2016, Indianapolis, IN, USA, 24–28 October 2016, pp. 2269–2274. ACM (2016)

17. Link, S., Prade, H.: Possibilistic functional dependencies and their relationship to possibility theory. IEEE Trans. Fuzzy Syst. **24**(3), 757–763 (2016)

18. Link, S., Prade, H.: Relational database schema design for uncertain data. In: Mukhopadhyay, S., et al. (eds.) Proceedings of the 25th ACM International conference on Information and Knowledge Management, CIKM 2016, Indianapolis, 24–28 October, pp. 1211–1220 (2016)

19. Meliou, A., Roy, S., Suciu, D.: Causality and explanations in databases. PVLDB **7**(13), 1715–1716 (2014)

20. González, A., Marín, N., Pons, O., Vila, M.A.: Qualification of fuzzy statements under fuzzy certainty. In: Melin, P., Castillo, O., Aguilar, L.T., Kacprzyk, J., Pedrycz, W. (eds.) IFSA 2007. LNCS (LNAI), vol. 4529, pp. 162–170. Springer, Heidelberg (2007). https://doi.org/10.1007/978-3-540-72950-1_17

21. Pivert, O., Prade, H.: Detecting suspect answers in the presence of inconsistent information. In: Lukasiewicz, T., Sali, A. (eds.) FoIKS 2012. LNCS, vol. 7153, pp. 278–297. Springer, Heidelberg (2012). https://doi.org/10.1007/978-3-642-28472-4_16

22. Pivert, O., Prade, H.: Querying uncertain multiple sources. In: Straccia, U., Calì, A. (eds.) SUM 2014. LNCS (LNAI), vol. 8720, pp. 286–291. Springer, Cham (2014). https://doi.org/10.1007/978-3-319-11508-5_24

23. Pivert, O., Prade, H.: A certainty-based model for uncertain databases. IEEE Trans. Fuzzy Syst. **23**(4), 1181–1196 (2015)

24. Pivert, O., Prade, H.: Database querying in the presence of suspect values. In: Morzy, T., Valduriez, P., Bellatreche, L. (eds.) ADBIS 2015. CCIS, vol. 539, pp. 44–51. Springer, Cham (2015). https://doi.org/10.1007/978-3-319-23201-0_6

25. Pivert, O., Prade, H.: Possibilistic conditional tables. In: Gyssens, M., Simari, G. (eds.) FoIKS 2016. LNCS, vol. 9616, pp. 42–61. Springer, Cham (2016). https://doi.org/10.1007/978-3-319-30024-5_3

26. Prade, H.: Lipski's approach to incomplete information databases restated and generalized in the setting of Zadeh's possibility theory. Inf. Syst. 9(1), 27–42 (1984)

27. Prade, H., Testemale, C.: Generalizing database relational algebra for the treatment of incompleteuncertain information and vague queries. Inf. Sci. 34, 115–143 (1984)

28. Roblot, T.K., Link, S.: Possibilistic cardinality constraints and functional dependencies. In: Comyn-Wattiau, I., Tanaka, K., Song, I.-Y., Yamamoto, S., Saeki, M. (eds.) ER 2016. LNCS, vol. 9974, pp. 133–148. Springer, Cham (2016). https://doi.org/10.1007/978-3-319-46397-1_11

29. Suciu, D., Olteanu, D., Ré, C., Koch, C.: Probabilistic Databases. Synthesis Lectures on Data Management. Morgan & Claypool Publishers, San Rafael (2011)

30. Zadeh, L.: Fuzzy sets as a basis for a theory of possibility. Fuzzy Sets Syst. 1, 3–28 (1978)

An Argumentative Recommendation Approach Based on Contextual Aspects

Juan Carlos Lionel Teze[1,2,3](✉), Lluis Godo[4], and Guillermo Ricardo Simari[1]

[1] Institute for Computer Science and Engineering (ICIC),
Departamento de Ciencias e Ing. de la Computación, Universidad Nacional del Sur,
Alem 1253, 8000 Bahía Blanca, Buenos Aires, Argentina
`{jct,grs}@cs.uns.edu.ar`
[2] Agents and Intelligent Systems Area, Faculty of Management Sciences,
Universidad Nacional de Entre Ríos, Tavella 1424, 3200 Concordia, E. R., Argentina
[3] Consejo Nacional de Investigaciones Científicas y Técnicas (CONICET),
Buenos Aires, Argentina
[4] Artificial Intelligence Research Institute (IIIA-CSIC), Campus UAB, Bellaterra,
08193 Barcelona, Spain
`godo@iiia.csic.es`

Abstract. Argumentation-based recommender systems constitute an interesting tool to provide reasoned recommendations in complex domains with unresolved contradictory information situations and incomplete information. In these systems, the use of contextual information becomes a central issue in order to come up with personalized recommendations. An argumentative recommender system that offers mechanisms to handle contextual aspects of the recommendation domain provides an important ability that can be exploited by the user. However, in most of existing works, this issue has not been extensively studied. In this work, we propose an argumentation-based formalization for dealing with this issue. We present a general framework that allows the design of recommender systems capable of handling queries that can include (possibly inconsistent) contextual information under which recommendations should be computed. To answer a query, in the proposed argumentation-based approach, the system first selects alternative instances according to the user's supplied contextual information, and then makes recommendations, in both cases through a defeasible argumentative analysis.

Keywords: Recommenders · Argumentation · Contextual information

1 Introduction and Motivation

Recommender systems represent an interesting specialized tool to assist users in domains where users are confronted with many choices. Although recommenders systems have been efficiently applied in different real-world scenarios [5,9,10, 16], the dynamic nature of user's preferences and knowledge usually leads to conflicting situations and incomplete domain information. In such a setting,

© Springer Nature Switzerland AG 2018
D. Ciucci et al. (Eds.): SUM 2018, LNAI 11142, pp. 405–412, 2018.
https://doi.org/10.1007/978-3-030-00461-3_31

defeasible argumentation can be a useful alternative to tackle these issues since its reasoning mechanism can effectively deal with incomplete and conflicting information. Several research lines have shown that defeasible argumentation can be integrated with recommendation technologies as underlying inference model, providing an interesting possibility for the development of architectures for recommender systems [6–8,19].

A central question in recommendation systems is to anticipate the user's needs in order to provide personalized recommendations. An approach to deal with this situation is the use of contextual information. This is not a new idea, it has been for a long the focus in a variety of theories and implementations [1,17]. One of the first approaches to take contextual information into consideration is the work by Herlocker and Kostan [13], where they proposed a task-focused approach in which user's task profile is used for making recommendations. Other approaches, like [20], allow recommendation systems to work with dynamic information and answer queries based on facts received as contextual information. Recently, the authors of [2] claim that user's contextual information is in fact not fully exploited in recommendation systems, and they propose a methodology that incorporates contextual factors in the process of evaluating the recommendations in order to make them more personalized according to the user situation. Finally, it is important to note that contextual information is a multifaceted concept that has been commonly used in several areas of computer science. For example, the representation of textual contexts by vector space models and its variants has been used to improve the performance of several natural language processing (NLP) applications, including machine translation, sentiment analysis and summarization, see e.g. [4,14,15,18].

In this work, we propose an approach where argumentation is used both for a first selection process based on contextual information and for the recommendation process itself. Moreover, it presents a special type of conditional query that includes the alternatives to be considered in the recommendations and elements of the context that should be satisfied for each of the instances of such alternatives. Next, we present an example that will serve to motivate the main ideas of our proposal, as well as a running example.

Example 1. *Consider a recommendation system to help to users regarding places to go in their spare time in a given city. The system considers two types of alternatives: go to pubs and go to cinemas. Additionally, the system considers certain contextual aspects that will help in selecting among different available options for each of the two kinds of alternatives considered, e.g., the kind of area where places are located, which are classified as safe/unsafe, or touristic/non-touristic zones. Considering the alternatives and contextual information, two users Luis and Pilar request recommendations to the system. Luis prefers pubs and cinemas in touristic areas, while Pilar prefers pubs in safe zones but cinemas in touristic areas.*

Despite their clear interest, mechanisms to dynamically select instances of a recommendation alternative depending on contextual aspects have not seemingly been extensively studied for argumentative recommender systems. In this short paper, we present work in progress towards a generic argumentative recommendation framework able to deal with this question. In particular, we propose a system architecture based on this framework that uses Defeasible Logic Programming (DeLP) as its underlying argumentative reasoning system.

This paper is structured as follows. After overviewing in Sect. 2 the DeLP formalism, in Sect. 3 we present the components of the proposed framework and a DeLP-based recommender system. Finally, we finish in Sect. 4 with some conclusions and future research lines.

2   Defeasible Logic Programming: An Overview

In this section, we summarize the main concepts of DeLP [12]. A DeLP-program $\mathcal{P}$ is denoted as (Π, Δ) distinguishing the subset Π of facts and strict rules, and the subset Δ of defeasible rules. *Facts* are ground literals representing atomic information, or its negation using a strong negation "$\sim$". *Strict rules* represent non-defeasible information and are denoted as $L_0 \leftarrow L_1, \ldots, L_n$, where $\{L_i\}_{0<i\leq n}$ are ground literals. *Defeasible Rules* represent tentative information that may be used if there is nothing against, and are denoted as $L_0 \prec L_1, \ldots, L_n$, where again $\{L_i\}_{0<i\leq n}$ are ground literals. A defeasible rule expresses that *reasons to believe* in the antecedent give *reasons to believe* in the consequent.

Example 2. *The following are examples of what* DeLP*-programs representing information of the application domain introduced in Example 1 could look like. The program* $\mathcal{P}_{cont}$ *is meant to represent contextual information to decide whether a location (of a possible option) is either in a touristic or safe zone of the city, while the program* $\mathcal{P}_{rec}$ *is meant to represent information for recommending possible options to go.*

$$\mathcal{P}_{cont} = \left\{ \begin{array}{l} safeZone(D) \prec commercialZone(D) \\ \sim safeZone(D) \prec theftsZone(D) \\ touristicZone(D) \prec nearbyShops(D) \\ [\ldots] \end{array} \right\}$$

$$\mathcal{P}_{rec} = \left\{ \begin{array}{l} recPub(D, U) \prec pub(D), karaoke(D), likeKaraoke(U) \\ \sim recPub(D, U) \prec pub(D), karaoke(D), likeKaraoke(U), \sim craftBeer(D) \\ recCinema(D, U) \prec cinema(D), movie(M, D), genre(G, M), like(U, G) \\ [\ldots] \end{array} \right\}$$

The three rules shown in the program $\mathcal{P}_{cont}$ consider two contextual aspects: touristic zones and safe zones. The first two rules are used for suggesting whether a destination option is in a safe zone: to be near a commercial zone is a reason for believing it is in safe zone, whereas to be in a zone with many reported thefts is a reason for not believing it is in a safe zone. The last rule expresses that being in a zone with many shops is a reason to believe it is in a touristic area.

The rules shown in the program $\mathcal{P}_{rec}$ deal with recommendation alternatives: cinemas and pubs. The first two defeasible rules represent reasons for and against *recommending pubs: if the user likes to sing karaoke and the pub is a karaoke pub then these are good reasons for recommending such a pub, but if there is not craft beer in the pub then this is a reason for not recommending it. Finally, the last rule expresses reasons for recommending to go to a cinema: if the user likes the genre of the movie at the cinema, then this is a reason for recommending that cinema.*

To deal with contradictory information, DeLP builds arguments supporting the different pieces of the conflictive information. An *argument* for a literal L, denoted $\langle A, L \rangle$, where $A \subseteq \Delta$ is a minimal and non-contradictory set such that together with Π allows a derivation of L. To establish if $\langle A, L \rangle$ is a non-defeated argument, arguments that could be defeaters for $\langle A, L \rangle$ are considered, counter-arguments that according to some criterion are preferred to $\langle A, L \rangle$. As each defeater could in turn be defeated, a sequence of arguments called *argumentation line* is constructed. Clearly, there can be more than one defeater for a particular argument. Therefore, many acceptable argumentation lines could arise from this argument, leading to a tree structure called *dialectical tree* where each argument defeats its predecessor. The prevailing argument provides a warrant for the information it supports. A ground literal L is *warranted* if there exists an non-defeated argument for L. More details about the warrant procedure can be found in [12]. Finally, a DeLP-interpreter will return YES if the literal L is warranted, NO if the complement of L is warranted, UNDECIDED if neither L nor its complement are warranted, or UNKNOWN if L is not in the signature of the program.

The recommender system proposed in this paper use DeLP-servers [11], an implementation of DeLPthat provides a reasoning service for multi-agent systems. A query for a DeLP-server is a pair $[C, Q]$, where Q is a literal to be consulted to DeLP and C is the context for that query. The context is any DeLP-program.

3   A Recommendation Framework and a DeLP Instantiation

In this section, first we propose a recommendation framework providing a defeasible context-dependent way for selecting recommendation options, and then we sketch a particular instance of this framework: a defeasible logic programming recommender system (DeLP-recommender) based on contextual aspects. This particular system will answer queries using an argumentative reasoning and will allow users to express their preferences indicating the context aspects upon which consulted recommendations should be solved. In what follows, we will explain each of these proposals.

The recommendation framework consists of four main components, see Fig. 1(A). The *domain data repository* stores the necessary domain knowledge for making recommendations. In particular, the system uses the information from this repository both to evaluate the contextual aspects satisfying the user' requirements and to come up with the recommendations themselves. The *context module* is in charge of evaluating whether the different options in the domain data repository satisfy the users' contextual preferences. The *recommendation module* receives the available options from the context module and it computes the final answers. Finally, the *preference manager* takes as input the user's contextual preferences for the different alternatives and supplies the filtered options to the recommendation module. This component acts as the communication channel among the other system modules.

In the rest of the section we sketch a particular DeLP-based instantiation of proposed framework, whose architecture is represented in Fig. 1(B). As shown in the figure, we use DeLP-servers (that facilitate the implementation of reasoning services that work over knowledge bases represented as DeLP-programs) for the implementation of both the context and recommendation modules. Moreover, our approach assumes two finite sets of literals: a set *SAlts* representing recommendation alternatives, and a set *SContextA* with the available contextual aspects for such alternatives. For instance, from the application domain presented in Example 1 the following sets could be considered:

$$SAlts = \{cinema(X), pub(X)\}, SContextA = \{safeZone(X), touristicZone(X)\}.$$

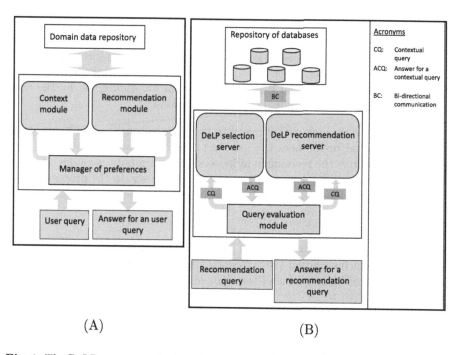

(A)   (B)

Fig. 1. The DeLP-recommender based on contextual aspects (B) as particular instance of the proposed recommendation framework (A).

Thus, having defined these sets, and following the conceptual schema in Fig. 1(A), the proposed system will contain the following two DeLP-servers (see Fig. 1(B)):

- The DeLP-*selection server* uses information about available alternatives (obtained from database repository) for determining, by means of an argumentative inference mechanism, which contextual elements are satisfied for particular alternatives. This will allow for filtering those options of a particular alternative that the system could suggest. An interesting feature of the proposed formalism is that the description of new contextual elements can be introduced by means of a set of DeLP-rules. Adopting this rule-based representation, the DeLP-inference mechanism allows the server to decide whether a given option D can be warranted to satisfy a particular contextual aspect. For instance, the DeLP-program $\mathcal{P}_{cont}$ in Example 2 shows defeasible rules for two contextual aspects, *safeZone(D)* and *touristicZone(D)*.

- The DeLP-*recommendation server* uses DeLP-rules to offer warranted recommendations to the user. To do this, it tries to build arguments for and against those options selected by the selection server. In order to provide personalized recommendations, we propose a knowledge-based approach where general preference criteria over available alternatives are encoded by DeLP-rules. One of the main advantages of this approach is its flexibility. Using rules provides the possibility of easily changing or expanding the criteria the server can use. For instance, considering the DeLP-program $\mathcal{P}_{rec}$ introduced in Example 2, several defeasible rules have been encoded to represent different general preference criteria for and against recommending cinemas and pubs. Besides those rules, the DeLP-server can use information given by the user about her interests to output its recommendations.

An extended recommendation query to the system consists of a query itself together contextual preferences the user can specify as part of the query. This type of queries allows the recommendation system to compute recommendations using context information specifically required by the user. In the proposed approach, an extended *recommendation query* consists of a sequence

$$[(Alt_1, ContextA_1), \ldots, (Alt_n, ContextA_n)],$$

where $Alt_i \in SAlts$ and $ContextA_i$ is a Boolean expression built from literals of $SContextA$, in which each pair $(Alt_i, ContextA_i)$ expresses the contextual aspects $ContextA_i$ required to recommend the elements of Alt_i. For instance, a query

$$[(cinema(X), turistZone(X)), (pub(X), turistZone(X) \vee safeZone(X))]$$

can be interpreted in the following way: *"the user prefers cinemas that are in touristic zones and pubs in either touristic or safe zones"*.

For answering a recommendation query, the system considers all those options (i.e. instantiations) for each alternative included in the query and that are warranted by the DeLP-selection server to satisfy the users' contextual preferences.

The final output presented to the user is a list *RList* with all recommendations for each alternative included in the query and that have been warranted by the DeLP-recommendation server.

4 Conclusions and Future Work

In this short paper we have presented preliminary work about an enhanced argumentation-based approach for the development of recommenders, by providing them additional reasoning capacities to handle contextual preference information explicitly declared by the user. One of the main objectives of this proposal has been to introduce a concrete way of filtering possible user's recommendations through the evaluation of this contextual information using a defeasible argumentative reasoning. As for future work, we plan to develop an instance of the proposed framework allowing to incorporate priorities in the defeasible rules, both in the DeLP-selection and recommendation servers. For instance, we can introduce priorities as weights in the style of Possibilistic-DeLP [3], and/or to allow for gradual evaluation of the contextual conditions (e.g. some areas may be evaluated as being safer than others). Both features would lead to provide rankings of recommendations to the user rather than plain sets. Another future goal is to make an analysis of time efficiency and computational complexity.

Acknowledgements. This work has been partially supported by EU H2020 research and innovation programme under the Marie Sklodowska-Curie grant agreement No. 690974 for the project MIREL: MIning and REasoning with Legal texts, and by funds provided by CONICET, Universidad Nacional del Sur by PGI-UNS (grant 24/N040), and Universidad Nacional de Entre Ríos. Godo acknowledges the Spanish FEDER/MINECO project TIN2015-71799- C2-1-P.

References

1. Adomavicius, G., Tuzhilin, A.: Context-aware recommender systems. In: Ricci, F., Rokach, L., Shapira, B. (eds.) Recommender Systems Handbook, pp. 191–226. Springer, Boston, MA (2015). https://doi.org/10.1007/978-1-4899-7637-6_6
2. Afzal, M., et al.: Personalization of wellness recommendations using contextual interpretation. Expert Syst. Appl. **96**, 506–521 (2018)
3. Alsinet, T., Chesñevar, C.I., Godo, L., Simari, G.R.: A logic programming framework for possibilistic argumentation: formalization and logical properties. Fuzzy Sets Syst. **159**(10), 1208–1228 (2008)
4. Baroni, M., Dinu, G., Kruszewski, G.: Don't count, predict! A systematic comparison of context-counting vs. context-predicting semantic vectors. In: Proceedings of the 52nd Annual Meeting of the Association for Computational Linguistics (Volume 1: Long Papers), vol. 1, pp. 238–247 (2014)
5. Bobadilla, J., Serradilla, F., Hernando, A.: Collaborative filtering adapted to recommender systems of e-learning. Knowl.-Based Syst. **22**(4), 261–265 (2009)
6. Briguez, C.E., Budán, M.C., Deagustini, C.A.D., Maguitman, A.G., Capobianco, M., Simari, G.R.: Argument-based mixed recommenders and their application to movie suggestion. Expert Syst. Appl. **41**(14), 6467–6482 (2014)

7. Briguez, C.E., Budán, M.C., Deagustini, C.A., Maguitman, A.G., Capobianco, M., Simari, G.R.: Towards an argument-based music recommender system. COMMA **245**, 83–90 (2012)
8. Briguez, C.E., Capobianco, M., Maguitman, A.G.: A theoretical framework for trust-based news recommender systems and its implementation using defeasible argumentation. Int. J. Artif. Intell. Tools **22**(4), 1350021 (2013)
9. Carrer-Neto, W., Hernández-Alcaraz, M.L., Valencia-García, R., Sánchez, F.G.: Social knowledge-based recommender system. Application to the movies domain. Expert Syst. Appl. **39**(12), 10990–11000 (2012)
10. Castro-Schez, J.J., Miguel, R., Vallejo, D., López-López, L.M.: A highly adaptive recommender system based on fuzzy logic for B2C e-commerce portals. Expert Syst. Appl. **38**(3), 2441–2454 (2011)
11. García, A.J., Rotstein, N.D., Tucat, M., Simari, G.R.: An argumentative reasoning service for deliberative agents. In: Zhang, Z., Siekmann, J. (eds.) KSEM 2007. LNCS (LNAI), vol. 4798, pp. 128–139. Springer, Heidelberg (2007). https://doi.org/10.1007/978-3-540-76719-0_16
12. García, A.J., Simari, G.R.: Defeasible logic programming: an argumentative approach. Theory Pract. Log. Program. (TPLP) **4**(1–2), 95–138 (2004)
13. Herlocker, J.L., Konstan, J.A.: Content-independent task-focused recommendation. IEEE Internet Comput. **5**(6), 40–47 (2001)
14. Iacobacci, I., Pilehvar, M.T., Navigli, R.: Embeddings for word sense disambiguation: an evaluation study. In: Proceedings of the 54th Annual Meeting of the Association for Computational Linguistics (Volume 1: Long Papers), vol. 1, pp. 897–907 (2016)
15. Kågebäck, M., Johansson, F., Johansson, R., Dubhashi, D.: Neural context embeddings for automatic discovery of word senses. In: Proceedings of the 1st Workshop on Vector Space Modeling for Natural Language Processing, pp. 25–32 (2015)
16. Lippi, M., Torroni, P.: MARGOT: a web server for argumentation mining. Expert Syst. Appl. **65**, 292–303 (2016)
17. Ricci, F., Rokach, L., Shapira, B.: Introduction to recommender systems handbook. In: Ricci, F., Rokach, L., Shapira, B., Kantor, P.B. (eds.) Recommender Systems Handbook, pp. 1–35. Springer, Boston, MA (2011). https://doi.org/10.1007/978-0-387-85820-3_1
18. Taghipour, K., Ng, H.T.: Semi-supervised word sense disambiguation using word embeddings in general and specific domains. In: Proceedings of the 2015 Conference of the North American Chapter of the Association for Computational Linguistics: Human Language Technologies, pp. 314–323 (2015)
19. Teze, J.C., Gottifredi, S., García, A.J., Simari, G.R.: Improving argumentation-based recommender systems through context-adaptable selection criteria. Expert Syst. Appl. **42**(21), 8243–8258 (2015)
20. Tucat, M., García, A.J., Simari, G.R.: Using defeasible logic programming with contextual queries for developing recommender servers. In: AAAI Fall Symposium: The Uses of Computational Argumentation (2009)

Author Index

Antonucci, Alessandro   35
Augustin, Thomas   351

Baroni, Pietro   243
Bauters, Kim   333
Bellodi, Elena   78
Ben Amor, Nahla   359
Bertossi, Leopoldo   368
Blaas, Arno   373
Bueno, Marcos L. P.   93
Butz, Raphaela   50

Cabitza, Federico   64
Calliess, Jan-Peter   373
Castelltort, Arnaud   152
Ciucci, Davide   64
Cobb, Adam D.   373
Cota, Giuseppe   78

da Costa Pereira, Célia   380
Divari, Maria   321
Doria, Serena   108
Dubois, Didier   3, 124, 359

Facchini, Alessandro   35
Faux, Francis   124

Godo, Lluis   387, 405

Hommersom, Arjen   50, 93

Imoussaten, Abdelhak   140

Kaminski, Benjamin L.   290
Katoen, Joost-Pieter   290

Lamma, Evelina   78
Laurent, Anne   152
Leray, Philippe   276
Lesot, Marie-Jeanne   152
Liu, Weiru   333
Lobo, Mariana   93
Lucas, Peter J. F.   93

Malchiodi, Dario   380
Marsala, Christophe   152
Martinez, Maria Vanina   387
Matheja, Christoph   290
McAreavey, Kevin   333

Nickles, Matthias   164

Ozaki, Ana   181

Peñaloza, Rafael   181
Perrussel, Laurent   228
Pfeifer, Niki   196
Pivert, Olivier   396
Potyka, Nico   212
Pozos-Parra, Pilar   228
Prade, Henri   3, 124, 359, 396

Rago, Antonio   243
Rico, Agnès   18, 124
Rifqi, Maria   152
Riguzzi, Fabrizio   78
Roberts, Stephen J.   373
Rodrigues, Pedro P.   93

Saidi, Syrine   359
Sanfilippo, Giuseppe   196, 260
Simari, Gerardo I.   387
Simari, Guillermo Ricardo   405

Tettamanzi, Andrea G. B.   380
Teze, Juan Carlos Lionel   405
Thévenin, Jean Marc   228
Toni, Francesca   243

van der Gaag, Linda C.   276
van Eekelen, Marko   50
van Keulen, Maurice   290
Viappiani, Paolo   306
Vivona, Doretta   321

Xu, Mengwei   333

Zese, Riccardo   78

Printed in the United States
By Bookmasters